```python
import numpy as np

index_value = np.random.random([3, 3, 3])
print('原数据', index_value)
print('####################')
print(index_value[:1, 1:3, 2])
```

```
D:\anaconda3\python.exe D:/DLAI/qwenAILearn/untitled9/studyNumpy/数组的索引2.py
原数据 [[[0.96735713 0.20008986 0.93431251]
  [0.39629697 0.40814864 0.22539278]
  [0.52097874 0.38204418 0.91030824]]

 [[0.81261282 0.92788045 0.74221152]
  [0.5017542  0.80595748 0.72478601]
  [0.48712281 0.25715609 0.51998573]]

 [[0.56078442 0.35462929 0.23275688]
  [0.81109904 0.3736148  0.58315035]
  [0.08836854 0.25946715 0.80110368]]]
####################
[[0.22539278 0.91030824]]
```

图 1-25　index_value 的取值过程

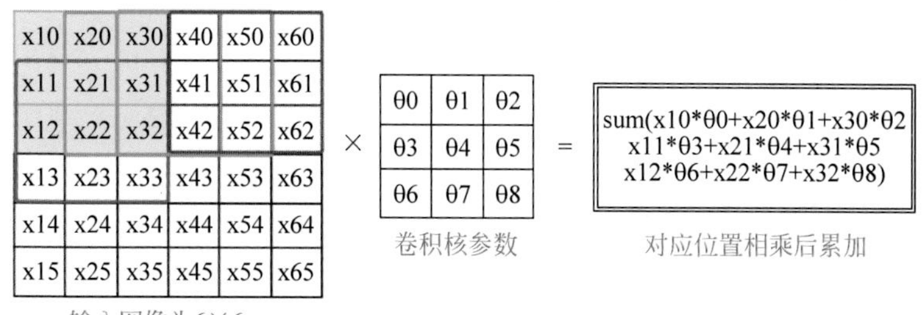

图 4-1　卷积计算($s=1$)

图 4-2　卷积计算($s=3$)

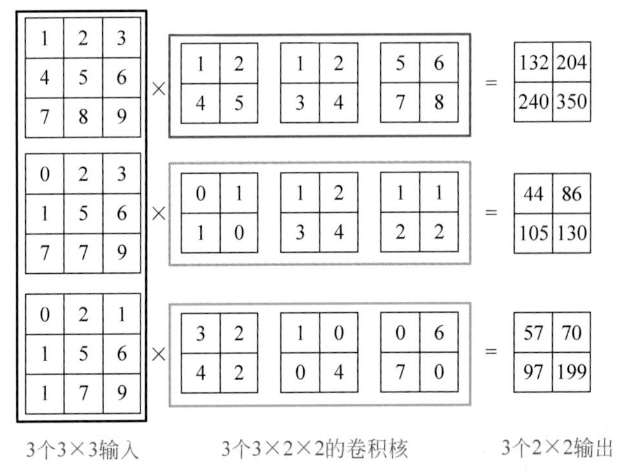

图 4-8　多通道输入与多通道输出

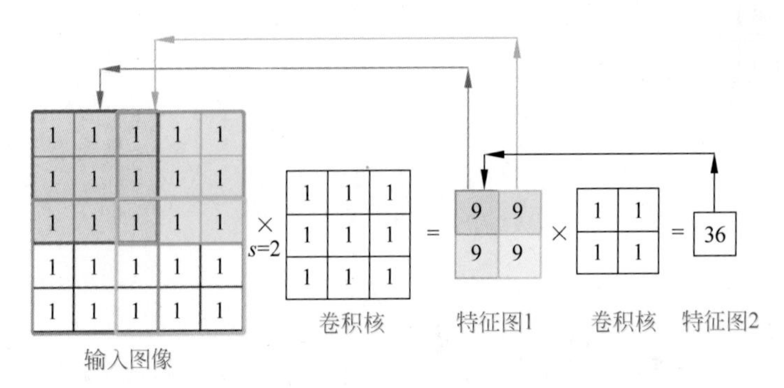

图 4-13　感受野的映射

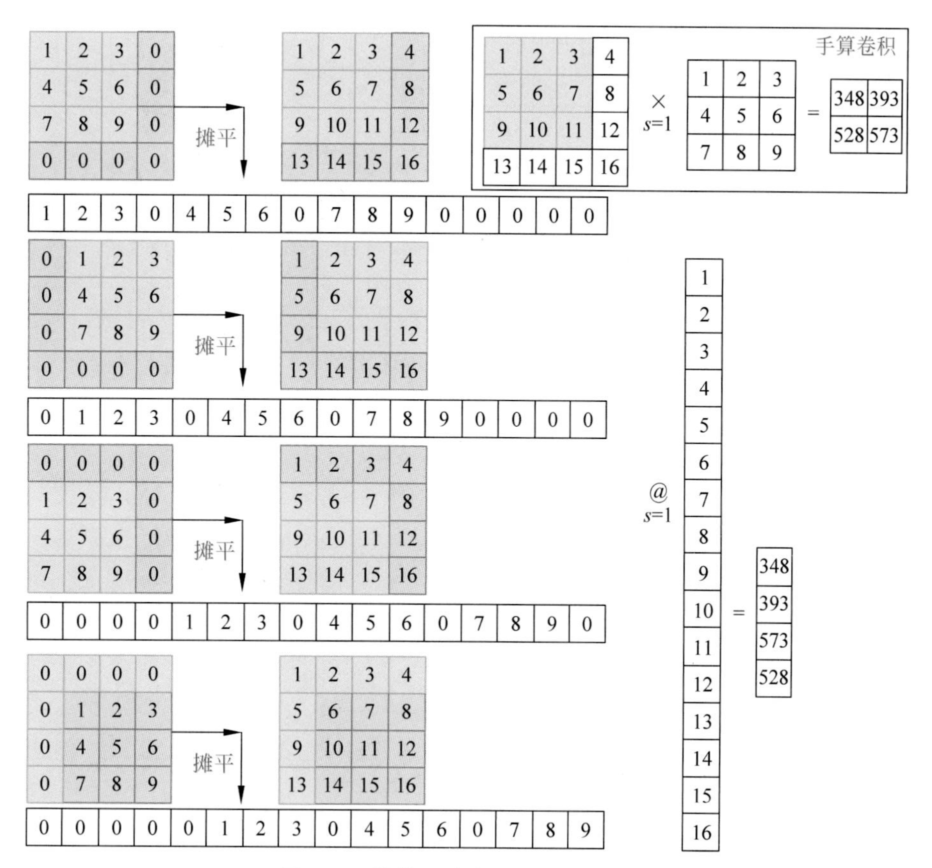

图 4-14　计算机中卷积的计算($s=1$)

1	3	2	1	0
2	9	1	1	0
1	3	2	3	0
5	6	1	2	0
0	0	0	0	0

kernel_size=3×3
$s=2\times2$

9	3
6	6

图 4-19　平均池化(2)

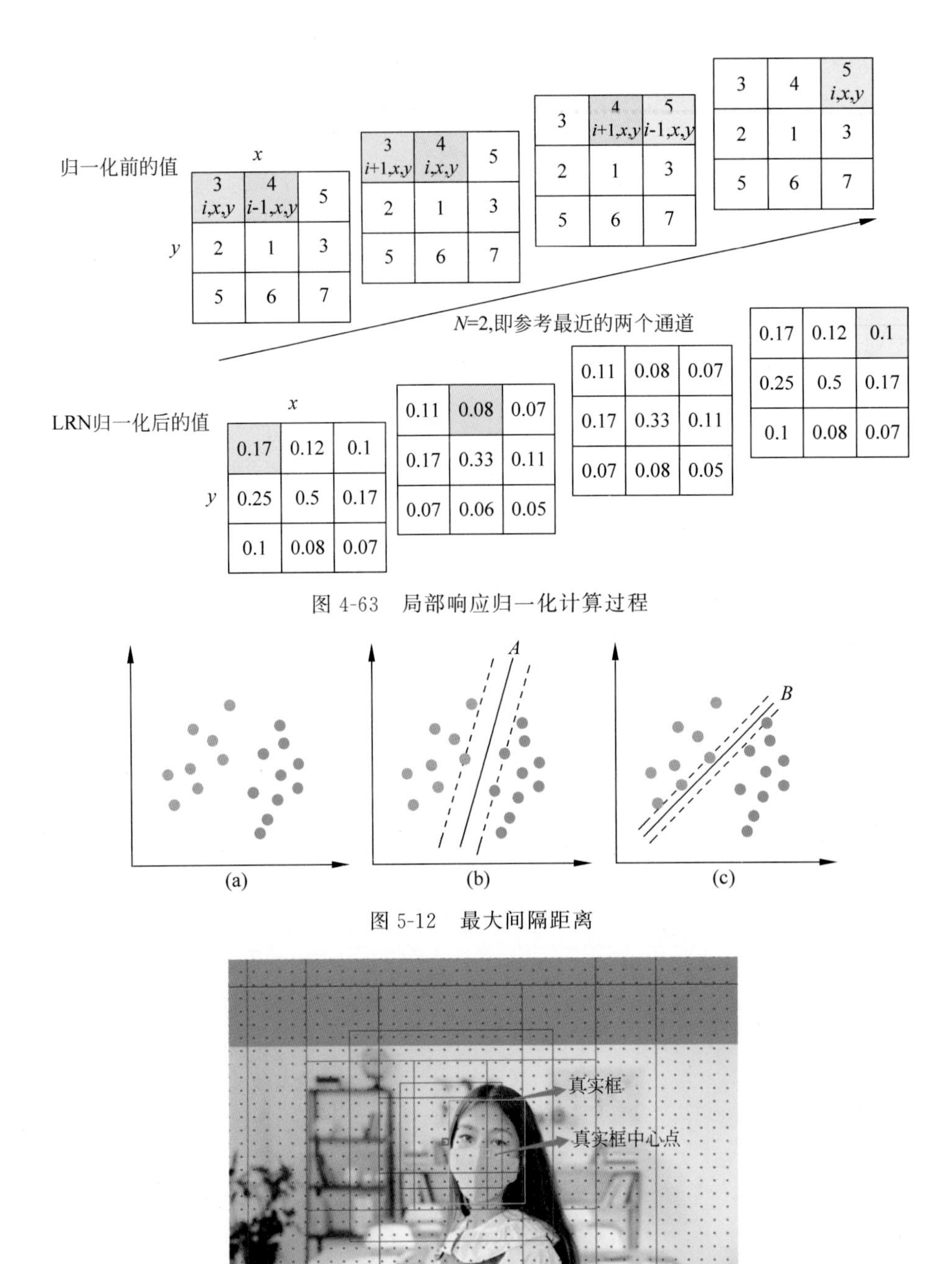

图 4-63　局部响应归一化计算过程

图 5-12　最大间隔距离

图 5-19　Faster R-CNN 建议框的生成

图 5-20　Faster R-CNN 正样本的选择

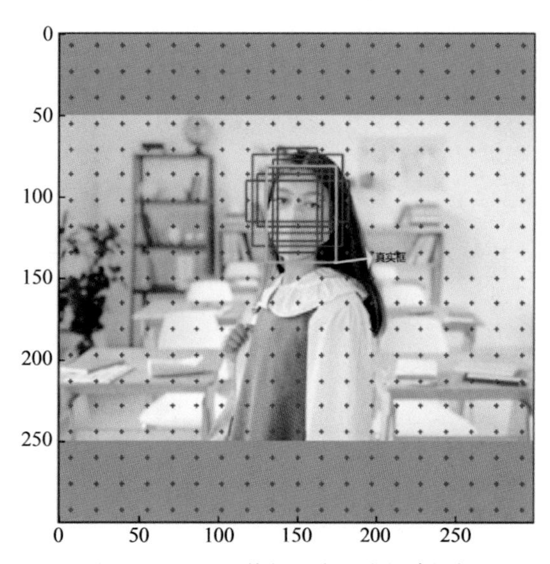

图 5-26　Conv7 特征图中生成的建议框

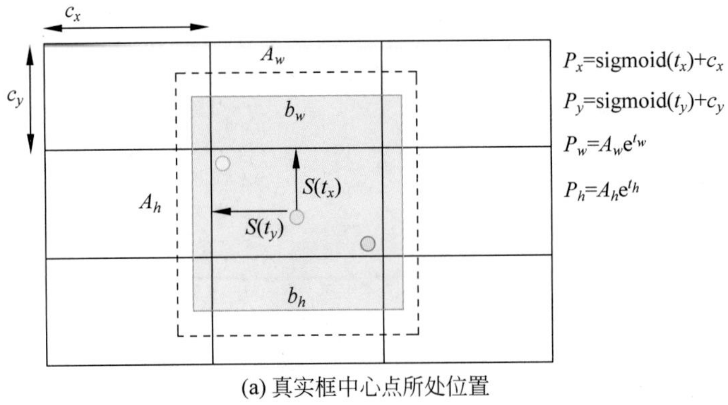

(a) 真实框中心点所处位置

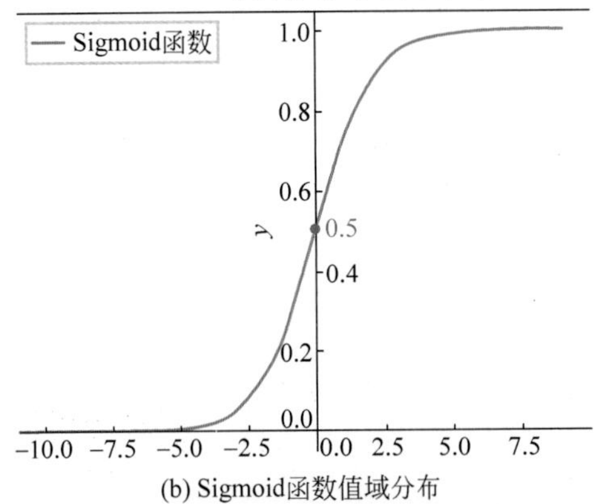

(b) Sigmoid函数值域分布

图 5-79 Grid 敏感度

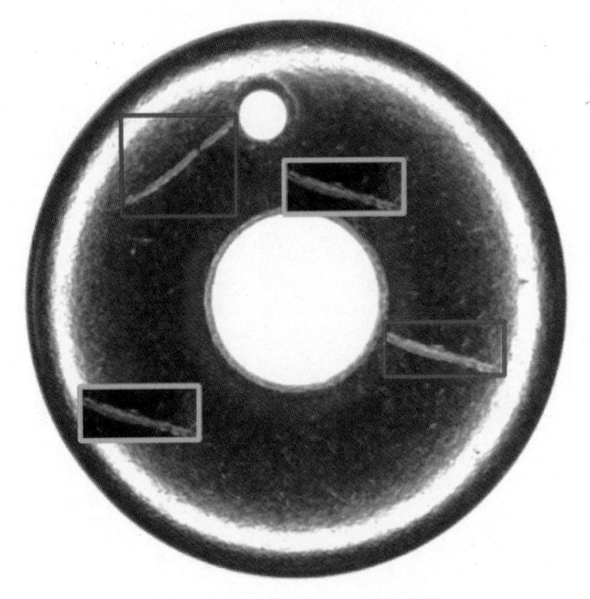

图 5-110 Replicate 数据增强

跟我一起学人工智能

深度学习

从零基础快速入门到项目实践

文青山◎编著

清华大学出版社
北京

内容简介

本书从Python基础入手，循序渐进地讲解机器学习、深度学习等领域的算法原理和代码实现，在学习算法理论的同时也使读者的代码工程能力逐步提高。

本书共6章，第1章从零基础介绍Python基础语法、Python数据处理库NumPy、Pandas、Matplotlib、OpenCV的使用；第2章主要介绍机器学习算法的原理并配有代码实例，方便在理解原理的同时也能写出代码；第3章主要介绍深度学习框架TensorFlow、Keras、PyTorch的API和网络模型的搭建方法，力保读者能够掌握主流深度学习框架的使用；第4章主要介绍卷积神经网络各种卷积的特性，并通过代码实现了多个经典分类网络；第5章介绍目标检测领域中的多个经典算法的原理，并配套展现了代码调试的过程，将算法原理与代码进行结合，方便读者更深入地理解算法原理；第6章分享深度学习项目的分析和实现过程。

本书精心设计的算法原理讲解、代码实现，不仅适合对深度学习感兴趣的初学者，同时对高校学生、教师、相关技术人员、研究人员及从事深度学习的工程师都有较高的参考价值。

图书在版编目（CIP）数据

深度学习：从零基础快速入门到项目实践 / 文青山编著. -- 北京：清华大学出版社，2024. 7. --（跟我一起学人工智能）. -- ISBN 978-7-302-66860-2

Ⅰ. TP181

中国国家版本馆CIP数据核字第2024SM1055号

责任编辑：赵佳霓
封面设计：吴　刚
责任校对：时翠兰
责任印制：沈　露

出版发行：清华大学出版社
网　　址：https://www.tup.com.cn，https://www.wqxuetang.com
地　　址：北京清华大学学研大厦A座　　邮　　编：100084
社 总 机：010-83470000　　邮　　购：010-62786544
投稿与读者服务：010-62776969，c-service@tup.tsinghua.edu.cn
质量反馈：010-62772015，zhiliang@tup.tsinghua.edu.cn
课件下载：https://www.tup.com.cn，010-83470236
印 装 者：三河市人民印务有限公司
经　　销：全国新华书店
开　　本：186mm×240mm　　印　张：33.75　　彩　插：3　　字　数：763千字
版　　次：2024年8月第1版　　印　次：2024年8月第1次印刷
印　　数：1～2000
定　　价：129.00元

产品编号：100987-01

前言

PREFACE

如何学习深度学习呢?

针对这个问题也许每个人都有不同的观点,本书编排的目录也就是笔者的学习过程。

大概是在2022年的某一天,很偶然接触到这个领域,刚开始时没有什么信心,不知自己能否入门,因为对这个领域十分陌生,听起来也非常高端,随便翻翻书,密密麻麻的数学公式看得人头晕眼花,一筹莫展。

本着复杂的知识如果工作需要,则一定要搞懂的心态,立志后坚定地开始了密集的学习,学习的过程是有点忙碌的,每天研究原理和看代码直到深夜,周末也没有休息时间。好在经过一个多月的努力,原来复杂的原理似乎懂了一些,根据原理看开源的代码也没有那么痛苦了,也可以根据工作需要修改开源的代码实现某些功能,虽然还有一点云里雾里的感觉,但至少已经迈出了重要的一步。

赶巧2022年年底在家被闭关半个月。闲着无聊突发奇想重写一些经典模型的代码,以加深对于经典算法的理解,于是完成了本书大部分代码的主体。

然后又在机缘巧合之下跟清华大学出版社的编辑加上了好友,聊着聊着就讨论到了如何学习深度学习,如何用最短的时间入门深度学习,后来也就有了写本书的想法。

因为定位零基础入门,所以从Python基础知识入手并逐步深入机器学习、卷积神经网络、目标检测等经典算法。为了更好地理解这些知识,并在学习过程中逐步提高代码编写能力,本书没有调用知名第三方库实现算法或直接解析开源代码,这是因为第三方库和开源代码往往将细节隐藏得很好,不利于初学者掌握基本原理及代码实现的技巧。本书中的每个例子的每行代码都经过测试,很有参考意义。

至于阅读本书是否需要很强的数学背景,个人感觉读者只要会求导就行。本书涉及的数学知识有些看起来不够"严谨",因为基本上对复杂的数学公式已经通过数值代入进行了"破坏性"计算,有着很详细的步骤,这样做的目的是使读者更好地理解背后的原理。在编程的世界里,复杂的数学公式很多时候就是一个Python函数。作为应用方,个人觉得怎么算、严不严谨,不是很重要的,重要的是理解它背后做了什么事,以及意义是什么。

本书主要内容

第1章从零基础介绍Python基础语法、Python数据处理库NumPy、Pandas、Matplotlib、OpenCV的使用。

第 2 章主要介绍机器学习算法的原理并配有代码实例，方便在理解原理的同时也能写出代码。

第 3 章主要介绍深度学习框架 TensorFlow、Keras、PyTorch 的 API 和网络模型的搭建方法，力保读者能够掌握主流深度学习框架的使用。

第 4 章主要介绍卷积神经网络各种卷积的特性，并通过代码实现了多个经典分类网络。

第 5 章主要介绍目标检测领域中的多个经典算法的原理，并配套展现了代码调试的过程，将算法原理与代码进行结合，方便读者更深入地理解算法原理。

第 6 章主要分享深度学习项目的分析和实现过程。

阅读建议

本书是一本深入浅出的深度学习入门指南，旨在帮助读者快速入门深度学习，所以本书从 Python 的基础开始讲解，逐步引导读者步入机器学习和深度学习的世界。为了确保每个知识点都能被充分理解和掌握，本书也提供了详细的示例代码供读者调试运行，以期更好地理解算法，同时也提供了配套 PPT 和源码供读者参考，通过阅读及调试这些代码可以更好地理解算法原理，并加深对知识的理解。扫描目录上方二维码，可获取本书配套资源。

以下是一些阅读建议。

(1) 全面理解：从第 1 章开始，确保对 Python 的基础语法和数据处理库有足够的了解，这是学习后续章节的基础。

(2) 理论与实践并重：第 2 章深入讲解机器学习算法的原理，并通过配套的代码实例进行练习，帮助读者更好地掌握这些概念。

(3) 熟悉深度学习框架：第 3 章介绍 TensorFlow、Keras 和 PyTorch 等深度学习框架的 API 和模型搭建的方法，这些库的应用是构建深度学习代码的关键工具。

(4) 深化 CNN 知识：第 4 章深入讲解卷积神经网络的各种特性，并通过实战项目来进一步理解分类网络的工作原理。

(5) 目标检测的应用：第 5 章详细描述目标检测领域经典算法的原理，以及算法原理关键的实现代码，尝试理解并调试这些代码，可以加深对于算法的理解与应用。

(6) 项目分析与实现：第 6 章讲解如何分析和实施深度学习项目，这将帮助读者将前面所学的知识应用于实际业务场景。

(7) 定期复习：定期回顾之前的学习内容是非常重要的。可以创建一个学习计划，包括定期复习和实践编程任务。也可参考每节的总结和练习任务。

(8) 扩展阅读：如果可能，则可以利用其他资源(如书籍、在线课程和博客文章)来补充本书的内容，以获得更丰富的视角和深入的见解。

无论你是初学者还是有一定基础的学习者都可以通过阅读本书，并结合本书提供的学习路线、代码资源等更好地提升自己。

致谢

感谢清华大学出版社赵佳霓编辑的耐心和支持！感谢我的家人，尤其是我的爱人祁超超女士在本书写作过程中的支持和鼓励！

由于时间仓促，书中难免存在不妥之处，请读者见谅并指正。

文青山

2024年6月

目录
CONTENTS

教学课件(PPT)

本书源码

第1章 Python 编程基础

深度学习实战的基础是编程，而 Python 是深度学习的主流编程语言，掌握 Python 编程基础，并同时掌握一些机器学习、深度学习常用库的使用方法，有助于理解机器学习原理及算法的代码实现。本章从 Python 基础语法讲起，将详细地介绍学习深度学习需要掌握的 NumPy、Pandas、Matplotlib 库的重要 API，并就计算机视觉领域中需要用到的 OpenCV 常用函数进行介绍，只有学习并熟练掌握本章知识，才可以为后面学习深度学习框架及深度学习算法代码打下坚实的基础。

1.1 环境搭建

万事开头难，所有事情都需要从零开始，怎么从零开始呢？

当然是安装环境，每台计算机，每个程序员，任何一门编程语言第 1 步需要做的都是安装开发环境，Python 环境怎么安装呢？

建议使用 Anaconda 集成开发环境进行开发环境的管理，访问 Anaconda 网站，然后单击 Download 按钮进行下载，如图 1-1 所示。

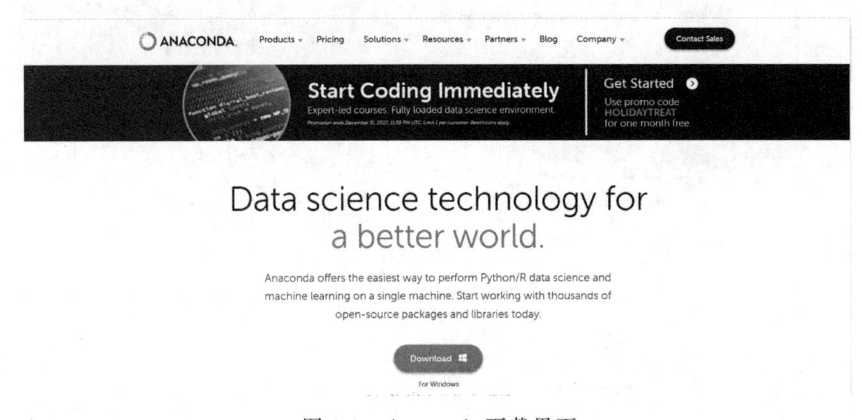

图 1-1　Anaconda 下载界面

一路默认单击 Next 按钮，当安装到如图 1-2 所示时，需要选中 Add Anaconda3 to my PATH environment variable，将 Anaconda3 添加到环境变量（环境变量提供了一个快捷访问的方式）。

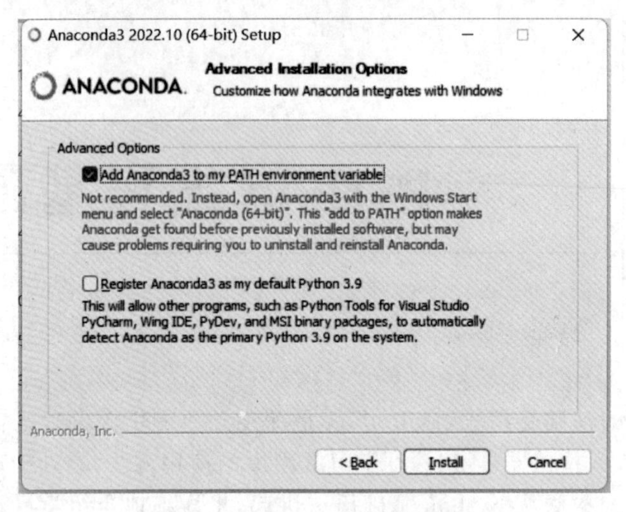

图 1-2 Anaconda 选中环境变量

在 Windows 计算机的开始菜单，单击 Anaconda Prompt 应用，如图 1-3 所示。

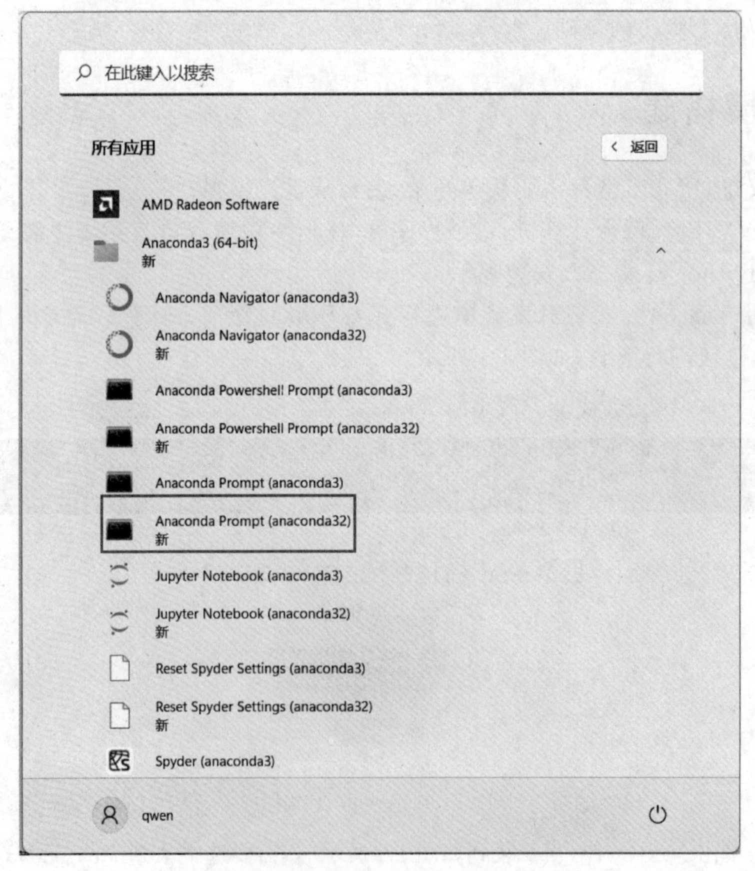

图 1-3 单击 Anaconda Prompt

在 Anaconda Prompt 命令行管理界面中，使用 conda create -n StudyDNN python=3.8 命令创建一个开发环境，命名为 StudyDNN，并输入 y 进行相关库的安装，如图 1-4 所示。

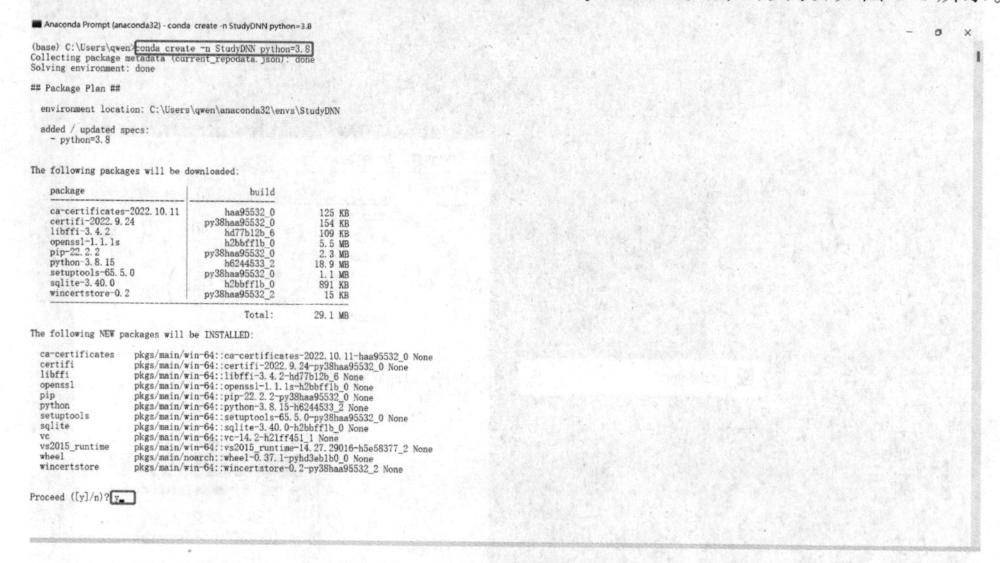

图 1-4　conda 命令创建开发环境

接下来，双击 PyCharm-community-2020.1.exe，在弹出的窗口中连续单击 Next 按钮，如图 1-5 所示。

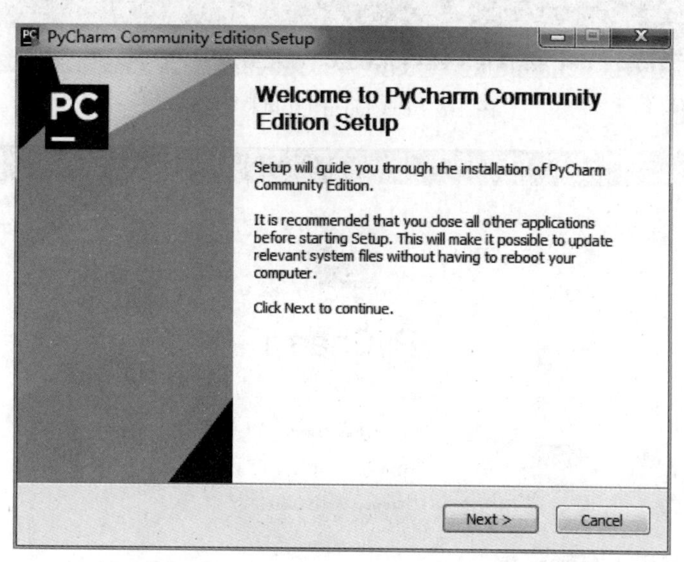

图 1-5　PyCharm 初始安装界面

安装成功后，在 Windows 的开始菜单中，打开 PyCharm 后有两种 UI 风格供选择，如图 1-6 所示，根据自己的喜好，可以选择黑夜模式或者白天模式。

单击 Creat New Project 创建一个新的项目，如图 1-7 所示。

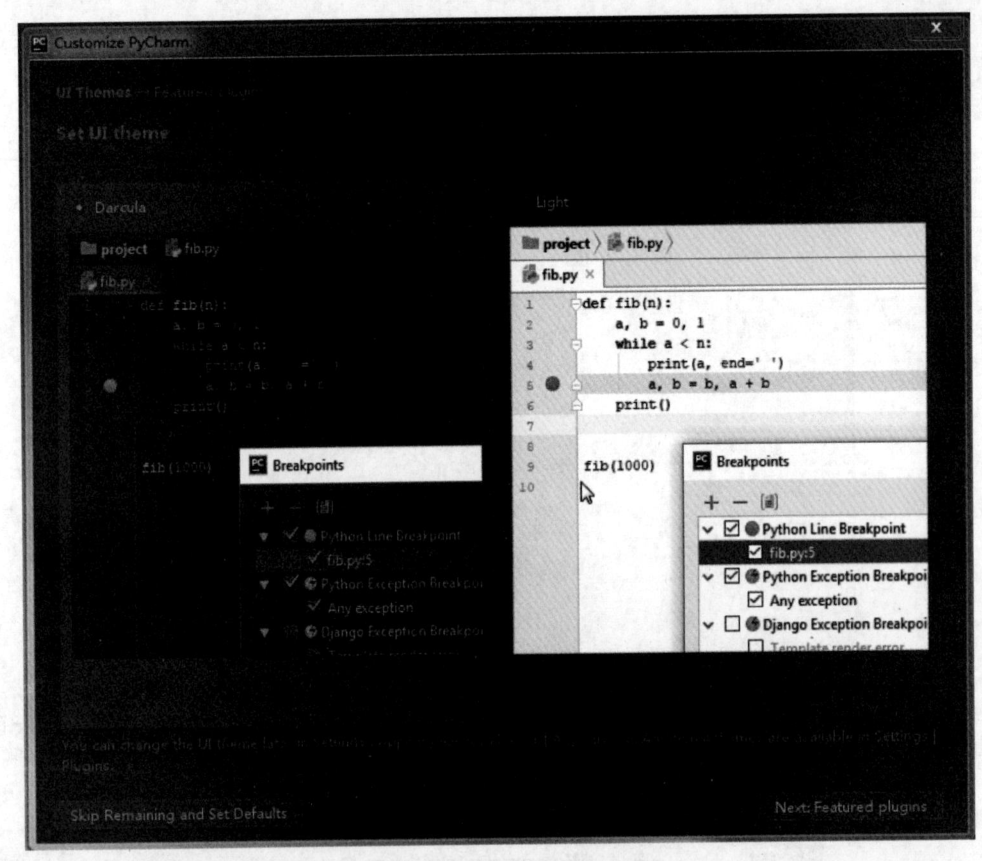

图 1-6　PyCharm UI 风格选择

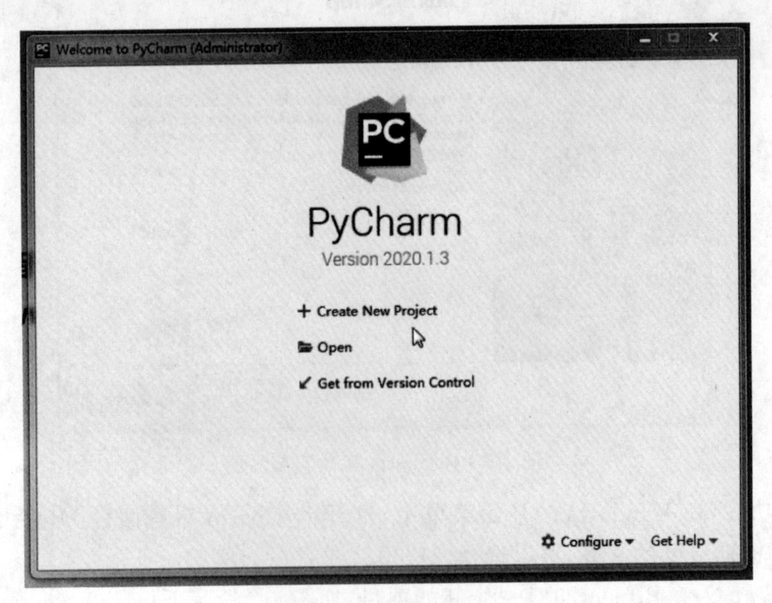

图 1-7　创建项目

在弹出的新页面中选择 Existing interpreter 模式，而不是使用 Virtualenv 模式，因为 Virtualenv 环境更适合多项目多人协作而不是学习时使用，如图 1-8 所示。

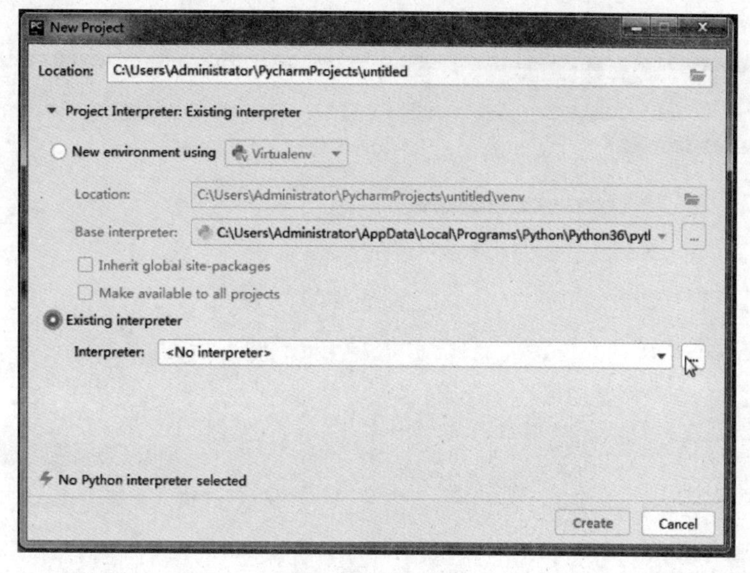

图 1-8 项目模式

单击“…”按钮并在弹出的窗口中单击 Conda Environment，在 Interpreter 右侧选择 StudyDNN 环境下的 python. exe，然后单击 OK 按钮，如图 1-9 所示。

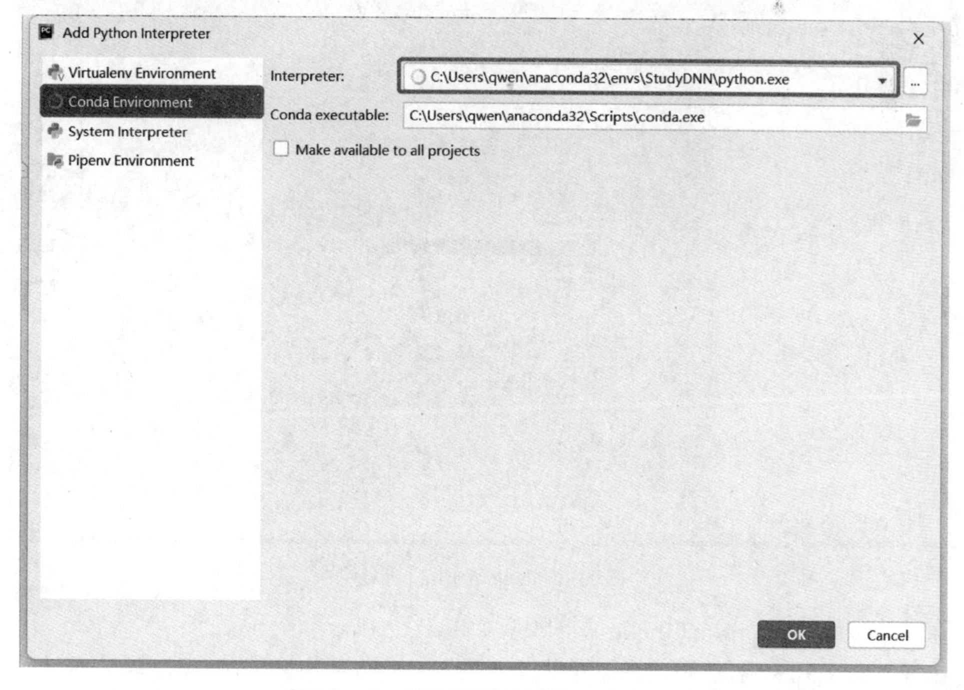

图 1-9 StudyDNN 环境下的 python. exe

项目创建成功后，默认的名称是 untitled，此时需要创建一个.py 文件，用来存放写的代码。选中 untitled 后右击菜单并依次选择 New→Python File，如图 1-10 所示，然后输入文件名称 helloWorld.py，如图 1-11 所示。

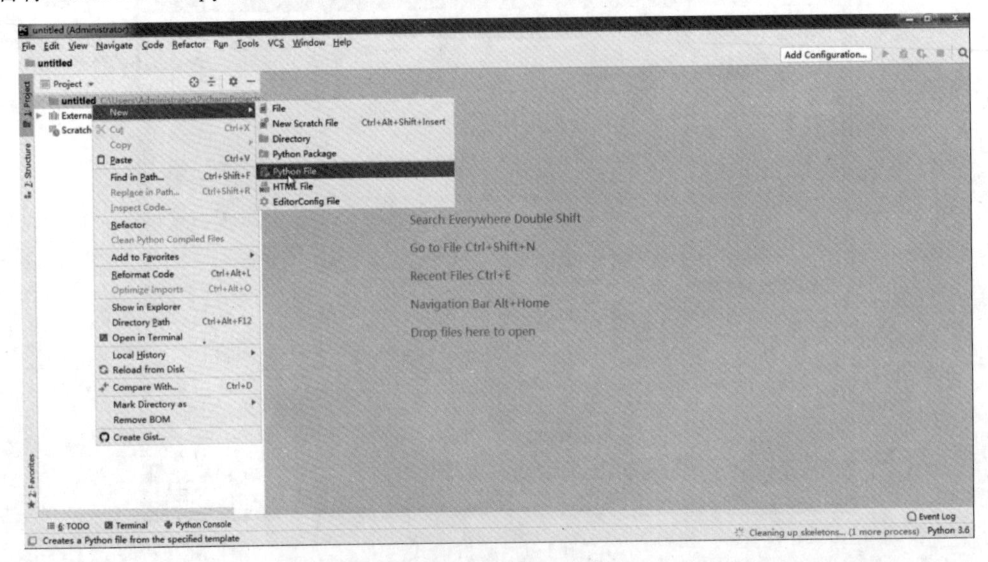

图 1-10　创建.py 文件

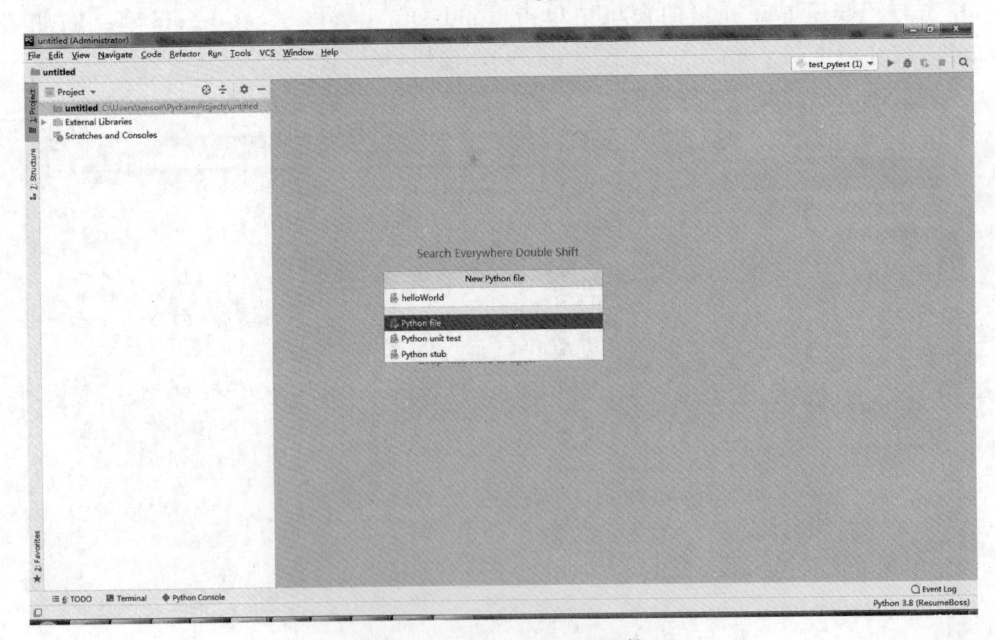

图 1-11　helloWorld.py 文件

在空白区域，输入 print('hello world')，如图 1-12 所示。

然后右击空白区域，在弹出的菜单中选择 Run 'helloWorld.py'，程序就会在计算机中运行，如图 1-13 所示。

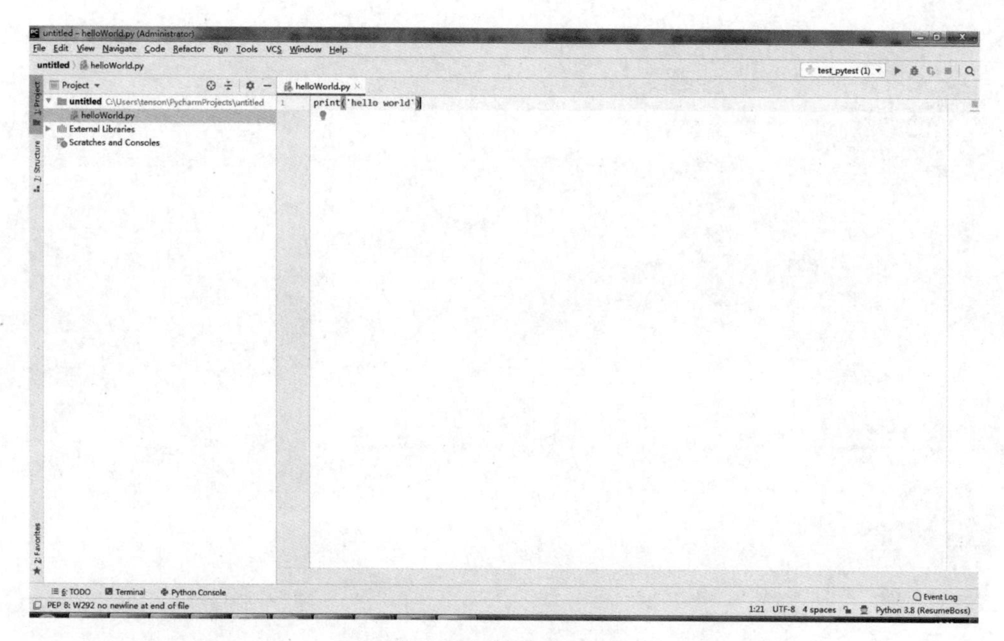

图 1-12　helloWorld. py

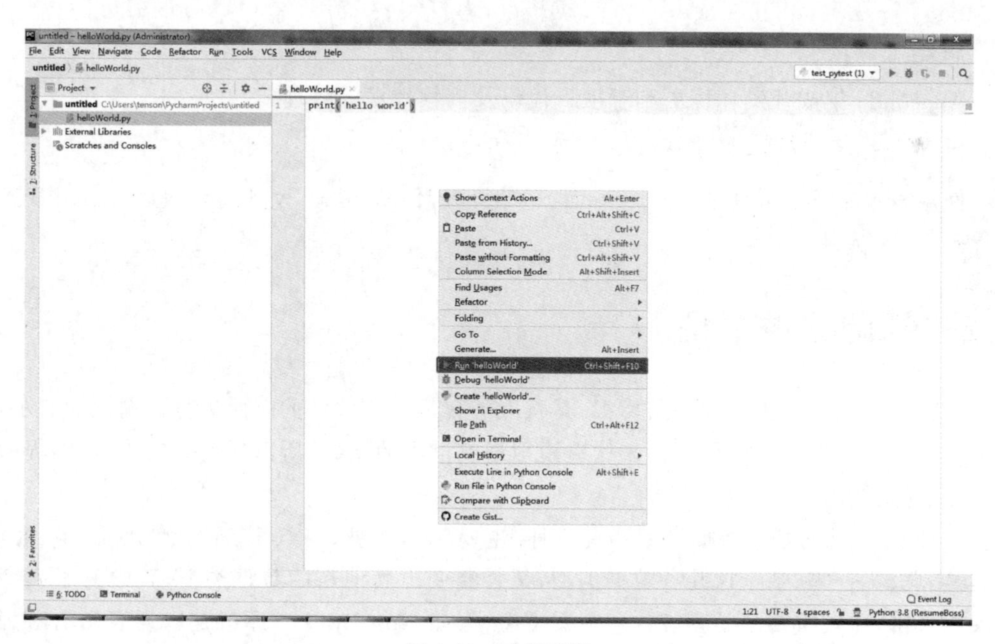

图 1-13　运行程序

运行程序后得到的结果如图 1-14 所示，在 3 号区域输出了一个字符串 hello world。1 号区域用来管理 Python 源文件，即编写的代码文件；2 号区域为代码的内容；3 号区域为代码执行结果，至此 Python 开发环境就搭建好了。

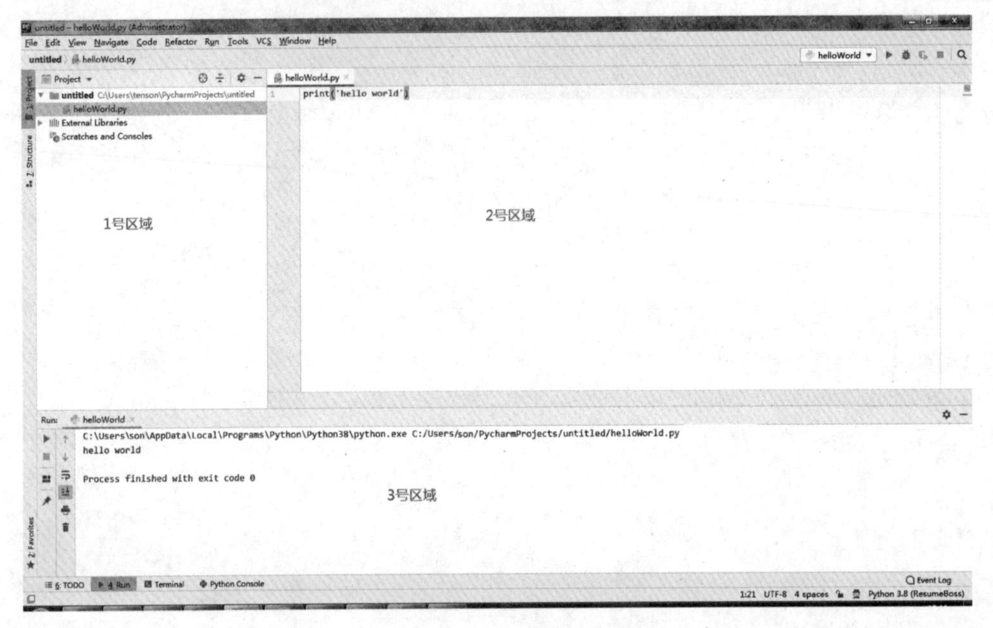

图 1-14　运行结果

总结

Anaconda 集成开发环境更利于项目代码及依赖环境的管理。

练习

使用 conda 命令创建一个 Python 3.9 的开发环境，并在 PyCharm 中运行 helloWorld 的代码。

1.2　基础数据类型

编程是对现实世界的抽象与模拟，是将人类智慧一些共有的东西进行数据化的展示，而数据类型就是在编程的世界中用来构造模拟现实世界的基础，所以任何一门语言，编程的基础都是数据类型。

Python 常见的基本数据类型共有 8 种，主要包括整型、浮点型、布尔型、字符串、元组、列表、字典、集合等，基于这几种数据类型，又会演变出其他新的数据类型，其中对于人工智能编程来讲，比较重要的是列表这种数据类型，例如 NumPy 这个库就是以列表的语法为基础的。

1.2.1　数值型

数值型主要包括 3 种数据类型，即整数、浮点数、复数。Python 中的数字支持一般的数学运算，如“+”代表加法、“-”代表减法、“*”代表乘法、“**”代表乘方。

在进行算术运算之前，需要了解怎样定义一个变量，以及怎样定义一个变量并直接使用，如“变量名称＝值”，等号为赋值的意思，即将右边的值赋给左边的变量。同时 print(变量)为计算机输出当前变量值的方法。数值型常见的操作，代码如下：

```
#第 1 章/数值型操作.py
#定义一个整型变量 a
a = 12
#定义一个整型变量 b
b = 14
#定义一个变量 c,用来存储 a + b,即存储加法计算的结果
c = a + b
#输出变量 c 的结果
print(c)
print('**********************************')
#定义一个变量 c,用来存储 a - b,即存储减法计算的结果
c = a - b
print(c)
print('**********************************')
#定义一个变量 c,用来存储 a *b,即存储乘法计算的结果
c = a *b
print(c)
print('**********************************')
#定义一个变量 c,用来存储 a / b,即存储除法计算的结果
c = a / b
print(c)
print('**********************************')
#定义一个变量 c,用来存储 a //b,即存储整除的结果
c = a //b
print(c)
print('**********************************')
#定义一个变量 c,用来存储 a % b,即存储求余的结果
c = a % b
print(c)
d = complex(3)              #定义一个复数
print(d, type(d))
print('**********************************')
#常用的数字函数
print(abs(-3.) + 3e-3)     #取-3.0 的绝对值, 3e-3 为科学记数法,即 0.003
print(pow(3, 2))            #3 的 2 次方
print(round(3.1496, 2))     #对 3.1496 进行四舍五入,并保留两位小数
```

运行结果如下：

```
26
**********************************
-2
**********************************
168
**********************************
0.8571428571428571
**********************************
```

```
0
**********************************
12
(3+0j) <class 'complex'>
**********************************
3.003
9
3.15
```

注意：type(var)函数用于输出某个变量的数据类型。浮点数的表示方法可以为0.03或者3e-2。abs(number)、round(number, ndigits)、pow(number, exp)等均为Python内置函数。

1.2.2 字符串

字符串即文本，在Python中用单引号、双引号、三引号包含的内容都视为字符串，大多数程序使用单引号或者双引号，示例代码如下：

```
#第1章/字符串1.py
#(1)使用单引号
V1 = 'qwentest123'
#(2)使用双引号
V2 = "qwentest123"
#(3)使用三引号,例如今天的热搜
hotSrh = """
1一个班17人被录取为空军飞行学员
2中国电竞用户规模达4.84亿人
"""
```

注意：单引号里不能有单引号，双引号里不能有双引号，三引号里不能有三引号，引号的个数是成对的。

并不是输入什么字符串就会输出什么样的字符串，某些特殊格式的字符串输出时会有变化，这些字符串也就是常说的转义字符，代码如下：

```
#第1章/字符串2.py
filePath = "D:\\课程ppt\\510"                #这是一个转义之后的文件路径
filePath2 = 'D:\课程ppt\510'                 #这是从计算机中复制的文件路径
say = "路漫漫其修远兮\n吾将上下而求索"
print(filePath)
print(filePath2)
print(say)
```

运行结果如下：

```
D:\课程ppt\510
D:\课程pptň
```

```
路漫漫其修远兮
吾将上下而求索
```

注意：\\被转义成\，而\510 被转义成 ñ，\n 被转义成换行符。一般说来当遇到\这个字符时可能会有变化。

字符串还包括很多常用的函数，这些函数把字符串常用的方法进行了封装，只需调用它们便可以完成相关工作，代码如下：

```
#第 1 章/字符串 3.py
poetry = "最是那一低头的温柔，像一朵水莲花不胜凉风的娇羞"
author = '徐志摩'
print('作者：' + author + '\n' + poetry)                #+为字符串拼接
print('***********************************')
poerty1 = poetry.replace('娇羞', '。')                  #对指定的字符串进行替换
print(poerty1)
print('***********************************')
v = ' qwen test123 '
print(v.strip(' '))                                     #去除两边指定的空格字符
print(v.rstrip(' '))                                    #去除左边的指定字符
print(v.lstrip(' '))                                    #去除右边指定的字符
print('***********************************')
jobInfo = '山哥|男|18 岁|3 年'
print(jobInfo.split('|'))                               #按指定的字符串有规律地进行分割
```

运行结果如下：

```
作者：徐志摩
最是那一低头的温柔，像一朵水莲花不胜凉风的娇羞
***********************************
最是那一低头的温柔，像一朵水莲花不胜凉风的。
***********************************
qwen test123
  qwen test123
qwen test123
***********************************
['山哥', '男', '18 岁', '3 年']
```

注意：str.split(char)会将源字符串分割成多个，从而组成一个列表，[]列表中每个逗号为一个元素，在上述代码中源字符串被分割成 4 部分。

字符串还提供了一个下标取值和切片的操作，例如需要从“最是那一低头的温柔，像一朵水莲花不胜凉风的娇羞”中提取“一”这个字符，代码如下：

```
#第 1 章/字符串 4.py
poetry = "最是那一低头的温柔，像一朵水莲花不胜凉风的娇羞"
print(poetry[3])
```

运行代码后的结果如下：

```
一
```

从左往右数为正整数。同样地，也可以从右往左数，例如取“风”字，则代码如下：

```
#第 1 章/字符串 4.py
poetry = "最是那一低头的温柔,像一朵水莲花不胜凉风的娇羞"
print(poetry[-4])
```

运行结果如下：

```
风
```

注意： 从左往右数为正整数。从后往前数则是从－1 开始。这在 Python 中叫下标取值。下标取值可以在元组、列表等数据类型中使用，同时也可以在很多自定义类型中使用。

还是上面的例子，如果想取“一低头的温柔”，则又该怎么取呢？可以想象一下，如果手里有两把标尺，则只需同时按下去取值，代码如下：

```
#第 1 章/字符串 4.py
poetry = "最是那一低头的温柔,像一朵水莲花不胜凉风的娇羞"
print(poetry[3:9])
```

运行结果如下：

```
一低头的温柔
```

为什么不是 poetry([3:8])呢？因为语法的规则是左闭右开。想取的字符，在右边需要＋1。如果想取“一朵水莲花”，则代码如下：

```
#第 1 章/字符串 4.py
poetry = "最是那一低头的温柔,像一朵水莲花不胜凉风的娇羞"
print(poetry[-12:-7])
```

运行结果如下：

```
一朵水莲花
```

如果想取“娇羞”，则可以省略右边的区间，代码如下：

```
#第 1 章/字符串 4.py
poetry = "最是那一低头的温柔,像一朵水莲花不胜凉风的娇羞"
print(poetry[-2:])
```

运行结果如下：

```
娇羞
```

同样，如果要取"最是那一低头的温柔"，则代码如下：

```
#第 1 章/字符串 4.py
poetry = "最是那一低头的温柔,像一朵水莲花不胜凉风的娇羞"
print(poetry[:9])
```

运行结果如下：

```
最是那一低头的温柔
```

如果取"是一头温"，则代码如下：

```
#第 1 章/字符串 4.py
print(poetry[1:9:2]) #从下标 1 到下标 8 为止,但是每两个下标取一个值
```

运行结果如下：

```
是一头温
```

如果字符串很多，并且肉眼无法获取是第几个位置，则只需结合 str.index(string)函数获取位置，代码如下：

```
#第 1 章/字符串 4.py
poetry = "最是那一低头的温柔,像一朵水莲花不胜凉风的娇羞"
start_index = poetry.index('一朵')
end_index = poetry.index('不胜')
print(poetry[start_index:end_index])
```

运行结果如下：

```
一朵水莲花
```

注意：下标取值从左往右数是从 0 开始的；从后往前数是−1 开始的；

切片取值是左闭右开，右边的下标值需要+1；

切片的完整语法为 list[startIndex:endIndex:sep]。

1.2.3 元组

元组是 Python 中的一种数据类型，满足(object,object)的格式即为元组。object 代指任意数据类型。

元组也支持下标取值和切片的语法，并同时支持+号拼接，示例代码如下：

```
#第 1 章/元组 1.py
v2 = ('qwentest123', 1, 2, 3, 99.99, ('a', 'b', 'c'))
print(type(v2))
print(v2[1])            #获取第 1 个值
print(v2[1:4])
```

```
#如果元组内仍然是元组,则可以通过链式表达,获取相应的值
print(v2[-1][:2])
v1 = (12, 2, 33, 47, 59, 60, 7)
v2 = ('qwentest123', 1, 2, 3, 99.99, ('a', 'b', 'c'))
print(v1 + v2)              #元组的拼接
print(v1.index(12))         #获取元素值在元组中的位置
print(v1.count(60))         #获取元素值在元组中出现的次数
```

运行结果如下：

```
<class 'tuple'>
1
(1, 2, 3)
('a', 'b')
(12, 2, 33, 47, 59, 60, 7, 'qwentest123', 1, 2, 3, 99.99, ('a', 'b', 'c'))
0
1
```

注意：不能对元组中的元素值进行修改、删除、增加操作，这是元组的一大特性。

1.2.4 列表

列表是 Python 中一种可以增加、修改、删除的数据类型，满足[object,object]格式。列表支持下标取值和切片，对同样的两个列表进行拼接使用+号，示例代码如下：

```
#第 1 章/列表 1.py
anyList = [1, 3.14, 'qwentest123', (11, 12), [110, 111]]
print(type(anyList))
print(anyList[1])               #列表的下标取值
print(anyList[1:3])             #列表的切片
#可以通过链式表达,获取 anyList 列表中的 11 或者 111
print(anyList[-2][0])
print(anyList[-1][1])
l1 = [1, 2, 3, 4, 5, 6, 7, 8, 9, 10]
l2 = [11, 12, 13, 14, 15, 16, 17, 18, 19, 20]
#列表的拼接
print(l1 + l2)
```

运行结果如下：

```
<class 'list'>
3.14
[3.14, 'qwentest123']
11
111
[1, 2, 3, 4, 5, 6, 7, 8, 9, 10, 11, 12, 13, 14, 15, 16, 17, 18, 19, 20]
```

列表的增加可以使用 list.append(object)、list.insert(index,object)、list.extend(iterable)方法，示例代码如下：

```
#第 1 章/列表 2.py
spaceList = []
spaceList.insert(0, 1)              #向 0 这个位置插入 1
print(spaceList)
spaceList.append(2)                 #向列表的最后位置插入 2
print(spaceList)
spaceList.extend([3, 4, 5])         #向列表一次插入多个值
print(spaceList)
```

运行结果如下：

```
[1]
[1, 2]
[1, 2, 3, 4, 5]
```

注意：list. extend(iterable)中的 iterable 是一个可迭代类型，例如字符串、元组、列表；iterable 是一个整体的参数，而不是多个参数。

列表的修改通过下标或切片之后在指定位置赋值即可，示例代码如下：

```
#第 1 章/列表 2.py
number = [0, 1, 2, 3, 4, 5, 6, 7, 8, 9]
number[0] = 11                    #修改列表中某个位置的值
print(number)
number[0:2] = ['a', 'b']          #修改列表中多个位置的值
print(number)

number = [0, 1, 2, 3, 4, 5, 6, 7, 8, 9]
number[10] = 11                   #本行代码会报错,因为不存在下标 10
print(number)
```

运行结果如下：

```
[11, 1, 2, 3, 4, 5, 6, 7, 8, 9]
['a', 'b', 2, 3, 4, 5, 6, 7, 8, 9]
Traceback (most recent call last):
  File "D:\DLAI\qwenAILearn\untitled9\列表 2.py", line 17, in <module>
    number[10] = 11 #本行代码会报错,因为不存在下标 10
IndexError: list assignment index out of range
```

注意：可以一次性修改多个位置的值，但是修改的总数量与赋值的总数量要相等。

列表的删除可以使用 list. pop(index)或者 del(list[index])方法，示例代码如下：

```
#第 1 章/列表 2.py
number = [0, 1, 2, 3, 4, 5, 6, 7, 8, 9]
number.pop(0)                   #pop 传的是位置
print(number)
del (number[1])                 #删除列表中的第几个
```

```
print(number)
newNum = [1,2,3,1,4]
newNum.remove(1)            #删除的是列表中的元素，但是只能删除重复的第 1 个
print(newNum)
```

运行结果如下：

```
[1, 2, 3, 4, 5, 6, 7, 8, 9]
[1, 3, 4, 5, 6, 7, 8, 9]
[2, 3, 1, 4]
```

对列表进行排序可以使用 list.sort()，对列表进行倒置可以使用 list.reverse()，示例代码如下：

```
#第 1 章/列表 2.py
newNum = [1, 2, 3, 1, 4, 5, 6, 7, 8, 9]
newNum.sort()                        #排序后从小到大排列
print(newNum)
newNum.sort(reverse=True)            #排序后从大到小排列
print(newNum)
newNum2 = [1, 2, (2, 3), [3, 4]]     #所谓倒置，是将列表最后的元素值与第 1 个进行交换
newNum2.reverse()
print(newNum2)
newNum2 = [1, 2, (2, 3), [3, 4]]
newNum2.sort()                       #报错，只能对相同数据类型进行排序
print(newNum2)
```

运行结果如下：

```
[1, 1, 2, 3, 4, 5, 6, 7, 8, 9]
[9, 8, 7, 6, 5, 4, 3, 2, 1, 1]
[[3, 4], (2, 3), 2, 1]
Traceback (most recent call last):
  File "D:\DLAI\qwenAILearn\untitled9\列表 2.py", line 41, in <module>
    newNum2.sort() #报错，只能对相同数据类型进行排序
TypeError: '<' not supported between instances of 'tuple' and 'int'
```

注意：列表的任何操作是对源列表进行的操作，改变的是列表本身；对列表进行排序时，列表中的元素的数据类型要相同，否则将会报错。

1.2.5 字典

字典是一组以{关键字：内容}表示的一组内容，类似于新华字典，其语法格式满足{'key1'：object，'key2'：object}。字典的 key 不能重名，如果重名，则后面的 key 会覆盖前面的 key。key 的数据类型只能为不可改变的数据类型，例如 int、float、str、tuple，不能为 list、dict 等可变数据类型，示例代码如下：

```
#第 1 章/字典 1.py
#字典的定义
me = {
    'name': '文山', 'age': 18, 'height': 185,
    'love': ['读书', '运动'], 'study': ('xxx 高中', 'xx 大学'),
    'experience': [
        {'time': '2010 年', 'job': 'C#开发工程师'},
        {'time': '2014 年', 'job': '高级开发工程师'}
    ]
}
print(type(me))
#key 不能重复,如果重复,则后面的 key 会覆盖前面的内容
me = {'keyname': 1, 'keyname': 2}
print(me)
#key 的类型可以是 int、float、str、tuple 等不可变内容类型
dict = {1: 2, 1.0: 2.0, 'str': 'string', (1, 2): (2, 1)}
print(type(dict))
#以下代码会报错,因为 key 不能为可变内容的数据类型
dict = {[1, 2]: [2, 3], {'u': '1'}: 1}
print(type(dict))
```

运行结果如下：

```
<class 'dict'>
{'keyname': 2}
<class 'dict'>
Traceback (most recent call last):
  File "D:\DLAI\qwenAILearn\untitled9\字典 1.py", line 19, in <module>
    dict = {[1, 2]: [2, 3], {'u': '1'}: 1}
TypeError: unhashable type: 'list'
```

注意：字典的键一般会被设置为字符串类型，方便统一应用与管理，不能为 list 或者 dict 等可变内容的数据类型。

字典中 key 值的获取可以使用 dict[key]或者 dict.get(key)方法，示例代码如下：

```
#第 1 章/字典 2.py
#字典的定义
me = {
    'name': '文山', 'age': 18, 'height': 185,
    'love': ['读书', '运动'], 'study': ('xxx 高中', 'xx 大学'),
    'experience': [
        {'time': '2010 年', 'job': 'C#开发工程师'},
        {'time': '2014 年', 'job': '高级开发工程师'}
    ]
}
print(me['study'])                    #dict[key]获取键值
print(me.get('name'))                 #dict.get(key)获取键值
print(me.get('love')[0])              #链式取值
print(me.get('study')[-1])
print(me.get('experience')[0].get('job'))
```

运行结果如下：

```
('xxx 高中', 'xx 大学')
文山
读书
xx 大学
C#开发工程师
```

注意：当 dict[key]获取键值时，如果 key 不存在，则会抛出 KeyError 异常，而当 dict.get(key)获取键值时，如果 key 不存在，则返回 None，所以推荐使用 dict.get(key)。

字典的拼接需要使用 dict.update(dict)方法，不能再使用＋号进行连接，示例代码如下：

```
#第 1 章/字典 2.py
dict1 = {'user': 'qwentest123'}
dict2 = {'pwd': '123456'}
dict1.update(dict2)            #把 dict2 更新到 dict1 中
print(dict1)
print(dict1 + dict2)           #会报错,因为字典拼接不能用+号
```

代码运行结果如下：

```
{'user': 'qwentest123', 'pwd': '123456'}
Traceback (most recent call last):
  File "D:\DLAI\qwenAILearn\untitled9\字典 2.py", line 21, in <module>
    print(dict1 + dict2)      #会报错
TypeError: unsupported operand type(s) for +: 'dict' and 'dict'
```

字典是一个可增加、修改、删除的数据类型，向字典增加新值只需 dict[key]＝object，当 key 不存在时，赋一个新值即为增加；如果 key 存在，则为修改。删除可以调用 dict.pop(key)或者 del 来完成，示例代码如下：

```
#第 1 章/字典 3.py
dict = {}
dict['newKey'] = '增加'
print('给 dict 新增了一个 key:value 值', dict)
dict['newKey'] = 2              #dict 修改
print('把 dict 中 key 为 newKey 的值进行了修改', dict)
k = {'u': 'qwentest', 'p': '1111'}
k.pop('u')                      #按 key 删除
print(k)
del (k['p'])                    #删除 dict 中的某个键和值
print(k)
```

运行结果如下：

```
给 dict 新增了一个 key:value 值 {'newKey': '增加'}
把 dict 中 key 为 newKey 的值进行了修改 {'newKey': 2}
{'p': '1111'}
{}
```

注意：当字典的 key 不存在时为增加，当字典的 key 存在时为修改，这是字典的一个重要特性。

字典还有一些高频使用函数，例如 dict. values()、dict. keys()、dict. items()等，示例代码如下：

```
#第 1 章/字典 3.py
#获取字典的值以组成一个序列
print(dict.values())
print('---------------------------------')
#获取字典的 key 以组成一个序列
print(dict.keys())
print('---------------------------------')
#获取字典的每个(key- value)以组成一个序列
print(dict.items())
print('---------------------------------')
#如果 name 存在，则为获取，否则将 name 的初始值设置为 None
print(dict.setdefault('vname'))
print('---------------------------------')
#将不存在的 myname 的初始值设置为 qwentest123
dict.setdefault('myvname', 'qwentest')
print(dict)
```

运行结果如下：

```
dict_values([2])
---------------------------------
dict_keys(['newKey'])
---------------------------------
dict_items([('newKey', 2)])
---------------------------------
None
---------------------------------
{'newKey': 2, 'vname': None, 'myvname': 'qwentest'}
```

1.2.6 集合

集合由无序不重复元素组成，用大括号{}来表示，但是如果创建一个空集合，则必须用 set()而不是{}。集合的主要功能是用来去重或者取两个集合的交集或并集等，示例代码如下：

```
#第 1 章/集合 1.py
#此时为空字典
s = {}
print(type(s))
#创建空的集合
s = set({})
print(type(s))
```

```
print('-----------------------')
s = {'apple', 'orange', 'apple', 'pear', 'orange', 'banana'}
print(s)
a = set('abracadabra')
b = set('alacazam')
#在集合 a 中的字母,但不在集合 b 中
print(a - b)
print('------------------------')
#在集合 a 或 b 中的字母
print(a | b)
print('------------------------')
#在集合 a 和 b 中都有的字母
print(a & b)
print('-------------------------')
#在集合 a 或 b 中的字母,但不同时在集合 a 和 b 中
print(a ^ b)
```

运行结果如下：

```
<class 'dict'>
<class 'set'>
-----------------------
{'apple', 'orange', 'banana', 'pear'}
{'b', 'r', 'd'}
------------------------
{'c', 'd', 'l', 'm', 'a', 'b', 'r', 'z'}
------------------------
{'c', 'a'}
-------------------------
{'m', 'b', 'r', 'z', 'd', 'l'}
```

注意：集合一旦参与运算就会自动去重，并且去重后的顺序是不固定的。

1.2.7 数据类型的转换

数据类型之间是可以互相转换的，其转换的通用方式是“数据类型(变量)”，示例代码如下：

```
#第 1 章/数据类型转换 1.py
#数字与字符串数值进行转换
s = 1
print(float(s), str(s))                #转换为浮点数,字符串
s = 3.14
print(int(s), str(s))                  #转换为 int 型的 3,字符串
s = '3.14'
print(int(float(s)))                   #将浮点数的字符串转换为 int 型时,
                                       #要先转换成 float,再转换成 int

print('########################')
#字符串与 list、tuple、set 转换
```

```
s = 'abcdefghijla'
print(list(s))                          #转换为列表
print(tuple(s))                         #转换为元组
print(set(s))                           #转换为集合,但是会自动去重
print('########################')
#将元组转换成 str、list、set
s = ('a', 'b', 'c', 'd', 'e', 'f', 'g', 'h', 'i', 'j', 'l', 'a')
print(''.join(s))                       #将元组转换为字符串
print(list(s))                          #转换为列表
print(set(s))                           #转换为集合,但是会自动去重
print('########################')
s = ['a', 'b', 'c', 'd', 'e', 'f', 'g', 'h', 'i', 'j', 'l', 'a']
print(''.join(s))                       #将列表转换为字符串
print(tuple(s))                         #转换为元组
print(set(s))                           #转换为集合,但是会自动去重
print('########################')
s = [1, 2, 3, 4, 5]
print(''.join(s))                       #报错,数字不能直接转换
```

代码的运行结果如下：

```
1.0 1
3 3.14
3
########################
['a', 'b', 'c', 'd', 'e', 'f', 'g', 'h', 'i', 'j', 'l', 'a']
('a', 'b', 'c', 'd', 'e', 'f', 'g', 'h', 'i', 'j', 'l', 'a')
{'c', 'j', 'e', 'l', 'a', 'f', 'd', 'g', 'i', 'h', 'b'}
########################
abcdefghijla
['a', 'b', 'c', 'd', 'e', 'f', 'g', 'h', 'i', 'j', 'l', 'a']
{'c', 'j', 'e', 'l', 'a', 'f', 'd', 'g', 'i', 'h', 'b'}
########################
abcdefghijla
('a', 'b', 'c', 'd', 'e', 'f', 'g', 'h', 'i', 'j', 'l', 'a')
{'c', 'j', 'e', 'l', 'a', 'f', 'd', 'g', 'i', 'h', 'b'}
########################
Traceback (most recent call last):
  File "D:\DLAI\qwenAILearn\untitled9\数据类型转换 1.py", line 28, in <module>
    print(''.join(s)) #报错,数字不能直接转换
TypeError: sequence item 0: expected str instance, int found
```

注意：将字符串浮点数转换 int 时需要先转换成 float，再转换成 int。''.join(iterable)函数是将序列拼接在一起以组成一个字符串，如果内容是 int 型，则需要''.join([str(i) for i in s])。

总结

Python 基础数据类型中列表、字典、字符串是核心类型，其中列表切片的语法最为重要。

练习

本书配套资源"Python 基础语法练习 1.txt"中有专门针对基础数据类型的题目,可以练习使用。

1.3 条件语句

Python 数据类型 int、float、str、tuple、list、dict、set 构成了数据结构的基础,对这些数据进行过滤就需要条件语句。例如陶喆《爱我还是他》中的一段歌词:

"你爱我还是他
是不是真的他有比我好
你为谁在挣扎
你爱我还是他
就说出你想说的真心话
你到底要跟我 还是他"

要来表现"爱我还是他",又该怎么进行呢？这时,就需要使用 Python 中的 if 语句,其语法结构如下:

```
if 条件表达式:
        语句块
elif 条件表达式:
        语句块
elif 条件表达式:
        语句块
else:
        语句块
```

翻译成可解释性语言为

```
如果满足条件:
     执行 语句块
否则如果满足条件:
     执行 语句块
否则如果满足条件:
     执行 语句块
都不满足:
     执行语句块
```

elif 为 else if 的缩写,elif 和 else 也可以省略:

```
if 条件表达式:
     语句块
```

也可以表达为

```
if 条件表达式:
     语句块
else:
     语句块
```

那么，怎么用代码实现“爱我还是他”呢？代码如下：

```
#第 1 章/条件 1.py
import time

youLove = input('告诉我，你到底喜欢谁(me 或者 him ?)：')
if youLove == 'him':
   print('如果喜欢他，则 3s 后播放歌曲《伤心太平洋》')
   time.sleep(3)
   print("""一波还未平息
           一波又来侵袭
           一波还来不及
           一波早就过去
           深深太平洋底深深伤心""")
elif youLove == 'me':
   print('如果喜欢俺，则 3s 后播放歌曲《咱们结婚吧！》')
   time.sleep(3)
   print("""好想和你拥有一个家
                这一生 最美的梦啊
                有你陪伴我同闯天涯
                哦 My Love 咱们结婚吧""")
else:
   print('如果都不是，则 3s 后播放歌曲《洗刷刷》')
   time.sleep(3)
   print("""冷啊冷 疼啊疼 哼啊哼
                我的心 哦
                等啊等 梦啊梦 疯啊疯""")
```

运行结果如下：

```
告诉我，你到底喜欢谁(me 或者 him ?)：me
如果喜欢俺，则 3s 后播放歌曲《咱们结婚吧！》
好想和你拥有一个家
                这一生 最美的梦啊
                有你陪伴我同闯天涯
                哦 My Love 咱们结婚吧
```

注意：代码中调用了 time 模块，time. sleep(3)即程序会在此暂停 3s。input()为 Python 中等候命令行输入的语句。当输入 me 回车，3s 后程序会输出《咱们结婚吧！》的歌词；当输入 him 回车，3s 后程序会输出《伤心太平洋》的歌词；当输入其他，3s 后输出《洗刷刷》的歌词。

条件表达式中的判断符号主要使用“＞”大于、“＜”小于、“＞＝”大于或等于、“＜＝”小于或等于、“！＝”不等于、“＝＝”等于、“in”存在。

“＞”大于、“＞＝”大于或等于、“＝＝”等于的用法，代码如下：

```
#第 1 章/条件 2.py
#>、>=、== (大于、大于或等于、等于)
result = 2 > 2
print(result)
```

```
result = 2 >= 2
print(result)
result = 2 == 2
print(result)
```

运行结果如下：

```
False
True
True
```

注意：result 是一个变量，这个变量用来存储右边比较的值，比较的结果只能为 True 或者 False。True、False 在 Python 中称为 bool 型，即真或者假，假为 0，真可以为任意整数。

“<”小于、“<=”小于或等于、“!=”不等于的用法，代码如下：

```
#第 1 章/条件 2.py
print(2 < 2)
print(2 <= 2)
print(2 != 2)
```

运行结果如下：

```
False
True
False
```

in 或者 not in 的用法，代码如下：

```
#第 1 章/条件 2.py
#字符串的比较
print('q' in 'qwentest123')
print('v' not in 'qwentest123')
print('******************')
#元组的比较
print('q' in tuple('qwentest123'))
print('v' not in tuple('qwentest123'))
print('******************')
#列表的比较
print('q' in list('qwentest123'))
print('v' not in list('qwentest123'))
print('******************')
#集合的比较
print('v' in set('qwentest123'))
print('t' not in set('qwentest123'))
print('******************')
#字典的比较 key
print('qwentest123' in {'v': 'qwentest123'})
print('v' in {'v': 'qwentest123'})
print('******************')
```

```
#字典比较值
print('qwentest123' in {'v': 'qwentest123'}.values())
print('v' in {'v': 'qwentest123'}.keys())
```

运行结果如下：

```
True
True
*****************
True
True
*****************
True
True
*****************
False
False
*****************
False
True
*****************
True
True
```

注意：a in b 指 a 是否在 b 这个对象中，b 是一个 iterable 对象，例如 str、tuple、list、set、dict 等具有多个内容的对象；set('qwentest123')会自动去重；字典默认比较时比较的是 key，比较值需要转换成 dict. values()。

条件表达符号有“＞”“＞＝”“＜”“＜＝”“!＝”“＝＝”“in”“not in”，有时需要在条件表达式中同时比较多个字段，这时就需要使用逻辑关系符，逻辑关系符主要有“or”或者、“and”并且、“not”否则，代码如下：

```
#第 1 章/条件 3.py
#or 语句，只要一个为真，则为真
print(1 >= 2 or 1 <= 3)
#and 语句，只有同时满足时，才能为真
print(1 >= 2 and 1 <= 3)
#not 语句，取反的意思
print(not True)
print(not False)
```

运行结果如下：

```
True
False
False
True
```

if 语句是很重要的条件过滤语句，例如“已知周一到周五的英文单词，根据用户输入的

首字母来判断是星期几，如果首字母有重复，则根据输入的第 2 个字母进行判断”，代码如下：

```
#第 1 章/条件 4.py
week = {
    'M': 'Monday',
    'TU': 'Tuesday',
    'W': 'Wednesday',
    'TH': 'Thursday',
    'F': 'Friday',
    'SA': 'Saturday',
    'SU': 'Sunday'
}

first, second = input('请输入首字母：').upper(), ''
if first in ['T', 'S']:
    second = input('请输入第 2 个字母:').upper()
print(week.get(first + second))
```

运行结果如下：

```
请输入首字母：t
请输入第 2 个字母:u
Tuesday
```

注意：这里巧妙地用了字典这种数据类型来描述源数据，然后充分使用 dict.get(key)的特点获取用户输入的字母信息，当 key 存在时输出，如果不存在，则输出为 None。

数据的描述推荐使用列表或者字典进行描述，可以简化代码。

总结

条件语句是过滤数据的基本用法，是程序的重要组成部分。

练习

本书配套资源“Python 基础语法练习 2. txt”中有专门针对条件语句的题目，可以练习使用。

1.4 循环语句

所谓循环，就是多次重复的意思。在多次重复中使用条件语句或者其他语句来过滤数据，在 Python 中主要使用 for 循环和 while 循环，代码如下：

```
#第 1 章/循环 1.py
#使用 for 循环
for x in range(3):
    print(x, 'for:love')
print('---------------')
```

```
#使用 while 语句来表达
i = 0
while i < 3:
    print(i, 'while:love')
    i += 1
```

代码的运行结果如下：

```
0 for:love
1 for:love
2 for:love
---------------
0 while:love
1 while:love
2 while:love
```

注意：range()函数是 Python 集成的常用内置函数，range(3)即产生 0～2 的数。

x 指将 range()函数中产生的数存储为 x 这个变量，x 即为每个具体的内容。for…in 是标准固定关键字。

range()函数有 3 个参数，即 range(start＝0,end,step＝1)，默认 start 为 0，step 为 1，每循环 1 次 step 会加 1。

while 循环后面跟的是条件表达式，当为 True 时才会执行循环。while 循环需要自己控制循环的次数和步长，而 for 循环在遍历完 range()函数所有的内容时结束循环。

---

break、continue 是循环语句中的常用关键词，不仅可以用到 while 语句，还可以用到 for…in 语句，当满足条件时 break 会退出整个循环，而 continue 只会结束本次循环，代码如下：

```
#第 1 章/循环 2.py
i = 1
while True:
    print('又是新的一天,开始打{}次怪吧!'.format(i))
    if i >= 2:            #当 i 大于或等于 2 时会结束整个循环,一共循环 2 次
        print('打怪结束,解放了!')
        break
    i += 1
print('---------------')

ii = 0
while True:
    ii += 1
    if ii <= 2:           #当 ii 小于或等于 2 时,结束本次循环,但是 ii 仍然会+1
        print('服务期限不够,还得继续服务{}!'.format(ii))
        continue
    print('又是新的一天,开始打{}次怪吧!'.format(ii))
    if ii >= 5:           #当 ii 大于或等于 5 时,结束整个循环语句,一共循环 5 次
        print('打怪结束,解放了!')
        break
```

运行结果如下：

```
又是新的一天,开始打 1 次怪吧!
又是新的一天,开始打 2 次怪吧!
打怪结束,解放了!
---------------
服务期限不够,还得继续服务 1!
服务期限不够,还得继续服务 2!
又是新的一天,开始打 3 次怪吧!
又是新的一天,开始打 4 次怪吧!
又是新的一天,开始打 5 次怪吧!
打怪结束,解放了!
```

for…in 不仅可以通过 range()函数产生多次循环,也能够用来循环其他数据类型,只要是一个可迭代的数据类型,例如字符串、元组、列表、集合等,代码如下：

```
#第 1 章/循环 3.py
#遍历字符串
for s in 'qwentest123': print(s)
print('------------------------------')
#使用 for…in 来循环元组
for t in (1, 2, 3, ): print(t)
print('------------------------------')
#使用 for…in 来循环列表
for l in [1, 2, 3,]: print(l)
print('------------------------------')
#使用 for…in 来循环集合
for se in {1, 2, 3,}: print(se)
print('------------------------------')
#使用 for…in 来循环字段,不过循环打印的是字典的 key
for u in {'vv': 'qwentest123', 'name': '文山'}: print(u)
print('------------------------------')
#如果想获取字段的值或者每个 key: value
for u in {'v': 'qwentest123', 'name': '文山'}.values(): print(u)
print('------------------------------')
for k, v in {'v': 'qwentest123', 'name': '文山'}.items(): print('{0}={1}'.format
(k, v))
```

运行结果如下：

```
q
w
e
n
t
e
s
t
1
2
3
------------------------------
```

```
1
2
3
------------------------------
1
2
3
------------------------------
1
2
3
------------------------------
vv
name
------------------------------
qwentest123
文山
------------------------------
v=qwentest123
name=文山
```

注意：字典的遍历 for var in dict 时，遍历的是 key。如果要遍历 dict 值或者每项，则需要调用 dict.values()或者 dict.items()方法。

大多数时候，使用一次或两次循环即可完成相关工作，但有时需要使用多次循环，例如打印九九乘法表，代码如下：

```
#第 1 章/循环 4.py
for i in range(1, 10):
   for j in range(1, i + 1):
      d = i *j
      print('%d *%d=%-2d' % (i, j, d), end=' ')
   print()
```

运行结果如下：

```
1*1=1
2*1=2 2*2=4
3*1=3 3*2=6 3*3=9
4*1=4 4*2=8 4*3=12 4*4=16
5*1=5 5*2=10 5*3=15 5*4=20 5*5=25
6*1=6 6*2=12 6*3=18 6*4=24 6*5=30 6*6=36
7*1=7 7*2=14 7*3=21 7*4=28 7*5=35 7*6=42 7*7=49
8*1=8 8*2=16 8*3=24 8*4=32 8*5=40 8*6=48 8*7=56 8*8=64
9*1=9 9*2=18 9*3=27 9*4=36 9*5=45 9*6=54 9*7=63 9*8=72 9*9=81
```

注意：'%d * %d=%-2d' % (i, j, d)是一种格式化字符串的方法，%d 代表填空的位置要输入 int 型，如果是%f，则代表填入浮点数，如果是%s，则代表填入字符串。

列表推导式提供了从序列创建列表的简单途径，其语法满足[变量 for 变量 in 可循环对象]，代码如下：

```
#第1章/循环4.py
vec = [2, 4, 6]
print([3 *x for x in vec])
print([[x, x **2] for x in vec])
print([3 *x for x in vec if x > 3]) #可以跟 if 语句，意为如果 x>3，则会*3
```

代码的运行结果如下：

```
[6, 12, 18]
[[2, 4], [4, 16], [6, 36]]
[12, 18]
```

总结

循环语句在 Python 中又称为遍历语句，与条件语句统称为控制语句，是程序的重要组成部分。

练习

本书配套资源“Python 基础语法练习 3.txt”中有专门针对循环语句的题目，可以练习使用。

1.5 函数

函数是可重复使用的用来实现单一功能的集合，它把前面一系列的代码集中在一起供其他地方调用。函数的定义，解决的主要是重复使用问题，使程序更具有可维护性，其语法如下：

```
def 函数的名称(参数列表):
    函数体，即代码的内容
```

函数的名称只能是英文、数字、下画线的组合，不能以数字开头，也不能与编程关键字重名。参数列表，也就是说有多个参数，例如 a、b、c、d 等变量，此时 a、b、c、d 称为函数的形参，也就是说它们等待传值，是形式上的参数，代码如下：

```
#第1章/函数1.py

#a 和 b 是形参，只有象征意义
def add(a, b):
   return a + b


def clc(a, b):
   #如果函数的返回值有多个，则默认返回的是元组
   return a + b, a - b, a *b, a / b


print(add(1, 1))
print(clc(2, 2))
```

运行结果如下：

```
2
(4, 0, 4, 1.0)
```

注意：可以把 Python 中的函数看成 f(a,b)＝a＋b 的数学函数，a 和 b 是数学函数中的形参，return 是函数运行后的结果；函数的参数在 Python 中可以有 0～N 个，并且没有类型限制；return 返回多个值的默认数据类型是 tuple。

函数的另一种定义方式是关键字参数函数，使用关键字参数能很好地给每个形参赋初值，同时如果参数名定义恰当，则能更好地描述参数的作用。

使用关键字参数允许函数调用时传参的顺序与声明时不一致，因为 Python 解释器能够自适应参数匹配，也就是说一旦使用关键字参数的形式，在调用函数时传参的顺序就可以灵活地变动，代码如下：

```
#第 1 章/函数 2.py
import random

#关键字参数必填在必填参数的后面
def gussNumber(a, start=1, end=10):
    """
    猜数游戏的函数。
    :param a: 用户输入的数字
    :param start: 随机数开始的位置
    :param end: 随机数结束的位置
    :return:
    """
    #是否为正整数字符串
    if a.isdigit():
        #把 a 这个字符串强制转换成 int 型
        int_a = int(a)
        #产生 1~ 10 的随机整数
        num = random.randint(start, end)
        if int_a == num:
            print('猜中了')
            #如果猜中了,则函数的返回值为 None,如果没有猜中,则会返回 None
            return True
        elif int_a < num:
            print('猜小了')
        else:
            print('猜大了')
        print("{0}{1}".format(num, "为产生的随机数"))
    else:
        print('请输入正整数')


i = 0
while True:
  a = input("请输入你猜的数字：")
  if gussNumber(a): break
```

```
    if i >= 2:
        print('3 次机会都用完了')
        break
    i += 1
```

运行结果如下：

```
请输入你猜的数字：3
猜小了
9 为产生的随机数
请输入你猜的数字：3
猜大了
1 为产生的随机数
请输入你猜的数字：3
猜大了
2 为产生的随机数
3 次机会都用完了
```

注意：函数当中如果执行到 return 语句，则整个函数就会结束，如果没有 return 语句，则函数会返回 None。

函数的参数还有一些常见形式，例如 * args 可以传入任意个参数、** kargs 可以传入任意个关键字参数、args:type 提示参数类型，代码如下：

```
#第 1 章/函数 3.py


def sumAnyNum(*args):
    """求任意个数字的和"""
    print('传递的参数值为 %s,数据类型为 %s' % (args, type(args)))
    return sum(args)


def sumAnyNum2(**kwargs):
    print('传递的参数值为 %s,数据类型为 %s' % (kwargs, type(kwargs)))
    return sum(kwargs.values())


def sumAnyNum3(x: list):
    print('传递的参数 x 的类型提示为 list 格式')
    return sum(x)

#这里是传入的任意个参数
print(sumAnyNum(11, 12, 13))
print(sumAnyNum(0, 1, 2, 3, 4, 5, 6, 7, 8, 9))
print('----------------------------------------')
#这里是传入的任意个关键字参数
print(sumAnyNum2(k1=11, k2=12, k3=13))
#调用参数类型提示的函数
print(sumAnyNum3([1, 2, 3]))
```

运行结果如下：

```
传递的参数值为 (11, 12, 13),数据类型为 <class 'tuple'>
36
传递的参数值为 (0, 1, 2, 3, 4, 5, 6, 7, 8, 9),数据类型为 <class 'tuple'>
45
----------------------------------------
传递的参数值为 {'k1': 11, 'k2': 12, 'k3': 13},数据类型为 <class 'dict'>
36
传递的参数 x 的类型提示为 list 格式
6
```

注意：sumAnyNum3(x：list)中的 x 虽然提示为 list 格式，但是在 Python 中传入元组仍然可以得到结果，这是因为 Python 是一个弱类型语言，对于参数的类型本身是没有限制的。

在 Python 中还可以创建 lambda 匿名函数，所谓匿名即不再使用 def 语句来定义一个函数，而是采用 lambda 表达式进行定义。lambda 的主体是一个表达式，而不是一个代码块，所以 lambda 表达式中只能封装有限的逻辑，代码如下：

```
#第 1 章/函数 3.py
#lambda 表达式
addsum = lambda arg1, arg2: arg1 + arg2
print("相加后的值为 ", addsum(10, 20))
print("相加后的值为 ", addsum(20, 20))
```

运行结果如下：

```
相加后的值为  30
相加后的值为  40
```

如果一个变量定义在函数内部，则其作用域仅在函数中，如果一个变量定义在函数外，则其作用域是整个程序，代码如下：

```
#第 1 章/函数 4.py
total = 0 #这是一个全局变量


#可写函数说明
def sum(arg1, arg2):
   #返回两个参数的和
   total = arg1 + arg2 #total 在这里是局部变量
   print("函数内是局部变量: ", total)
   return total


#调用 sum 函数
sum(10, 20)
print("函数外是全局变量: ", total)
```

运行结果如下：

```
函数内是局部变量: 30
函数外是全局变量: 0
```

装饰器是修改其他函数功能的函数,它可以让其他函数在不需要做任何代码变动的前提下增加额外的功能,代码如下:

```python
#第 1 章/函数 5.py
import time


#定义了一个装饰器,用来输出每个函数的运行时间
#func 是引用装饰器函数的地址
def getRunTime(func):
    #*args 和**kwargs 是用来接收 func 这个函数的参数
    def wrapper(*args, **kwargs):
        #在 func 之前执行
        start_time = time.time()
        #执行调用装饰器函数,即 func 函数
        result = func(*args, **kwargs)
        #在 func 之后执行
        end_time = time.time()
        print(end_time - start_time)
        return result

    return wrapper


#在一个函数中使用装饰器
@getRunTime
def func1():
    time.sleep(0.8)


#在类的方法中使用装饰器
class Method(object):
    @getRunTime
    def func2(self):
        time.sleep(0.1)



func1()
Method().func2()
```

代码的运行结果如下:

```
0.8003103733062744
0.11751770973205566
```

注意:getRunTime 这个装饰器的参数 func 是 func1 在内存中的地址,这个地址指向 func1 函数,而 * args 和 ** kwargs 是 func1 函数的参数,return wrapper 返回的是一个地址。

总结

函数式编程又称面向过程的编程，常用的函数定义方式有必填参数、关键字参数。

练习

本书配套资源"Python 基础语法练习 4. txt"中有专门针对循环语句的题目，供练习使用。

1.6 类

常用的编程组织方式有两种，即面向过程和面向对象。面向过程是早期程序员使用的方法，其主要方法是将问题分解成多个步骤，然后用函数逐个实现，在运行时依次调用函数；而面向对象是把构成问题的事物组合抽象成类，通过类创建对象来完成各个功能。在面向对象编程中主要涉及以下几个概念。

(1) 类：用来描述具有相同属性和方法的对象的集合。它定义了该集合中每个对象所共有的属性和方法。

(2) 方法：类中定义的函数。

(3) 类变量：类变量在整个实例化的对象中是公用的。类变量定义在类中且在函数体之外。类变量通常不作为实例变量使用。

(4) 数据成员：类变量或者实例变量用于处理类及其实例对象的相关数据。

(5) 方法重写：如果从父类继承的方法不能满足子类的需求，则可以对其进行改写，这个过程叫方法的覆盖，也称为方法的重写。

(6) 局部变量：定义在方法中的变量，只作用于当前实例的类。

(7) 实例变量：在类的声明中，属性是用变量来表示的，这种变量就称为实例变量，实例变量就是一个用 self 修饰的变量。

(8) 继承：即一个派生类继承基类的字段和方法。继承也允许把一个派生类的对象作为一个基类对象对待。

(9) 实例化：创建一个类的实例，即类的具体对象。

(10) 对象：通过类定义的数据结构实例。

通过类关键字定义方法，代码如下：

```
#第 1 章/类 1.py
class ClassName():
    pass
```

class 是关键字，通常类的名称的首字母需要大写。

类对象支持两种操作，即属性引用和实例化。类对象创建后，类命名空间中所有的命名都是有效属性名，代码如下：

```
#第1章/类1.py
class MyClass():
    """一个简单的类实例"""
    i = 12345   #类的变量,又称为类的属性

    #类中的函数,又称为方法
    def f(self):
        return 'hello world'


#实例化类
x = MyClass()

#访问类的属性和方法
print("MyClass 类的属性 i 为", x.i)
print("MyClass 类的方法 f 输出为", x.f())
```

运行结果如下：

```
MyClass 类的属性 i 为 12345
MyClass 类的方法 f 输出为 hello world
```

如何描述类的属性及成员函数的定义、继承、重载、扩展方面的应用，代码如下：

```
#第1章/类2.py
class Person():
    def __init__(self):
        """
        构造函数,即类的初始函数,可以赋初始值
        """
        #__var 表示私有属性,也就是只能在 Person 类中进行访问
        self.__name = '文山'
        self.__age = 18
        self.__sex = 'man'
        self.__tops = 178
        self.__weight = 70
        self.__weiXin = 'qwenTest123'

    #读属性
    @property
    def name(self): return self.__name

    #写属性
    @name.setter
    def name(self, value): self.__name = value

    @property
    def age(self): return self.__age

    @age.setter
    def age(self, value): self.__age = value
```

```
    @property
    def sex(self): return self.__sex

    @sex.setter
    def sex(self, value): self.__sex = value

    @property
    def tops(self): return self.__tops

    @tops.setter
    def tops(self, value): self.__tops = value

    #实例类所拥有的一些方法,即对象所拥有的行为
    def eat(self):
        print('首要任务是吃饭')

    def think(self):
        print('思考是人之所在')

    def see(self):
        print('看花花世界')
```

注意：__var 表示类的私有属性，只能在 Person 类中访问，但是在 def name 上面增加 @property 便设置了可读属性，可以此去访问 self.__name 属性的值。@name.setter 表示将该属性设置为可写。eat()、think()、see()用于描述 Person 类的行为，所以定义成了方法。

继承 Person()类并定义一个 PoliticalStatus()类来描述人所具有的社会属性，代码如下：

```
#第 1 章/类 2.py

#继承,是这个世界赋予人的政治面貌、国家、党派等
class PoliticalStatus(Person):
    def __init__(self):
        """
        构造函数,继承 Person 类所拥有的属性
        """
        Person.__init__(self)
        self.__country = "CN"
        self.__party = None

    @property
    def country(self): return self.__country

    @country.setter
    def country(self, value): self.__country = value

    @property
    def party(self): return self.__party
```

```
    @party.setter
    def party(self, value): self.__party = value

    def take_party(self):
        print('加入')
```

注意：PoliticalStatus()类继承自 Person 类，并通过 Person. __init__(self)继承父类的构造函数，但是不能访问 self. __name、self. __age、self. __sex，然后 PoliticalStatus()类就拥有 name、age、sex、tops 等父类属性，以及 eat()、think()、see()等父类方法，并同时拥有子类属性 country()和子类方法 take_party()。

再定义职业类 Profession()来继承 Person()类，代码如下：

```
#第 1 章/类 2.py
#继承,使人具备职业信息
class Profession(Person):
    def __init__(self):
        Person.__init__(self)
        self.__occupation = 'tester'
        self.__salary = 1000

    @property
    def occupation(self): return self.__occupation

    @occupation.setter
    def occupation(self, value): self.__occupation = value

    @property
    def salary(self): return self.__salary

    @salary.setter
    def salary(self, value): self.__salary = value

    def work_content(self):
        print("工作内容就是'哈哈哈...'")
```

然后定义一个 Me 类，同时继承 PoliticalStatus()和 Profession()类，并在 describeMe()方法中调用父类中的各种方法进行扩展，代码如下：

```
#第 1 章/类 2.py
class Me(PoliticalStatus, Profession):
    """
    #重载
    继承 PoliticalStatus、Profession 类
    并重写 Profession 类中的 work_content 方法
    """

    def work_content(self):
        print('work work work')
```

```
    def describeMe(self):
        """
        调用父类,扩展
        :return:
        """
        self.eat()
        self.see()
```

注意：Me()同时继承了 PoliticalStatus()类和 Profession()类，而这两个类又继承了 Person()类，那么此时 Me()就同时拥有 Person()、PoliticalStatus()、Profession()类可访问的属性和方法，需要注意的是 Me()中有一个 work_content()会重写 Profession()类中的 workContent()方法，当实例化 Me()时调用的是 Me()中的 work_content()，当实例化 Profession()时，调用的是 Profession()中的 workContent()。

最后分别实例化不同的类并调用相关方法，代码如下：

```
#第 1 章/类 2.py
"""
调用相关类
"""
p = PoliticalStatus()
#读取默认的人的政治情况
print("name = %s, party=%s" % (p.name, p.party))
#新加入的人的政治情况
p.name = 'qcc'
p.party = "pragmatism"
print("name = %s, party=%s" % (p.name, p.party))
p.take_party()
print('==============')
p2 = Profession()
#读取默认的人的职业信息
print("name = %s" % (p2.name))
#新加入的人的职业信息
p2.name = 'qcc'
p2.sex = 'fmale'
print("name = %s" % (p2.name))
p2.work_content()              #此时调用的是 Profession()类中的 work_content
print('==============')
m = Me()
m.describeMe()
m.work_content()               #此时调用的是 Me()类中的 work_content
```

运行结果如下：

```
name = 文山, party=None
name = qcc, party=pragmatism
加入
==============
name = 文山
```

```
name = qcc
工作内容就是'哈哈哈...'
==============
首要任务是吃饭
看花花世界
work work work
```

注意：在拥有多个类之间的继承关系时，需要注意实例化对象所对应的类中的继承关系。

总结

面向对象的编程是一种更高级的编程方式，适合大规模的程序编写。

练习

基于 Person 类继承并创建一个学生类。

1.7 文件处理

在 Python 中处理文件的基本操作可以使用内置 open()函数，代码如下：

```
#第 1 章/文件处理 1.py
#写一个 TXT 文件。如果当前目录下面有此文件，则覆盖。如果没有此文件，就创建一个文本文件
f = open('文山.txt', mode='w')
#写一行内容
f.write('大家好，我是文山。')
#一次写多行内容
f.writelines(['第 1 行\n', '第 2 行\n'])
f.close()
print('####################')
#读内容
f1 = open('文山.txt', mode='r')
#读多行，返回的是一个 list
print(f1.readlines())
f1.close()
```

运行结果如下：

```
####################
['大家好，我是文山。第 1 行\n', '第 2 行\n']
```

注意：mode 中的 w 表示覆盖写、a 表示追加写、r 表示只读。

总结

在 Python 中读取文件时使用 open()函数即可打开一个文件。

练习

以二进制文件打开一个 JPG 图片并另存为 PNG 格式的图片。

1.8 异常处理

所谓异常就是指程序在运行的过程中出现错误，异常处理就是当程序出现异常时，程序捕获这些异常并进行相应处理，其语法格式如下：

```
try:
    从这一段代码中捕获异常
except 异常类型 1:
    当出现指定异常类型时执行的代码
except 异常类型 2:
    当出现指定异常类型时执行的代码
except Exception as e:
    当出现未知异常时的处理
finally:
    不管有没有异常都需要执行
```

except 至少需要 1 个，finally 表示不管前面有没有异常都会被执行，代码如下：

```
#第 1 章/异常处理.py
try:
    print(1 / 0)
except ZeroDivisionError:
    print('0 不能作为除数')
finally:
    print('有异常时会执行')
print('hello world')
```

运行结果如下：

```
0 不能作为除数
有异常时会执行
hello world
```

总结

try-except-finally 结构能够增强代码的可靠性，使异常信息能够被捕获并得到相应的处理。

练习

尝试使用 try-except-finally 结构打开一个不存在的文件，并观察其提示错误的提示语。

1.9 模块与包

模块是最高级别的程序组织单元，它将程序代码和数据封装起来以便重复使用，从简单使用的角度来看，模块往往对应于 Python 程序文件，即. py 文件。

模块的使用语法有两种，一种是使用 import 包名. 模块名，另一种是 from 包名. 模块名

import 模块中的函数、类或者变量。先新建一个 mod1. py 文件，代码如下：

```
#第 1 章/mod1.py
Name = '文山'


def anySum(*args):
   return sum(args)


class Test():
   pass
```

再新建一个 mod2. py 文件，并且使用 import 模块的方法导入 Name、anySum()函数、Test()类，代码如下：

```
#第 1 章/mod2.py
import mod1

#使用 mod1.py 文件中的 Name 变量
print(mod1.Name)
#使用 mod1.py 文件中的 anySum 函数
print(mod1.anySum(1, 2, 3))
#也可以使用 mod1.py 文件中的 Test 类
print(mod1.Test())
```

运行结果如下：

```
文山
6
<mod1.Test object at 0x000001812D72A880>
```

当使用 import 导入模块时还可以使用 as 重命名关键字，例如新建一个 mod3. py 文件，代码如下：

```
#第 1 章/mod3.py
import mod1 as p

#使用 mod1.py 文件中的 Name 变量
print(p.Name)
#使用 mod1.py 文件中的 anySum 函数
print(p.anySum(1, 2, 3))
#也可以使用 mod1.py 文件中的 Test 类
print(p.Test())
```

运行结果如下：

```
文山
6
<mod1.Test object at 0x000001812D72A880>
```

import 是整个模块的名称，而 from 模块 import 使用的是具体对象的名称，例如新建

一个 mod4. py 文件,代码如下:

```
#第 1 章/mod4.py

from mod1 import Name, anySum, Test

print(Name)
#使用 mod1.py 文件中的 anySum 函数
print(anySum(1, 2, 3))
#也可以使用 mod1.py 文件中的 Test 类
print(Test())
```

运行结果如下:

```
文山
6
<mod1.Test object at 0x000001812D72A880>
```

注意:import 导入的是模块的名称,而 from…import…导入的是具体的对象名称。

Python 中还有一个包的概念,其实就是文件夹,更确切地说是一个包含__init__. py 文件的文件夹,因此,如果想手动创建一个包,则只需新建一个文件夹,然后在该文件夹中创建一个__init__. py 文件,该文件可以不编写任何代码。在 PyCharm 中右击并选择新建一个 modPackage 包,然后新建 mod5. py 文件,代码如下:

```
#第 1 章/modPackage/mod5.py

Name = '文山'


def anySum(*args):
   return sum(args)


class Test():
   pass
```

然后新建 mod6. py 文件,代码如下:

```
#第 1 章/mod6.py

from modPackage.mod5 import Name, anySum, Test

print(Name, anySum(1, 2, 3), Test())
```

运行结果如下:

```
文山 6 <modPackage.mod5.Test object at 0x000002A3C23F9FA0>
```

以上文件的结构如图 1-15 所示。

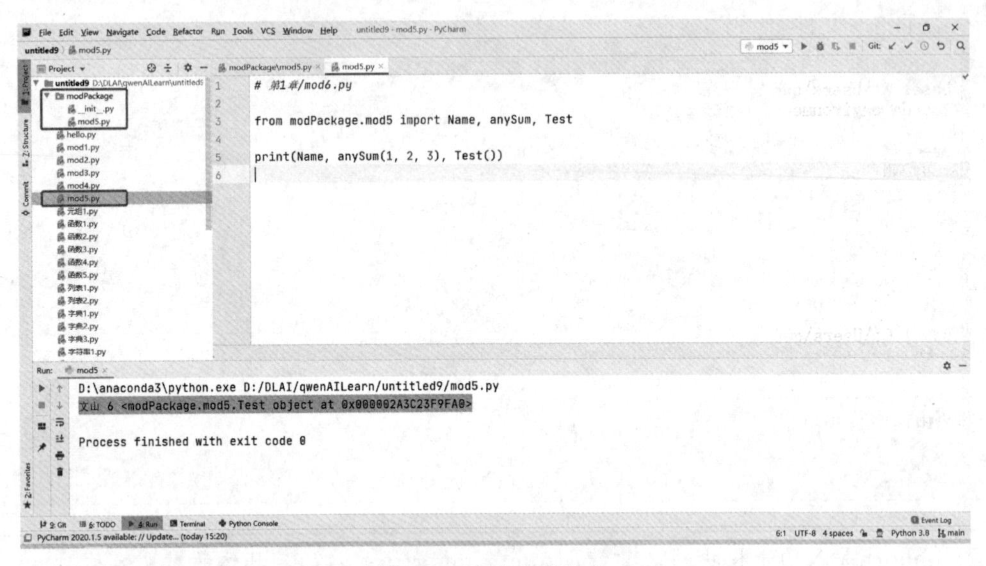

图 1-15　包的文件结构

总结

模块是最高级别的程序组织单元，它将程序代码和数据封装起来以便重复使用，模块分为 Python 自带模块及第三方模块。

练习

在项目目录中新建具有 3 个层级的文件夹，在第 3 级文件夹中新建 a. py 文件并定义一个函数，然后在项目根目录中调用 a. py 文件中的函数。

1.10　包的管理

Python 提供了大量的第三方包，可以通过 pip 或者 conda 命令对第三方包进行下载、更新、删除。

首先在 Windows 开始菜单中单击 Anaconda Prompt. exe，然后输入命令 conda env list 查看当前环境，如图 1-16 所示。

然后输入命令 conda activate StudyDNN 进入当前开发环境，如图 1-17 所示。

然后输入命令"conda install 包名"，进行第三方包的安装，如图 1-18 所示。

conda 会自动搜索要安装包的依赖项，并询问用户是否继续下载，当输入 y 确认后就会自动下载，如图 1-19 所示。

也可以使用命令"pip install 包名"的方式进行安装，如图 1-20 所示。

删除包使用命令"conda uninstall 包名"，如图 1-21 所示。

pip 删除包的命令，如图 1-22 所示。

查看目前安装了哪些包，可以使用命令 conda list 或者 pip list，如图 1-23 所示。

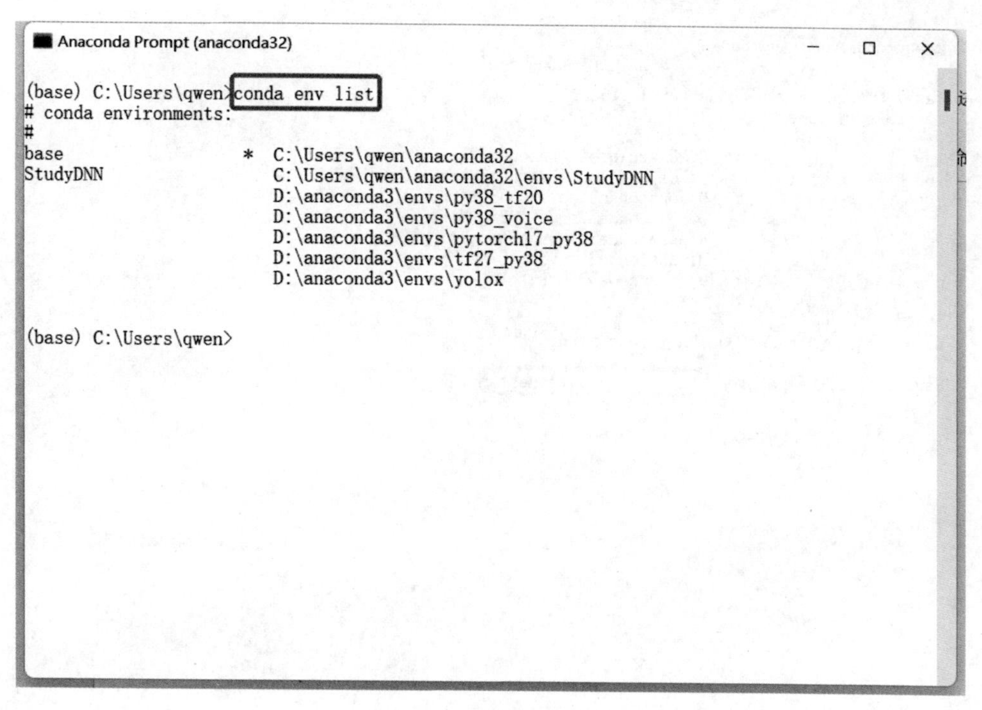

图 1-16 查看 conda 环境

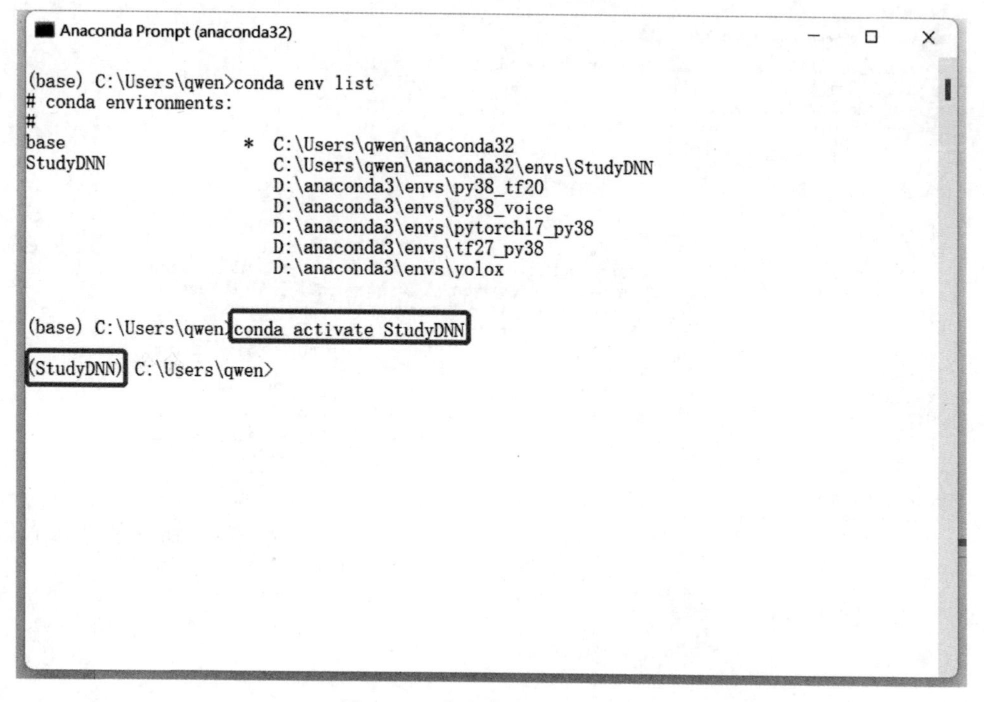

图 1-17 进入某个 conda 环境

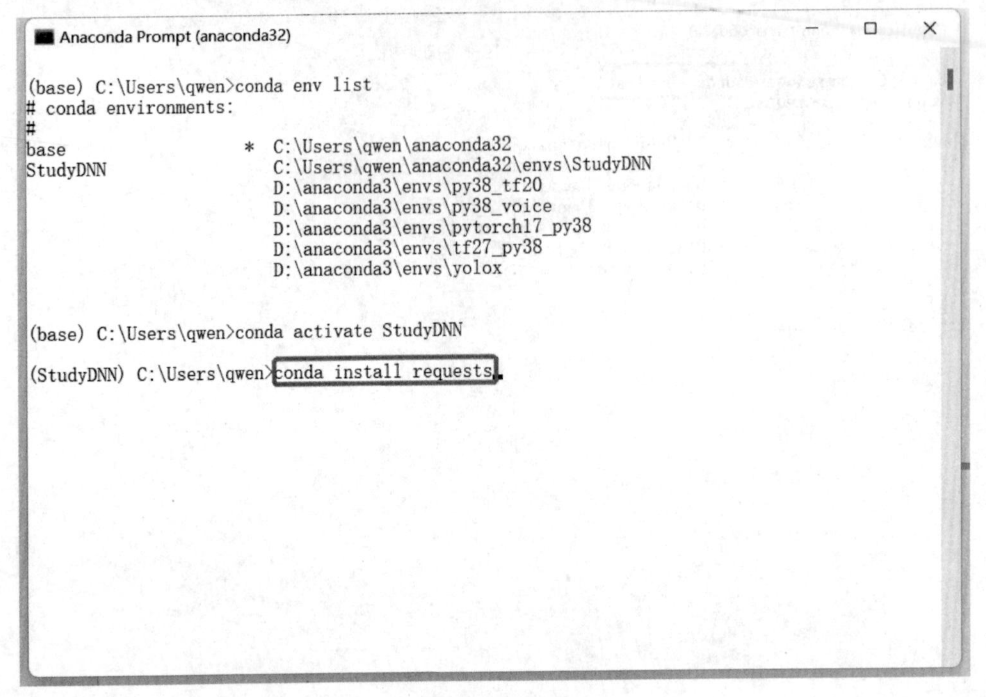

图 1-18　conda 安装包

```
Anaconda Prompt (anaconda32) - conda install requests
   requests-2.28.1              |    py38haa95532_0         100 KB
   urllib3-1.26.12              |    py38haa95532_0         185 KB
   win_inet_pton-1.1.0          |    py38haa95532_0          35 KB
   ------------------------------------------------------------
                                          Total:          2.0 MB

The following NEW packages will be INSTALLED:

  brotlipy           pkgs/main/win-64::brotlipy-0.7.0-py38h2bbff1b_1003 None
  cffi               pkgs/main/win-64::cffi-1.15.1-py38h2bbff1b_3 None
  charset-normalizer pkgs/main/noarch::charset-normalizer-2.0.4-pyhd3eb1b0_0 None
  cryptography       pkgs/main/win-64::cryptography-38.0.1-py38h21b164f_0 None
  idna               pkgs/main/win-64::idna-3.4-py38haa95532_0 None
  pycparser          pkgs/main/noarch::pycparser-2.21-pyhd3eb1b0_0 None
  pyopenssl          pkgs/main/noarch::pyopenssl-22.0.0-pyhd3eb1b0_0 None
  pysocks            pkgs/main/win-64::pysocks-1.7.1-py38haa95532_0 None
  requests           pkgs/main/win-64::requests-2.28.1-py38haa95532_0 None
  urllib3            pkgs/main/win-64::urllib3-1.26.12-py38haa95532_0 None
  win_inet_pton      pkgs/main/win-64::win_inet_pton-1.1.0-py38haa95532_0 None

Proceed ([y]/n)? y
```

图 1-19　conda 确认下载

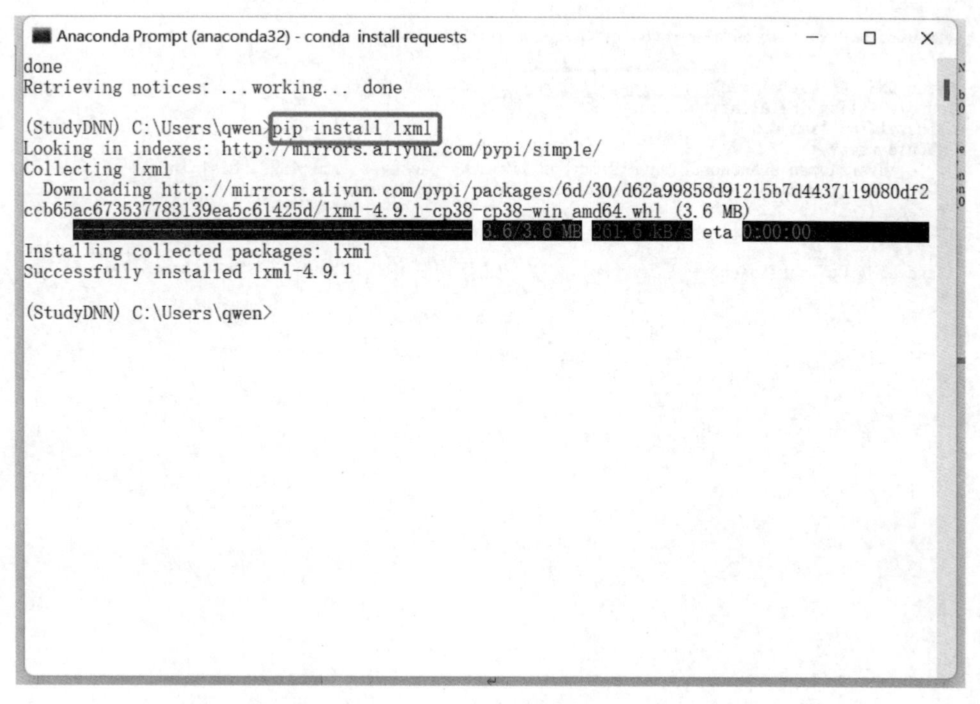

图 1-20　pip 下载安装包

```
Anaconda Prompt (anaconda32) - conda install requests - conda uninstall requests
(StudyDNN) C:\Users\qwen>conda uninstall requests
Collecting package metadata (repodata.json): done
Solving environment: done


==> WARNING: A newer version of conda exists. <==
  current version: 22.9.0
  latest version: 22.11.1

Please update conda by running

    $ conda update -n base -c defaults conda



## Package Plan ##

  environment location: C:\Users\qwen\anaconda32\envs\StudyDNN

  removed specs:
    - requests


The following packages will be REMOVED:

  brotlipy-0.7.0-py38h2bbff1b_1003
  cffi-1.15.1-py38h2bbff1b_3
  charset-normalizer-2.0.4-pyhd3eb1b0_0
  cryptography-38.0.1-py38h21b164f_0
  idna-3.4-py38haa95532_0
```

图 1-21　conda 删除包

```
Anaconda Prompt (anaconda32) - conda install requests - conda uninstall requests

(StudyDNN) C:\Users\qwen>pip uninstall lxml
Found existing installation: lxml 4.9.1
Uninstalling lxml-4.9.1:
  Would remove:
    c:\users\qwen\anaconda32\envs\studydnn\lib\site-packages\lxml-4.9.1.dist-info\*
    c:\users\qwen\anaconda32\envs\studydnn\lib\site-packages\lxml\*
Proceed (Y/n)? y
  Successfully uninstalled lxml-4.9.1

(StudyDNN) C:\Users\qwen>
```

图 1-22　pip 删除包

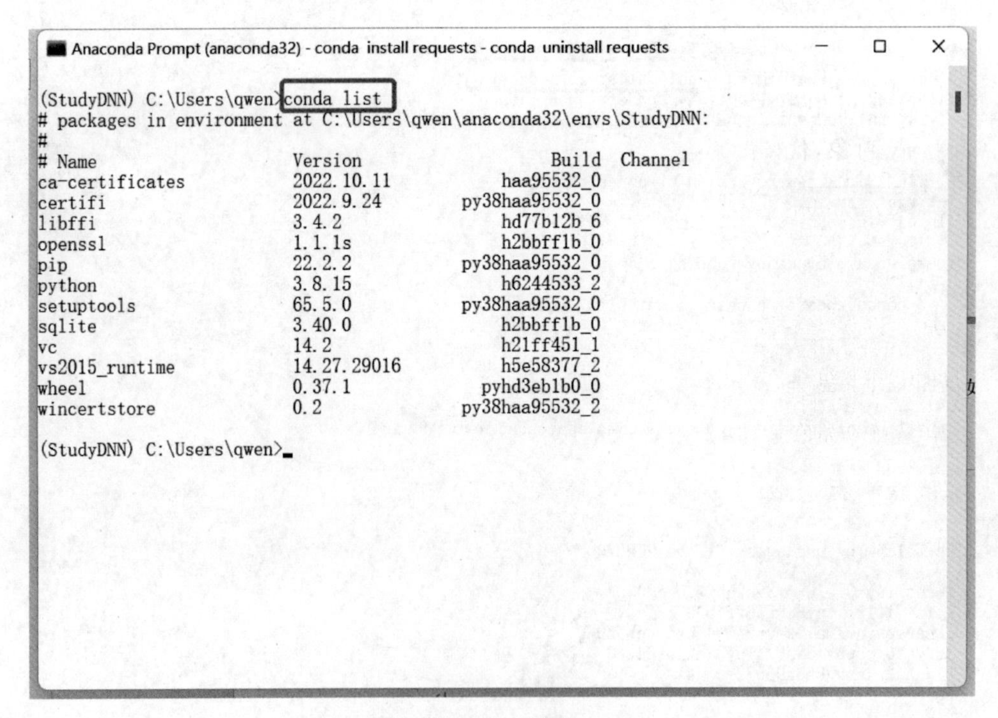

图 1-23　查看包

注意：推荐使用 conda 命令，因为 conda 会根据当前环境依赖的相关包去下载对应的版本，而 pip 下载的包有可能会与当前环境不兼容，从而导致出现一些问题。

总结

安装第三方包可以使用 conda 或者 pip 命令，推荐使用 conda 命令。

练习

使用 conda 命令在创建的 Python 3.9 环境中安装 NumPy、Pandas、Matplotlib、OpenCV 等第三方包，然后删除这些第三方包。

1.11 NumPy 简介

NumPy 是一个科学计算的基础包，提供了对于多维数组操作的 API，如数学计算函数、排序、离散傅里叶变换、基本线性代数、基本统计运算和随机模拟等功能，并能够与 Pandas、Matplotlib、OpenCV、TensorFlow、Pytorch 等包互相转换，NumPy 包的核心是 NDArray 对象。

1.11.1 NDArray 的创建

可以将列表或者元组的内容转换成 NDArray 的形式，此时只需调用 numpy.array() 函数，同时 NumPy 中提供了 numpy.zeros()、numpy.ones()、numpy.arange() 等函数来初始化 NDArray 对象，代码如下：

```
#第 1 章/studyNumPy/数组的创建.py
#安装命令 conda install numpy
import numpy as np

#创建一维数组，并将 a 的数组类型指定为 float32
a =np.array([2, 3, 4], dtype=np.float32)
#创建二维数组
b =np.array([[1.5, 2, 3], [4, 5, 6]])
#创建一个初始值为 0 的数组
c =np.zeros([3, 4])
#创建初始值为 1 的三维数组，并将数据类型指定为 int16
d =np.ones([2, 3, 4], dtype=np.int16)
#创建一个从 1 到 10 的一维数组，步长是 0.5
e =np.arange(1, 10, 0.5)
#随机生成指定维度的数
f =np.random.random((2, 3))
#得到维度
print(f.ndim)
#得到数组的形状
print(f.shape)
```

进行断点调试，可以看到变量 a、b、c、d、e、f 全变成了 NDArray 对象，如图 1-24 所示。

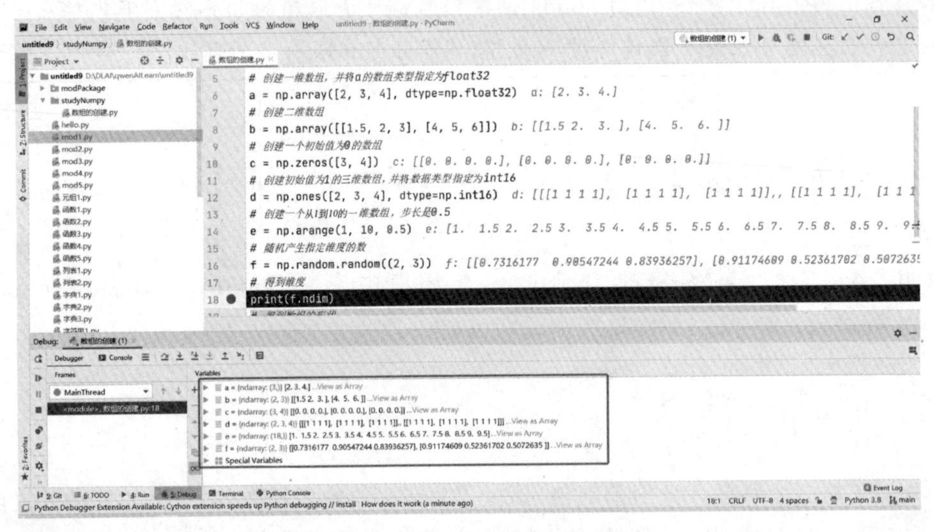

图 1-24　初始化 NDArray 对象

注意：float32、float16、float64、int16、int32 等表示的是 float 或者 int 型，但是其数值的范围不一样。

1.11.2　NDArray 索引与切片

NDArray 提供了像列表一样可以进行索引和切片的操作，其索引的语法满足 NDArray[index0,index1,index2,…]，其中 NDArray 指某个数组，index0 代表第 0 轴上的第几个下标的值，index1 代表第 1 轴的第几个下标的值，由外向内轴的数量+1，代码如下：

```
#第 1 章/studyNumPy/数组索引 1.py

import numpy as np

index_value =np.random.random([2, 2, 2])
print('原数据', index_value)
print('####################')
print(index_value[0, 1, 1])
```

运行结果如下：

```
原数据 [[[0.7121764 0.26118264]
  [0.12411949 0.42472087]]

[[0.35309219 0.63477725]
  [0.36814087 0.22884683]]]
####################
0.42472087
```

注意：index_value[0,1,1]指第 0 轴的第 0 个下标，第 1 轴的第 1 个下标，第 2 轴的第 1 个下标，即这个三维数组中的第 0 轴第 0 个位置是[0.7121764 0.26118264] [0.12411949 0.42472087]，在它的基础上取第 1 轴中的下标为 1 的值[0.12411949 0.42472087]，然后在它的基础上取第 2 轴中的下标为 1 的值 0.42472087。

NDArray 也可以使用切片的语法取多个值，其语法格式满足：

```
NDArray[start_index0:end_index0:step0,start_index1:end_index1:step1, …]
```

start_index0：end_index0：step0 指的第 0 轴上某个开始的下标（如果为 0，则可以省略），end_index0 指第 0 轴上某个结束的下标（如果为－1，则可以省略），step0 表示第 0 轴上每隔几个值取一个值，以此类推，可以在每个轴上进行类似取值，代码如下：

```
#第 1 章/studyNumPy/数组索引 2.py
import numpy as np

index_value = np.random.random([3, 3, 3])
print('原数据', index_value)
print('####################')
print(index_value[:1, 1:3, 2])
```

运行结果如下：

```
原数据 [[[0.96735713 0.20008986 0.93431251]
  [0.39629697 0.40814864 0.22539278]
  [0.52097874 0.38204418 0.91030824]]

[[0.81261282 0.92788045 0.74221152]
  [0.5017542 0.80595748 0.72478601]
  [0.48712281 0.25715609 0.51998573]]

[[0.56078442 0.35462929 0.23275688]
  [0.81109904 0.3736148 0.58315035]
  [0.08836854 0.25946715 0.80110368]]]
####################
[[0.22539278 0.91030824]]
```

如图 1-25 所示，index_value[:1,1:3,2]中“:1”的取值为红色区域，“1:3”的取值为绿色区域，“2”为蓝色区域中下标为 2 的值。

NumPy 允许使用 3 个点“...”表示产生完整索引元组所需的冒号，代码如下：

```
#第 1 章/studyNumPy/数组索引 3.py
import numpy as np

index_value = np.random.random([2, 2, 2])
print('原图', index_value)
print('#########################')
print(index_value[:1, :, :])            #第 0 轴第 0 个下标的值
```

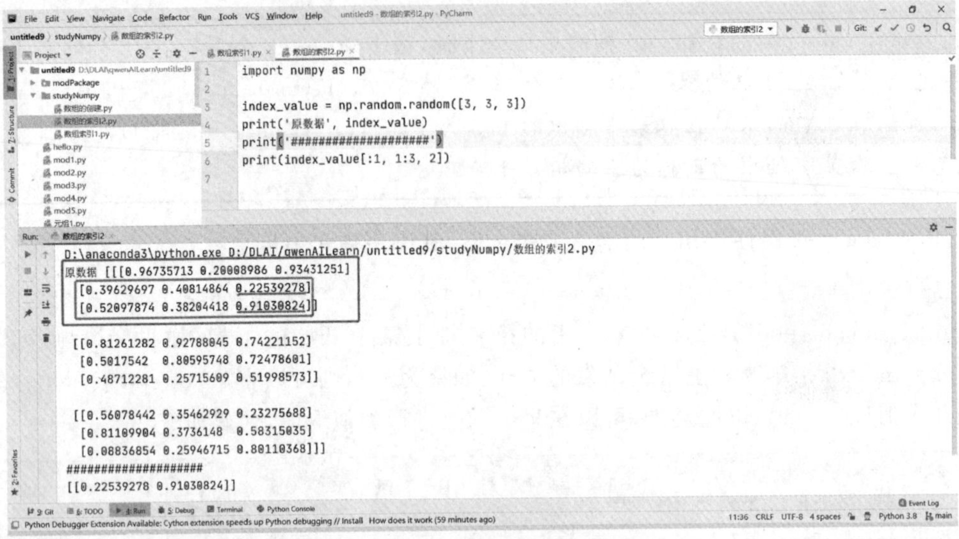

图 1-25　index_value 的取值过程(见彩插)

```
print(index_value[:1, ...])                 #与上面相同
print('##########################')
print(index_value[:, 1, :])
print(index_value[:, 1, ...])               #第 1 轴第 1 个下标的值
print('##########################')
print(index_value[:, :, 1])                 #最后一个轴,所有下标为 1 的值
print(index_value[..., 1])
```

运行结果如下:

```
原图 [[[0.26706571 0.49764647]
  [0.82469316 0.02161888]]

 [[0.1535775 0.07303183]
  [0.44224731 0.99207168]]]
##########################
[[[0.26706571 0.49764647]
  [0.82469316 0.02161888]]]
[[[0.26706571 0.49764647]
  [0.82469316 0.02161888]]]
##########################
[[0.82469316 0.02161888]
 [0.44224731 0.99207168]]
[[0.82469316 0.02161888]
 [0.44224731 0.99207168]]
##########################
[[0.49764647 0.02161888]
 [0.07303183 0.99207168]]
[[0.49764647 0.02161888]
 [0.07303183 0.99207168]]
```

当索引号为 NDArray 且内容又是单个数字时，下标取值取的是第 0 轴相应下标的内容，代码如下：

```
#第 1 章/studyNumPy/数组索引 4.py
import numpy as np

palette = np.array([[0, 0, 0],                #黑色
                    [255, 0, 0],              #红色
                    [0, 255, 0],              #绿色
                    [0, 0, 255],              #蓝色
                    [255, 255, 255]])         #白色
image = np.array([[0, 1, 2, 0],
                  [0, 3, 4, 0]])
#当目前标取值的索引号为 NumPy 且索引值是数字时，取的是第 0 轴的内容
#[0, 1, 2, 0]，即取 4 次下标分别为 0、1、2、0 的值
print(palette[image])
```

运行结果如下：

```
[[[  0   0   0]
  [255   0   0]
  [  0 255   0]
  [  0   0   0]]

 [[  0   0   0]
  [  0   0 255]
  [255 255 255]
  [  0   0   0]]]
```

当索引号为数组时，每个维度的索引数组必须具有相同的形状，代码如下：

```
#第 1 章/studyNumPy/数组索引 5.py
import numpy as np

a = np.arange(12).reshape(3, 4)
print('原值', a)
print('#############')
i = np.array([[0, 1], [1, 2]])
j = np.array([[2, 1], [3, 3]])
#i 表示取第 0 轴，j 表示取第 1 轴，所以当 i 为 0 时，j 为 2
print(a[i, j])
print('#############')
#在第 1 轴中，每个第 0 轴都取[2, 1]，[3, 3]次
print(a[:, j])
```

运行结果如下：

```
原值 [[ 0  1  2  3]
 [ 4  5  6  7]
 [ 8  9 10 11]]
#############
```

```
[[ 2  5]
 [ 7  11]]
#############
[[[ 2  1]
  [ 3  3]]

 [[ 6  5]
  [ 7  7]]

 [[10  9]
  [11  11]]]
```

注意：a[i,j]中 i 表示取第 0 轴而 j 表示取第 1 轴，所以当 i 为 0 时，j 为 2；a[:,j]表示在第 0 轴中每个值都要取[2,1]，[3,3]的下标值再组成新的数组。

NumPy 中可以使用布尔值过滤出想要的数据，代码如下：

```
#第 1 章/studyNumPy/数组索引 6.py

import numpy as np

a = np.arange(12).reshape(3, 4)
print('原值', a)
print('###############')
b1 = np.array([False, True, True])
b2 = np.array([True, False, True, False])
print(a[b1, :])              #按第 0 轴选择
print('1###############')
print(a[b1])                 #同上
print('2###############')
print(a[:,b2])               #按第 1 轴选择
```

运行结果如下：

```
原值 [[ 0  1  2  3]
 [ 4  5  6  7]
 [ 8  9 10 11]]
###############
[[ 4  5  6  7]
 [ 8  9  10 11]]
1###############
[[ 4  5  6  7]
 [ 8  9  10 11]]
2###############
[[ 0  2]
 [ 4  6]
 [ 8  10]]
```

注意：布尔取值，即所对应的下标为 False 时不取值，而下标为 True 时取值。

1.11.3 NDArray 常用运算函数

NumPy 提供了一些常用的数据运算函数，例如 numpy.sin()、numpy.cos()、numpy.exp()等，代码如下：

```
#第 1 章/studyNumPy/常用表达函数.py

import numpy as np

c = np.arange(12).reshape(3, 4)
print(np.exp(c))            #e 的 x 幂次方
print('###############')
print(np.sqrt(c))           #开方
print('###############')
print(np.argmax(c))         #求最大值的索引位置
print('###############')
print(np.floor(c))          #向下取整
print('###############')
print(c.T)                  #转置，即行列交换
```

运行结果如下：

```
[[1.00000000e+00 2.71828183e+00 7.38905610e+00 2.00855369e+01]
 [5.45981500e+01 1.48413159e+02 4.03428793e+02 1.09663316e+03]
 [2.98095799e+03 8.10308393e+03 2.20264658e+04 5.98741417e+04]]
###############
[[0.         1.         1.41421356 1.73205081]
 [2.         2.23606798 2.44948974 2.64575131]
 [2.82842712 3.         3.16227766 3.31662479]]
###############
11
###############
[[ 0.  1.  2.  3.]
 [ 4.  5.  6.  7.]
 [ 8.  9. 10. 11.]]
###############
[[ 0  4  8]
 [ 1  5  9]
 [ 2  6 10]
 [ 3  7 11]]
```

NumPy 对于两个数组可以进行基本的算术运算，代码如下：

```
#第 1 章/studyNumPy/基本算术运算.py
import numpy as np

x = np.array([[1, 3], [2, 5]])
w = np.array([[3, 6], [4, 7]])
print('x 值\n', x)
print('w 随机值\n', w)
print('###################')
```

```
print('点乘\n', x *w)            #矩阵点乘法
print('相加\n', x + w)           #矩阵相加
print('内积\n', x @w)            #内积
print('####################')
x2 = np.random.random((2, 3))
w2 = np.random.random((3, 4))
print(x2 @w2)                    #得到一个 2 *4 的矩阵
```

运行结果如下：

```
x 值
 [[1 3]
 [2 5]]
w 随机值
 [[3 6]
 [4 7]]
####################
点乘
 [[ 3 18]
 [ 8 35]]
相加
 [[ 4  9]
 [ 6 12]]
内积
 [[15 27]
 [26 47]]
####################
[[0.85383671 0.56970322 0.95527218 1.3576547 ]
 [1.0573956  0.77879435 1.13676695 1.67171272]]
```

关于矩阵内积，假设矩阵 $\boldsymbol{N}\times\boldsymbol{M}$，那么另外一个矩阵需要满足 $\boldsymbol{M}\times\boldsymbol{N}$，矩阵的内积运算过程如图 1-26 所示。

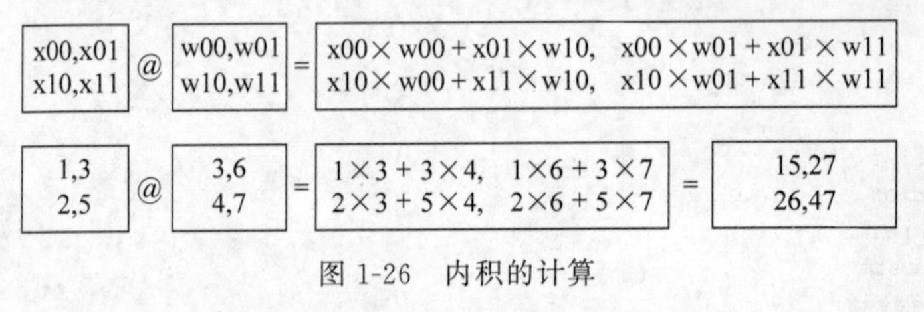

图 1-26　内积的计算

注意：矩阵点乘指两个矩阵对应位置相乘，矩阵内积指行列相乘后相加（只有行列的维度相等才能进行内积运算），通常说的矩阵乘法指的是内积运算。

1.11.4　NDArray 广播机制

广播机制是一种轻量级数组复制的手段，例如存储时是 np.random.random([3])，但是当它与其他数组进行数学运算时可以自动变成[N,3]，通过广播机制可以使存储空间变小以节省资源，代码如下：

```
#第 1 章/studyNumPy/广播机制.py
import numpy as np

x = np.random.random([2, 4])
w = np.random.random([4, 3])
b = np.random.normal([3])
result = x @w + b #b 自动扩维为(2,3)
print(result.shape)
```

运行结果如下：

```
(2, 3)
```

实现自动广播需要满足条件，两个数组的后缘维度的轴长度相等或者其中的一方的维度为 1 则认为可以广播，代码如下：

```
#第 1 章/studyNumPy/广播机制.py
#32, 32, 2
#32, 1 当最后一个维度为 1 时可以广播
x = np.random.random([32, 32, 2])
x = x + np.random.random([32, 1])
print(x.shape)
print('###########################')
#32, 32, 2
#32, 2 当最后两个维度相同时可以广播
x = np.random.random([32, 32, 2])
x = x + np.random.random([32, 2])
print(x.shape)
print('###########################')
#32, 32, 2
#32, 1 当最后一个维度为 1 时可以广播
x = np.random.random([32, 32, 2])
x = x + np.random.random([32, 1])
print(x.shape)
print('###########################')
#32, 32, 2
#1, 32, 1 当两边为 1 且中间(为 32)相等时可以广播
x = np.random.random([32, 32, 2])
x = x + np.random.random([1, 32, 1])
print(x.shape)
print('###########################')
#32, 32, 2
#3, 2 当倒数第 2 个维度不等时不能广播
x = np.random.random([32, 32, 2])
x = x + np.random.random([3, 2])
print(x.shape)
```

运行结果如下：

```
(32, 32, 2)
###########################
(32, 32, 2)
```

```
###########################
(32, 32, 2)
###########################
(32, 32, 2)
###########################
Traceback (most recent call last):
  File "D:\DLAI\qwenAILearn\untitled9\studyNumPy\广播机制.py", line 37, in <module>
    x = x + np.random.random([3, 2])
ValueError: operands could not be broadcast together with shapes (32,32,2) (3,2)
```

注意：x＝x＋np.random.random([3,2])广播失败的原因是因为3这个维度与原x的32没有对齐，所以不能进行广播。

总结

NumPy的索引取值与列表的语法基本一致，但更灵活、更强大，并支持内积运算。

练习

本书配套资源“NumPy基础练习1.txt”中有专门针对NumPy基础用法题目，供练习使用。

1.12 Pandas简介

Pandas是一个数据分析支持库，提供了快速、灵活、明确的数据结构，旨在简单、直观地处理关系型、标记型数据，适合处理与SQL或Excel表类似的含异构列的表格数据，其主要包含Series一维对象和DataFrame二维对象，对于深度学习的工作有时会从Excel或者SQL读取数据，所以主要还是使用DataFrame对象。Pandas基于NumPy开发，可以与其他第三方科学计算库完美集成。

1.12.1 Pandas对象的创建

Pandas对象的创建使用Series()或者DataFrame()放入相关数据即可，代码如下：

```
#第1章/Pandas对象的创建.py
import numpy as np
import pandas as pd

#初始化一维对象 np.nan,表示异常值
s = pd.Series([1, 2, 3, np.nan, 5, 6])
#初始化二维对象.行列
#以长度为4,长度为1的内容会自动复制。类似广播
df2 = pd.DataFrame({
    'A': 1.,
    'B': pd.Timestamp('20130102'),
    'C': pd.Series(1, index=list(range(4)), dtype='float32'),
```

```
    'D': np.array([3] * 4, dtype='int32'), #[3] * 4 表示复制 4 份[3]
    'E': pd.Categorical(["test", "train", "test", "train"]),
    'F': 'foo'
})
print(s)
print(df2)
```

运行结果如下：

```
0    1.0
1    2.0
2    3.0
3    NaN
4    5.0
5    6.0
dtype: float64
     A          B    C  D      E    F
0  1.0 2013-01-02  1.0  3   test  foo
1  1.0 2013-01-02  1.0  3  train  foo
2  1.0 2013-01-02  1.0  3   test  foo
3  1.0 2013-01-02  1.0  3  train  foo
```

在 DataFrame()中数字用来表示行号，A、B、C、D、E、F 表示列名，而 Series()中没有列名只有行号，如图 1-27 所示。

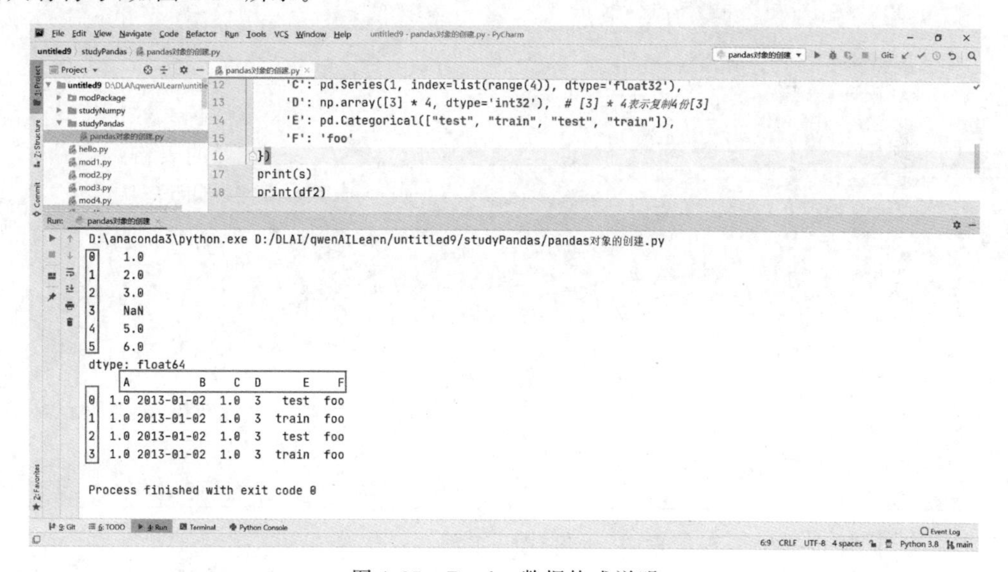

图 1-27 Pandas 数据构成说明

1.12.2 Pandas 的索引与切片

DataFrame 对象的索引和切片的语法类似于 NumPy 的语法，同时 Pandas 还提供了 loc[]和 iloc[]这两个对象以进行操作，代码如下：

```
#第 1 章/Pandas 索引与切片 iloc 用法.py
import numpy as np
import pandas as pd
#随机生成一个 3 行 3 列的数
rnd_value = np.random.random([3, 3])
#将 NumPy 数据转换为 DataFrame 对象
df = pd.DataFrame(rnd_value)
print('原值', df)
print('################')
#基本用法,下标取值,默认为行的索引
print(df.iloc[2])
print('################')
#分别在行取索引为 0、1 行,在列中取索引为 1 的列
print(df.iloc[:2, 1:2])
print('################')
#分别在行取索引为 0、2 行,在列中取索引为 0、2 列
print(df.iloc[[0, 2], [0, 2]])
print('################')
#取索引为 0 行且列索引为 0 列的值
print(df.iloc[0, 0])
```

运行结果如下：

```
原值          0         1         2
0  0.385156  0.982662  0.144847
1  0.275933  0.258540  0.836210
2  0.252520  0.533021  0.384131
################
0    0.252520
1    0.533021
2    0.384131
Name: 2, dtype: float64
################
          1
0  0.982662
1  0.258540
################
          0         2
0  0.385156  0.144847
2  0.252520  0.384131
################
0.385156
```

Pandas 同样也提供了布尔索引,示例代码如下：

```
#第 1 章/Pandas 布尔索引的用法.py

import numpy as np
import pandas as pd

#随机生成一个 3 行 3 列的数
rnd_value = np.random.random([3, 3])
```

```
#将 NumPy 数据转换为 DataFrame 对象
df = pd.DataFrame(rnd_value)
print('原值', df)
print('###########')
#如果 df 中的值大于 0.5,则为 True,否则为 False
print(df > 0.5)
print('###########')
#将 df 中大于 0.5 的值取出来,否则为 NaN 异常值
print(df[df > 0.5])
```

运行结果如下：

```
原值          0         1         2
0  0.247702  0.053419  0.965267
1  0.424877  0.312497  0.720876
2  0.631195  0.546980  0.331154
###########
       0      1      2
0  False  False   True
1  False  False   True
2   True   True  False
###########
          0        1         2
0       NaN      NaN  0.965267
1       NaN      NaN  0.720876
2  0.631195  0.54698       NaN
```

1.12.3　Pandas 常用统计函数

Pandas 的一大特点是提供了较多统计函数，例如 head()、tail()、describe()等函数，使用者可以快速地对数据的分布有一个总体上的了解，代码如下：

```
#第 1 章/Pandas 布尔索引的用法.py

import numpy as np
import pandas as pd

#随机生成一个 3 行 3 列的数
rnd_value = np.random.random([3, 3])
#将 NumPy 数据转换为 DataFrame 对象
df = pd.DataFrame(rnd_value)
#取第 1 行
print(df.head(1))
print('#'*35)
#取最后 1 行
print(df.tail(1))
print('#'*35)
#统计摘要信息
print(df.describe())
print('#'*35)
#转置
```

```
print(df.T)
print('#'*35)
#先按 1 列排序,再按 2 列排序
print(df.sort_values(by=[1, 2]))
```

运行结果如下：

```
          0         1         2
0  0.297752  0.468549  0.824418
###################################
          0         1         2
2  0.376324  0.400587  0.693341
###################################
              0         1         2
count  3.000000  3.000000  3.000000
mean   0.477736  0.295387  0.741207
std    0.246842  0.243451  0.072334
min    0.297752  0.017027  0.693341
25%    0.337038  0.208807  0.699602
50%    0.376324  0.400587  0.705862
75%    0.567727  0.434568  0.765140
max    0.759131  0.468549  0.824418
###################################
          0         1         2
0  0.297752  0.759131  0.376324
1  0.468549  0.017027  0.400587
2  0.824418  0.705862  0.693341
###################################
          0         1         2
1  0.759131  0.017027  0.705862
2  0.376324  0.400587  0.693341
0  0.297752  0.468549  0.824418
```

1.12.4 Pandas 文件操作

Pandas 另一个非常好用的功能就是集成了读取或者存储 CSV、Excel 等文件函数，并且支持连接各数据库，可以方便地与 Pandas 索引取值或者转换成 NumPy 进行数据处理，代码如下：

```
#第 1 章/pandas 文件操作.py

import numpy as np
import pandas as pd

#随机生成一个 3 行 3 列的数
rnd_value = np.random.random([3, 3])
#将 NumPy 数据转换为 DataFrame 对象
df = pd.DataFrame(rnd_value)
#写 CSV 文件,默认为当前文件路径
df.to_csv('test.csv')
```

```
#写 Excel 文件
df.to_excel('test.xlsx')
#读 CSV 文件
print(pd.read_csv('test.csv'))
print('#' *35)
print(pd.read_excel('test.xlsx'))
```

运行结果如下:

```
   Unnamed: 0         0         1         2
0           0  0.349515  0.862236  0.972058
1           1  0.610629  0.275613  0.593430
2           2  0.452929  0.383856  0.234271
###################################
   Unnamed: 0         0         1         2
0           0  0.349515  0.862236  0.972058
1           1  0.610629  0.275613  0.593430
2           2  0.452929  0.383856  0.234271
```

注意:在进行Excel存取操作时,可能会报No module named 'openpyxl'的错误或者No module named"某个包名"的错误,此时只需使用conda或者pip安装对应的包便可以解决环境问题。

Pandas也可以读取MySQL中的数据,但是需要安装pymysql、sqlalchemy等依赖包,具体的使用方法,代码如下:

```
#第 1 章/pandas 读取 mysql.py
import numpy as np
import pandas as pd

import pandas as pd
from sqlalchemy import create_engine

#初始化数据库连接,使用 pymysql 模块
engine = create_engine('mysql+pymysql://MySQL用户登录名:MySql; 登录密码:
//@MySql 地址:端口/库名')
#查询语句,选出 employee 表中的所有数据
sql = '''select *from testdf limit 2;'''
#read_sql_query 的两个参数: SQL 语句和数据库连接
df = pd.read_sql_query(sql, engine)
#输出 user 表的查询结果
print('查询数据: ', df)
print('#' *35)
#插入数据的用法
#新建 Pandas 中的 DataFrame, 只有 id 和 num 两列
df = pd.DataFrame({'id': [110], 'name': ['test'], 'score': [90]})
#将新建的 DataFrame 存储为 MySQL 中的数据表,不存储 index 列
#if_exists='append'表示追加
df.to_sql('testdf', engine, index=False, if_exists='append')
```

```
#查询新建的信息是否成功
sql = 'select * from testdf where id=110'
df = pd.read_sql_query(sql, engine)
print('查询插入的数据：', df)
```

运行结果如下：

```
查询数据：      id name  score
0  10   张三     80
1  27   李四     75
####################################
查询插入的数据：      id  name  score
0  110  test     90
```

注意：create_engine 函数中的 MySQL 服务器的信息需要自己先搭建，并创建一个拥有 id、name、score 共 3 个字段的表才可执行代码。

总结

Pandas 虽然只能处理一维、二维数据，但是能够很方便地读取 Excel、CSV、数据库等关系型、标记型数据。

练习

本书配套资源"Pandas 基础练习 1. txt"中有专门针对 Pandas 基础用法的题目，供练习使用。

1.13 Matplotlib 简介

Matplotlib 是最流行的 Python 底层绘图库，可以非常方便地构建二维或三维图表，可结合 NumPy 或者 Pandas 数据，绘制图形包括线性图、柱状图、直方图、密度图、散布图等，能够满足大部分的图表绘制需求。

1.13.1 Matplotlib 基本使用流程

Matplotlib 绘图需要设定标题、设定 x 轴和 y 轴的名称和范围，同时如果一张图里面包含几个图形，则还需要添加子图，当这些设定完毕后就可以调用 plot() 函数进行绘制了，其绘图的基本流程可以参考图 1-28 的说明。

按照上面的流程，绘制折线图的代码如下：

```
#第 1 章/studyMatplotlib/绘制折线图.py
#安装此库的方法
#conda install matplotlib
import matplotlib.pyplot as plt
import numpy as np
```

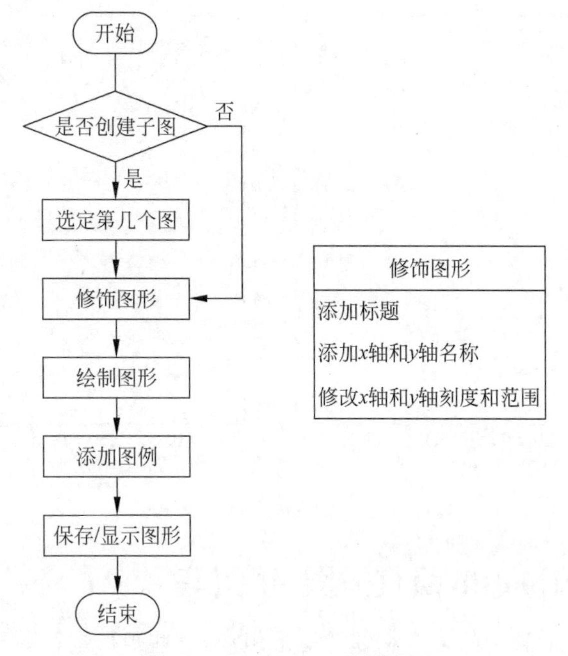

图 1-28 Matplotlib 绘图流程

```
#用来正常显示中文标签
plt.rcParams['font.sans-serif'] = ['SimHei']
plt.rcParams['axes.unicode_minus'] = False

#将画布的大小设置为 10×10, dpi 是图像质量
fig = plt.figure(figsize=(10, 10), dpi=80)
#设置子图 1
plt.subplot(2, 2, 1)              #表示图像是 2 行 2 列的,选取第 1 个位置
x1_value = [0.1, 0.2, 0.3, 0.4]
y_value = np.linspace(0, 1.0, len(x1_value))
plt.title('图 1')
plt.plot(x1_value, y_value)   #绘 x1
plt.legend(['x1'])                #图例说明
#设置子图 2
plt.subplot(2, 2, 2)
x2_value = [0.4, 0.5, 0.6, 0.7]
plt.title('图 2')
plt.plot(x2_value, y_value)   #绘 x2
plt.legend(['x2'])                #图例说明
#plt.show()                       #弹出来显示图像
plt.savefig('1.png')              #保存图像。当使用此语句时,需要注意 plt.show()
```

运行结果如图 1-29 所示。

注意:Matplotlib 默认不支持中文,需要设置 plt. rcParams。在执行 plt. show()之后再执行 plt. savefig()函数可能导致保存的图像为空。np. linspace()将指定范围内的数均分。

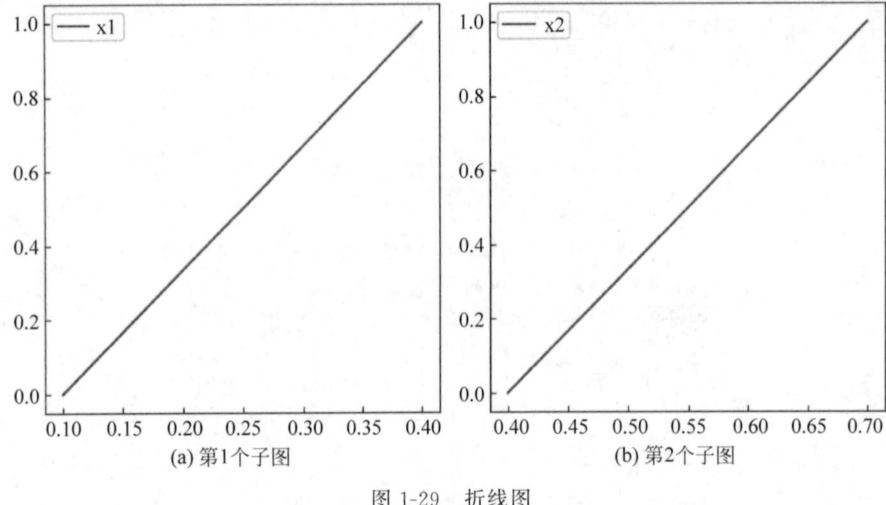

(a) 第1个子图　(b) 第2个子图

图 1-29　折线图

1.13.2　Matplotlib 绘直方图、饼图等

直方图由竖立在 x 轴上的多个相邻的矩形组成，这些矩形把 x 轴拆分为一段彼此不重叠的线段，矩形的面积跟落在其所对应的元素数量成正比。plt.hist()函数可以显示直方图，代码如下：

```
#第 1 章/studyMatplotlib/绘制直方图.py
import numpy as np
import matplotlib.pyplot as plt
#用来正常显示中文标签
plt.rcParams['font.sans-serif'] = ['SimHei']
plt.rcParams['axes.unicode_minus'] = False

#将画布的大小设置为 10×10, dpi 是图像质量
fig = plt.figure(figsize=(10, 10), dpi=80)

p = np.random.randint(0, 100, 100)
#直方图
plt.hist(p, bins=20, color='g')
#plt.show()
plt.savefig('直方图.png')
```

运行结果如图 1-30 所示。

条状图跟直方图有点类似，可以用 bar()函数生成条状图，代码如下：

```
#第 1 章/studyMatplotlib/条状图.py

import matplotlib.pyplot as plt

index = [1, 2, 3, 4, 5]
values = [10, 3, 19, 6, 4]
```

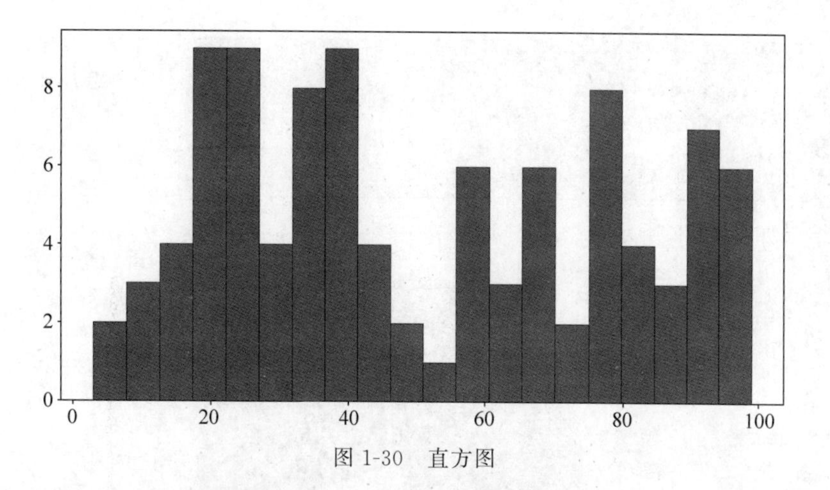

图 1-30 直方图

```
plt.bar(index, values, color='c')
#plt.show()
plt.savefig('条状图.png')
```

运行结果如图 1-31 所示。

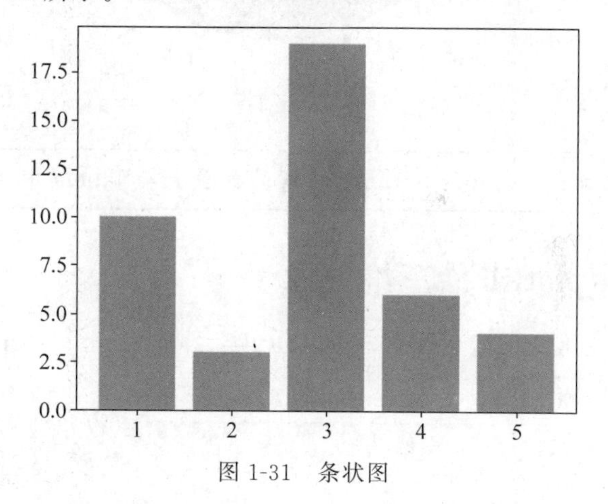

图 1-31 条状图

饼图函数为 pie(),代码如下:

```
#第 1 章/studyMatplotlib/饼图.py
import matplotlib.pyplot as plt

#用来正常显示中文标签
plt.rcParams['font.sans-serif'] = ['SimHei']
plt.rcParams['axes.unicode_minus'] = False

labels = ['小目标', '中目标', '大目标']
values = [30, 25, 24]
colors = ['green', 'blue', 'red']
#画饼图 plt.pie(值,标签,颜色,显示数字,是否有阴影,比例)
plt.pie(values, labels=labels, colors=colors,
```

```
        autopct='%1.1f%%', shadow=False, startangle=90)
plt.axis('equal')
#plt.legend(loc='upper right')
plt.show()
plt.savefig('饼图.png')
```

运行结果如图 1-32 所示。

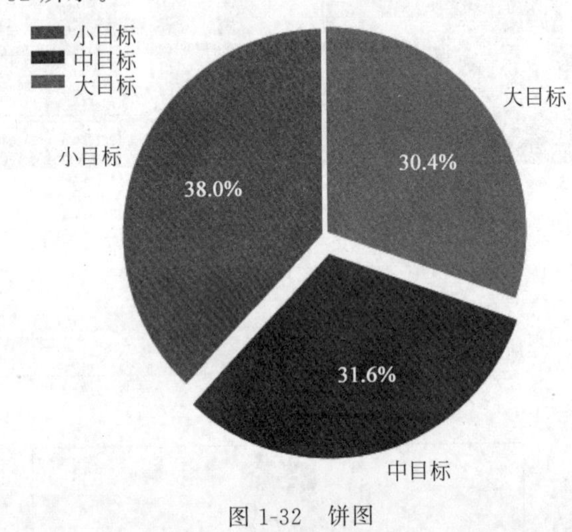

图 1-32　饼图

注意：color='c'代表青色。colors=colors 参数的数量要与 labels 中参数的数量保持一致。

1.13.3　Matplotlib 绘三维图像

Matplotlib 绘制三维图像需要调用 Axes3D()类，并使用 plot_surface()函数传入 X、Y、Z 这 3 个参数的值，代码如下：

```
#第 1 章/studyMatplotlib/3D 图像.py
import matplotlib.pyplot as plt
import numpy as np
from mpl_toolkits.mplot3d import Axes3D

#用来正常显示中文标签
plt.rcParams['font.sans-serif'] = ['SimHei']
plt.rcParams['axes.unicode_minus'] = False

fig = plt.figure()
ax = fig.add_axes(Axes3D(fig))                #初始化
X = np.arange(-1, 1, 0.01)
Y = np.arange(-1, 1, 0.01)
X, Y = np.meshgrid(X, Y)
Z = (X **2 + Y **2 - (2 *X *Y))               #(x-y)**2,x - y 的平方
#三维图像数据需要 x、y、z 这 3 个坐标
#rstride     数组行距(步长大小)
```

```
#cstride     数组列距(步长大小)
#color           曲面块颜色映射
ax.plot_surface(X, Y, Z, rstride=1, cstride=1, cmap='rainbow')
#plt.show()
plt.savefig('3d图像')
```

运行结果如图 1-33 所示。

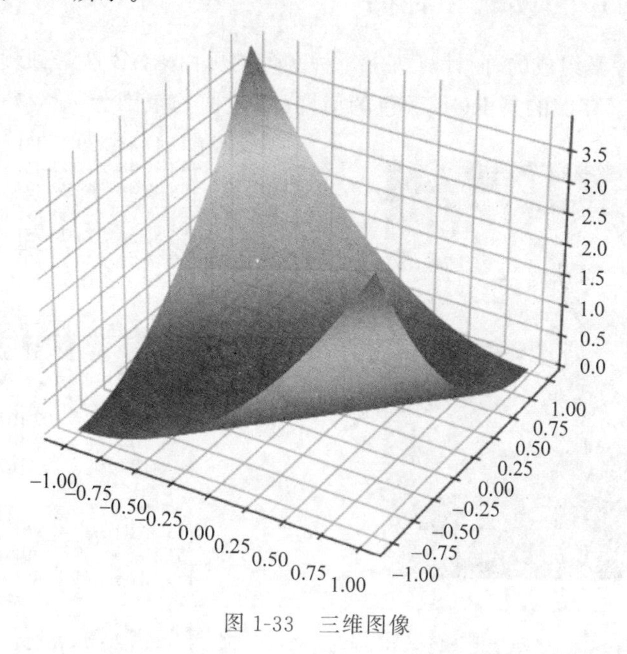

图 1-33　三维图像

注意：np. meshgrid()函数的作用是生成基于 X、Y 的坐标系。假设 X 为 np. array([1,2,3]),Y 为 np. array([7,8]),np. meshgrid(X,Y)后得到的 X 内容为 np. array([1,2,3],[1,2,3]),Y 的内容为 np. array([7,7,7],[8,8,8])。

总结

Matplotlib 可以很方便地绘出折线图、直方图、饼图等,并可结合 NumPy 等数据快速地出图。

练习

本书配套资源“Matplotlib 基础练习 1. txt”中有专门针对 Matplotlib 绘图的练习任务。

1.14　OpenCV 简介

OpenCV 是一个跨平台的计算机视觉和机器学习软件库,可以运行在 Linux、Windows、Android 和 mac OS 等操作系统上,同时提供了 Python、C++ 等多语言调用的接口,并且也实现了图像处理和计算机视觉方面的很多通用算法。在后面的深度学习算法中，需要用到 OpenCV 的一些基本函数以实现图像的读取、存储，以及一些数据增强部分的内容，因此本节的重点在于介绍一些 OpenCV 基本函数的使用，而关于 OpenCV 的一些通用算法并不在此范围内。

1.14.1 图片的读取和存储

图片是由一些像素组成的，而计算机中每种颜色均可由 RGB 这三种颜色合成，所以对于计算机来讲每个像素存储的是 RGB 颜色的值，其颜色值的范围为 0～255，如图 1-34 所示。

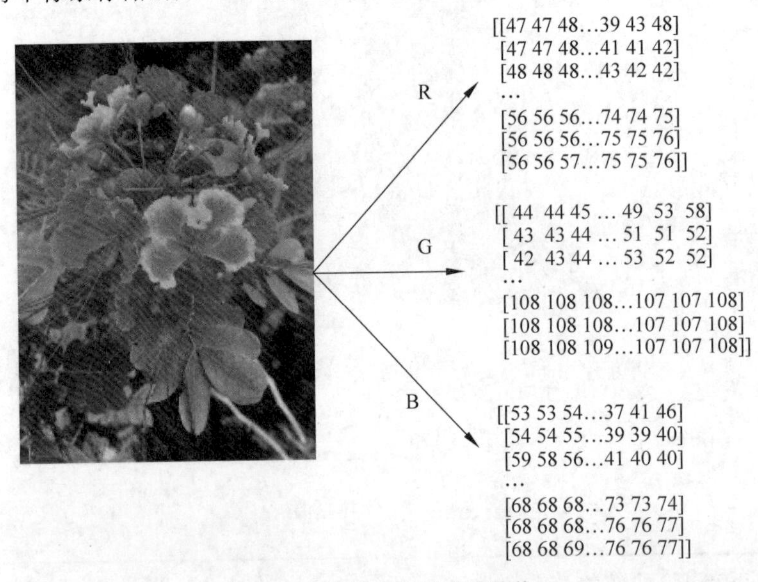

图 1-34 读取对象对应的数值

图像的读取使用 cv2.imread()函数，而存储使用 cv2.imwrite()函数，示例代码如下：

```
#第 1 章/studyOpenCV/cv_ex1.py
#环境安装命令：conda install opencv
import cv2

#读取图片 cv2.imread(图片位置,图像格式)
img = cv2.imread("test1.jpg", cv2.COLOR_BGR2RGB)
#将图像的尺寸改为 432,416
img = cv2.resize(img, [432, 416])
#在弹出窗口中显示图片
cv2.imshow('img', img)
#窗口停留的时间
cv2.waitKey(0)
#释放所有窗口句柄
cv2.destroyAllWindows()
#将 img 写入某个图像中
cv2.imwrite('test1_2.jpg', img)
```

注意：在弹出的窗口中按任意键后，图像将会被保存到当前目录下面的 test1_2.jpg 中。如果不想弹出窗口，则将相关窗口函数注释即可。

cv2.COLOR_BGR2RGB 用于将 BGR 图像转换为 RGB 格式的图像，常见的图像格式有灰度图、BGR 格式、RGB 格式、HSV 格式等。

1.14.2 画矩形、圆形等

在计算机视觉领域尤其是目标检测领域，需要对识别出来的目标进行突出显示，而显示的方法就包括把目标用一个矩形、圆形、直线等表示出来，因此有必要对以下方法进行掌握，代码如下：

```
#第 1 章 studyOpenCV/基本绘图功能.py
import cv2
import numpy as np
import math

#读取图片 cv2.imread(图像位置,图像格式)
img =cv2.imread("test1.jpg", cv2.COLOR_BGR2RGB)
#在图像区域中画矩形
#cv2.rectangle(图像,(起点 x,起点 y), (终点 x,终点 y),颜色,线条粗细)
cv2.rectangle(img, (100, 100), (200, 200), (255, 0, 0), 2)
#在图像中的某个位置画圆、画点,即半径为 0
#cv2.circle(图像,中心点,半径, 颜色,线条粗细)
cv2.circle(img, (150, 150), 50, (0, 255, 0), 2)
#在图像中画线
#cv2.circle(图像,(起点 x,起点 y), (终点 x,终点 y), 颜色,线条粗细)
cv2.line(img, (100, 100), (200, 200), (0, 0, 255), 2)
#生成一串连续的坐标点
t =np.linspace(0, math.pi, 1000)
x =np.sin(t)
y =np.cos(t) +np.power(x, 2.0 / 3)
pts =[[[int(x *255), int(y *255)] for x, y in zip(x, y) if x >0 and y >0]]
#画多边形
cv2.polylines(img, np.array(pts), True, (0, 255, 255), 2)
cv2.imshow('img', img)
#窗口停留的时间
cv2.waitKey(0)
#释放所有窗口句柄
cv2.destroyAllWindows()
```

运行结果如图 1-35 所示。

1.14.3 在图中增加文字

利用文字对于识别的目标进行标识是目标检测的重要结果，而 OpenCV 提供的 cv2.putText()函数只能显示英文，如果需要显示中文，则需要借助 Pillow 库，代码如下：

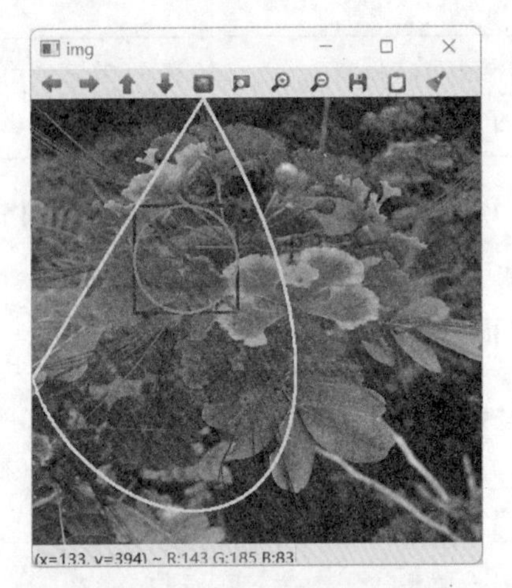

图 1-35　画矩形、圆形的结果

```
#第 1 章 studyOpenCV/增加文字描述.py
import cv2
import numpy as np
import math
#conda install pillow
from PIL import Image, ImageDraw, ImageFont

#读取图片 cv2.imread(图像位置,图像格式)
img = cv2.imread("test1.jpg", cv2.COLOR_BGR2RGB)
#在图像区域中画矩形
#cv2.rectangle(图像,(起点 x,起点 y), (终点 x,终点 y),颜色,线条粗细)
cv2.rectangle(img, (100, 100), (200, 200), (255, 0, 0), 2)
#OpenCV 自带函数,实现文字描述,只能显示英文
#cv2.putText(图,文字,位置,字体,大小,颜色,线条粗细)
cv2.putText(img, 'flower', (100, 100), cv2.FONT_HERSHEY_SIMPLEX, 1, (255, 255,
255), 1)
#显示中文的处理
#转换为 Image 对象需要的格式
img = Image.fromarray(cv2.cvtColor(img, cv2.COLOR_BGR2RGB))
#使用 Image 重绘图片
draw = ImageDraw.Draw(img)
#设置字体
fontText = ImageFont.truetype("SIMYOU.TTF", 24, encoding="utf-8")
#写中文
draw.text((200, 200), '花儿', (255, 255, 255), font=fontText)
#从 Image 格式再换回 OpenCV 的图像格式
img = cv2.cvtColor(np.array(img), cv2.COLOR_BGR2RGB)
#窗口展示
cv2.imshow('img', img)
#窗口停留的时间
cv2.waitKey(0)
#释放所有窗口句柄
cv2.destroyAllWindows()
```

运行结果如图 1-36 所示。

图 1-36　图中写文字的效果

注意：如果绘图时要显示中文，则需要调用 Pillow 库实现，其中字体的选择即 SIMYOU.TTF 是从 Windows 计算机中的 C:\Windows\Fonts 中复制出来的。

1.14.4　读取视频或摄像头中的图像

视频的本质是连续的图片，即每秒能够播放的图片，在深度学习计算机视觉任务中有相当大一部分的操作是对摄像头中的图像或者视频进行相关识别，而 OpenCV 提供了 cv2.VideoCapture()来读取视频的内容，代码如下：

```
#第 1 章/studyOpenCV/读取视频.py

import cv2
import numpy as np

#读取视频文件或者摄像头中的图像
#如果是 RTMP 协议，则 path 值为 rtmp://ip:port/地址
#如果是 RTSP 协议，则 path 值为 rtsp://ip:port/地址
#如果是视频文件，则 path 为视频目录地址
#如果是计算机的摄像头，则 path 为 0
path = 0
vc = cv2.VideoCapture(0)
if vc.isOpened():
    #读取 1 帧的图像，open 是否有图像，frame 这一帧的图像值
    isOpen, frame = vc.read()
else:
    isOpen = False

while isOpen:
    ret, frame = vc.read()
```

```
    if frame is None:
        break
    if ret == True:
        #灰色图像格式读取
        img = cv2.cvtColor(frame, cv2.COLOR_RGB2GRAY)
        cv2.imshow('result', img)
        if cv2.waitKey(0):
            break
vc.release()
cv2.destroyAllWindows()
```

注意：cv2. VideoCapture(0)表示打开当前计算机的摄像头，如果要换视频文件，则需要传入文件地址。

总结

OpenCV 能够对图像或者视频进行读写、再加工等操作，是计算机视觉领域中比较重要的第三方库。

练习

打开摄像头或视频后在每帧图中画一个矩形，并显示出来。

第 2 章

机器学习基础

在介绍了 Python 基础、数据分析常用库及 OpenCV 基础操作后，本章将介绍机器学习的部分知识，理解并掌握机器学习的一些概念是学习深度学习的一个重要过程，因此本章结合学习深度学习的理论要求将重点讲解 KNN、梯度下降、线性回归、逻辑回归、聚类、神经网络等算法，并使用 Python 基础代码而不是调用（如 scikit-learn）第三方机器学习库来完成算法代码的实现，通过本章的训练有助于提高初学者的理论和代码编写水平。

2.1 HelloWorld 之 KNN 算法

2.1.1 KNN 算法原理

假设小文到了一定的年纪并想找一个女朋友成家，但是一直没有找到心仪的对象，迫于无奈找到万婆婆让其介绍一个相亲对象，然后万婆婆拿出相亲记录表，见表 2-1。

表 2-1 相亲记录表

编号	姓名	年龄	性别	月收入	学历	相亲次数	成功与否
1	文 *	25	男	5000	本科	3	否
2	祁 *	30	女	20 000	本科	2	是
3	杨 *	28	男	11 000	高中	4	否
4	万 *	30	女	40 000	博士	5	是
5	余 *	32	男	6000	本科	3	否
6	张 *	21	女	7000	大专	2	否
7	陈 *	27	男	4000	高中	99	否
8	利 *	26	男	14 000	本科	10	是
9	杨 *	34	男	24 000	本科	10	是
10	钱 *	36	男	7000	高中	99	否
11	小文	26	男	13 000	本科	0	?

那么小文这次能够相亲成功吗？

通过以上数据，需要预测出小文相亲成功或者相亲失败，这本质上是一个分类任务。分类任务是机器学习和深度学习的一个重要任务。

假设只参考年龄、月收入两个字段,把它展示在图中,如图 2-1 所示。

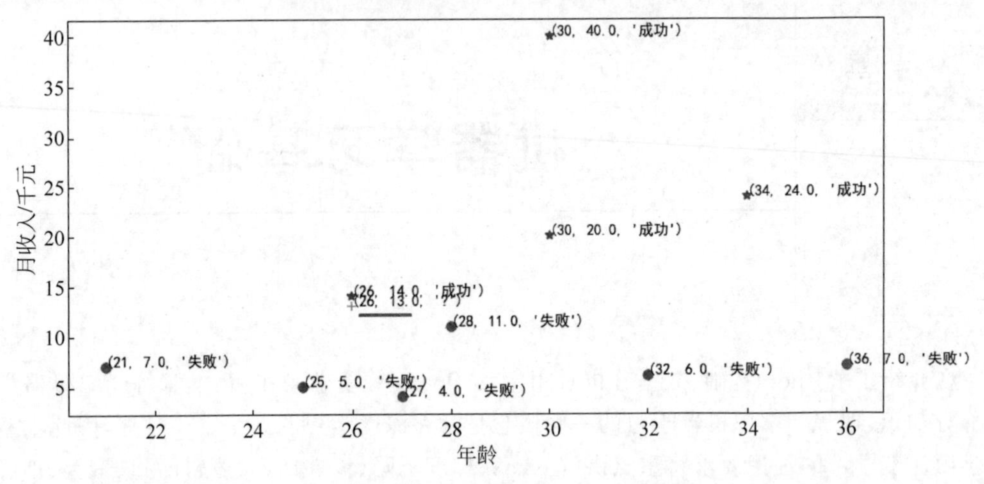

图 2-1　相关人员的年龄与收入关系

依据我们的生活经验很容易就知道小文是很有可能相亲成功的,因为根据小文的条件离他最近的是(26,14 000)的利＊,但是如果把参考范围扩大一些,例如最近的 5 个,则这次相亲可能失败,所以 KNN 算法的核心其实就是参考距离最近的样本,所参考距离样本的数量哪个类别最多,需要判别的样本(已知数据又称为样本数据)就是哪一个。如果参考样本数为 5,则其中 3 个失败 2 个成功,所以预测结果为相亲失败,如图 2-2 所示。

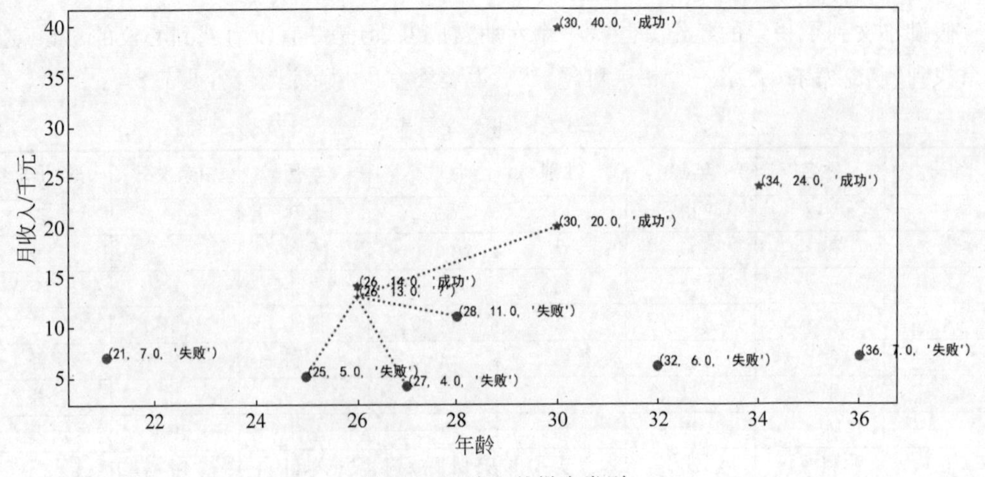

图 2-2　K 为 5 的样本类别

根据上面的分析,KNN 算法实现的基本过程如下:

(1) 计算所有样本数据与待预测数据之间的距离。

(2) 按照距离大小进行递增排序。

(3) 确定前 K 个样本所在类别出现的频率,并输出出现频率最高的类别。

其中,K 为超参数,即可调节的参数,K 值不同所得到的结果就可能不同。计算距离可以使用欧氏距离的公式：

$$d(x,y)=\sqrt{(x_1-y_1)^2+(x_2-y_2)^2+\cdots+(x_i-y_i)^2}=\sqrt{\sum_{i=1}^{n}(x_i-y_i)^2} \quad (2\text{-}1)$$

另一个问题是如何确定 K 值，一个比较好的办法就是将样本数据划分成训练集和验证集，然后设置不同的 K 值(例如 1、3、5、7)并统计验证集中预测错误数，然后取错误率较低到错误率较高的拐点值，$K=3$ 或 $K=5$ 的结果较佳，如图 2-3 所示。

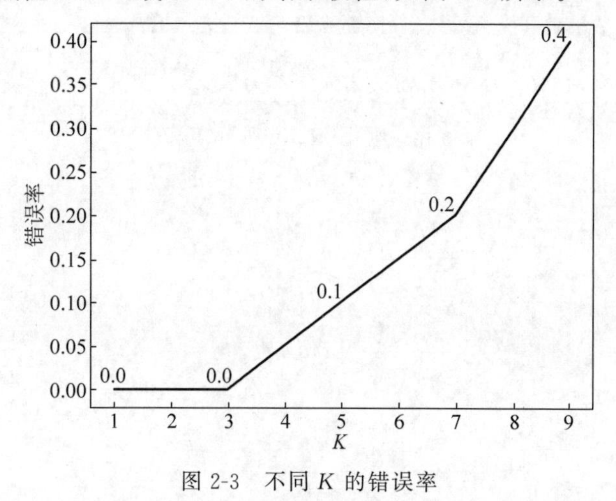

图 2-3　不同 K 的错误率

2.1.2　KNN 算法代码实现

依据 2.1.1 节介绍的算法原理，设计程序流程图，如图 2-4 所示。

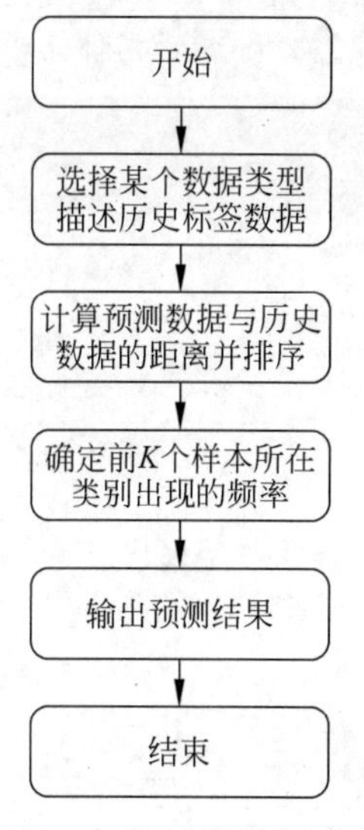

图 2-4　KNN 算法代码实现流程图

按照程序流程图的设计，代码如下：

```
#第 2 章/machineLearn/KNN 算法的实现.py
import math

amount_ratio = 1000

#历史相亲数据的描述
def historical_data():
    #将所有金额缩小到以 K 为单位，方便计算
    data = {
        #姓名：[年龄，收入，是否成功(0 为失败，1 为成功)]
        '文*': [25, 5000 / amount_ratio, 0],
        '祁*': [30, 20000 / amount_ratio, 1],
        '杨*1': [28, 11000 / amount_ratio, 0],
        '万*': [30, 40000 / amount_ratio, 1],
        '余*': [32, 6000 / amount_ratio, 0],
        '张*': [21, 7000 / amount_ratio, 0],
        '陈*': [27, 4000 / amount_ratio, 0],
        '利*': [26, 14000 / amount_ratio, 1],
        '杨*2': [34, 24000 / amount_ratio, 1],
        '钱*': [36, 7000 / amount_ratio, 0],
    }
    return data


def calculate_european_distance(input_x: list, data: dict):
    """
    返回欧氏距离后的结果，并以[[人名，距离]]的格式返回
    :param input_x:
    :param data:
    :return:
    """
    distance_list = []
    for key, v in data.items():
        #计算欧氏距离
        d = math.sqrt((input_x[0] - v[0]) **2 + (input_x[1] - v[1]) **2)
        distance_list.append([key, round(d, 2)])
    #对计算后的距离按升序进行排列
    #假设有 10 条数据，那么计算后的结果就是 10 条，即每个数据都有一个距离
    return sorted(distance_list, key=lambda l: l[1])


def predict_ByK(distance_list, raw_data, k=5):
    #确定前 K 个样本所在类别出现的频率，并输出出现频率最高的类别
    labels = {"相亲失败": 0, "相亲成功": 0}
    for s in distance_list[:k]:
        #根据上一步距离计算出来的人名，从原始数据中取出样本信息
        #即[25, 5000 / amount_ratio, 0]
        sample_information = raw_data[s[0]]
        if sample_information[-1] == 0:
            labels['相亲失败'] += 1
```

```
        elif sample_information[-1] == 1:
            labels['相亲成功'] += 1
    #根据相亲结果出现的次数按升序进行排列
    return sorted(labels.items(), key=lambda l: l[1], reverse=True)


if __name__ == "__main__":
    raw_data = historical_data()
    input_x = ['小文', 26, 13000 / amount_ratio]
    distance_list = calculate_european_distance(input_x[1:], raw_data)
    result = predict_ByK(distance_list, raw_data, k=5)
    print(f'预测{input_x[0]}的相亲结果为{result[0][0]}, '
          f'概率为{round(result[0][1] / sum([y for x, y in result]), 2)}')
```

在 historical_data()函数中，使用字典来描述历史样本数据，并对收入进行了 1000 倍缩放(方便计算)，然后在 calculate_european_distance()函数中，调用 math 数学库来计算预测数据与所有历史样本数据的距离，并使用 sorted()函数按照距离大小进行递增排序，然后在 predict_ByK()函数中，根据超参数 K 统计标签出现的频率，并选择出现最多的标签作为预测结果。

运行结果如下：

```
预测小文的相亲结果为相亲失败，概率为 0.6
```

从算法实现的过程可以发现，KNN 算法简单且易于实现，但是该算法存在两个问题：第 1 个是 KNN 没有模型训练的过程，训练和预测是结合在一起的，当数据量大时内存开销大、分类慢；第 2 个是业务可解释性差，无法给出业务上可理解的业务规则。

总结

KNN 通过计算预测数据最短距离所在数据的类别，采用计数投票的方式进行预测，是一种简单有效的分类算法。

练习

更换成 Pandas、NumPy 库实现上述代码的功能。

2.2　梯度下降

2.2.1　什么是梯度下降

梯度下降的基本思想可以类比为下山的过程，假设山上浓雾很大，可视度很低，下山的路无法确定，而小文又被困在了山顶，小文要下山就必须利用周围的信息来帮助自己下山，这时就可以采用梯度下降的算法，具体来说就是以小文所处的位置为基准，寻找这个位置最陡峭的地方，朝着山高度下降的地方走，然后每隔一段距离反复采用此方法，这样就能成功地抵达山底。

这座山最陡峭的地方是无法通过肉眼立马观察出来的，因此就需要一个工具进行测量，假设小文有这个工具，所以小文每走一段距离都需要时间来测量所在位置的最陡峭的方向，这是比较耗时的，而为了在太阳下山前到达山底，就要尽可能地减少测量方向的次数，所以找到一个合适的频率来确保下山的方向正确，并同时又不至于耗时太多是比较重要的。

梯度下降的过程就和下山的场景类似，首先需要一个可微分的函数，这个函数就代表着一座山。下山的目的就是找到这个函数的最小值，也就是山底处。根据场景设定，最快的下山的方式就是找到当前位置最陡峭的方向，然后沿着该方向向下走，对应到函数中就是找到给定点的梯度，然后朝着梯度相反方向就能让函数值下降最快，因为梯度的方向就是函数变化最快的方向。

看待微分可以从函数的斜率或者函数的变化率的角度，例如单变量微分：

$$\frac{\partial x^2}{\partial x}=2x \tag{2-2}$$

当一个函数有多个变量时就有了多变量的微分，即分别对每个变量求微分：

$$\frac{\partial(\theta_1+2\theta_2-3\theta_3)}{\partial\theta_2}=2 \tag{2-3}$$

梯度实际上就是多变量微分的一般化：

$$\Delta J(\theta)=\left[\frac{\partial J}{\partial\theta_1},\frac{\partial J}{\partial\theta_2},\frac{\partial J}{\partial\theta_3}\right]=[1,2,-3] \tag{2-4}$$

从上文可知梯度就是分别对每个变量进行微分，然后用逗号分隔在一个向量中。在多变量函数中，梯度就是一个向量，而向量是有方向的，梯度的方向就是函数在给定点的上升最快的方向。

所以为了下山需要千方百计地求取梯度，观测最陡峭的地方，梯度的方向是函数在给定点上升最快的方向，那么梯度的反方向就是函数在给定点下降最快的方向，所以只要沿着梯度的方向一直走，就能走到局部的最低点，如图 2-5 所示。

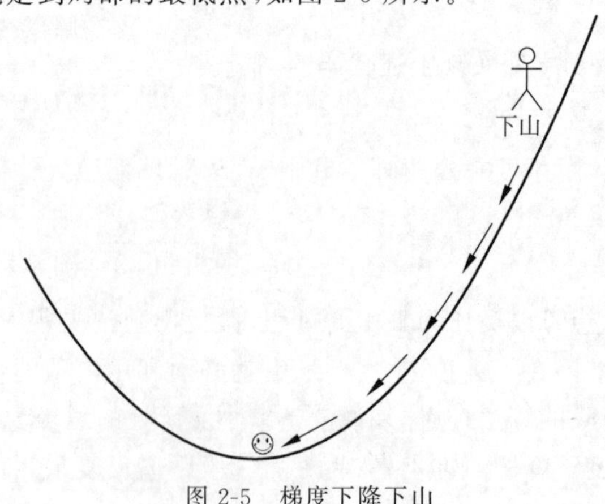

图 2-5　梯度下降下山

本节花了较多篇幅介绍梯度下降算法的基本思想和场景设定，如果从数学角度出发，则梯度下降的公式可以表示为

$$\theta^{k+1}=\theta^{k}-\alpha * \nabla_{\theta}J(\theta) \tag{2-5}$$

J 是关于 θ 变量的一个函数，当前所处的位置为 θ^k，要从这个点走到 J 的最小值，也就是山底，首先要确定前进的方向，也就是梯度的方向，然后走一段距离的步长 α，走完这段步长就到达了 θ^{k+1} 这个点。

在梯度下降算法中 α 称为学习率，通过 α 来控制每步走的距离，如果 α 设置的值较大，则有可能错过最低点，如果 α 设置得太小，则需要更多的时间，α 的设置在梯度下降算法中是非常重要的，如图 2-6 所示。

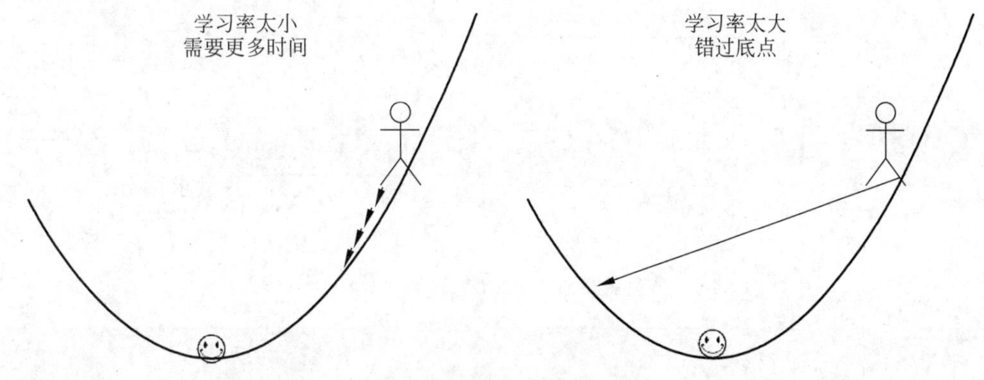

图 2-6　学习率 α 设置的影响

梯度前加一个负号，就意味着朝着梯度相反的方向前进。梯度的方向实际上就是函数在此点上升最快的方向，而为了下山就需要朝着下降最快的方向走，所以需要加上负号。

2.2.2　梯度下降的代码实现

假设有一个目标函数如下：

$$J(\theta)=(\theta_1-10)^2+(\theta_2-10)^2 \tag{2-6}$$

根据梯度下降算法的原理设 $\theta^0=(20,20)$，初始学习率 $\alpha=0.1$，函数的梯度为 $\nabla J(\theta)=[2(\theta_1-10),2(\theta_2-10)]$，则通过本算法计算目标函数的最小值的过程如下：

$\theta^1=(20,20)-0.1*[2*(20-10),2*(20-10)], J(\theta)=128$

$\theta^2=(18,18)-0.1*[2*(18-10),2*(18-10)], J(\theta)=81.92$

$\theta^3=(16.4,16.4)-0.1*[2*(16.4-10),2*(16.4-10)], J(\theta)=52.42$

$\theta^4=(15.12,15.12)-0.1*[2*(15.12-10),2*(15.12-10)], J(\theta)=33.55$

……

$\theta^{30}=(10.01,10.01)-0.1*[2*(10.01-10),2*(10.01-10)], J(\theta)=0.0003$

将上面的计算过程转换成代码，代码如下：

```
#第 2 章/machineLearn/梯度下降 1.py

def compute_loss(theta1, theta2):
    #目标函数
    return (theta1 - 10) **2 + (theta2 - 10) **2


def compute_gradient_sgd(theta1, theta2, learning_rate):
    #计算梯度的变化
    theta1 = theta1 - learning_rate *2 *(theta1 - 10) #梯度下降
    theta2 = theta2 - learning_rate *2 *(theta2 - 10)
    return theta1, theta2


if __name__ == "__main__":
    #迭代次数
    epochs = 31
    #初始权重值
    theta1, theta2 = 20, 20
    #学习率
    learning_rate = 0.1
    for x in range(1, epochs):
        #计算梯度
        theta1, theta2 = compute_gradient_sgd(
            theta1, theta2, learning_rate
        )
        #计算目标函数的值
        loss = compute_loss(theta1, theta2)
        #仅在第 1、第 2、第 3、第 4、第 30 次时输出计算过程信息
        if x in [1, 2, 3, 4, 30]:
            print(f"({round(theta1, 2)},"
                  f"{round(theta2, 2)})"
                  f"-{learning_rate}*[2*({round(theta1, 2)}-10),"
                  f"2*({round(theta2, 2)}-10)],J(θ)={round(loss, 4)}")
```

运行结果如下：

```
(18.0,18.0)-0.1*[2*(18.0-10),2*(18.0-10)],J(θ)=128.0
(16.4,16.4)-0.1*[2*(16.4-10),2*(16.4-10)],J(θ)=81.92
(15.12,15.12)-0.1*[2*(15.12-10),2*(15.12-10)],J(θ)=52.4288
(14.1,14.1)-0.1*[2*(14.1-10),2*(14.1-10)],J(θ)=33.5544
(10.01,10.01)-0.1*[2*(10.01-10),2*(10.01-10)],J(θ)=0.0003
```

通过梯度下降目标函数已靠近最小值，如图 2-7 所示。

根据上面的分析，梯度下降算法的实现过程如下：

(1) 初始设置 θ^0，迭代次数 $k=0$，学习率为 α。

(2) 计算 $\theta^{k+1}=\theta^k-\alpha\times\nabla J(\theta)$。

(3) 判断目标函数是否达到最小值，如果达到最小值，则转至步骤(5)。

(4) 设置 $k=k+1$，转至步骤(2)。

(5) 输出 θ，结束。

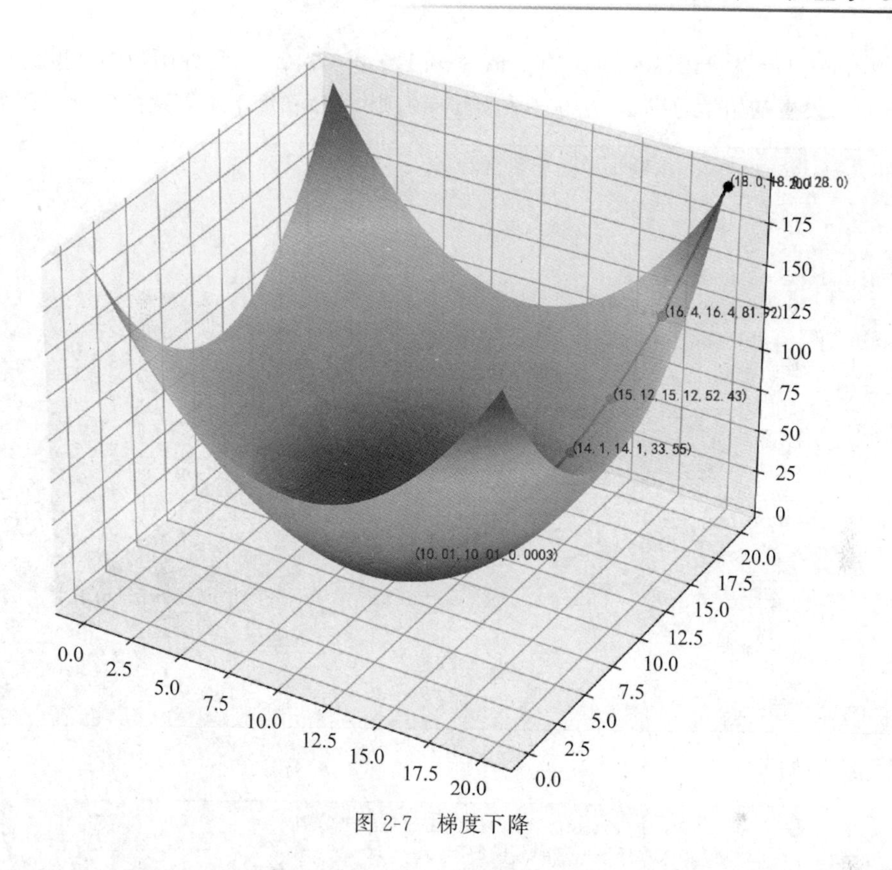

图 2-7 梯度下降

2.2.3 SGD、BGD 和 MBGD

梯度下降算法根据每次学习更新参数使用的样本数，可以分为随机梯度下降(Stochastic Gradient Descent，SGD)、全量梯度下降(Batch Gradient Descent，BGD)、小批量梯度下降(Mini-Batch Gradient Descent，MBGD)。

假设存在线型预测函数 $f(x)=\theta x$，x 拥有的训练样本为$\{x1,x2,x3,x4,\cdots\}$，已知训练样本的 y 值为$\{y1,y2,y3,y4,\cdots\}$，构造一个目标函数让 $f(x)$ 预测值接近 y 值，即 $f(\text{loss})=\frac{1}{2}(f(x)-y)^2$。

SGD 梯度下降算法每次从训练集中随机选择一个样本进行学习，即

$$\theta=\theta-\alpha\nabla\theta J(\theta;x_i;y_i) \tag{2-7}$$

可以通过伪代码来表示 SGD 梯度下降算法的计算过程，代码如下：

```
#第 2 章/machineLearn/随机梯度下降 SGD.py

for i in range(epoch):
    for i in range(m):                         #m 为样本的数量
        x_i = data[i]                          #data 为所有的数量
        gradient = ((theat *x_i) - y) *x_i
        theat = theat - alpha *gradient        #alpha 为学习率
```

假设 train_data 为{1,2,3,4}对应的 y 值为{0.1,0.2,0.3,0.4}，目标函数的梯度为 $\nabla J(\theta)=(\theta x_i-y)x_i$，设置初始学习率 $\alpha=0.003$，初始 $\theta=0$，则 SGD 梯度下降算法的实现代码如下：

```
#第 2 章/machineLearn/随机梯度下降 SGD.py
theta = 0
alpha = 0.003
epoch = 60
train_data = [1, 2, 3, 4]
y = [0.1, 0.2, 0.3, 0.4]
loss_list = []
epoch_list = []
item = 0
for j in range(epoch):
    for i in range(len(train_data)):
        x_i = train_data[i]
        params_grad = (theta *x_i - y[i]) *x_i
        theta = theta - alpha *params_grad
        loss = 0.5 * (theta *x_i - y[i]) **2
        if item % 50 == 0:
            print(f'epoch={item},theta={theta},loss={round(loss, 6)}')
        loss_list.append(loss)
        epoch_list.append(item)
        item += 1
print(f'预测结果={theta *4}')
```

运行结果如下：

```
epoch=0,theta=0.00030000000000000003,loss=0.00497
epoch=50,theta=0.06808552961444406,loss=0.004583
epoch=100,theta=0.0899134546122745,loss=5.1e-05
epoch=150,theta=0.0967712461979024,loss=4.7e-05
epoch=200,theta=0.09897955468546994,loss=1e-06
预测结果=0.398362529500783
```

SGD 梯度下降算法运算时间较慢，每次迭代更新时有可能会波动，迭代次数也会增多，其损失与迭代次数的收敛变化如图 2-8 所示。

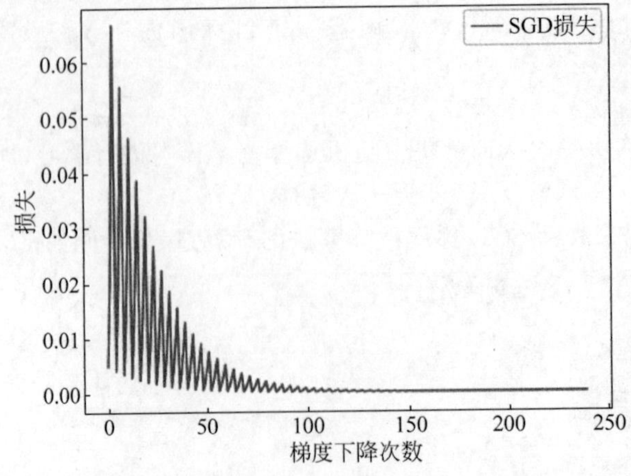

图 2-8　SGD 梯度下降迭代次数与损失

BGD 梯度下降算法每次学习都使用整个训练集，如果训练集较大，则需要消耗大量的内存和显存空间，即

$$\theta=\theta-\alpha\nabla\theta J(\theta) \tag{2-8}$$

BGD 梯度下降算法的伪代码计算过程，代码如下：

```
#第 2 章/machineLearn/全量梯度下降 BGD.py
for in range(epoch):
    #m 为样本的数量
    #x_i 为每个样本
    gradient = 1/m * Σ(theta *x_i - y) *x_i
    #alpha 为学习率
    theat = theat - alpha *gradient
```

BGD 梯度下降算法相对于 SGD 梯度下降算法计算了所有样本的梯度，因此 BGD 梯度下降算法的代码如下：

```
#第 2 章/machineLearn/全量梯度下降 BGD.py
def compute(theta, x_i, y_i):
    #目标函数
    return 0.5 *(theta *x_i - y_i) **2


def pre(theta, x_i):
    #预测函数
    return theta *x_i


def coumpute_all_loss(theta, train_data, y_data):
    #求所有样本的平均损失
    loss = 0
    for x_i, y_i in zip(train_data, y_data):
        loss += compute(theta, x_i, y_i)
    return loss / len(train_data)


def grad_all_params(theta, train_data, y_data):
    #求所有样本的平均梯度
    grad = 0
    for x_i, y_i in zip(train_data, y_data):
        grad += (theta *x_i - y_i) *x_i
    return grad / len(train_data)


if __name__ == "__main__":
    theta = 0
    alpha = 0.003
    epoch = 100
    train_data = [1, 2, 3, 4]
    y = [0.1, 0.2, 0.3, 0.4]
    loss_list = []
```

```
    epoch_list =[]

    for j in range(epoch):
        #求梯度
        grad = grad_all_params(theta, train_data, y)
        #梯度下降
        theta = theta - alpha *grad
        #求 loss
        loss = coumpute_all_loss(theta, train_data, y)
        if j % 20 == 0:
            print(f'epoch={j},theta={theta},loss={round(loss, 6)}')
            loss_list.append(loss)
            epoch_list.append(j)
print(f'预测结果={theta *4}')
```

运行结果如下：

```
epoch=0,theta=0.0022500000000000003,loss=0.035831
epoch=20,theta=0.03799137697558632,loss=0.014419
epoch=40,theta=0.060664252384820004,loss=0.005802
epoch=60,theta=0.07504700209459667,loss=0.002335
epoch=80,theta=0.08417083334582977,loss=0.00094
预测结果=0.35890995677530885
```

BGD梯度下降算法的迭代次数与损失变化，如图2-9所示。

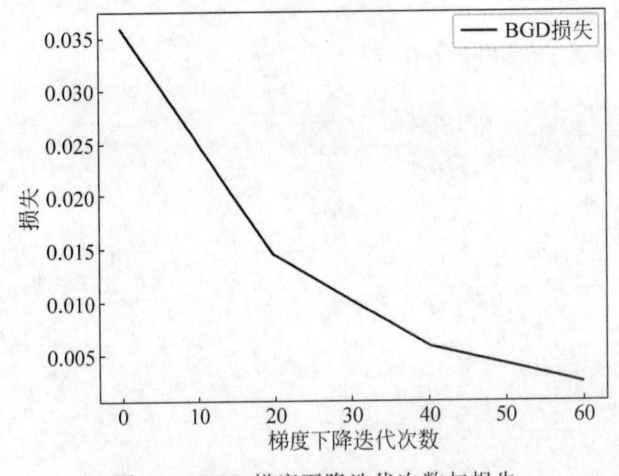

图2-9　BGD梯度下降迭代次数与损失

MBGD梯度下降算法综合了SGD梯度下降算法和BGD梯度下降算法，每次从训练集中选择一定数量的样本进行训练，即

$$\theta=\theta-\alpha\nabla\theta J(\theta;x_{i:i+m};y_{i:i+m}) \tag{2-9}$$

MBGD梯度下降算法的伪代码计算过程，代码如下：

```
#第2章/machineLearn/小批量梯度下降 MBGD.py
#伪代码实现过程
```

```
for in range(epoch):
   #m 为样本的数量
   #x_i 为每个样本,每次统计 10 个样本的梯度
   for i in range(0,m,step=10):
                          10
      gradient = 1/10 * Σ (theta *x_i - y) *x_i
                          i
    #alpha 为学习率
theat = theat - alpha *gradient
```

MBGD 梯度下降算法相对于 BGD 梯度下降算法只计算了指定样本数量的梯度,因此 MBGD 梯度下降算法的代码如下:

```
#第 2 章/machineLearn/小批量梯度下降 MBGD.py
def compute(theta, x_i, y_i):
   #目标函数
   return 0.5 *(theta *x_i - y_i) **2


def pre(theta, x_i):
   #预测函数
   return theta *x_i


def coumpute_all_loss(theta, x_i, y_i):
   #求样本的损失
   loss = compute(theta, x_i, y_i)
   return loss


def grad_all_params(theta, x_i, y_i):
   #求样本的梯度
   grad = (theta *x_i - y_i) *x_i
   return grad


if __name__ == "__main__":
   theta = 0
   alpha = 0.003
   batch_size = 2 #每次计算梯度的样本数
   epoch = 50
   train_data = [1, 2, 3, 4]
   y = [0.1, 0.2, 0.3, 0.4]
   loss_list = []
   epoch_list = []
   m = len(train_data)
   item = 0
   for j in range(epoch):
       loss = 0
       grad = 0
       #按 batch size 每次累加 grad 和 loss
       for i in range(0, m, batch_size):
```

```
            for x in range(i, i + batch_size):
                #求梯度
                grad += grad_all_params(theta, train_data[x], y[x])
            #梯度下降
            theta = theta - alpha *grad/batch_size
            #求 loss
            loss += coumpute_all_loss(theta, train_data[i], y[i])

            if j % 30 == 0:
print(f'epoch={item}, batch_size={batch_size}, theta={theta}, loss={round
(loss/batch_size, 6)}')
            loss_list.append(loss/batch_size)
            epoch_list.append(item)
        item += 1
print(f'预测结果={theta *4}')
```

运行结果如下：

```
epoch=0,batch_size=2,theta=0.00075,loss=0.001231
epoch=0,batch_size=2,theta=0.005221875,loss=0.011337
epoch=30,batch_size=2,theta=0.08014024443570612,loss=4.9e-05
epoch=30,batch_size=2,theta=0.08103505898889657,loss=0.000454
预测结果=0.3726179707299525
```

梯度下降算法的迭代次数与损失变化，如图 2-10 所示。

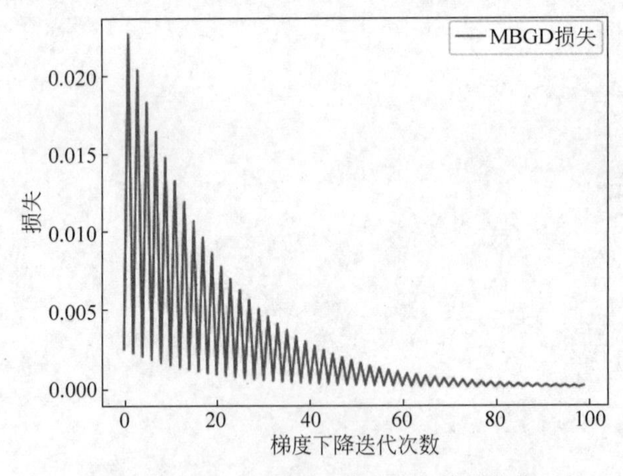

图 2-10　MBGD 梯度下降迭代次数与损失

相对于 SGD 随机梯度下降算法，MBGD 梯度下降算法降低了收敛波动性，使更新参数更加稳定。相对于 BGD 全量梯度下降算法，MBGD 梯度下降算法减少了内存占用，相对于 SGD 提高了学习的速度，并且不用担心内存瓶颈问题。MBGD 梯度下降算法常用于神经网络的训练。

2.2.4　Momentum 优化算法

在梯度下降的最开始需要对 θ 进行初始化，但是这个初始化有可能不是最佳的，从而导致目标函数在训练的过程中出现在局部最小值处并振荡，导致收敛缓慢，而引入 Momentum

动量优化算法，可以在一定程度上优化该问题，如图 2-11 所示。

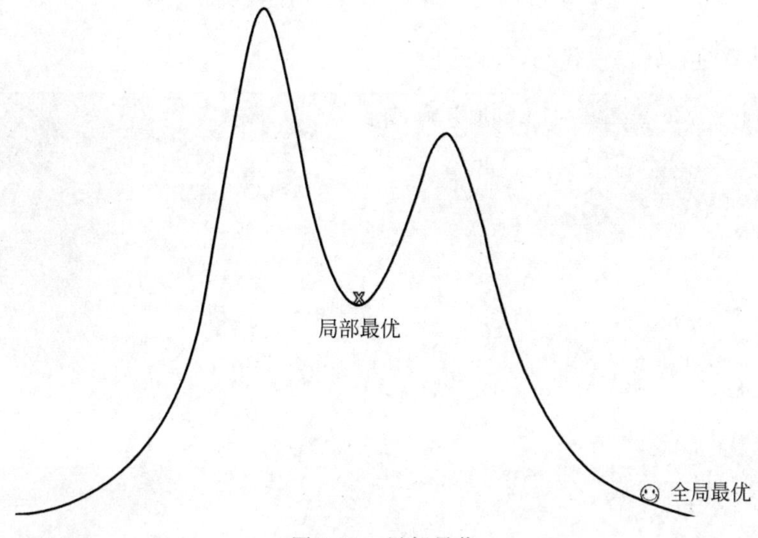

图 2-11　局部最优

Momentum 动量优化算法在原理上模拟了物理学中的动量，当一个小球从山上滚下来且没有阻力时它的动量会越来越大，速度会越来越快，但是如果遇到了阻力，速度就会变慢。动量优化算法就是借鉴此思想，使梯度方向在不变的维度上，参数更新变快，当梯度有所改变时，更新参数使其变慢，这样就能够加快收敛并且减少振荡，图 2-12 展示了有无动量的收敛。

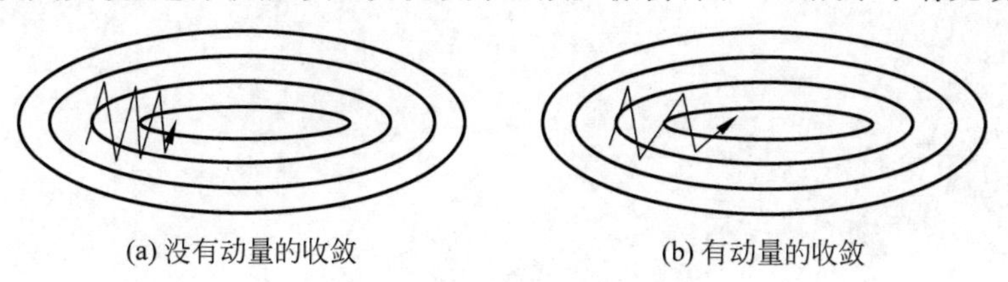

图 2-12　动量收敛的区别

Momentum 动量优化算法的公式为

$$v_t = \gamma v_{t-1} + \alpha \times \nabla_\theta J(\theta)$$
$$\theta = \theta - v_t \tag{2-10}$$

假设有一个目标函数如下：

$$J(\theta) = (\theta - 20)^2 + 5 \tag{2-11}$$

根据 Momentum 动量优化算法的公式设 $\theta^0 = 2$，初始学习率 $\alpha = 0.1$，权重 $\gamma = 0.5$，$v_0 = 0$，函数的梯度为$\nabla J(\theta) = 2(\theta - 20)$，则通过本算法计算目标函数的最小值的过程如下：

$$v_1 = \gamma v_0 + \alpha \nabla J(\theta) = 0.5 \times 0 + 0.1 \times (2 \times (2 - 20)) = -3.60$$
$$\theta^1 = \theta^0 - v_1 = 2 - (-3.6) = 5.6$$
$$v_2 = \gamma v_1 + \alpha \nabla J(\theta) = 0.5 \times (-3.6) + 0.1 \times (2 \times (5.6 - 20)) = -4.68$$

$$\theta^2 = \theta^1 - v_2 = 5.6 - (-4.68) = 5.6 = 10.28$$

…

将以上计算过程转换成代码，代码如下：

```
#第 2 章/machineLearn/动量法梯度下降 2.py

def compute_loss(theta):
    #目标函数
    return (theta - 20) **2 + 5


def compute_gradient_momentum(theta, gamma, v, learning_rate):
    #增加动量的计算
    #γv_0+α∇J(θ)
    v = gamma *v + learning_rate *2 *(theta - 20)
    #θ^0-v_1
    theta = theta - v
    return theta, v


def compute_gradient(theta, learning_rate):
    #没有动量的计算
    theta = theta - learning_rate *2 *(theta - 20)
    return theta


if __name__ == "__main__":
    theta = 2 #初始 θ
    gamma = 0.5 #γ
    v = 0 #初始 v
    learning_rate = 0.1 #学习率
    epochs = 20

    for x in range(1, epochs):
        theta = compute_gradient(theta, learning_rate)
        loss = compute_loss(theta)
        if x % 4 == 0 or x > 16:
            print(f'没有 momentum 动量时,'
                  f'第{x}次 theta ={round(theta, 4)},'
                  f'J(θ)={round(loss, 4)}')
    print('#' *50)
    theta = 2
    for x in range(1, epochs):
        theta, v = compute_gradient_momentum(
            theta, gamma, v, learning_rate
        )
        loss = compute_loss(theta)
        if x % 4 == 0:
            print(f'有 momentum 动量时,'
                  f'第{x}次,v={round(v, 4)},'
                  f'theta={round(theta, 4)},'
                  f'J(θ)={round(loss, 4)}')
```

运行结果如下：

```
没有 momentum 动量时,第 4 次 theta =12.6272,J(θ)=59.3582
没有 momentum 动量时,第 8 次 theta =16.9801,J(θ)=14.1198
没有 momentum 动量时,第 12 次 theta =18.763,J(θ)=6.53
没有 momentum 动量时,第 16 次 theta =19.4933,J(θ)=5.2567
没有 momentum 动量时,第 17 次 theta =19.5947,J(θ)=5.1643
没有 momentum 动量时,第 18 次 theta =19.6757,J(θ)=5.1051
没有 momentum 动量时,第 19 次 theta =19.7406,J(θ)=5.0673
##################################################
有 momentum 动量时,第 4 次,v=-3.2292,theta=17.7932,J(θ)=9.87
有 momentum 动量时,第 8 次,v=0.0772,theta=21.1777,J(θ)=6.3871
有 momentum 动量时,第 12 次,v=0.2,theta=20.1098,J(θ)=5.0121
有 momentum 动量时,第 16 次,v=-0.0096,theta=19.9238,J(θ)=5.0058
```

从运行结果来看，加了 Momentum 后只运行了 12 次 $J(\theta)$ 就接近最小，而没有加 Momentum，则需要运行 19 次，所以 Momentum 动量优化算法可以使梯度下降收敛更快，迭代次数更少，如图 2-13 所示。

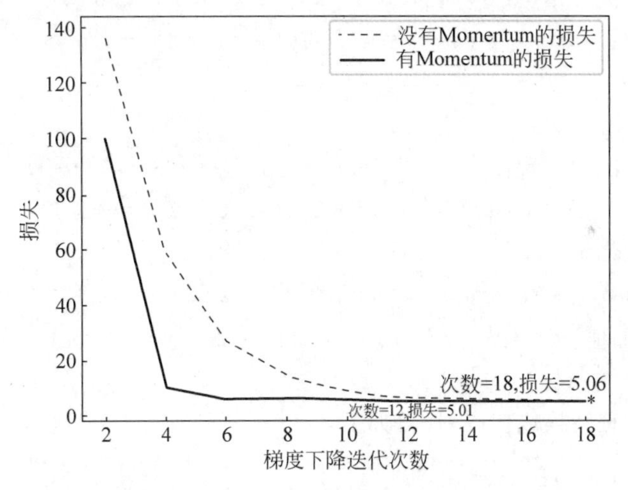

图 2-13 有无动量算法收敛的区别

2.2.5 NAG 优化算法

从山顶往下滚的球会盲目地选择斜坡，更好的方式应该是在遇到倾斜向上之前减慢速度。NAG(Nesterov Accelerated Gradient)不仅增加了动量项，并且在计算参数的梯度时，在目标函数中减去了动量项，即计算 $\nabla_\theta J(\theta-\gamma v_{t-1})$，这种方式预估了下一次参数所在的位置。

$$
\begin{aligned}
v_t &= \gamma v_{t-1} + \alpha \times \nabla_\theta J(\theta - \gamma v_{t-1}) \\
\theta &= \theta - v_t
\end{aligned} \tag{2-12}
$$

在图 2-14 中，Momentum 算法假定动量因子参数 $\gamma=0.9$，首先计算当前梯度项(如“蓝 1”向量)，然后加上动量项，这样便得到了大的跳跃(如“蓝 2”向量)，而在 NAG 算法中，首先来一个大的动量跳跃，然后加上一个小的使用了动量计算的当前梯度（如“红 1”）进行修正以得到绿色的向量，这样便可以阻止过快更新而提高稳定性。

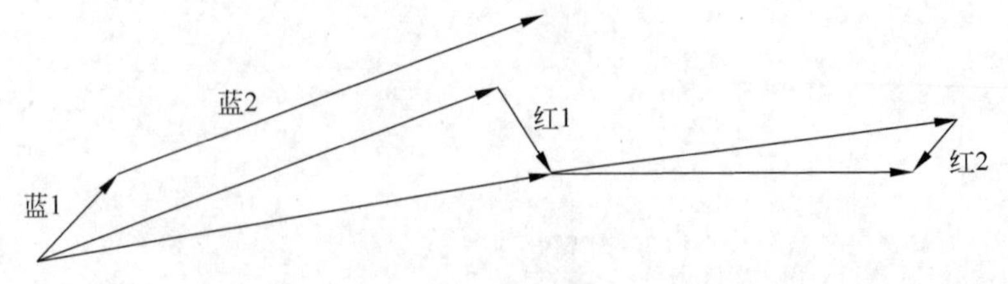

图 2-14 NAG 更新

假设有一个目标函数 $J(\theta)=(\theta-20)^2+5$，根据 NAG 优化算法的公式设 $\theta^0=2$，初始学习率 $\alpha=0.1$，权重 $\gamma=0.5$，$v_0=0$，函数的梯度为 $\nabla_\theta J(\theta)=2(\theta-20)$，则通过本算法计算目标函数的最小值的过程如下。

（1）第 1 次迭代。

计算 $\alpha\times\nabla_1$ 本次变化量：$\alpha\times\nabla_1=0.1\times2\times(2-20)=-3.6$

计算总变化量：$v_1=\gamma v_0+\alpha\times\nabla_1=0.5\times0+(-3.6)=-3.6$

更新权重 θ^1 的值：$\theta^1=\theta^0-v_1=2-3.6=5.6$

（2）第 2 次迭代。

预估 θ 的下一次的位置：$\theta^1-\gamma v_1=5.6-0.5\times(-3.6)=7.40$

计算 $\alpha\times\nabla_2$ 本次变化量：$\alpha\times\nabla_2=0.1\times2\times(7.40-20)=-2.52$

计算上一次变化量：$\gamma v_1=0.5\times(-3.6)=-1.8$

计算总变化量：$v_2=\gamma v_1+\alpha\times\nabla_2=-1.8+(-2.52)=-4.32$

更新权重 θ^2 的值：$\theta^2=\theta^1-v_2=5.6-4.32=9.92$

…

将以上计算过程转换成代码，代码如下：

```
#第 2 章/machineLearn/NAG 梯度下降 3.py

def compute_loss(theta):
    #目标函数
    return (theta - 20) **2 + 5


def compute_gradient_momentum(theta, gamma, v, learning_rate):
    #增加动量的计算
    #γv_0+α ∇J(θ)
    v = gamma *v + learning_rate *2 * (theta - 20)
    #θ^0-v_1
    theta = theta - v
    return theta, v


def compute_gradient_nag(theta, gamma, v, learning_rate):
```

```
    #NAG 梯度下降
    #预估 θ 的下一次位置
    theta_next = theta - gamma *v
    #预估下一次的梯度
    theta_gradient = 2 * (theta_next - 20)
    #计算本次变化量
    current_delta = learning_rate *theta_gradient
    #计算上次变化量的折扣
    last_delta_discount = gamma *v
    #计算总变化量,利用上一次梯度和下一次梯度来修正本次变化量,含大小和方向
    v_current = last_delta_discount + current_delta
    #计算 θ 的下一次位置
    theta_new = theta - v_current
    return theta_new, theta_gradient, v_current


if __name__ == "__main__":
    theta = 2                              #初始 θ
    gamma = 0.5                            #γ
    v = 0                                  #初始 v
    learning_rate = 0.1                    #学习率
    epochs = 20

    for x in range(1, epochs):
        theta, v = compute_gradient_momentum(
            theta, gamma, v, learning_rate
        )
        loss = compute_loss(theta)
        if x % 2 == 0:
            print(f'有 momentum 动量时,'
                  f'第{x}次,v={round(v, 4)},'
                  f'theta={round(theta, 4)},'
                  f'J(θ)={round(loss, 4)}')
    print('#' *50)
    theta = 2
    v = 0
    for x in range(1, epochs):
        theta, theta_gradient, v = compute_gradient_nag(
            theta, gamma, v, learning_rate
        )
        loss = compute_loss(theta)
        if x % 2 == 0:
            print(f'使用 nag 算法时,'
                  f'第{x}次,v={round(v, 4)},'
                  f'第{x}次 theta={round(theta, 4)},'
                  f'J(θ)={round(loss, 4)}')
```

运行结果如下：

```
有 momentum 动量时,第 2 次,v=-4.68,theta=10.28,J(θ)=99.4784
有 momentum 动量时,第 4 次,v=-3.2292,theta=17.7932,J(θ)=9.87
有 momentum 动量时,第 6 次,v=-1.0581,theta=20.9073,J(θ)=5.8232
```

```
有 momentum 动量时,第 8 次,v=0.0772,theta=21.1777,J(θ)=6.3871
有 momentum 动量时,第 10 次,v=0.3178,theta=20.5858,J(θ)=5.3432
有 momentum 动量时,第 12 次,v=0.2,theta=20.1098,J(θ)=5.0121
有 momentum 动量时,第 14 次,v=0.0585,theta=19.9293,J(θ)=5.005
有 momentum 动量时,第 16 次,v=-0.0096,theta=19.9238,J(θ)=5.0058
有 momentum 动量时,第 18 次,v=-0.0213,theta=19.9651,J(θ)=5.0012
##################################################
使用 nag 算法时,第 2 次,v=-4.32,第 2 次 theta=9.92,J(θ)=106.6064
使用 nag 算法时,第 4 次,v=-2.7648,第 4 次 theta=16.4288,J(θ)=17.7535
使用 nag 算法时,第 6 次,v=-1.0783,第 6 次 theta=19.3272,J(θ)=5.4526
使用 nag 算法时,第 8 次,v=-0.2477,第 8 次 theta=20.1408,J(θ)=5.0198
使用 nag 算法时,第 10 次,v=0.014,第 10 次 theta=20.1978,J(θ)=5.0391
使用 nag 算法时,第 12 次,v=0.0486,第 12 次 theta=20.104,J(θ)=5.0108
使用 nag 算法时,第 14 次,v=0.0289,第 14 次 theta=20.0349,J(θ)=5.0012
使用 nag 算法时,第 16 次,v=0.0107,第 16 次 theta=20.0057,J(θ)=5.0
使用 nag 算法时,第 18 次,v=0.0022,第 18 次 theta=19.9981,J(θ)=5.0
```

从代码结果来看 Momentum 算法运行了 12 次 $J(\theta)$接近最小,而 NAG 算法只运行了 8 次,NAG 算法收敛更快,并且在靠近极值点时有减速,如图 2-15 所示。

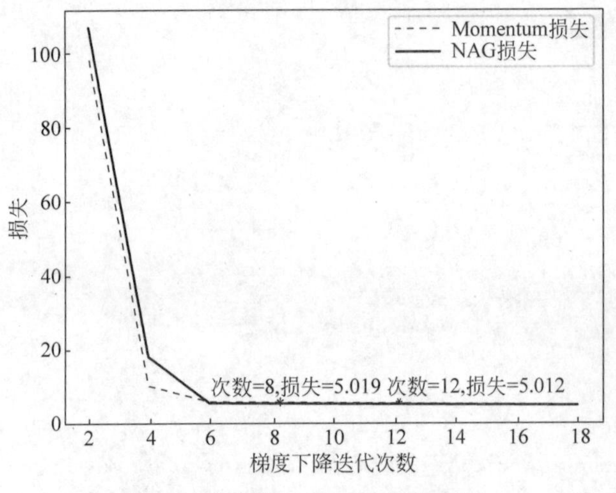

图 2-15 动量与 NAG 损失的对比

2.2.6 AdaGrad 优化算法

AdaGrad 算法也是一种基于梯度的优化算法,它能够对每个参数自适应不同的学习速率,如果一个参数的梯度一直都非常大,则其对应的学习率就小一点以防止震荡,而一个参数的梯度一直非常小,那么这个参数的学习率就变得大一点,使其能够更快地更新。

在 2.2.1 节中每个模型参数 θ_i 使用相同的学习速率 α,而 AdaGrad 在每个更新步骤中对于每个模型参数 θ_i 使用不同的学习速率 α_i,设在第 t 次更新步骤中,目标函数的参数 θ_i 梯度为 $g_{t,i}$,即

$$g_{t,i} = \nabla_\theta J(\theta_i) \tag{2-13}$$

则梯度下降的公式变为

$$\theta_{t+1,i}=\theta_{t,i}-\alpha\times g_{t,i} \tag{2-14}$$

而 AdaGrad 算法对每个参数使用不同的学习速率，所以公式变为

$$\theta_{t+1,i}=\theta_{t,i}-\frac{\alpha}{\sqrt{G_{t,ii}+\epsilon}}\times g_{t,i} \tag{2-15}$$

其中，$G_t\in R^{d\times d}$ 是一个对角矩阵，其中第 i 行的对角元素 e_{ii} 为过去到当前第 i 个参数 θ_i 的梯度的平方和，ϵ 是一个平滑参数，为了使分母不为 0。进一步将所有 $G_{t,ii}$ 和 $g_{t,i}$ 的元素写成向量 $\boldsymbol{G}_t$，$\boldsymbol{g}_t$，这样便可以使用向量点乘操作，公式就变为

$$\theta_{t+1}=\theta_t-\frac{\alpha}{\sqrt{G_t+\epsilon}}\odot g_t \tag{2-16}$$

假设有一个目标函数 $J(\theta_1,\theta_2)=0.5\times(\theta_1-10)^2+5\times(\theta_2+30)^2$，根据 AdaGrad 算法的要求，设置初始化学习 $\alpha=0.1$，初始权重 $(\theta_1,\theta_2)=(0,0)$，初始历史权重 $(g_{\theta_1},g_{\theta_2})=(0,0)$，函数的梯度为 $(\theta_1-10,10\times(\theta_2+30))$，则通过本算法计算目标函数最小值的代码如下：

```
#第 2 章/machineLearn/AdaGrad 梯度下降 3.py
import numpy as np


def compute_loss(theta1, theta2):
    #目标函数
    return 0.5 * (theta1 - 10) **2 + 5 * (theta2 + 30) **2


def compute_gradient_adagrad(
        theta1, theta2,
        g_theta1, g_theta2, learning_rate
):
    #AdaGrad 梯度下降
    theta1_gradient = theta1 - 10
    theta2_gradient = 10 * (theta2 + 30)
    #计算历史梯度平方的和
    g_theta1 = g_theta1 + np.square(theta1_gradient)
    g_theta2 = g_theta2 + np.square(theta2_gradient)
    #计算 theta 的下一个位置
    epsilon = 1e-8                          #平滑参数
    lr_theta1 = learning_rate / np.sqrt(g_theta1 + epsilon)
    lr_theta2 = learning_rate / np.sqrt(g_theta2 + epsilon)
    #梯度下降
    theta1 = theta1 - lr_theta1 *theta1_gradient
    theta2 = theta2 - lr_theta2 *theta2_gradient
    return [
        theta1, theta2, g_theta1,
        g_theta2, theta1_gradient,
        theta2_gradient, lr_theta1, lr_theta2
    ]
```

```
if __name__ == "__main__":
    theta1, theta2 = 0, 0            #初始θ
    g_theta1, g_theta2 = 0, 0        #初始历史θ
    learning_rate = 0.1              #学习率
    epochs = 20

    for x in range(1, epochs):
        result = compute_gradient_adagrad(
            theta1, theta2, g_theta1, g_theta2, learning_rate
        )
        theta1 = result[0]
        theta2 = result[1]
        g_theta1 = result[2]
        g_theta2 = result[3]
        loss = compute_loss(theta1, theta2)
        if x % 2 == 0:
            print(f'step={x},'
                  #f'theta1={round(theta1, 4)},'
                  #f'theta2={round(theta2, 4)},'
                  f'loss={round(loss, 4)},'
                  f'theta1_gradient={round(result[4], 4)},'
                  f'lr_theta1={round(result[6], 4)},\n'
                  f'theta2_gradient={round(result[5], 4)},'
                  f'lr_theta2={round(result[7], 6)}')
```

运行结果如下：

```
step=2,loss=4497.2787,theta1_gradient=-9.9,lr_theta1=0.0071
theta2_gradient=299.0,lr_theta2=0.000236
step=4,loss=4464.2542,theta1_gradient=-9.7724,lr_theta1=0.0051
theta2_gradient=297.7183,lr_theta2=0.000167
step=6,loss=4438.1587,theta1_gradient=-9.6787,lr_theta1=0.0042
theta2_gradient=296.7746,lr_theta2=0.000137
step=8,loss=4415.9317,theta1_gradient=-9.6013,lr_theta1=0.0036
theta2_gradient=295.9923,lr_theta2=0.000119
step=10,loss=4396.2618,theta1_gradient=-9.5338,lr_theta1=0.0033
theta2_gradient=295.3094,lr_theta2=0.000106
step=12,loss=4378.444,theta1_gradient=-9.4733,lr_theta1=0.003
theta2_gradient=294.6958,lr_theta2=9.7e-05
step=14,loss=4362.0477,theta1_gradient=-9.418,lr_theta1=0.0028
theta2_gradient=294.134,lr_theta2=9e-05
step=16,loss=4346.787,theta1_gradient=-9.3667,lr_theta1=0.0026
theta2_gradient=293.6127,lr_theta2=8.4e-05
step=18,loss=4332.4604,theta1_gradient=-9.3188,lr_theta1=0.0025
theta2_gradient=293.1245,lr_theta2=8e-05
```

AdaGrad 的主要优势在于它能够为每个参数自适应不同的学习速率，梯度越大学习率越小，梯度越小学习率越大，但是从训练开始时累积平方梯度值会越来越大，从而使学习率过早和过量减少，从而导致迭代后期由于学习率过小而使模型收敛缓慢，如图 2-16 所示。

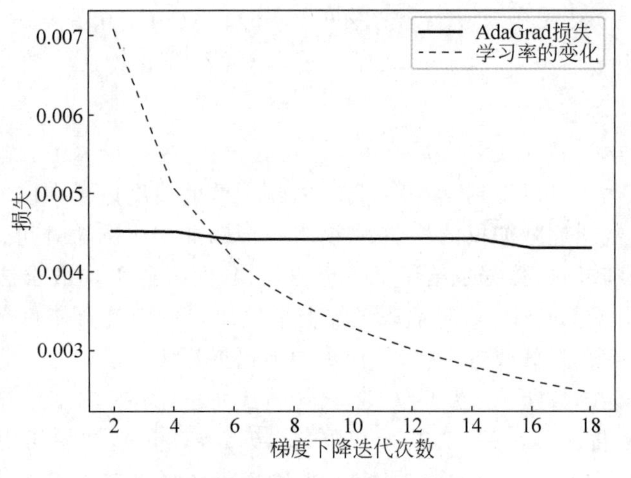

图 2-16　AdaGrad 损失与学习率的变化

2.2.7　RMSProp 优化算法

针对 AdaGrad 算法每个元素的学习率在迭代中一直在降低或不变，在迭代后期由于学习率过小，所以可能较难找到最优解的问题，RMSProp 算法对 AdaGrad 算法做了一点修改，即不同于 AdaGrad 算法里状态变量 $E[g^2]_t$ 是截止时间步 t 所有小批量随机梯度 g_t 按元素的平方和，RMSProp 算法将这些梯度按元素的平方做了指数加权移动平均，加了一个衰减系数，以此来控制历史信息的获取。

给定参数 $0 \leqslant \gamma < 1$，RMSProp 算法在时间步 $t > 0$ 计算：

$$E[g^2]_t = \gamma E[g^2]_{t-1} + (1-\gamma) g_t^2 \tag{2-17}$$

RMSProp 算法将目标函数自变量中每个元素的学习率通过按元素运算重新调整，然后更新公式就变为

$$\theta_{t+1} = \theta_t - \frac{\alpha}{\sqrt{E[g^2]_t + \epsilon}} \odot g_t \tag{2-18}$$

关于指数加权平均，假设最近 3 天的温度值分别为 24、25、24，则这 3 天平均温度为 sum([24,25,24])/3≈24.33，而指数加权平均本质上是一种近似求平均的方法，假设 $\gamma =$ 0.9 则根据公式：

$$v_t = \gamma v_{t-1} + (1-\gamma)\theta_t \tag{2-19}$$

其中，v_t 代表第 t 天的平均温度值，θ_t 代表第 t 天的温度值，γ 是一个超参数，则根据指数平均公式求法如下：

$$v_3 = 0.9 \times v_2 + (1-0.9)\theta_3$$

$$v_2 = 0.9 \times v_1 + (1-0.9)\theta_2$$

$$v_1 = 0.9 \times v_0 + (1-0.9)\theta_1$$

化简后得到：

$$v_3=0.9\times(0.9\times(0.9\times v_0+0.1\times\theta_1)+0.1\times\theta_2)+0.1\times\theta_3$$
$$=0.9\times0.9\times0.9\times v_0+0.9\times0.9\times0.1\times\theta_1+0.9\times0.1\times\theta_2+0.1\times\theta_3$$

而 v_0 的初始状态为 0，则

$$v_3=0.9^2\times0.1\times\theta_1+0.9^1\times0.1\times\theta_2+0.9^0\times0.1\times\theta_3$$

通过上面的表达式可以得到 v_3 等于每个时刻天数的温度值再乘以一个权值，所以指数加权平均就是以指数式递减加权的移动平均，各数值的加权比例随时间而指数式递减，越近的数据加权越重，但较旧的数据也给予了一定的加权，而普通平均数求法，它的每项的权值都是一样的，因此对于 RMSProp 算法来讲加上指数加权平均就会使每个变量 θ_t 的学习率在迭代过程中不再一直降低或者不变，从而更快地逼近最优解。

假设有一个目标函数 $J(\theta_1,\theta_2)=0.5\times(\theta_1-10)^2+5\times(\theta_2+30)^2$，根据 RMSProp 算法的要求，设置初始化学习 $\alpha=0.1$，初始权重 $(\theta_1,\theta_2)=(0,0)$，初始历史权重 $(g_{\theta_1},g_{\theta_2})=(0,0)$，函数的梯度为 $(\theta_1-10,10\times(\theta_2+30))$，则通过本算法计算目标函数最小值的代码如下：

```
#第 2 章/machineLearn/RMSProp 梯度下降 4.py
import numpy as np


def compute_loss(theta1, theta2):
    #目标函数
    return 0.5 * (theta1 - 10) **2 + 5 * (theta2 + 30) **2


def compute_gradient_rmsprop(
        theta1, theta2,
        g_theta1, g_theta2,
        learning_rate, gamma
):
    #RMSProp 梯度下降
    theta1_gradient = theta1 - 10
    theta2_gradient = 10 * (theta2 + 30)
    #RMSProp 相对于 AdaGrad 增加了一个系数，以此来计算历史梯度平方的和
    g_theta1 = gamma *g_theta1 + (1 - gamma) *np.square(theta1_gradient)
    g_theta2 = gamma *g_theta2 + (1 - gamma) *np.square(theta2_gradient)
    #计算 theta 的下一个位置
    epsilon = 1e-8                          #平滑参数
    lr_theta1 = learning_rate / np.sqrt(g_theta1 + epsilon)
    lr_theta2 = learning_rate / np.sqrt(g_theta2 + epsilon)

    #梯度下降
    theta1 = theta1 - lr_theta1 *theta1_gradient
    theta2 = theta2 - lr_theta2 *theta2_gradient
    return [
        theta1, theta2, g_theta1,
        g_theta2, theta1_gradient,
        theta2_gradient, lr_theta1, lr_theta2
    ]
```

```
if __name__ == '__main__':
    theta1, theta2 = 0, 0           #初始θ
    g_theta1, g_theta2 = 0, 0       #初始历史θ
    learning_rate = 0.1             #学习率
    epochs = 20
    gamma = 0.9

    for x in range(1, epochs):
        result = compute_gradient_rmsprop(
            theta1, theta2,
            g_theta1, g_theta2,
            learning_rate, gamma
        )
        theta1 = result[0]
        theta2 = result[1]
        g_theta1 = result[2]
        g_theta2 = result[3]
        loss = compute_loss(theta1, theta2)
        if x % 2 == 0:
            print(f'step={x},'
                  #f'theta1={round(theta1, 4)},'
                  #f'theta2={round(theta2, 4)},'
                  f'loss={round(loss, 4)},'
                  f'theta1_gradient={round(result[4], 4)},'
                  f'lr_theta1={round(result[6], 4)},\n'
                  f'theta2_gradient={round(result[5], 4)},'
                  f'lr_theta2={round(result[7], 6)}')
```

运行结果如下：

```
step=2,loss=4382.8616,theta1_gradient=-9.6838,lr_theta1=0.0233,
theta2_gradient=296.8377,lr_theta2=0.000769
step=4,loss=4274.3999,theta1_gradient=-9.2705,lr_theta1=0.0178,
theta2_gradient=292.6497,lr_theta2=0.000577
step=6,loss=4185.4115,theta1_gradient=-8.9551,lr_theta1=0.0156,
theta2_gradient=289.4184,lr_theta2=0.000498
step=8,loss=4106.7348,theta1_gradient=-8.6834,lr_theta1=0.0145,
theta2_gradient=286.6142,lr_theta2=0.000455
step=10,loss=4034.5335,theta1_gradient=-8.4371,lr_theta1=0.0138,
theta2_gradient=284.0558,lr_theta2=0.000428
step=12,loss=3966.7946,theta1_gradient=-8.2074,lr_theta1=0.0134,
theta2_gradient=281.6563,lr_theta2=0.000411
step=14,loss=3902.3213,theta1_gradient=-7.9893,lr_theta1=0.0132,
theta2_gradient=279.3668,lr_theta2=0.000399
step=16,loss=3840.343,theta1_gradient=-7.7798,lr_theta1=0.0131,
theta2_gradient=277.1568,lr_theta2=0.00039
step=18,loss=3780.3356,theta1_gradient=-7.5769,lr_theta1=0.0131,
theta2_gradient=275.006,lr_theta2=0.000385
```

在相同的参数的情况下，将 RMSProp 算法中的损失及学习率与 AdaGrad 算法进行图示化对比，可以发现 RMSProp 算法的 loss 下降更快，学习率在迭代的后期移动幅度更小，如图 2-17 所示。

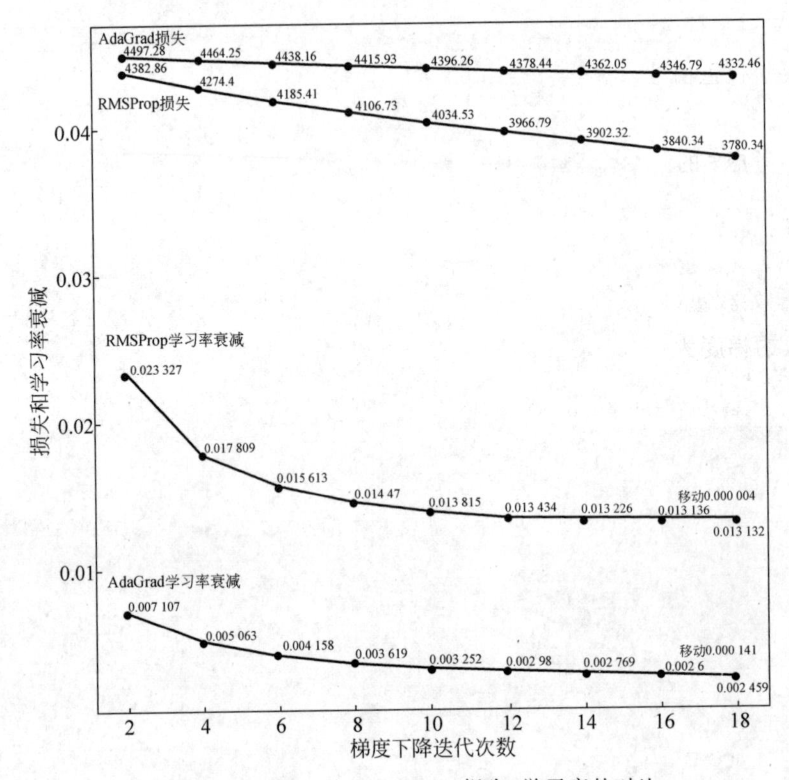

图 2-17　AdaGrad 与 RMSProp 损失、学习率的对比

2.2.8　AdaDelta 优化算法

RMSProp 算法虽然可以对不同的权重参数自适应地改变学习率，但仍要指定超参数学习率 α，AdaDelta 算法则不需要指定。

同 RMSProp 算法一样，AdaDelta 算法也使用了小批量随机梯度 g_t 按元素的平方的指数加权移动平均来计算历史梯度，给定超参数 $0 \leqslant \gamma < 1$，在时间步 $t > 0$ 时计算：

$$E[g^2]_t = \gamma E[g^2]_{t-1} + (1-\gamma) g_t^2 \tag{2-20}$$

然后计算历史梯度的均方根 $\mathrm{RMS}[g]_t$：

$$\mathrm{RMS}[g]_t = \sqrt{E[g^2]_t + \epsilon} \tag{2-21}$$

不同于 RMSProp 算法，AdaDelta 算法还计算了一个额外的状态变量 Δ_{θ_t}，并使用 Δ_{θ_t-1} 来计算变量的变化：

$$\mathrm{RMS}[\Delta\theta]_{t-1} = \sqrt{E[\Delta\theta^2]_{t-1} + \epsilon}$$

$$\Delta_{\theta_t} = \frac{\mathrm{RMS}[\Delta\theta]_{t-1}}{\mathrm{RMS}[g]_t} \odot g_t \tag{2-22}$$

计算历史 Δ_{θ_t} 平方和的加权平均 $E[\Delta\theta^2]_t$，以便下次迭代使用：

$$E[\Delta\theta^2]_t = \gamma \times E[\Delta\theta^2]_{t-1} + (1-\gamma) \times \Delta\theta_t^2 \tag{2-23}$$

然后更新权重值：

$$\theta_{t+1} = \theta_t - \Delta_{\theta_t} \tag{2-24}$$

如果不考虑ϵ的影响，则 AdaDelta 算法相对于 RMSProp 算法的不同之处在于使用了Δ_{θ_t-1}来代替学习率α。

假设有一个目标函数$J(\theta_1,\theta_2)=0.5\times(\theta_1-10)^2+5\times(\theta_2+30)^2$，根据 AdaDelta 算法的要求，初始权重$(\theta_1,\theta_2)=(0,0)$，初始历史权重$(g_{\theta_1},g_{\theta_2})=(0,0)$，初始$(\Delta_{\theta_1},\Delta_{\theta_2})=(0,0)$，函数的梯度为$(\theta_1-10,10\times(\theta_2+30))$，则通过本算法计算目标函数最小值的代码如下：

```
#第 2 章/machineLearn/AdaDelta 梯度下降 4.py
import numpy as np


def compute_loss(theta1, theta2):
    #目标函数
    return 0.5 * (theta1 - 10) **2 + 5 * (theta2 + 30) **2


def compute_gradient_adadelta(
        theta1, theta2,
        g_theta1, g_theta2,
        delta_theta1, delta_theta2,
        gamma
):
    #AdaDelta 梯度下降
    epsilon = 1e-8                              #平滑参数
    theta1_gradient = theta1 - 10
    theta2_gradient = 10 * (theta2 + 30)
    #(1) 计算历史梯度平方和的加权平均
    g_theta1 = gamma *g_theta1 + (1 - gamma) *np.square(theta1_gradient)
    g_theta2 = gamma *g_theta2 + (1 - gamma) *np.square(theta2_gradient)
    #(2) 计算历史梯度的平方根
    rms_g_theta1 = np.sqrt(g_theta1 + epsilon)
    rms_g_theta2 = np.sqrt(g_theta2 + epsilon)
    #(3) 计算 delta_θ 的均方根
    rms_delta_theta1 = np.sqrt(delta_theta1 + epsilon)
    rms_delta_theta2 = np.sqrt(delta_theta2 + epsilon)
    #(4) 计算 delta_Δθ_(t-1)
    x_delta_theta1 = (rms_delta_theta1 / rms_g_theta1) *theta1_gradient
    x_delta_theta2 = (rms_delta_theta2 / rms_g_theta2) *theta2_gradient
    #(5) 计算 Δ_(θ_t )的平方和的加权平均,以便下一次迭代使用
    delta_theta1 = gamma *delta_theta1 + \
                (1 - gamma) *np.square(x_delta_theta1)
    delta_theta2 = gamma *delta_theta2 + \
                (1 - gamma) *np.square(x_delta_theta2)
    #(6) 计算 θ 的下一个位置
    theta1 = theta1 - x_delta_theta1
```

```
    theta2 = theta2 - x_delta_theta2
    return [
        theta1, theta2,                          #权重
        theta1_gradient, theta2_gradient,        #梯度
        g_theta1, g_theta2,                      #历史梯度的加权平均
        delta_theta1, delta_theta2               #Δ_θ 的平方和的加权平均
    ]


if __name__ == '__main__':
    theta1, theta2 = 0, 0                        #初始 θ
    g_theta1, g_theta2 = 0, 0                    #初始历史 θ
    delta_theta1, delta_theta2 = 0, 0            #初始 Δ_θ
    epochs = 18616
    gamma = 0.9
    for x in range(1, epochs):
        result = compute_gradient_adadelta(
            theta1, theta2,
            g_theta1, g_theta2,
            delta_theta1, delta_theta2,
            gamma
        )

        theta1 = result[0]
        theta2 = result[1]
        theta1_gradient = result[2]
        theta2_gradient = result[3]
        g_theta1 = result[4]
        g_theta2 = result[5]
        delta_theta1 = result[6]
        delta_theta2 = result[7]

        loss = compute_loss(theta1, theta2)
        if x % 3000 == 0:
            print(f'step={x},'
                  f'theta1={round(theta1, 6)},'
                  f'theta2={round(theta2, 6)},'
                  f'loss={round(loss, 6)},\n'
                  f'theta1_gradient={round(theta1_gradient, 6)},'
                  f'theta2_gradient={round(theta2_gradient, 6)},'
                  )
```

运行结果如下：

```
step=3000,theta1=3.288102,theta2=-3.522077,loss=3527.926692,
theta1_gradient=-6.713327,theta2_gradient=264.795727,
step=6000,theta1=7.500645,theta2=-9.165646,loss=2173.475,
theta1_gradient=-2.500612,theta2_gradient=208.363949,
step=9000,theta1=10.0,theta2=-15.42624,loss=1061.972441,
theta1_gradient=0.0,theta2_gradient=145.758464,
step=12000,theta1=10.0,theta2=-21.432474,loss=367.012484,
theta1_gradient=0.0,theta2_gradient=85.694056,
```

```
step=15000,theta1=10.0,theta2=-26.478751,loss=61.995976,
theta1_gradient=0.0,theta2_gradient=35.226928,
step=18000,theta1=10.0,theta2=-29.737688,loss=0.344037,
theta1_gradient=0.0,theta2_gradient=2.629193,
```

训练次数与损失的变化,如图 2-18 所示。

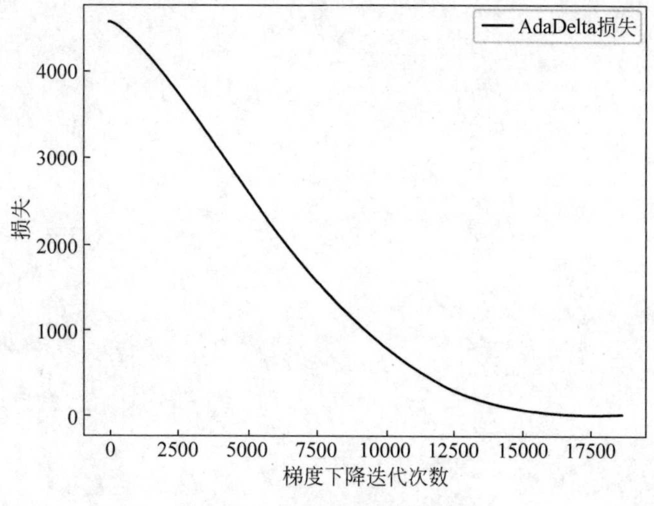

图 2-18 AdaDelta 损失的变化

2.2.9 Adam 优化算法

Adam 算法在 RMSProp 算法的基础上对小批量随机梯度也做了指数加权移动平均,但是与计算历史梯度衰减方式 RMSProp 算法不同,它不再使用历史平方衰减,其衰减方式类似于动量的方法,其计算公式如下:

$$m_t=\beta_1 m_{t-1}+(1-\beta_1)g_t$$

$$v_t=\beta_2 v_{t-1}+(1-\beta_2)g_t^2 \tag{2-25}$$

m_t 与 v_t 分别是梯度第 1 个时刻的平均值和第 2 个时刻估计值的梯度,其中 β_1、β_2 的取值范围都为[0,1],建议 $\beta_1=0.9$,$\beta_2=0.999$,如果将 m_0 和 v_0 都设置为 0,Adam 算法的作者则发现 m_t 与 v_t 更倾向于 0,为了解决这个问题算法的原作者对此进行了修正:

$$\hat{m}=\frac{m_t}{1-\beta_1^t}$$

$$\hat{v}=\frac{v_t}{1-\beta_2^t} \tag{2-26}$$

最终,Adam 算法的更新方程为

$$\theta_{t+1}=\theta_t-\frac{\alpha}{\sqrt{\hat{v}_t}+\epsilon}\hat{m}_t \tag{2-27}$$

假设有一个目标函数 $J(\theta_1,\theta_2)=0.5\times(\theta_1-10)^2+5\times(\theta_2+30)^2$，根据 Adam 算法的要求，初始权重$(\theta_1,\theta_2)=(0,0)$，初始化学习率 $\alpha=0.1$，初始历史权重$(g_{\theta_1},g_{\theta_2})=(0,0)$，初始$(m_0,v_0)=(0,0)$，函数的梯度为$(\theta_1-10,10\times(\theta_2+30))$，则通过本算法计算目标函数最小值的代码如下：

```
#第 2 章/machineLearn/AdaDelta 梯度下降 4.py
import numpy as np


def compute_loss(theta1, theta2):
    #目标函数
    return 0.5 * (theta1 - 10) **2 + 5 * (theta2 + 30) **2


def compute_gradient_adam(
        t,
        theta1, theta2,
        m_theta1, m_theta2,
        v_theta1, v_theta2,
        beta1, beta2,
        lr_rate, is_correct=True
):
    #AdaDelta 梯度下降
    epsilon = 1e-8                                      #平滑参数
    theta1_gradient = theta1 - 10
    theta2_gradient = 10 * (theta2 + 30)

    #(1)计算 m
    m_theta1 = beta1 *m_theta1 + (1 - beta1) *theta1_gradient
    m_theta2 = beta1 *m_theta2 + (1 - beta1) *theta2_gradient

    #(2)计算 v
    v_theta1 = beta2 *v_theta1 + (
            1 - beta2) *np.square(theta1_gradient)
    v_theta2 = beta2 *v_theta2 + (
            1 - beta2) *np.square(theta2_gradient)
    t += 1
    #(3.1)计算 m^
#t=2 时,1-0.81=0.18, -1.89 / 0.18=-9.94
#如果没有修正,则为-1.89,更接近 0,有修正时为-9.94
    hat_m1 = m_theta1 / (1 - np.power(beta1, t) if is_correct else 1)
    hat_m2 = m_theta2 / (1 - np.power(beta1, t) if is_correct else 1)
    #(3.2)计算 v^
#t=2 时 1-0.99=0.001, 0.19 / 0.001=99
#如果没有修正,则为 0.19,更接近 0,有修正时为 99.00
    hat_v1 = v_theta1 / (1 - np.power(beta2, t) if is_correct else 1)
    hat_v2 = v_theta2 / (1 - np.power(beta2, t) if is_correct else 1)
    #(4)更新权重
    theta1 = theta1 - lr_rate *hat_m1 / (np.sqrt(hat_v1) + epsilon)
    theta2 = theta2 - lr_rate *hat_m2 / (np.sqrt(hat_v2) + epsilon)
    return [
```

```
        theta1, theta2,
        m_theta1, m_theta2,
        v_theta1, v_theta2,
        theta1_gradient, theta2_gradient
    ]


if __name__ == '__main__':
    lr_rate = 0.1
    beta1 = 0.9
    beta2 = 0.999
    m_theta1, m_theta2 = 0, 0                          #初始 m_0
    v_theta1, v_theta2 = 0, 0                          #初始 v_0
    theta1, theta2 = 0, 0                              #初始 θ
    epochs = 700
    for t in range(0, epochs):
        result = compute_gradient_adam(
            t,
            theta1, theta2,
            m_theta1, m_theta2,
            v_theta1, v_theta2,
            beta1, beta2,
            lr_rate, True
        )
        theta1, theta2 = result[0:2]
        m_theta1, m_theta2 = result[2:4]
        v_theta1, v_theta2 = result[4:6]
        theta1_gradient, theta2_gradient = result[6:]
        loss = compute_loss(theta1, theta2)
        if t % 100 == 0:
            print(f'step={t},'
                  f'loss={round(loss, 4)},'
                  f'theta1={round(theta1, 4)},'
                  f'theta2={round(theta2, 4)},\n'
                  f'm_theta1={round(m_theta1, 4)},'
                  f'm_theta2={round(m_theta2, 4)},'
                  f'v_theta1={round(v_theta1, 4)},'
                  f'v_theta2={round(v_theta2, 4)},'
                  #f'theta1_gradient={round(theta1_gradient, 4)},'
                  #f'theta2_gradient={round(theta2_gradient, 4)},'
                  )
```

运行结果如下：

```
step=0,loss=4519.055,theta1=0.1,theta2=-0.1,
m_theta1=-1.0,m_theta2=30.0,v_theta1=0.1,v_theta2=90.0,
step=100,loss=2111.8123,theta1=7.8005,theta2=-9.4603,
m_theta1=-2.7099,m_theta2=214.0407,v_theta1=3.4842,v_theta2=6126.7331,
step=200,loss=854.0795,theta1=9.8495,theta2=-16.9304,
m_theta1=-0.2197,m_theta2=137.3467,v_theta1=3.2523,v_theta2=8215.068,
step=300,loss=289.9812,theta1=9.9983,theta2=-22.3845,
m_theta1=-0.0036,m_theta2=80.8128,v_theta1=2.943,v_theta2=8439.1463,
```

```
step=400,loss=80.9139,theta1=10.0,theta2=-25.9772,
m_theta1=0.0,m_theta2=43.148,v_theta1=2.6628,v_theta2=7951.4784,
step=500,loss=18.327,theta1=10.0,theta2=-28.0855,
m_theta1=-0.0,m_theta2=20.769,v_theta1=2.4093,v_theta2=7275.6971,
step=600,loss=3.3473,theta1=10.0,theta2=-29.1818,
m_theta1=0.0,m_theta2=8.9808,v_theta1=2.1799,v_theta2=6599.9759,
```

训练次数与损失的变化，如图 2-19 所示。

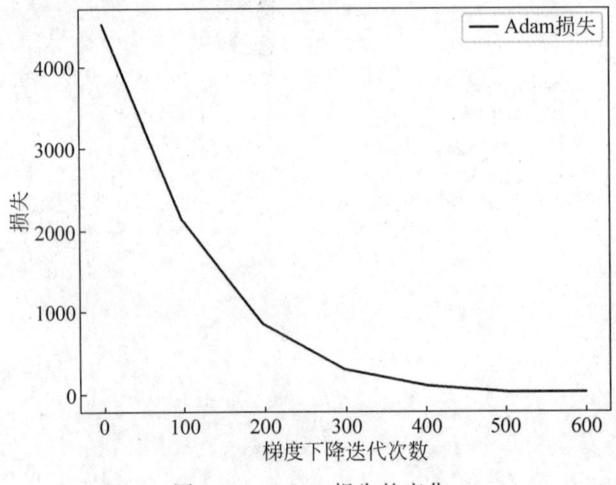

图 2-19　Adam 损失的变化

2.2.10　学习率的衰减

学习率的衰减是一种调整学习率的策略，在训练的早期学习的步长可以大一些，但是随着训练次数的增加为了避免梯度下降无法收敛到全局最优点，学习的步长最好越来越小，从而使学习能够逼近最优解。学习率衰减的策略，主要包括指数衰减、分段衰减、余弦衰减、余弦退火衰减等。

指数衰减学习率是先使用较大的学习率来快速地得到一个较优的解，然后随着迭代的次数的增加逐步减小学习率，使模型在训练后期更加稳定，即

$$\alpha_{t+1}=\alpha_0 e^{t/s} \tag{2-28}$$

设初始学习率为 α_0，t 是为迭代次数，e 为衰减系数，α_{t+1} 是随着迭代次数 t 的递增而衰减的，s 用来控制衰减的速度，如果 s 大一些，$\frac{t}{s}$就会增长缓慢，从而使 α_{t+1} 衰减得慢，否则学习率很快就会衰减为趋近于 0。

设初始学习率 $\alpha_0=0.1$，衰减系数 $e=0.9$，$s=10$ 每隔 10 个 epoch 进行衰减，迭代次数 epoch=100，则目标函数 $J(\theta)=(\theta-20)^2+5$ 梯度下降的过程，代码如下：

```
#第 2 章/machineLearn/梯度下降学习率系数衰减.py
if __name__ == "__main__":
    #初始学习率
```

```
learning_rate = 0.1
#衰减系数
decay_rate = 0.9
#decay_steps 控制衰减速度
decay_steps = 10
#迭代轮数
global_steps = 100
#初始权重
theta = 0
#随着训练次数的增加,学习率在降低
for global_step in range(global_steps):
    decayed_learning_rate = learning_rate *decay_rate **(
            global_step / decay_steps
    )
    theta = theta - decayed_learning_rate *(2 *(theta - 20))
    loss = (theta - 20) **2 + 5
    if global_step % 20 == 0:
        print(f'迭代={global_step},损失={loss},'
              f'学习率={decayed_learning_rate},'
              f'衰减系数={global_step / decay_steps}')
```

运行结果如下：

```
迭代=0,损失=261.0,学习率=0.1,衰减系数=0.0
迭代=20,损失=5.093818509932878,学习率=0.08100000000000002,衰减系数=2.0
迭代=40,损失=5.000175337443015,学习率=0.06561,衰减系数=4.0
迭代=60,损失=5.0000011696755955,学习率=0.05314410000000001,衰减系数=6.0
迭代=80,损失=5.000000021245625,学习率=0.04304672100000001,衰减系数=8.0
```

指数衰减时迭代次数、学习率、衰减系数的变化如图 2-20 所示。

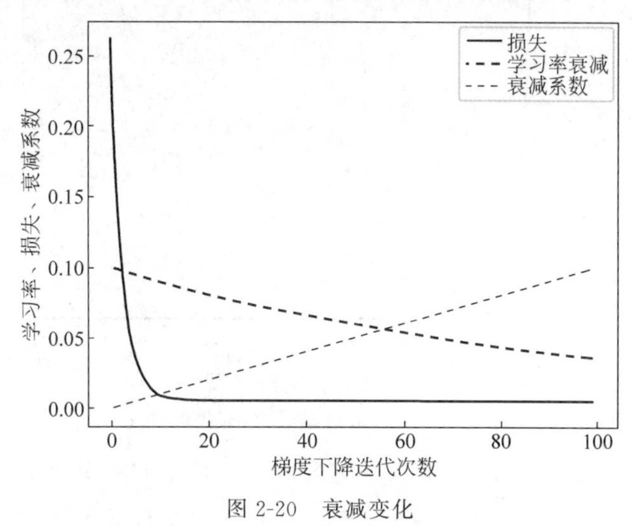

图 2-20 衰减变化

分段衰减即在指定的迭代次数时，学习率按给定的参数做乘法衰减，当指定迭代次数 $s=\{10,20,40,50,60\}$时，学习率衰减 0.5，则代码如下：

```
#第 2 章/machineLearn/梯度下降分段衰减.py
if __name__ == "__main__":
    #初始学习率
    learning_rate = 0.1
    #迭代轮数
    epoch = 100
    #初始权重
    theta = 0
    loss_list = []
    rate_list = []
    for step in range(epoch):
        if step in [10, 20, 40, 50, 60]:
            learning_rate = 0.5 * learning_rate
        theta = theta - learning_rate * (2 * (theta - 20))
        loss = (theta - 20) **2 + 5
        if step in [10, 20, 40, 50, 60]:
            print(f'迭代={step},损失={loss},学习率={learning_rate}')
```

运行结果如下：

```
迭代=10,损失=8.735465674926182,学习率=0.05
迭代=20,损失=5.506007705723608,学习率=0.025
迭代=40,损失=5.06849570860331,学习率=0.0125
迭代=50,损失=5.042346805488465,学习率=0.00625
迭代=60,损失=5.033345928511552,学习率=0.003125
```

分段衰减时迭代次数、学习率的变化如图 2-21 所示。

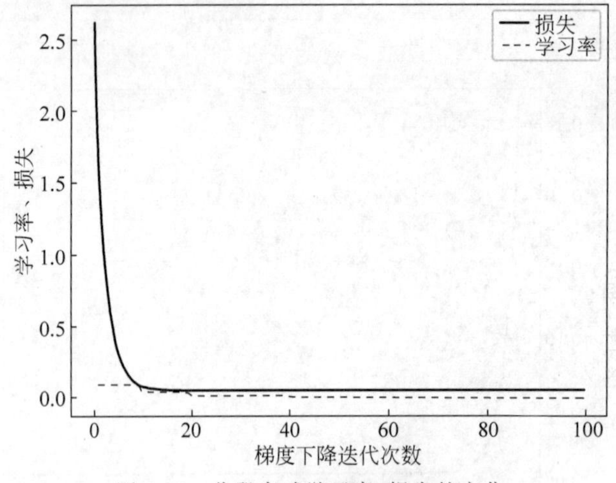

图 2-21 分段衰减学习率、损失的变化

余弦衰减就是采用余弦函数的相关方式进行学习率的衰减，其更新机制如下：

$$\alpha_t = \alpha_T + 0.5 \times (\alpha_0 - \alpha_T)\left(1 + \cos\left(\pi \frac{t}{T}\right)\right) \tag{2-29}$$

其中，α_0 是初始学习率，t 为当前迭代的次数，T 为总共迭代的次数，对于 $t > T$ 时学习率将固定在 α_T，代码如下：

```
#第 2 章/machineLearn/梯度下降学习率余弦衰减.py
import math
alpha = 0.1                     #初始学习率
last_lr = 1e-5                  #最小学习率
T = 30                          #T 步之前会衰减
epoch = 100                     #衰减全局步数
theta = 0                       #初始权重

decay_rate = alpha              #初始余弦学习率
for t in range(epoch):
    if t <= T:
        #当 step>max_steps 时,其学习率为设定的最后的学习率
        #当 step>max_steps 时,math.cos 为-1
        decay_rate = last_lr + (
                decay_rate - last_lr
        ) *(1 + math.cos(math.pi *(t / T))) *0.5

    theta = theta - decay_rate *(2 *(theta - 20))
    loss = (theta - 20) **2 + 5
    if (t <= T and t % 8 == 0) or t == epoch-1:
        print(f'迭代={t},损失={loss},学习率={decay_rate}')
```

运行结果如下：

```
迭代=0,损失=261.0,学习率=0.1
迭代=8,损失=18.204684755037157,学习率=0.056517582635429386
迭代=16,损失=12.53422654068735,学习率=0.0011665695498189522
迭代=24,损失=12.511414010499156,学习率=1.0004907422520221e-05
迭代=99,损失=12.488913301100498,学习率=1e-05
```

余弦衰减迭代次数、学习率的变化如图 2-22 所示。

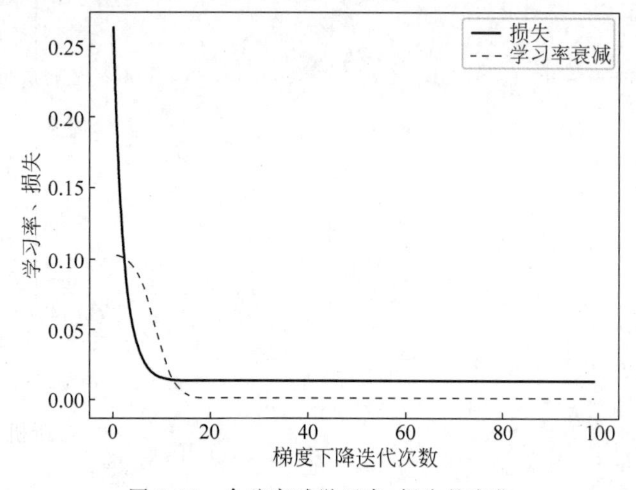

图 2-22　余弦衰减学习率、损失的变化

余弦退火衰减,学习率会先上升再下降。上升时使用线性上升,下降时使用余弦下降。上升的过程称为学习率预热,使用预热的原因是由于刚开始迭代时模型的权重可能是随机

初始化的，此时选择一个较大的学习率，可能会带来模型的不稳定。学习率预热在刚开始训练时先使用一个较小的学习率，然后使用较大的学习率来加快学习的过程，其更新的公式如下：

$$\alpha_{w_t}=\alpha_{w_0}+\frac{(\alpha_0-\alpha_{w_0})\times t}{w_t}$$

$$\alpha_t=\alpha_T+0.5\times(\alpha_0-\alpha_T)\left(1+\cos\left(\pi\frac{t-w}{T}\right)\right) \tag{2-30}$$

其中，α_{w_0} 为初始预热学习率，α_0 是初始学习率，t 为当前迭代的次数，T 为总共迭代的次数，w 为预热的迭代次数，对于 $t\leqslant w$ 时使用预热学习率 α_{w_t}，对 $t>w$ 时将使用余弦学习率，当 $t>T$ 时学习率将固定在最后学习率 α_T，代码如下：

```python
#第 2 章/machineLearn/梯度下降余弦退火衰减.py
import matplotlib.pyplot as plt
import math
import random

if __name__ == "__main__":
    alpha = 0.1                      #初始学习率
    last_lr = 0.01                   #最小学习率
    T = 40                           #T 步之前会衰减
    epoch = 100                      #衰减全局步数

    warmup_t = 20                    #设置预热的迭代次数
    warmup_lr = 0.0001               #设置预热的初始学习率
    theta = random.random()          #将参数值设置为初始值
    decay_rate = alpha               #初始余弦学习率

    for t in range(epoch):
        is_warmup = False
        if t <= warmup_t:
            #预热学习率
            warmup_lr = warmup_lr + (last_lr - warmup_lr) *t / warmup_t
            is_warmup = True
        elif t > warmup_t and t <= T:
            #余弦学习率
            #当 step>max_steps 时，其学习率为设定的最后的学习率
            #当 step>max_steps 时，math.cos 为-1
            decay_rate = last_lr + (
                    decay_rate - last_lr
            ) *(1 + math.cos(math.pi *((t - warmup_t) / T))) *0.5
        if is_warmup:
            theta = theta - warmup_lr *(2 *(theta - 20))
            if t % 5 == 0:
                print(f'迭代={t},预热学习率={warmup_lr}')
        else:
            theta = theta - decay_rate *(2 *(theta - 20))
            if (t <= T and t % 15 == 0) or t == epoch - 1:
                print(f'迭代={t},学习率={decay_rate}')
        #求损失
        loss = (theta - 20) **2 + 5
```

运行结果如下：

```
迭代=0,预热学习率=0.0001
迭代=5,预热学习率=0.005683105000000001
迭代=10,预热学习率=0.009675909107875
迭代=15,预热学习率=0.00999846867053471
迭代=20,预热学习率=0.01
迭代=30,学习率=0.05919118592560402
迭代=99,学习率=0.010775441191547337
```

余弦退火衰减迭代次数、学习率的变化如图 2-23 所示。

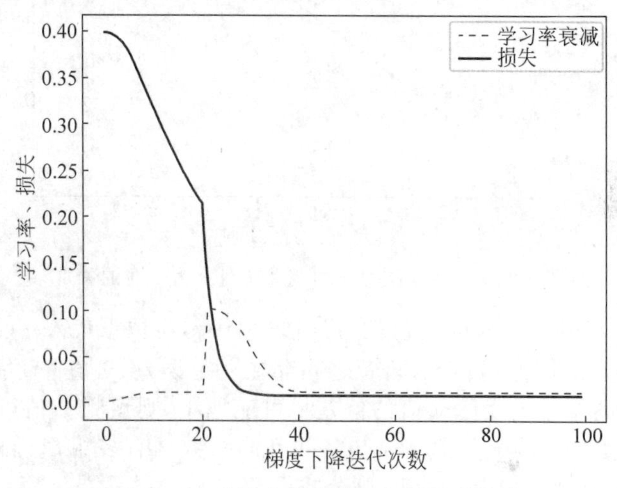

图 2-23　余弦退火衰减学习率、损失的变化

总结

梯度下降算法是深度学习的基础，不同的梯度下降算法是为了更快、更好地接近全局最优。优化器的选择在深度学习网络的训练过程中较为重要。

练习

可考虑将以上优化算法的函数实现封装成 1 个类进行管理；总结不同优化算法的优缺点。

2.3　线性回归

2.3.1　梯度下降求解线性回归

1889 年 K. Pearson 收集了大量父亲身高与儿子身高的资料，其中 10 组数据见表 2-2。

表 2-2　父亲身高与儿子身高

父亲身高为 x 时	60	62	64	65	66	67	68	70	72	74
儿子身高为 y 时	63.6	65.2	66	65.5	66.9	67.1	67.4	68.3	70.1	70

为了描述 y 与 x 之间的关系可以建立一个线性函数，其形式为 $f(x)=\theta x+b$，其中 θ、b 为要用实验数据确定的线性回归参数，图 2-24 展示了用一条线性函数来拟合表中的实验数据。

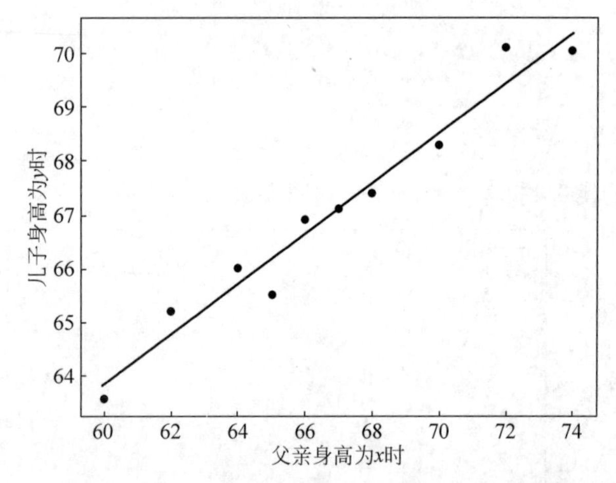

图 2-24　父亲身高为 x 时与儿子身高为 y 时的关系

当 x 为 $\{x_1,x_2,\cdots\}$ 时对应的 y 值就为 $\{y_1,y_2,\cdots\}$，将数据代入 $f(x_i)=\theta x_i+b$ 中会发现预测值 $\hat{y}$ 与实际 y 值并不相等，两者之间存在一定误差，$\hat{y}$ 值并不能准确地落在直线上，但是，如果预测的 $\hat{y}$ 值与实际 y 值的偏差尽可能地小，则预测 $\hat{y}$ 值可以认为是准确的，所以得到目标函数 $L(\theta)=(y-(\theta x+b))^2$，并且测试数据有多组，它们大小不一、有正有负，所以实际上只能希望总的误差最小，则公式更新如下：

$$L(\theta)=\frac{1}{2N}\sum_{i=1}^{N}((\theta x_i+b)-y_i)^2 \tag{2-31}$$

最小二乘法的中心思想就是寻找参数 θ 和 b 使预测值 $\hat{y}$ 与真实值 y 误差的平方和最小，其中 $L(\theta)$ 函数又称为损失函数。求目标函数最小值问题就转变为求最优化问题，根据 2.2 节梯度下降中的内容可以使用梯度下降算法进行求解，求解过程如下。

设初始权重值：

$$(\theta_0,b_0)=(0,0)$$

$$\frac{\partial J}{\partial\theta}=\frac{1}{N}\sum_{i=1}^{N}(y-(\theta x_i+b))\times x_i,\quad \frac{\partial J}{\partial b}=\frac{1}{N}\sum_{i=1}^{N}(y-(\theta x_i+b))$$

根据梯度下降公式：

$$(\theta,b)=\theta-\alpha\times\frac{\partial J}{\partial\theta},\quad b-\alpha\times\frac{\partial J}{\partial b}$$

设学习率 $\alpha=0.01$，则计算过程如下：

$$\theta_1=\theta_0-0.01\times\left\{\frac{1}{2}((60\times\theta_0-b_0)-63.6)\times 60+\frac{1}{2}((62\times\theta_0-b_0)-65.2)\times 62+\cdots\right\}$$

$$b_1=b_0-0.01\times\left\{\frac{1}{2}((60\times\theta_0-b_0)-63.6)+\frac{1}{2}((62\times\theta_0-b_0)-65.2)+\cdots\right\}$$

第 1 次迭代后的值为

$$\theta_1 = 0 - 0.01 \times \left(\frac{1}{2} \times ((60 \times 0 + 0) - 63.6) \times 60 + \frac{1}{2} \times ((65.2 \times 0 + 0) - 65.2) \times 62\right)$$

$$b_1 = 0 - 0.01 \times \left(\frac{1}{2} \times ((60 \times 0 + 0) - 63.6) + \frac{1}{2} \times ((65.2 \times 0 + 0) - 65.2)\right)$$

则

$$(\theta_1, b_1) = (39.292, 0.644)$$

第 2 次迭代后的值如下：

$$\theta_2 = -39.292 - 0.01 \times \left(\frac{1}{2} \times ((60 \times (-39.292) + (-0.644)) - 63.6) \times 60 + \frac{1}{2} \times ((65.2 \times (-39.292) + (-0.644)) - 65.2) \times 62\right)$$

$$b_2 = -0.644 - 0.01 \times \left(\frac{1}{2} \times ((60 \times (-39.292) + (-0.644)) - 63.6) + \frac{1}{2} \times ((65.2 \times (-39.292) + (-0.644)) - 65.2)\right)$$

则

$$(\theta_2, b_2) = (1501.81, 24.60)$$

将以上求解过程转换为代码，代码如下：

```
#第 2 章/machineLearn/一元线性回归梯度下降求解.py
#预测函数
def prediction(w, x, b):
    return w *x + b


#损失函数
def sum_loss(X, Y, w, b):
    total_loss = 0
    for x, y in zip(X, Y):
        total_loss += (y - prediction(w, x, b)) **2
    return 1 / (2 *len(X)) *total_loss


#w 的梯度,使用了全量梯度
def sum_gradient_w(X, Y, w, b):
    total_w = 0
```

```
    for x, y in zip(X, Y):
        total_w += (prediction(w, x, b) - y) *x
    return 1 / len(X) *total_w


#b的梯度,使用了全量梯度
def sum_gradient_b(X, Y, w, b):
    total_b = 0
    for x, y in zip(X, Y):
        total_b += (prediction(w, x, b) - y) *1
    return 1 / len(X) *total_b


if __name__ == "__main__":
    epoch = 21                          #迭代次数
    init_wb = 10, 10                    #初始参数值
    temp_w, temp_b = init_wb[0], init_wb[1]
    learning = 0.0001                   #学习率
    X = [60, 62, 64, 65, 66, 67, 68, 70, 72, 74]
    Y = [63.6, 65.2, 66, 65.5, 66.9, 67.1, 67.4, 68.3, 70.1, 70]
    result_wb = []                      #存储过程中的梯度
    for item in range(epoch):
        #梯度下降
        w = temp_w - learning *sum_gradient_w(X, Y, temp_w, temp_b)
        b = temp_b - learning *sum_gradient_b(X, Y, temp_w, temp_b)
        loss = sum_loss(X, Y, w, b)
        if item % 5 == 0:
            print('第{}次梯度,(w,b)参数为({:6f},{:6f }),总损失{:6f }'.format
(item + 1, w, b, loss))
        #更新后再赋值给temp,这样能保证w、b同时更新
        temp_w = w
        temp_b = b
          result_wb.append([temp_w,temp_b])
```

运行结果如下：

```
第1次梯度,(w,b)参数为(5.902224,9.938901),总损失57104.833861
第6次梯度,(w,b)参数为(1.112617,9.867525),总损失151.260033
第11次梯度,(w,b)参数为(0.867236,9.863916),总损失1.774189
第16次梯度,(w,b)参数为(0.854664,9.863778),总损失1.381829
第21次梯度,(w,b)参数为(0.854019,9.863819),总损失1.380794
```

根据result_wb这个列表的内容，将第4次和第21次的梯度下降得到的参数代入图像中会发现随着每次梯度下降学习而得到的参数可以更好地对现有数据进行线性拟合，代码如下：

```
#第2章/machineLearn/一元线性回归梯度下降求解.py

def plt_gradient_linear_process(X, Y, result_wb):
    #用来正常显示中文标签
    plt.rcParams['font.sans-serif'] = ['SimHei']
```

```
    plt.rcParams['axes.unicode_minus'] = False
    #将 x 轴的刻度线方向设置为向内
    plt.rcParams['xtick.direction'] = 'in'
    #将 y 轴的刻度线方向设置为向内
    plt.rcParams['ytick.direction'] = 'in'
    plt.scatter(X, Y)
    length = len(result_wb)
    num = int(length * 0.2)
    mid_wb = result_wb[num]
    last_wb = result_wb[-1]
    plt.plot(X, [mid_wb[0] * x + mid_wb[1] for x in X], label='学习第{}次拟合直线'.
format(num))
    plt.plot(X, [0.4645 * x + 35.97 for x in X], label='最佳拟合直线')
    plt.plot(X, [last_wb[0] * x + last_wb[1] for x in X], label='学习第{}次拟合直线'.
format(length))
    plt.title("父亲身高为 x 时与儿子身高为 y 时的关系\n 梯度下降线性拟合过程")
    plt.xlabel('父亲身高为 x 时')
    plt.ylabel('儿子身高为 y 时')
    plt.legend(loc='upper left')
    plt.text(x=71,
             y=60,
             s=f'temp_w={init_wb[0]}\n'
               f'temp_b={init_wb[1]}\n'
               f'learning={learning}\n'
               f'epoch={epoch}\n'
               f'loss={round(loss, 4)}',
             fontdict=dict(fontsize=8, family='monospace', ), #字体属性字典
             #添加文字背景色
             bbox={'facecolor': '#74C476',                          #填充色
                   'edgecolor': 'b',                                #外框色
                   'alpha': 0.8,                                    #框透明度
                   'pad': 0.9,                                      #文字与框周围距离
                   }
             )
    plt.savefig('loss_lr.png')
    plt.show()
plt_gradient_linear_process(X, Y, result_wb)
```

运行结果如图 2-25 所示。

运行程序并调整初始参数 temp_w 和 temp_b、学习率 learning 及 epoch 的设置可以发现不同的初始值，对于梯度下降学习的过程影响是巨大的，线性拟合的效果也不尽相同。

初始设置 temp_w=0、temp_b=0、learning=0.0001、epoch=21 时损失为 2.55，需要少量地增加 epoch 的次数，如图 2-26 所示。

初始设置 temp_w=0、temp_b=0、learning=0.001、epoch=21 时损失得到一个很大的值，说明已越过最优解，需要降低学习率，如图 2-27 所示。

初始设置 temp_w=0、temp_b=0、learning=0.00001、epoch=200 时损失为 2.55 比较接近图 2-26 的效果，如图 2-28 所示。

初始设置 temp_w=1、temp_b=40、learning=0.0001、epoch=21 时已接近最佳拟合直线，如图 2-29 所示。

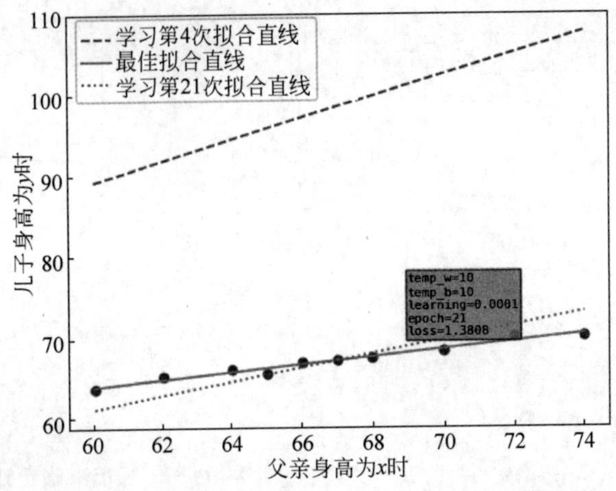

图 2-25　父亲身高为 x 时与儿子身高为 y 时的关系-梯度下降线性拟合过程

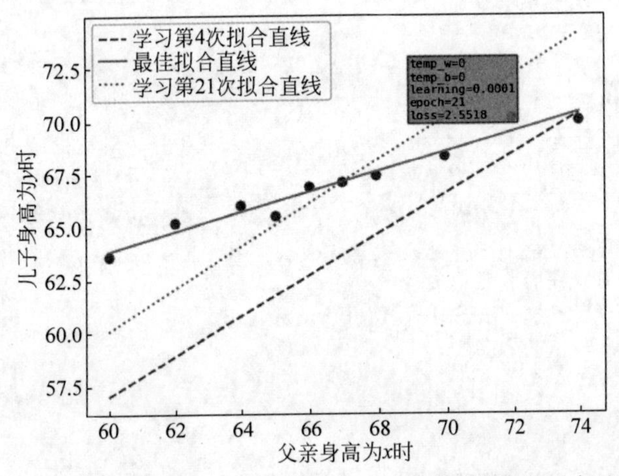

图 2-26　设置 temp_w＝0、temp_b＝0 时梯度下降线性拟合过程

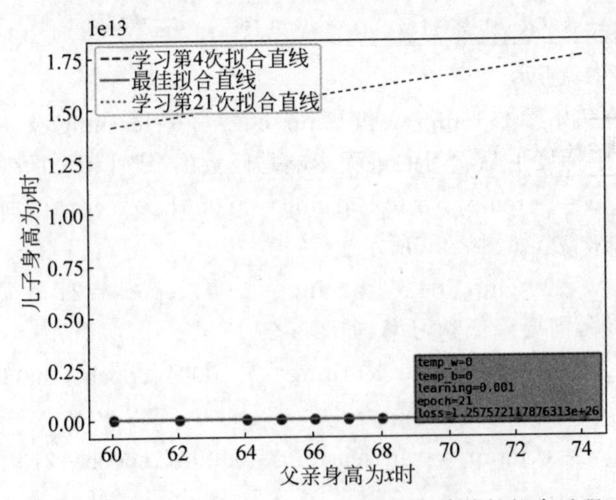

图 2-27　设置 learning＝0.001 时梯度下降线性拟合过程

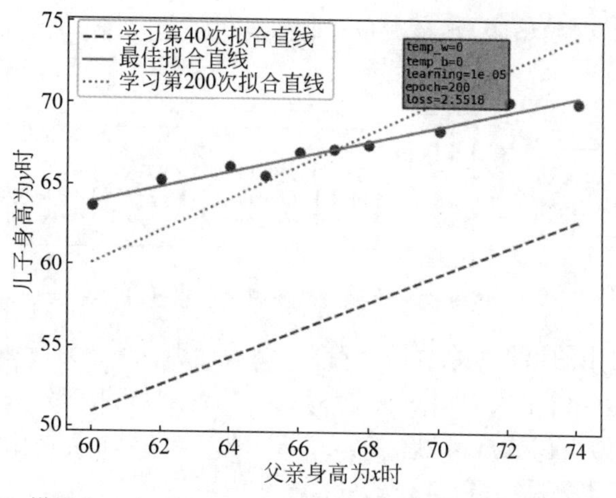

图 2-28　设置 learning＝0.00001、epoch＝200 时梯度下降线性拟合过程

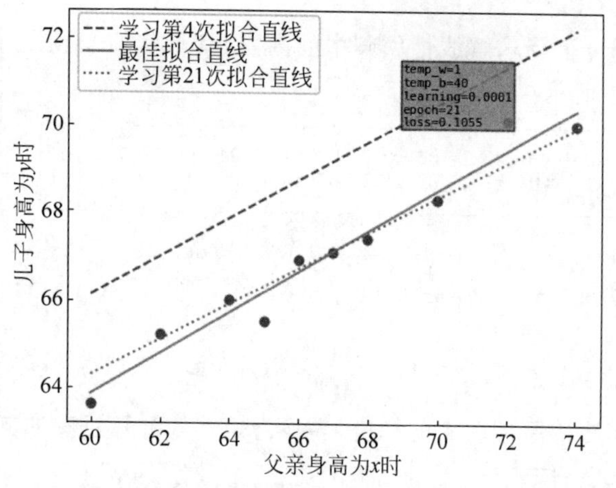

图 2-29　设置 temp_w＝1、temp_b＝40、learning＝0.0001、epoch＝21 时梯度下降线性拟合过程

综上可知，初始权重、学习率、学习次数的设置均可影响梯度下降学习的速度及学习成果，这些是机器学习中调节的主要参数。这些可调节参数又称为超参数。

注意：最佳拟合直线的参数 θ＝0.4645 和 b＝35.97 是通过正规方程求解线性回归获得的。

2.3.2　梯度下降求解多元线性回归

存在以下数据，见表 2-3。

表 2-3　年龄、性别、区域、婚否、小孩数与收入关系表

编号	年龄	性别	区域	婚否	小孩数量/个	收入/元
1	48	女	大城市	否	1	17 546
2	40	男	乡镇	是	3	30 085.1

假设年龄、性别、区域、婚否、小孩数量与收入存在线性关系，则线性回归预测的过程如图 2-30 所示。

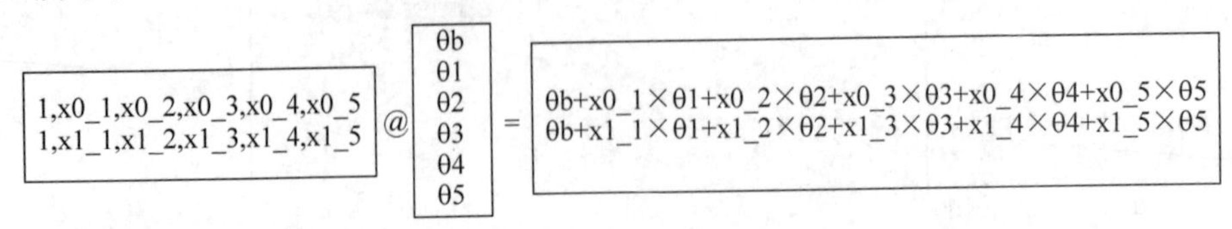

图 2-30 多元线性回归的预测过程

x0_1 表示第 1 行中的第 1 列数据即年龄、x0_2 为性别、x0_3 为区域、x0_4 为婚否、x0_5 为小孩数量。θ1 表示年龄字段的预测权重参数、θ2 为性别预测参数、θ3 为区域预测参数、θ4 为婚否预测参数、θ5 为小孩数量预测参数。在 X 中插入数据 1 与偏置参数 θb，根据矩阵内积的运算法则得到等式关系，其数学公式可以表示为

$$\hat{Y}=X@\Theta \tag{2-32}$$

根据 2.3.1 节梯度下降求解线性回归中的内容，其真实值与预测值的误差转换为矩阵求解的过程如图 2-31 所示。

θb+x0_1×θ1+x0_2×θ2+x0_3×θ3+x0_4×θ4+x0_5×θ5
θb+x1_1×θ1+x1_2×θ2+x1_3×θ3+x1_4×θ4+x1_5×θ5
−
y0
y1
=
Δ0
Δ1

图 2-31 矩阵损失求解

因此，损失函数的数学公式如下：

$$L(\theta)=\frac{1}{2N}\Sigma^{N}(\hat{Y}-Y)^{2} \tag{2-33}$$

根据梯度下降算法的原理求 θ 的梯度，转换为矩阵的计算过程如图 2-32 所示。

1,1
x0_1,x1_1
x0_2,x1_2
x0_3,x1_3
x0_4,x1_4
x0_5,x1_5
@
Δ0
Δ1
=
1×Δ0+1×Δ1
x0_1×Δ0+x1_1×Δ1
x0_2×Δ0+x1_2×Δ1
x0_3×Δ0+x1_3×Δ1
x0_4×Δ0+x1_4×Δ1
x0_5×Δ0+x1_5×Δ1

图 2-32 矩阵求梯度

其数学公式如下：

$$\frac{\partial L}{\partial \Theta}=\boldsymbol{X}^{\mathrm{T}}@(X@\Theta-Y)$$

$$\Theta=\Theta-\alpha\times\frac{\partial L}{\partial \Theta} \tag{2-34}$$

将以上分析过程使用 NumPy 库实现多元线性回归的梯度下降算法，代码如下：

```
#第 2 章/machineLearn/多元线性回归梯度下降求解.py
import numpy as np
import pandas as pd
```

```
import matplotlib.pyplot as plt


#数据预处理
def data_preprocessing(path, Debug=False):
    data = pd.read_csv(path, encoding='gbk')
    if Debug:
        data = data.loc[:50, :]
    #将收入转换为以千元为单位
    y = data['收入'] / 1000
    x1 = data['年龄']
    #如果性别为男,则将值设置为1,如果性别为女,则将值设置为2
    x2 = data.apply(lambda x: 1 if x['性别'] == '男' else 2, axis=1)
    area = {'大城市': 1, '乡镇': 2, '县城': 3, '农村': 4}
    #如果区域为大城市,则设置为1(乡镇为2,县城为3,农村为4)
    x3 = data.apply(lambda x: area[x['区域']] if x['区域'] in area else 0, axis=1)
    #如果已婚,则为1,如果未婚,则为2
    x4 = data.apply(lambda x: 1 if x['婚否'] == '是' else 2, axis=1)
    x5 = data['小孩']
    #追加1列并将初始值设置为1作为矩阵运算的偏置项
    data['偏置项'] = 1
    x = np.column_stack([data['偏置项'], x1, x2, x3, x4, x5])
    return x, y


#用三维图像来展示拟合情况
def gradient_descent_3d(X, y, theta):
    plt.rcParams['font.sans-serif'] = ['SimHei']
    plt.rcParams['axes.unicode_minus'] = False
    ax = plt.axes(projection='3d')
    x1 = X[:, 1]          #取年龄数据
    x2 = X[:, 2]          #取性别数据
    #将x1与x2的值组成坐标系数据
    X1, X2 = np.meshgrid(x1, x2)
    #得到最后的权重
    theta = theta.T
    #根据权重得到预测值
    #图中只展示了年龄、性别与收入的关系
    Y1 = (X1 *theta[0, 1] + X2 *theta[0, 2] + theta[0, 0])
    #三维散点图
    ax.scatter(x1, x2, y, c='b', marker='o', label='年龄,收入')
    #根据参数获得的结果绘制平面图,用来表示拟合情况
    ax.plot_surface(X1, X2, Y1, rstride=1, cstride=1, cmap='rainbow')

    ax.text(20, 2, 32,
            '学习第{}次拟合'.format(epoch),
            color='b', horizontalalignment='left')
    ax.set_xlabel(u'年龄')
    ax.set_ylabel(u'性别')
    ax.set_zlabel(u'收入')
    ax.set_title("年龄、性别、收入多元线性回归拟合过程")
    plt.savefig('年龄、性别、收入三维关系图')
```

```
    plt.show()


if __name__ == '__main__':
    #读取数据源,共计 600 条数据
    data = data_preprocessing("多元线性回归数据.csv")
    #X 矩阵的数据
    X = data[0]
    #Y 矩阵的数据
    Y = data[1].values
    #将 shape 转置为(6,1)
    Y = np.reshape(Y, [-1, 1])
    #设置学习率
    alpha = 1e-6
    #设置迭代次数
    epoch = 100
    #初始权重
    theta = np.zeros([6, 1])
    all_loss = []
    for i in range(epoch):
        #使用公式 X^T@(Y_pred-Y_true)求梯度
        gradient = np.matmul(np.transpose(X), np.matmul(X, theta) - Y)
        #梯度下降
        theta = theta - alpha *gradient
        #求损失
        loss = np.mean(np.square(np.matmul(X, theta) - Y)) / 2
        #供画图用
        all_loss.append(loss)
        if i % 20 == 0 or i ==epoch-1:
            print("第{}次损失={}".format(i + 1, loss))
    #使用三维图像来展示梯度下降后最后一次值的拟合过程
    gradient_descent_3d(X, data[1].values, theta)
```

运行结果如下:

```
第 1 次损失=54.547319224187035
第 21 次损失=36.03552930295452
第 41 次损失=36.02828819740307
第 61 次损失=36.021198957256594
第 81 次损失=36.01425679858055
第 100 次损失=36.007793793175814
```

其最后 1 次梯度下降得到的权重参数拟合效果的三维效果如图 2-33 所示。

注意:在函数 data_preprocessing()中对收入数据进行了等比例缩小,并且将性别、区域、婚否字段中的汉字用数字进行了替换。

总结

线性回归通过寻找一条最佳直线方程拟合现有数据。当损失函数 $L(\theta)$ 最小时,表明找到了最佳参数使线性方程能够较好地拟合现有数据。

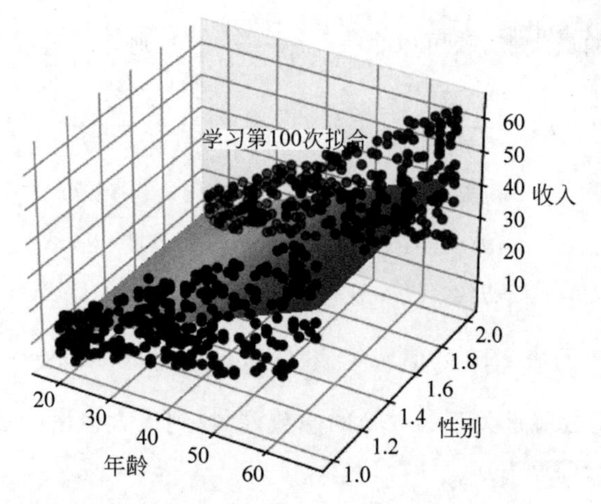

图 2-33　年龄、性别、收入多元线性回归拟合过程

练习

重写并调试多元线性回归的代码，并使用不同的学习率和初始权重值，以此观察拟合情况的变化。

2.4　逻辑回归

2.4.1　最大似然估计

如图 2-34 所示，在 1 个箱子里存放着 7 个黑球和 2 个白球，则拿出来的球最有可能是什么颜色呢？

根据生活常识推断可知“拿出来的球最有可能是黑色的”，“最有可能”就是最大似然的意思，这种想法常称为“最大似然原理”，总结起来“最大似然估计”的目的就是根据已观测到的样本结果反推最有可能(最大概率)导致这样结果的参数值 θ。

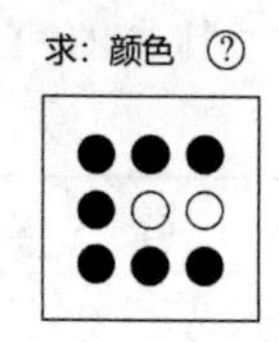

图 2-34　求未知颜色类型

最大似然估计提供了一种用给定的观察数据来评估模型参数的方法，即“模型已定，参数未知”。通过若干次试验，观察其结果，利用试验结果得到参数 θ 使样本出现的概率为最大。

假设 H 表示黑色、T 表示白色，共计取值 9 次，根据取值结果 $X=\{H,H,H,H,T,T,H,H,H\}$ 求最大概率 $P(x|\theta)$ 的过程如下：

$$
\begin{aligned}
P(x \mid \theta) &= P(H,H,T,T,H,H,H,H,H) \\
&= P(H \mid \theta) \times P(H \mid \theta) \times P(H \mid \theta) \times P(H \mid \theta) \times P(T \mid \theta) \times P(T \mid \theta) \times \\
&\quad P(H \mid \theta) \times P(H \mid \theta) \times P(H \mid \theta) \\
&= \theta \times \theta \times \theta \times \theta \times (1-\theta) \times (1-\theta) \times \theta \times \theta \times \theta \\
&= \theta^7 \times (1-\theta)^2
\end{aligned}
$$

要找到 $P(x|\theta)$的最大概率，可以令$\frac{\partial(P(x|\theta))}{\partial\theta}=0$，则计算过程如下：

$$
\begin{aligned}
L(\theta \mid x) &= \frac{\partial(\theta^7)}{\partial\theta} \times (1-\theta)^2 + \theta^7 \times \frac{\partial((1-\theta)^2)}{\partial\theta} \\
&= 7\theta^6 \times (1-\theta)^2 + \theta^7 \times 2 \times (1-\theta) \times (-1) \\
&= \theta^6 \times (1-\theta) \times (7 \times (1-\theta) - 2\theta) \\
&= \theta^6 \times (1-\theta) \times (7 - 9\theta)
\end{aligned}
$$

因为是求 $P(x|\theta)$的概率，θ 的值域应该在(0,1)之间，所以 $\theta=\frac{7}{9}$。

综上可知，似然函数就是关于参数 θ 的函数，要找到最大概率的参数，即找到$L(\theta|x)$最大概率值时的 θ，见公式：

$$L(\theta \mid x) = P(x \mid \theta) = \prod_{i=1}^{N} P(x_i;\theta) \tag{2-35}$$

由于式(2-35)是连乘的形式，可能导致 $L(\theta|x)$的值接近 0，所以借助对数函数来简化，见公式：

$$l(\theta \mid x) = \underset{\theta}{\operatorname{argmax}} \sum_{i=1}^{N} \log L(\theta \mid x) \tag{2-36}$$

式(2-36)又称为对数似然函数。

2.4.2 梯度下降求解逻辑回归

假设某学校根据语文、数学的考试分数进行录取工作，1 表示录取、0 表示不录取，现有某考生的分数语文 80 分、数学 60 分，问是否能够录取。历史样本数据共有 100 条，表 2-4 仅摘录了其中的两条。

表 2-4 某学校录取分数记录表摘录

语文	数学	录取结果
34.6	78.0	0
60.1	86.3	1

数据样本中的录取结果，如图 2-35 所示。

很显然该问题是一个二分类问题，需要根据已知样本数据的分布预估该学生被录取与否的概率，该分类方法称为逻辑回归。

根据 2.4.1 节最大似然估计可知，逻辑回归的求解服从"根据已知样本数据分布反推最有可能导致出现样本结果分布的参数值 θ"的任务，即

$$L(\theta \mid x) = \prod_{i=1}^{N} P(y_i \mid x_i;\theta) \tag{2-37}$$

由于 $P(y_i|x_i;\theta)$是一个概率问题，其值域应该在(0,1)之间，并以 0.5 为区分，大于 0.5 为分类 1，小于 0.5 为分类 0，而线性回归中的预测函数 $f(x_i)=\theta x_i+b$ 的值域可能大于 1

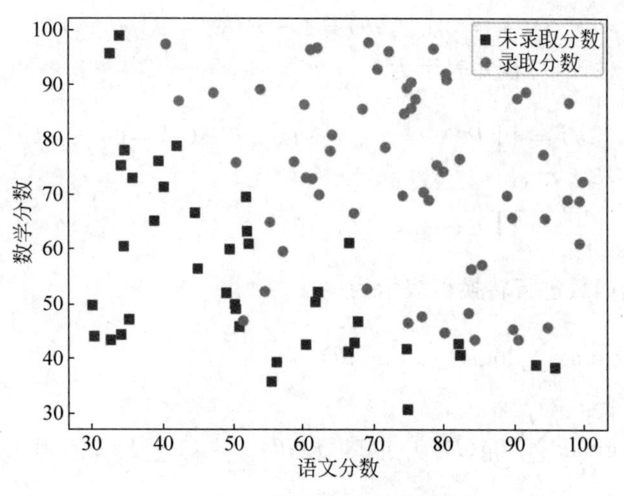

图 2-35 某录取结果分布

或者小于 0,已不再适合,故更换预测函数为 Sigmoid 函数,其公式为

$$P(y_i \mid x_i;\theta)=\frac{1}{1+\mathrm{e}^{-(\theta x_i+b)}} \tag{2-38}$$

Sigmoid 函数其值域在 0～1,并且当 x 趋近 $+\infty$ 时 $y=1$,当 x 趋近 $-\infty$ 时 $y=0$,其曲线图如 2-36 所示。

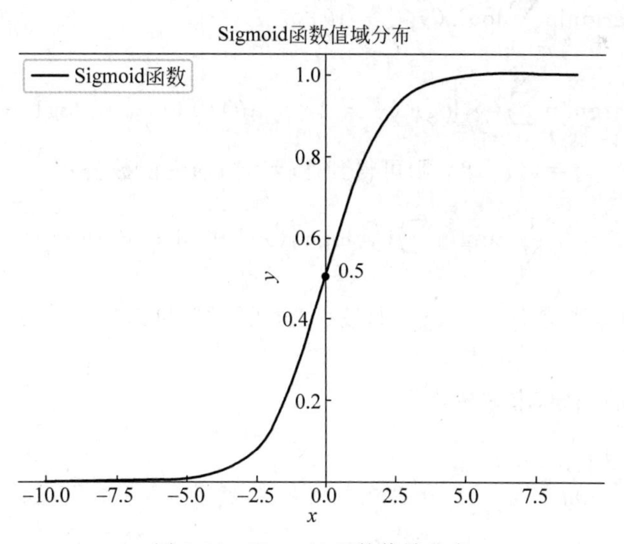

图 2-36 Sigmoid 函数值域分布

当分类结果 $y=1$ 时,其概率分布可以表示为

$$P(y_i=1 \mid x_i;\theta)=\frac{1}{1+\mathrm{e}^{-(\theta x_i+b)}}$$

当 $y=0$ 时表示为

$$P(y_i=0 \mid x_i;\theta)=1-P(y_i=1 \mid x_i;\theta)$$

根据最大似然估计，可合并表示为

$$\begin{aligned}L(y_i \mid x_i;\theta)&=\prod_{i=1}^{N}P(y_i=1 \mid x_i;\theta)^{y_i}\times P(y_i=0 \mid x_i;\theta)^{1-y_i}\\&=\prod_{i=1}^{N}P(y_i=1 \mid x_i;\theta)^{y_i}\times(1-P(y_i=1 \mid x_i;\theta))^{1-y_i}\end{aligned}$$

根据对数似然函数进行转换可表示为

$$\begin{aligned}l(\theta \mid y_i;x_i)&=\underset{\theta}{\operatorname{argmax}}\sum_{i=1}^{N}\log L(y_i \mid x_i;\theta)\\&=\underset{\theta}{\operatorname{argmax}}\sum_{i=1}^{N}\log(P(y_i=1 \mid x_i;\theta)^{y_i}\times(1-P(y_i=1 \mid x_i;\theta))^{1-y_i})\\&=\underset{\theta}{\operatorname{argmax}}\sum_{i=1}^{N}(y_i\log P(y_i=1 \mid x_i;\theta)+(1-y_i)\log 1-P(y_i=1 \mid x_i;\theta))\end{aligned}$$

加负号转换为求最小值问题：

$$\begin{aligned}l(\theta \mid y_i;x_i)&=\underset{\theta}{\operatorname{argmax}}\sum_{i=1}^{N}\log L(y_i \mid x_i;\theta)\\&=-\underset{\theta}{\operatorname{argmin}}\sum_{i=1}^{N}\log L(y_i \mid x_i;\theta)\\&=-\underset{\theta}{\operatorname{argmin}}\sum_{i=1}^{N}(y_i\log P(y_i=1 \mid x_i;\theta)+(1-y_i)\log 1-P(y_i=1 \mid x_i;\theta))\end{aligned}$$

令 $h_\theta(x_i)=P(y_i=1|x_i;\theta)$，则可得到逻辑回归损失函数公式：

$$l(\theta \mid y_i;x_i)=-\frac{1}{N}\underset{\theta}{\operatorname{argmin}}\sum_{i=1}^{N}(y_i\log h_\theta(x_i)+(1-y_i)\log(1-h_\theta(x_i))) \tag{2-39}$$

根据梯度下降算法求 $l(\theta|y_i;x_i)$ 的最小值，需要满足 $(\theta,b)=\theta-\alpha*\frac{\partial J}{\partial\theta},b-\alpha*\frac{\partial J}{\partial b}$，则关于 $\frac{\partial l(\theta|y_i;x_i)}{\partial\theta}$ 的偏导，求解过程如下：

$$\begin{aligned}\frac{\partial l(\theta \mid y_i;x_i)}{\partial\theta}=-\frac{1}{N}\underset{\theta}{\operatorname{argmin}}\sum_{i=1}^{N}\Bigg(&y_i\times\frac{1}{h_\theta(x_i)}\times\frac{\partial h_\theta(x_i)}{\partial\theta}+\\&(1-y_i)\times\frac{1}{1-h_\theta(x_i)}\times\frac{\partial(1-h_\theta(x_i))}{\partial\theta}\Bigg)\end{aligned}$$

又 $\frac{\partial h_\theta(x_i)}{\partial\theta}=\frac{\partial\frac{1}{1+e^{-(\theta x_i+b)}}}{\partial\theta}$，令 $z=\theta x_i+b$，则

$$
\begin{aligned}
\frac{\partial h_\theta(x_i)}{\partial \theta} &= \frac{\partial(1+\mathrm{e}^{-z})^{-1}}{\partial \theta} = -1 \times (1+\mathrm{e}^{-z})^{-1-1} \times \frac{\partial(1+\mathrm{e}^{-z})}{\partial \theta} \\
&= -1 \times (1+\mathrm{e}^{-z})^{-2} \times \left(\frac{\partial(1)}{\partial \theta_j} \times \mathrm{e}^{-z} + 1 \times \frac{\partial(\mathrm{e}^{-z})}{\partial \theta}\right) \\
&= -1 \times (1+\mathrm{e}^{-z})^{-2} \times \left(0 \times \mathrm{e}^{-z} + 1 \times \frac{\partial\left(\frac{1}{\mathrm{e}^{z}}\right)}{\partial \theta}\right) \\
&= -1 \times (1+\mathrm{e}^{-z})^{-2} \times \left(1 \times \frac{\frac{\partial(1)}{\partial \theta} \times \mathrm{e}^{z} - 1 \times \frac{\partial(\mathrm{e}^{z})}{\partial \theta}}{\mathrm{e}^{z} \times \mathrm{e}^{z}}\right) \\
&= -1 \times (1+\mathrm{e}^{-z})^{-2} \times \left(1 \times \frac{0 - \mathrm{e}^{z} \times \frac{\partial(\theta x_i + b)}{\partial \theta}}{\mathrm{e}^{z} \times \mathrm{e}^{z}}\right) \\
&= (1+\mathrm{e}^{-z})^{-2} \times \frac{1 \times x_i}{\mathrm{e}^{z}} \\
&= \frac{1}{(1+\mathrm{e}^{-z})^2} \times \mathrm{e}^{-z} \times x_i \\
&= \frac{1+\mathrm{e}^{-z}-1}{(1+\mathrm{e}^{-z})^2} \times x_i \\
&= \frac{1}{1+\mathrm{e}^{-z}} \times \left(1 - \frac{1}{1+\mathrm{e}^{-z}}\right) \times x_i \\
&= \frac{1}{1+\mathrm{e}^{-(\theta x_i + b)}} \times \left(1 - \frac{1}{1+\mathrm{e}^{-(\theta x_i + b)}}\right) \times x_i \\
&= h_\theta(x_i) \times (1 - h_\theta(x_i)) \times x_i
\end{aligned}
$$

又

$$
\frac{\partial(1-h_\theta(x_i))}{\partial \theta} = \frac{\partial(1)}{\partial \theta} - \frac{\partial h_\theta(x_i)}{\partial \theta} = -h_\theta(x_i) \times (1-h_\theta(x_i)) \times x_i
$$

故

$$
\begin{aligned}
\frac{\partial l(\theta \mid y_i; x_i)}{\partial \theta} = &-\frac{1}{N} \underset{\theta}{\operatorname{argmin}} \sum_{i=1}^{N} \left(y_i \times \frac{1}{h_\theta(x_i)} \times h_\theta(x_i) \times (1-h_\theta(x_i)) \times \right. \\
&\left. x_i + (1-y_i) \times \frac{1}{1-h_\theta(x_i)} \times -h_\theta(x_i) \times (1-h_\theta(x_i)) \times x_i\right) \\
= &-\frac{1}{N} \underset{\theta}{\operatorname{argmin}} \sum_{i=1}^{N} (y_i \times (1-h_\theta(x_i)) \times x_i + (1-y_i) \times -h_\theta(x_i) \times x_i) \\
= &-\frac{1}{N} \underset{\theta}{\operatorname{argmin}} \sum_{i=1}^{N} (y_i \times 1 \times x_i - h_\theta(x_i) \times y_i \times x_i -
\end{aligned}
$$

$$
\begin{aligned}
&1\times h_\theta(x_i)\times x_i + y_i \times h_\theta(x_i)\times x_i)\\
&= -\frac{1}{N}\underset{\theta}{\operatorname{argmin}}\sum_{i=1}^{N}(y_i\times x_i - h_\theta(x_i)\times x_i)\\
&= -\frac{1}{N}\underset{\theta}{\operatorname{argmin}}\sum_{i=1}^{N}((y_i\times - h_\theta(x_i))\times x_i)\\
&= \frac{1}{N}\underset{\theta}{\operatorname{argmin}}\sum_{i=1}^{N}((h_\theta(x_i)-y_i)\times x_i)\ .
\end{aligned}
$$

对参数 b 进行求导可得

$$
\frac{\partial l(\theta \mid y_i;x_i)}{\partial b} = \frac{1}{N}\underset{\theta}{\operatorname{argmin}}\sum_{i=1}^{N}((h_\theta(x_i)-y_i))\times 1
$$

综上可知关于 θ 和 b 的偏导数的结果如下：

$$
\begin{aligned}
\frac{\partial l(\theta \mid y_i;x_i)}{\partial \theta} &= \frac{1}{N}\underset{\theta}{\operatorname{argmin}}\sum_{i=1}^{N}((h_\theta(x_i)-y_i)\times x_i)\\
\frac{\partial l(\theta \mid y_i;x_i)}{\partial b} &= \frac{1}{N}\underset{\theta}{\operatorname{argmin}}\sum_{i=1}^{N}((h_\theta(x_i)-y_i))
\end{aligned}
\tag{2-40}
$$

式(2-40)只是对 1 个参数 θ 和 b 进行求偏导，而逻辑回归有可能存在多个 θ 参数，并且在代码实现时一般会选择 NumPy 进行求解，故矩阵化求解的过程如下：

首先将 $\theta x_i + b$ 矩阵化，计算过程如图 2-37 所示。

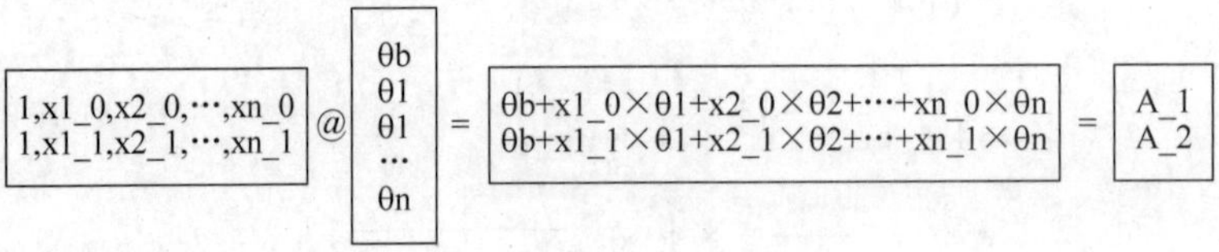

1,x1_0,x2_0,…,xn_0
1,x1_1,x2_1,…,xn_1
@
θb
θ1
θ1
…
θn
=
θb+x1_0×θ1+x2_0×θ2+…+xn_0×θn
θb+x1_1×θ1+x2_1×θ2+…+xn_1×θn
=
A_1
A_2

图 2-37　矩阵计算 $X@\Theta$

图 2-37 中 x1_0 表示第 0 行第 1 列、x2_1 表示第 1 行第 2 列，$\theta 1$ 表示与 x1 列相对应的参数进行内积运算后得 A_1 的结果，可以记为 $A=X@\Theta$。

根据式(2-40)中的 $h_\theta(x_i)-y_i$，其矩阵化计算过程如图 2-38 所示。

图 2-38 中 y1 为真实结果，e1 为 $h_\theta(x_i)-y_i$ 结果，可以记为 $E=H(A)-Y$。

根据式(2-38)中的 $(h_\theta(x_i)-y_i)\times x_i$，其矩阵化计算过程如图 2-39 所示。

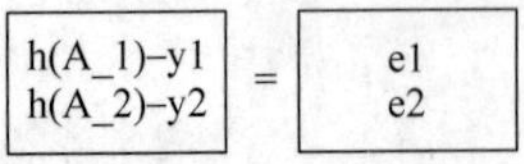

h(A_1)-y1
h(A_2)-y2
=
e1
e2

图 2-38　矩阵计算 $h_\theta(x_i)-y_i$

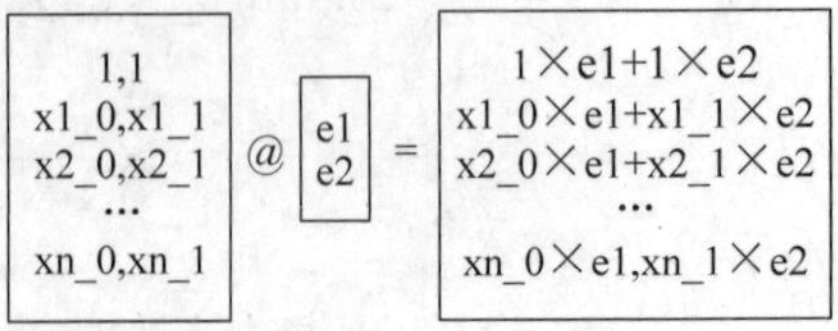

1,1
x1_0,x1_1
x2_0,x2_1
…
xn_0,xn_1
@
e1
e2
=
1×e1+1×e2
x1_0×e1+x1_1×e2
x2_0×e1+x2_1×e2
…
xn_0×e1,xn_1×e2

图 2-39　矩阵计算 $(h_\theta(x_i)-y_i)\times x_i$

将矩阵计算过程整理后可得公式：

$$\frac{\partial l(\theta \mid y_i;x_i)}{\partial \theta}=\frac{1}{N}\underset{\theta}{\operatorname{argmin}}\sum_{i=1}^{N}(\boldsymbol{X}^{\mathrm{T}}@(H(X@\Theta)-Y)) \tag{2-41}$$

综上，梯度下降求解逻辑回归的代码如下：

```
#第 2 章/machineLearn/逻辑回归梯度下降.py
import numpy as np
import matplotlib.pyplot as plt
import pandas as pd

#数据预处理
def data_preprocessing(path, frac=0.8):
    #读取数据
    data = pd.read_csv(path, encoding='gbk')
    #打乱数据并取 80%的内容
    data = data.sample(frac=frac).reset_index(drop=True)
    data.insert(0, '偏置项', 1)
    return data


#Sigmoid()函数
def sigmoid(z):
    return 1 / (1 + np.exp(-z))


#逻辑回归的损失函数
def cost_loss(X, Y, theta):
    #X 为样本,Y 为真实值
    #根据权重 theta 得到预测值
    pre_Y = sigmoid(np.matmul(X, theta))
    #损失计算可参考式(2-37)
    result = -np.sum(Y *np.log(pre_Y) +
                    (1 - Y) *np.log(1 - pre_Y)) / len(Y)
    return result
#根据获得的结果进行图形化验证
def verfiy_result(final_theta, theta2):
    x1 = np.arange(100, step=1)
    x2 = -(final_theta[0] + x1 *final_theta[1]) / final_theta[2]
    data = pd.read_csv('某学校录取结果.csv', encoding='gbk')
    #图形化展示其数据
    positive = data[data['录取结果'] == 1]      #1
    negetive = data[data['录取结果'] == 0]      #0
    fig, ax = plt.subplots(figsize=(8, 5))
    ax.scatter(positive['语文'], positive['数学'], label='未录取分数')
    ax.scatter(negetive['语文'], negetive['数学'], label='录取分数')
    ax.plot(x1, x2, label='最后决策边界')
    x3 = -(theta2[0] + x1 *theta2[1]) / theta2[2]
    ax.plot(x1, x3, label='中间决策边界')
    ax.set_xlabel('语文')
    ax.set_ylabel('数学')
    ax.set_title('某学校录取结果参数拟合过程')
    plt.legend(loc='upper right')
    plt.show()
```

```
if __name__ == "__main__":
    data = data_preprocessing('某学校录取结果.csv')
    X = data.iloc[:, :3].values
    Y = np.reshape(data.iloc[:, -1].values, [-1, 1])
    #设置学习率
    alpha = 1e-5
    #设置迭代次数
    epoch = 250000
    #初始权重
    theta = np.zeros([3, 1])
    loss_list = []
    theta_list = []
    for i in range(epoch):
        gradient = np.matmul(X.T, sigmoid(np.matmul(X, theta)) - Y)
        #梯度下降
        theta = theta - alpha *gradient
        #求损失
        loss = cost_loss(X, Y, theta)
        if i % 20 == 0 or i == 99:
            print('第{}次,损失={}'.format(i + 1, loss))
        loss_list.append(loss)
        theta_list.append(theta)
    #对 loss 进行图像绘制
    describe_plt_loss(epoch, loss_list)
    #某学生的考试成绩
    pre_x = np.array([[1, 80, 60]])
    #获取最后 1 次权重值来预测学生的录取结果
    result = sigmoid(np.matmul(pre_x, theta))
    #阈值自定义
    threshold = 0.5
    print("该学生被录取") if np.squeeze(result) > threshold else print('该学生未
被录取')
    #图形化验证最后 1 次的参数的分隔结果
    print('last_theta:', theta)
    verfiy_result(theta, theta_list[int(len(theta_list) *0.3)])
```

运行结果如下：

```
第 1 次,损失=0.6626861294398477
第 21 次,损失=0.6361135074795172
第 41 次,损失=0.635851561777337
第 61 次,损失=0.635755784494316
第 81 次,损失=0.6356720862541302
第 100 次,损失=0.6355934288822633
该学生被录取
```

其损失曲线如图 2-40 所示。

设置超参数 epoch＝250 000，并将最后得到的参数 theta 进行图形化展示，可观察到 theta＝[－7.716 896 8，0.071 046 23，0.057 944 98]能较好地拟合样本数据，但是并不能将所有的样本区分开，有能够被录取的分数被划分为未被录取，如图 2-41 所示。

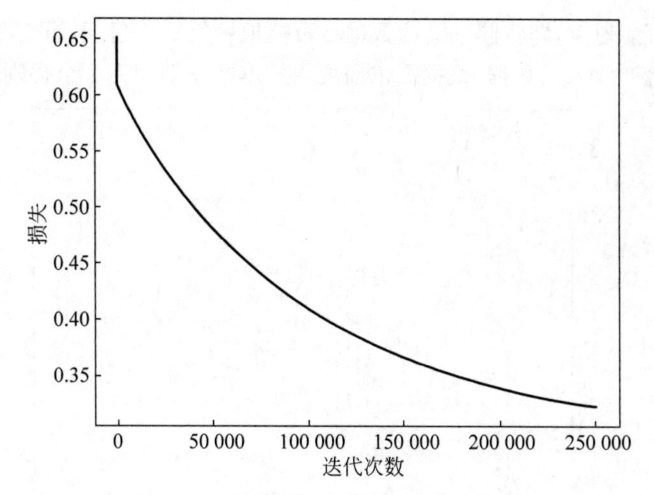

图 2-40 逻辑回归迭代次数与损失变化

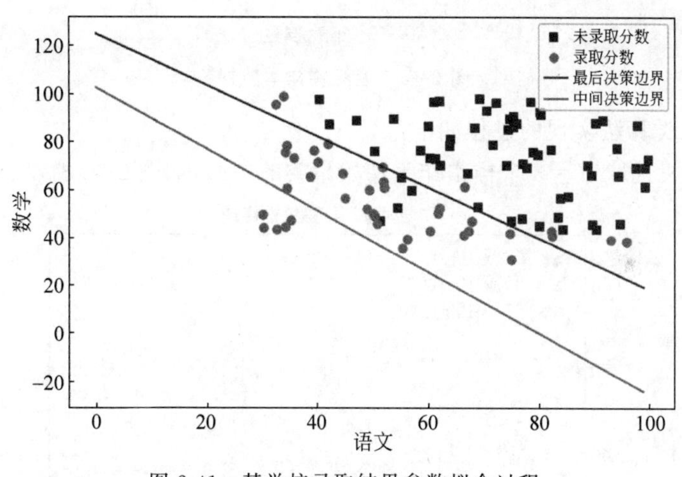

图 2-41 某学校录取结果参数拟合过程

总结

基于最大似然函数可推导出逻辑回归的损失函数。逻辑回归的分类,本质上是一个概率的分布。

练习

重写并调试逻辑回归的代码,并使用不同的学习率和初始权重值,观察逻辑回归拟合情况的变化。

2.5 聚类算法

聚类(K-Means)算法是一种无监督学习算法(只有样本数据没有标签),主要用于将相似的样本数据分割成不同的类或簇,使 1 个簇内的样本数据相似度尽可能大,使不同簇中的样本数据差异性也尽可能地大，即将同一类的数据尽可能地聚集到一起，不同类别数据尽量分离。

取“表 2-3”年龄、收入的数据，根据需求需要将样本数据划分为 3 个类别，如图 2-42 所示。

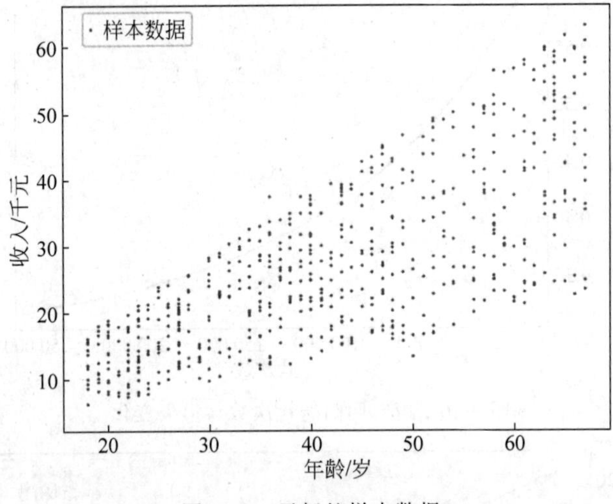

图 2-42　无标签样本数据

那么使用聚类算法实现的过程如下：

(1) 假设随机选取 $K=3$ 个样本作为初始聚类的中心，如图 2-43 所示。

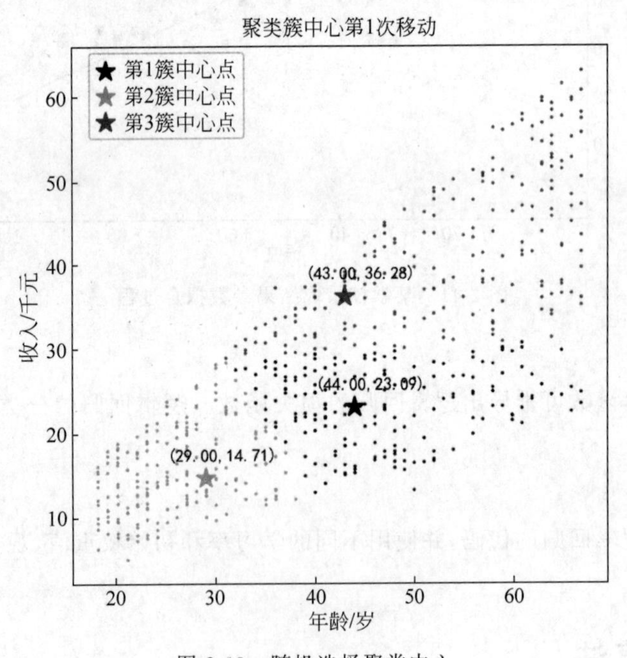

图 2-43　随机选择聚类中心

(2) 将每个样本到聚类中心做欧氏距离，哪个样本距离中心点最近就将该样本数据标记为某个类别，计算距离可参考欧氏距离的公式：

$$d(x,y)=\sqrt{\sum_{i=1}^{n}(x_i-y_i)^2} \tag{2-42}$$

经过计算归类后得到类别数据，如表 2-5 所示。

表 2-5　每个类别最近距离数据

类别(*K*)	数据 1	数据 2	数据 3	数据 *n*
0	48.,17.546	40.,30.0851	51.,16.5754	38.,26.6716
1	23.,20.3754	22.,8.87707	…	…
2	57.,50.5763	57.,37.8696	…	…

$K=0$ 代表第 1 个聚类中的数据包含[48.,17.546　40.,30.0851　51.,16.5754　38.,26.6716]，以此类推，每个类别都存储了离中心点最新的样本。

(3) 对每个类别中的数据求平均值，并作为新的聚类中心，如表 2-6 所示。

表 2-6　每个类别最近距离数据

类别(*K*)	第 1 次迭代中心	第 7 次迭代中心
0	43.00,36.28	53.36,41.76
1	44.00,23.09	47.00,23.37
2	29.00.,14.71	25.94.,15.67

观察图 2-44 可发现聚类中心点相对于图 2-43 已发生变化。

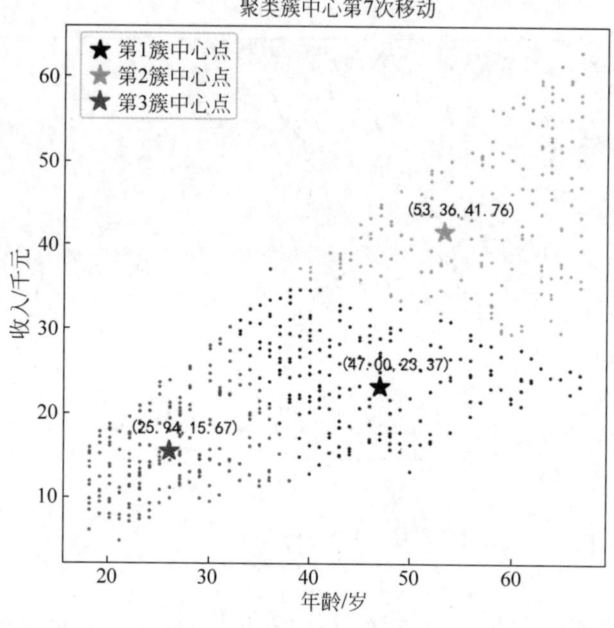

图 2-44　重新选择聚类中心

(4) 再次重复步骤 2，如果新的中心点与原中心点一样，则聚类过程结束。

由于步骤(1)中每次随机到的聚类中心点不同，所以需要进行中心点移动的次数也不同，最后得到的中心点的值也会有变化，如图 2-45 所示。

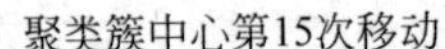

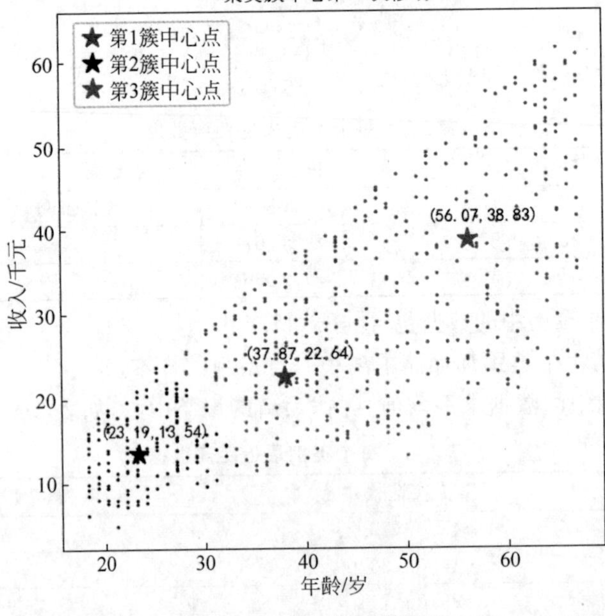

图 2-45　最后 1 次聚类中心

将以上过程使用代码来实现，代码如下：

```
#第 2 章/machineLearn/K_means 聚类.py
import numpy as np
import matplotlib.pyplot as plt
import pandas as pd
import random


#初始随机获取 K 个中心
def random_center(data: np, k: int):
    return np.array(random.choices(data, k=k))


#计算距离，欧氏距离
def calculate_european_distance(cluster_data, cluster_center: np):
    return np.sqrt(np.sum(np.power(cluster_data - cluster_center, 2)))


#根据每个簇的数据计算平均值，获得新的中心点
def calculate_new_center(cluster_new_data: dict):
    #初始新中心点
    return np.array([np.mean(cluster, axis=0) for _, cluster in cluster_new_
data.items()])


#聚类算法
def k_means(cluster_data: np, cluster_center: np):
```

```
    cluster_new_data = {}
    #簇分类标记:哪条数据与簇中心点最新,这条数据就属于哪个簇
    cluster_key = 0
    for cluster in cluster_data:
        #np.inf 表示+∞,没有确切的数值,类型为浮点型
        min_distance = float('inf')
        #本次 for 循环的目的是获取当前数据跟哪个簇的中心点最近
        for i in range(cluster_center.shape[0]):
            #获得某个簇的中心点
            center = cluster_center[i]
            #对每个数据与簇中心点进行距离计算
            european_distance = calculate_european_distance(cluster, center)
            #对本次距离与上一次的距离进行比较,如果本次小,则相应地对 flag 类别进行改变
            if european_distance < min_distance:
                min_distance = european_distance
                cluster_key = i
        #如果 flag 没有在字典中,则创建一个类似{1:[]}的字典
        if cluster_key not in cluster_new_data.keys():
            cluster_new_data[cluster_key] = list()
        #将本次比较数据放入簇中
        cluster_new_data[cluster_key].append(cluster)
    return cluster_new_data


#计算所有数据距离最新簇中心点的距离之和
def sum_distances_center(cluster_new_data: dict, cluster_center: np):
    loss = 0
    for i, cluster in cluster_new_data.items():
        #累计簇中的数据距离中心的距离之和
        cluster_loss = 0
        for row in cluster:
            cluster_loss += calculate_european_distance(row, cluster_center[i])
        #计算所有簇的距离
        loss += cluster_loss
return loss

#中心点距离所有距离的平方和
def plt_evaluation_distance(data):
    plt.plot(range(len(data)), data, '-', color='r', label='距离变化')
    plt.title('聚类中心离所有样本的距离变化')
    plt.legend()
    plt.xlabel('迭代次数')
    plt.ylabel('距离')
plt.show()

#读取收入数据
def describe_plt_data():
    data = pd.read_csv("多元线性回归数据.csv", encoding='gbk')
    data['收入'] = data['收入'] / 1000
    return data[['年龄', '收入']].to_numpy()


if __name__ == "__main__":
    #plt_original_data()
```

```
    k = 3
    #读取年龄和收入数据
    data = describe_plt_data()
    #初始簇中心
    init_center = random_center(data, k)
    #根据初始的簇中心,将数据分配到簇中
    cluster_new_data = k_means(data, init_center)
    #计算分配簇中的数据与中心点之间的距离之和
    sum_distances = sum_distances_center(cluster_new_data, init_center)
    #记录簇中心的变化
    old_distances = 0
    evaluation_distance_list = []
    evaluation_distance_list.append(sum_distances)
    #当上一次中心点与本次中心点没有变化时,停止移动中心点
    while abs(sum_distances - old_distances) > 0.00001:
        #得到新的中心点
        new_center = calculate_new_center(cluster_new_data)
        #根据新的中心点分配簇中的数据
        cluster_new_data = k_means(data, new_center)
        old_distances = sum_distances
        #新的中心点与簇中数据的距离
        sum_distances = sum_distances_center(cluster_new_data, new_center)
        #记录簇中心值及簇中数据的变化
        evaluation_distance_list.append(sum_distances)
    #查看聚类距离的变化
plt_evaluation_distance(evaluation_distance_list)
print('最后的中心点={}'.format(new_center))
```

运行结果如下：

```
最后的中心点=[[44.2601626  25.99936707]
 [25.68586387  15.44264728]
 [59.1595092  43.98177178]]
```

从图 2-46 可以发现,随着迭代次数的增加,聚类中心点离样本数据的距离在减小。

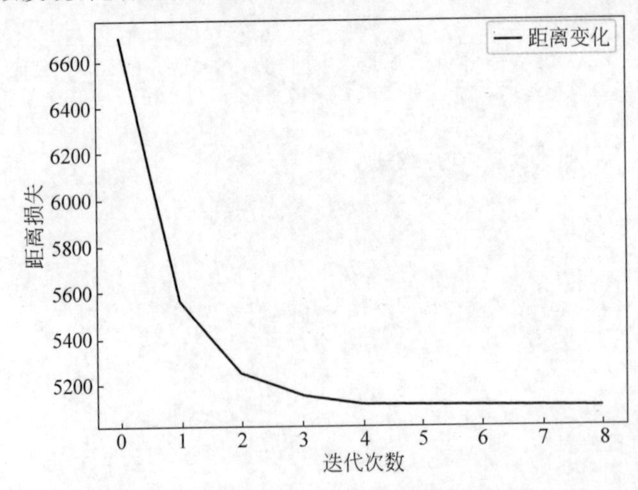

图 2-46　聚类中心离所有样本的距离变化

注意：由于每次都需要计算所有样本与聚类中心之间的距离，故在大规模数据集上 K-Means 算法的收敛速度较慢；如果 K 值不同，并且第 1 次随机聚类中心点的值不同，则最后迭代次数及得到的聚类中心点也可能不同。

总结

K-Means 计算的是某个簇内最小平均距离（离簇内数据都最近），而 KNN 是求 K 个最近距离数据的分类投票。一个是无监督学习，另一个是有监督学习。

练习

重写并调试 K-Means 代码，观察簇中心点的变化。

2.6 神经网络

2.6.1 什么是神经网络

1904 年生物学家确定了神经元的结构如图 2-47 所示。

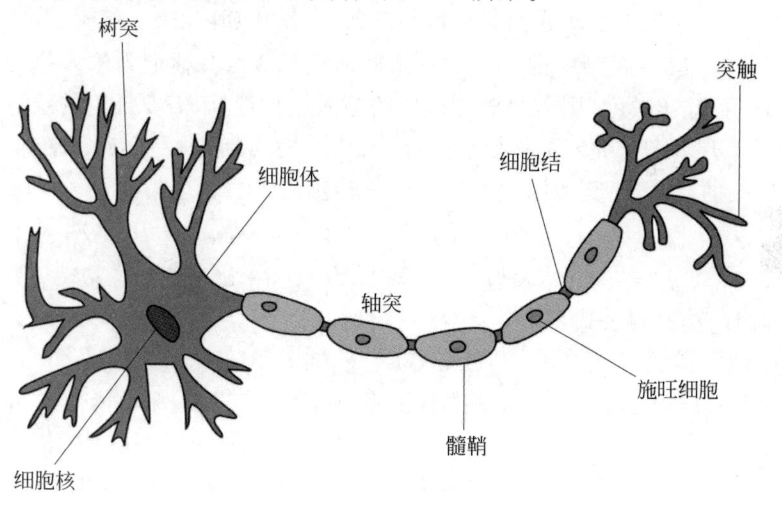

图 2-47 神经元结构

从生物学基础可知，信息输入树突后经过细胞体的加工处理，然后经过轴突传输到突触进行信息输出。树突可能会有多个，但是轴突只有 1 条。突触的主要作用是跟其他神经元的树突连接从而传递信号。

1943 年 Warren S. McCulloch 和 Walter Pitts 参考了生物神经元的结构，发表了抽象的神经元模型 M-P，其结构如图 2-48 所示。

输入 x 可以类比为神经元的树突，而输出可以类比为神经元的轴突，信息传播与处理的过程可以类比为细胞核。由于生物神经元具有不同的突触性质和突触强度，所以对神经元的影响不同，我们用 θ_i 权值来表示，其正负值模拟了生物神经元中突触的兴奋和抑制，其大小则代表了突触的不同连接强度。将其转换为数学表达式为 $y=\mathrm{sga}(x_1\times\theta_1+x_2\times\theta_2+x_3\times\theta_3)$。

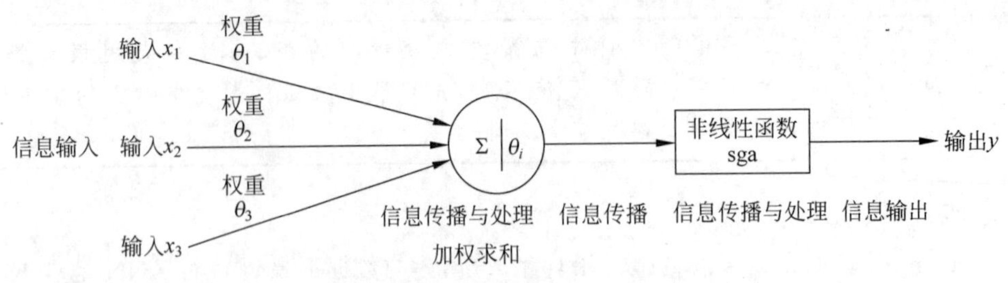

图 2-48 神经元模型 M-P 结构

从图 2-48 中可知，输出 y 是在输入 x 和权重 θ 的线性相加叠加后函数 sga 的值，当输入大于 0 时 sga 函数输出 1，否则输出 0。已知样本 x 称为特征，未知的结果 y 称为目标，sga 函数称为激活函数，只要找到特征 x 与目标 y 之间的线性关系 θ，神经元模型就可以预测出新样本的 y 值。

虽然 1943 年发布的 M-P 模型建立了神经网络大厦的地基，但是 M-P 模型中权重的值 θ 都是预先设置的，并不能通过学习获得。

1958 年心理学家 Rosenblatt 提出了由两层神经元组成的神经网络，称为感知机。感知机算法是首个可以学习到参数 θ 的人工神经网络，其结构如图 2-49 所示。

图 2-49 中，对 M-P 模型中的输入位置添加神经元节点并标记为输入单元，输入单元只负责传输数据，而输出单元，则需要对上一层的输入进行计算，需要计算的层称为计算层，如果只有一个计算层，则称为单层神经网络，其中 $\theta_{1\_1\_2}$ 表示后 1 层第 1 个神经元与上 1 层第 2 个神经元连接的权重值，其他以此类推。将其转换为数学表达式：

$$y_1 = g(a_{1\_1} \times \theta_{1\_1\_1} + a_{1\_2} \times \theta_{1\_1\_2} + a_{1\_3} \times \theta_{1\_1\_3})$$

$$y_2 = g(a_{1\_1} \times \theta_{1\_2\_1} + a_{1\_2} \times \theta_{1\_2\_2} + a_{1\_3} \times \theta_{1\_2\_3})$$

使用矩阵计算，则得公式：

$$Y = G(A_i @ \Theta_i) \tag{2-43}$$

输入 a 用矩阵 A_i 来表示，参数 θ 的转置用矩阵 $\boldsymbol{\Theta}_i$ 来表示，激活函数 g_i 用矩阵 $\boldsymbol{G}$ 来表示，则输出 y_i 用矩阵 $\boldsymbol{Y}$ 来表示，这个公式就是神经网络中从前一层计算后一层的矩阵运算，又称为前向传播。在此基础上 2 层或者多层神经网络的结构图，如图 2-50 所示。

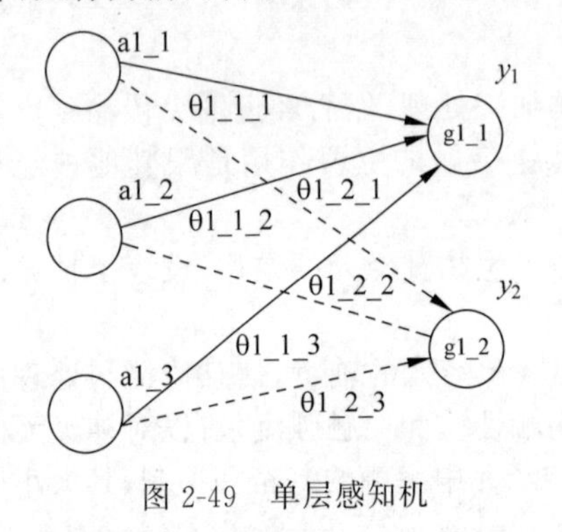

图 2-49 单层感知机

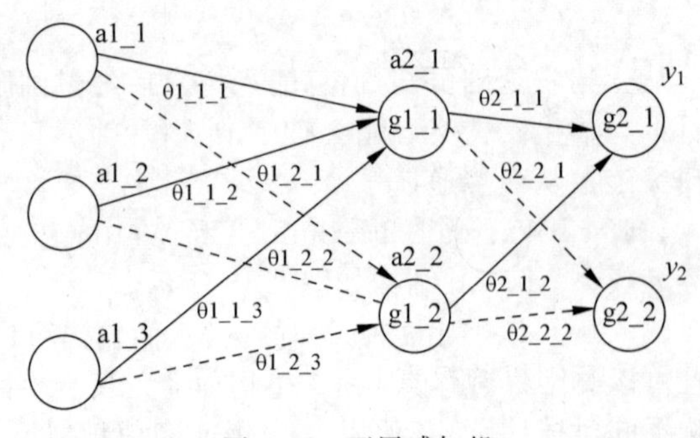

图 2-50 两层感知机

两层神经网络除了包含 1 个输入层和 1 个输出层以外，还增加了 1 个中间层（又称隐藏层），$a2\_1$ 表示第 2 层第 1 个神经元，其结果为 g1_1 计算的输出，并通过权重 $\theta_{2\_1\_1}$ 和 $\theta_{2\_1\_2}$ 连接并得到输出 g2_1。将其转换为数学表达为

$$
\begin{aligned}
&g_{1\_1}=g(a_{1\_1}\times\theta_{1\_1\_1}+a_{1\_2}\times\theta_{1\_1\_2})\\
&g_{1\_2}=g(a_{1\_1}\times\theta_{1\_2\_1}+a_{1\_2}\times\theta_{1\_2\_2})\\
&a_{2\_1}=g_{1\_1}\\
&a_{2\_2}=g_{1\_2}\\
&a_{2\_1}=g(a_{2\_1}\times\theta_{2\_1\_1}+a_{2\_2}\times\theta_{2\_1\_2})\\
&a_{2\_2}=g(a_{2\_1}\times\theta_{2\_2\_1}+a_{2\_2}\times\theta_{2\_2\_2})
\end{aligned}
$$

使用矩阵计算，则得公式：

$$
Y=G(G(A_i@\Theta_i)@\Theta_{i+1}) \tag{2-44}
$$

使用矩阵计算上一个节点的输出作为下一个节点的输入，并且增加了新的权重 Θ_{i+1}，然后经过计算得到新的输出 G_2，也就是预测值 Y。这里的 g 函数使用的是 Sigmoid 函数。

不管是 M-P 模型，还是感知机都没有提到偏置节点，但这些节点是默认存在的。在神经网络的每层中，除输出层都含有这样 1 个偏置单元。正如线性回归模型与逻辑回归模型中的一样，如图 2-51 所示。

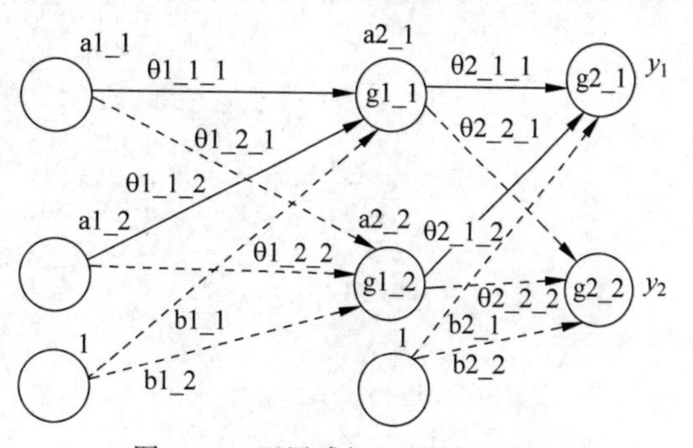

图 2-51　两层感知机增加偏置项

从图 2-51 可知，增加偏置项只需在每层中增加 1 个内容为 1 的神经元，其网络模型的权重参数为 2×2+2×2=8 个，算上偏置项后为 8+4=12 个。

如果是三层感知机，则其网络模型结构图如图 2-52 所示。

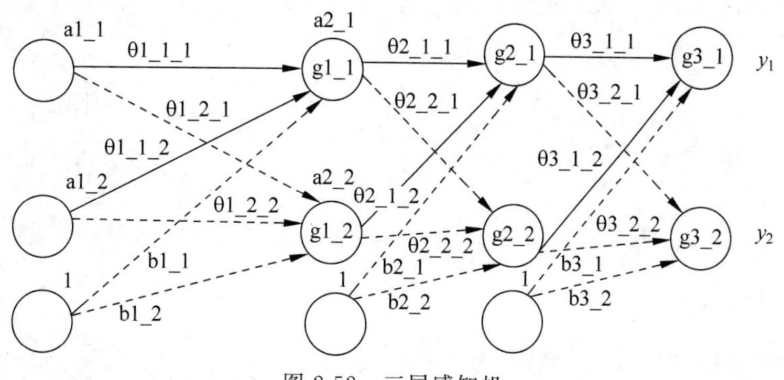

图 2-52　三层感知机

三层感知机的权重参数量为 2×2+2×2+2×2=12 个，外加偏置项为 6 个，共计 18 个。以此类推计算神经网络的权重参数量可使用以下公式。

$$\text{Total}_\theta = \sum_{i=1}^{n} M_i \times M_{i+1} + M_{i+1} \tag{2-45}$$

M_i 层神经元的个数与后一层 M_{i+1} 神经元的个数相乘，再加上 M_{i+1} 神经元个数的偏置项就是权重参数的个数。

权重参数的个数随着网络深度（层数）的增加，其网络的表达能力也在增加，假设输入的是 1 只猫的图像，第 1 个隐藏层可能学到的是猫的轮廓，也就是边缘特征，第 2 个隐藏层学到的可能是猫脸的特征，最后输出的是组成猫这个目标的特征，神经网络通过抽取更抽象的特征来对事物进行区分，从而获得更好的分类能力。

感知机算法中对于权重参数 Θ_i 的学习是通过反向传播算法实现的，其原理将会在后面章节讲解，先来看其前向传播的代码实现，代码如下：

```
#第 2 章/machineLearn/神经网络前向传播.py
import numpy as np

#Sigmoid()函数
def sigmoid(z):
    return 1 / (1 + np.exp(-z))

if __name__ == "__main__":
    #模拟 x 的值，当前只有 1 列，故为 x1
    X = np.array([[2.], [3.]])
    #插入偏置项后成为输入 a1 层，变为 3 个神经元
    a1 = np.insert(X, 0, 1, axis=1)
    #模拟输入层参数 theta1，其参数的个数与输入神经元的个数相等
    theta1 = np.array([[0.01], [0.02]])
    #输入与参数 theta1 进行线性计算
    a1_g1 = a1 @theta1
    #隐藏层 1，进行激活函数输出
    g2 = sigmoid(a1_g1)
    print('第 1 个层的输出变为下一层的输入:', a1_g1, g2)
    #模拟输出层参数 theta2，增加 1 个偏置项
    a2 = np.insert(g2, 0, 1, axis=1)
    #参数的个数与 a2 的神经元的个数相同
    theta2 = np.array([[0.03], [0.04]])
    #输入与参数 theta2 进行线性计算
    a2_g2 = a2 @theta2
    #输出层，进行激活函数输出
    g2 = sigmoid(a2_g2)
    print('最后输出层:', a2_g2, g2)
```

运行结果如下：

```
第 1 个层的输出变为下一层的输入: [[0.05]
 [0.07]] [[0.5124974 ]
 [0.51749286]]
最后输出层: [[0.0504999 ]
 [0.05069971]] [[0.51262229]
 [0.51267221]]
```

因为 X 的维度是(2,1)，值为[[2.],[3.]]，所以实际上在代码中仅代表有一个神经元，但是计算了两次。插入偏置项后其维度变为(2,2)，值为[[1. 2.],[1. 3.]]，由于每列对应 1 个参数 theta，所以需要两个 theta，其计算过程如图 2-53 所示。

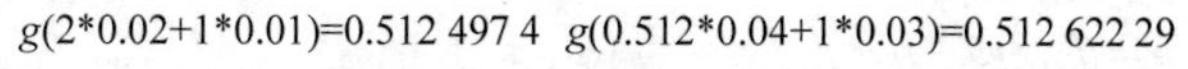

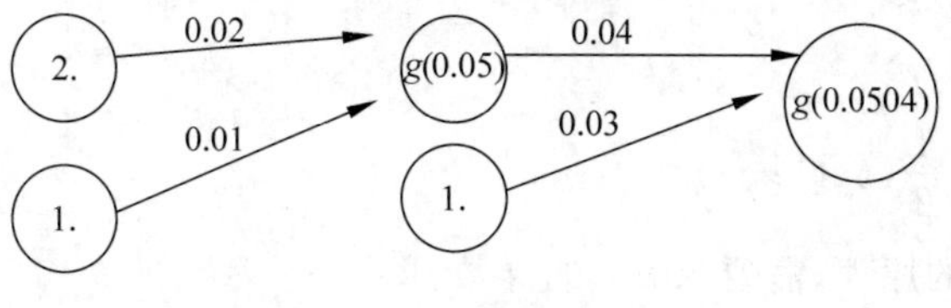

g(3*0.02+1*0.01)=0.517 492 86 g(0.517*0.04+1*0.03)=0.512 672 21

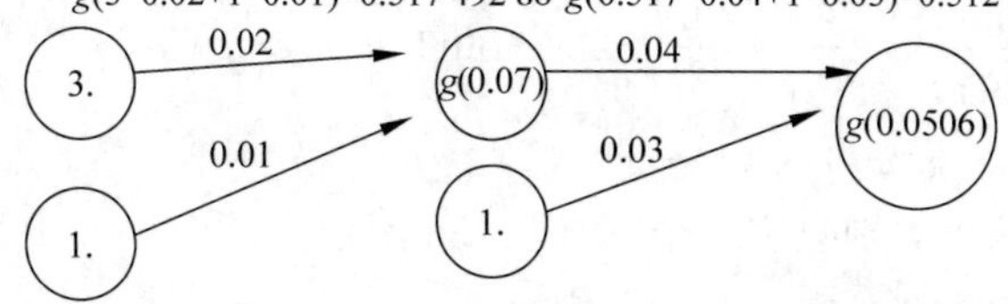

图 2-53　前向传播代码计算过程

2.6.2　反向传播算法

在 2.6.1 节中前向传播输出值是预测值 $\hat{y}$，根据 2.4 节逻辑回归的内容可知，模型函数的目标是使预测值与真实值之间的损失尽可能地小，因此神经网络可以转换为最优化问题，即“寻找最优参数 θ，使预测值 $\hat{y}$ 与真实值 y 之间的损失最小”。

根据前文可知，最优化问题结合梯度下降算法可以获得参数 θ，使损失函数最小，但是神经网络模型结构复杂，每次计算所有参数的梯度代价是昂贵的，因此还需要使用反向传播神经网络算法(Back Propagation Neural Network，BP)来计算梯度。反向传播算法不会一次计算所有参数的梯度，而是从网络的输出层反向逐层向前求梯度，即首先计算输出层的梯度，再计算倒数第 2 层的梯度，然后计算中间隐藏层的梯度，最后计算输入层的梯度，这样一层一层反向传播，直到获得所有网络层的梯度，其计算过程可参考图 2-54。

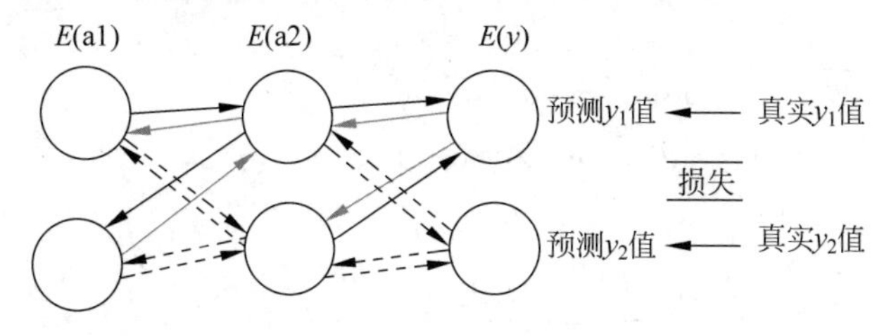

图 2-54　反向传播

其求解过程，可以总结为“一次前向传播，一次反向传播求导，然后由梯度下降更新参数”。图中 E 代表求偏导，正向实心箭头为前向传播，反向空心箭头为反向传播。

设损失函数为 $L(\theta)=\dfrac{1}{2N}\sum_{i=1}^{N}(\hat{y}_i-y_i)^2$，只有一个神经元且没有隐藏层，输出只有一个类别，如图 2-55 所示。

其前向传播过程如下：

设 $g(x_i)=\dfrac{1}{1+e^{-(\theta x_i+b)}}$

则连接过程 $a_1=\theta_{11}x_1+b_1$

激活函数 $\text{out1}=g(a_1)$

则损失函数 $L(\theta)=\dfrac{1}{2}(\text{out1}-y)^2$

根据反向传播算法的原理，需要求 θ_{11} 的导数，即

$$\frac{\partial L(\theta)}{\partial\theta_{11}}=\frac{\partial L(\theta)}{\partial \text{out1}}\times\frac{\partial \text{out1}}{\partial a_1}\times\frac{\partial a_1}{\partial\theta_{11}}=\frac{1}{2}\times 2\times(\text{out1}-y)\times g(a_1)\times(1-g(a_1))\times x_1$$
$$=(\text{out1}-y)\times g(a_1)\times(1-g(a_1))\times x_1$$

增加 1 个隐藏层，输出类别只有一个，如图 2-56 所示。

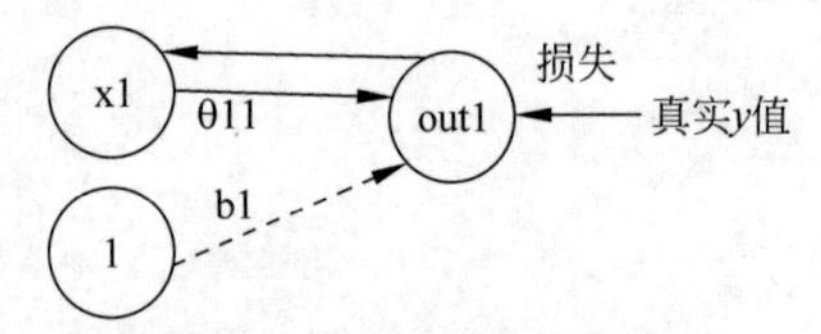

图 2-55　单个神经元没有隐藏层

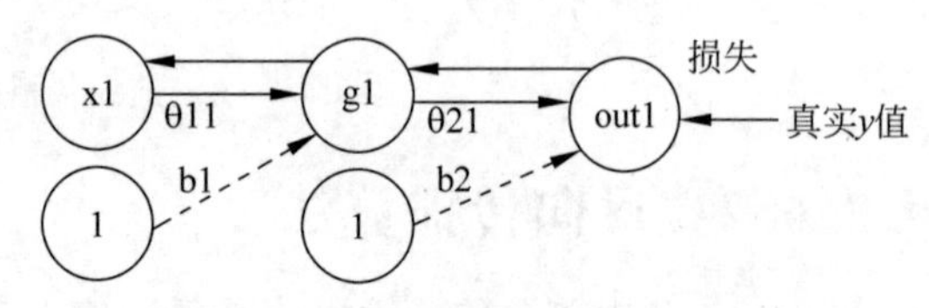

图 2-56　单个神经元有一个隐藏层

其前向传播过程如下：

连接过程 1，$a_1=\theta_{11}x_1+b_1$

激活函数 1，$\text{out}_{a_1}=g(\text{net}_1)$

连接过程 2，$a_2=\text{out}_{g_1}\times\theta_{21}+b_2$

激活函数 2，$\text{out1}=g(a_2)$

损失函数 $L(\theta)=\dfrac{1}{2}(\text{out1}-y)^2$，根据反向传播算法的原理，需要先求出 θ_{21} 的导数：

$$\frac{\partial L(\theta)}{\partial\theta_{21}}=\frac{\partial L(\theta)}{\partial \text{out1}}\times\frac{\partial \text{out1}}{\partial a_2}\times\frac{\partial a_2}{\partial\theta_{21}}$$
$$=(\text{out1}-y)\times g(a_2)\times(1-g(a_2))\times \text{out}_{a_1}$$

然后求出 θ_{11} 的导数：

$$\frac{\partial L(\theta)}{\partial\theta_{11}}=\frac{\partial L(\theta)}{\partial \text{out}_{a_1}}*\frac{\partial \text{out}_{a_1}}{\partial a_1}*\frac{\partial a_1}{\partial\theta_{11}}$$

从上式可知 θ_{11} 的导数受 out_{g_1} 的影响，所以需要先求出$\dfrac{\partial L(\theta)}{\partial \text{out}_{a_1}}$

因

$$\frac{\partial L(\theta)}{\partial \text{out}_{g_1}}=\frac{\partial L(\theta)}{\partial \text{out1}}\times\frac{\partial \text{out1}}{\partial a_2}\times\frac{\partial a_2}{\partial \text{out}_{a_1}}$$

故

$$\frac{\partial L(\theta)}{\partial \theta_{11}}=\frac{\partial L(\theta)}{\partial \text{out1}}\times\frac{\partial \text{out1}}{\partial a_2}\times\frac{\partial a_2}{\partial \text{out}_{a_1}}\times\frac{\partial \text{out}_{g_1}}{\partial \text{net}_1}\times\frac{\partial \text{net}_1}{\partial \theta_{21}}$$

$$=(\text{out1}-y)\times g(a_2)\times(1-g(a_2))\times\theta_{21}\times g(a_1)\times(1-g(a_1))\times x_1$$

增加 1 个神经元和 1 个隐藏层,输出为两个分类,如图 2-57 所示。

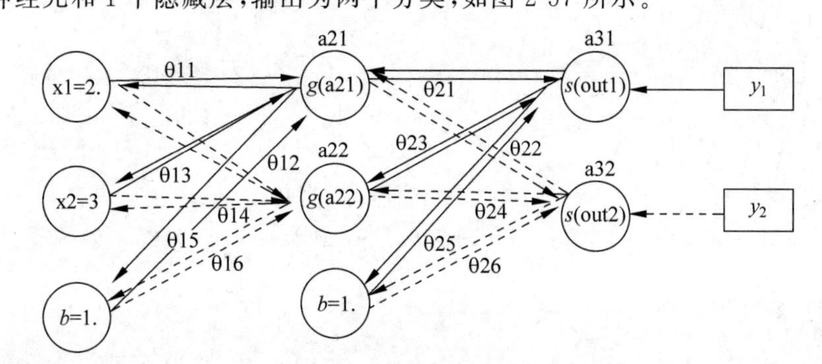

图 2-57 多个神经元有一个隐藏层

其前向传播过程如下:

连接过程 1_1,$a_{21}=\theta_{11}x_1+\theta_{13}x_2+b*\theta_{15}$

激活函数 1_1,$\text{out}_{a21}=g(a_{21})$

连接过程 1_2,$a_{22}=\theta_{12}x_1+\theta_{14}x_2+b*\theta_{16}$

激活函数 1_2,$\text{out}_{a22}=g(a_{22})$

连接过程 2_1,$a_{31}=\text{out}_{a21}\times\theta_{21}+\text{out}_{a22}\times\theta_{23}+b*\theta_{25}$

激活函数 2_1,$\text{out1}=g(a_{31})$

连接过程 2_2,$a_{32}=\text{out}_{a21}\times\theta_{22}+\text{out}_{a22}\times\theta_{24}+b\times\theta_{26}$

激活函数 2_2,$\text{out2}=g(a_{32})$

损失函数就变为 $L(\theta)=\frac{1}{2}((\text{out1}-y1)^2+(\text{out2}-y2)^2)$,根据反向传播算法的原理,先求隐藏层的梯度 θ_{21}:

$$\frac{\partial L(\theta)}{\partial \theta_{21}}=\frac{\partial L(\theta)}{\partial \text{out1}}\times\frac{\partial \text{out1}}{\partial a_{31}}\times\frac{\partial a_{31}}{\partial \theta_{21}}=(\text{out1}-y1)\times \text{out1}\times(1-\text{out1})*\text{out}_{a21}$$

以此类推可得

$$\frac{\partial L(\theta)}{\partial \theta_{23}}=(\text{out1}-y1)\times \text{out1}\times(1-\text{out1})*\text{out}_{a22}$$

$$\frac{\partial L(\theta)}{\partial \theta_{25}}=(\text{out1}-y1)\times \text{out1}\times(1-\text{out1})\times 1$$

$$\frac{\partial L(\theta)}{\partial \theta_{22}}=(\text{out2}-y2)\times \text{out2}\times(1-\text{out2})\times \text{out}_{a21}$$

$$\frac{\partial L(\theta)}{\partial \theta_{24}}=(\text{out2}-y2)\times \text{out2}\times(1-\text{out2})\times \text{out}_{a22}$$

$$\frac{\partial L(\theta)}{\partial \theta_{26}}=(\text{out2}-y2)\times \text{out2}\times(1-\text{out2})*1$$

然后从隐藏层反向求输入层 θ_{11} 的导数：

$$\frac{\partial L(\theta)}{\partial \theta_{11}}=\left(\frac{\partial L(\theta)}{\partial \text{out1}}\times\frac{\partial \text{out1}}{\partial a_{31}}\times\frac{\partial a_{31}}{\partial \text{out}_{a21}}+\frac{\partial L(\theta)}{\partial \text{out2}}\times\frac{\partial \text{out2}}{\partial a_{32}}\times\frac{\partial a_{21}}{\partial \text{out}_{a21}}\right)\times\frac{\partial \text{out}_{a21}}{\partial a_{21}}\times\frac{\partial a_{21}}{\partial \theta_{11}}$$

代入后 θ_{11} 的梯度为

$$\frac{\partial L(\theta)}{\partial \theta_{11}}=((\text{out1}-y1)\times \text{out1}\times(1-\text{out1})\times\theta_{21}+(\text{out2}-y2)\times \text{out2}\times(1-\text{out2})\times\theta_{22})\times \text{out}_{a21}\times(1-\text{out}_{a21})\times x_1$$

以此类推可得

$$\frac{\partial L(\theta)}{\partial \theta_{13}}=((\text{out1}-y1)\times \text{out1}\times(1-\text{out1})\times\theta_{21}+(\text{out2}-y2)\times \text{out2}\times(1-\text{out2})\times\theta_{22})\times \text{out}_{a21}\times(1-\text{out}_{a21})\times x_2$$

$$\frac{\partial L(\theta)}{\partial \theta_{15}}=((\text{out1}-y1)\times \text{out1}\times(1-\text{out1})\times\theta_{21}+(\text{out2}-y2)\times \text{out2}\times(1-\text{out2})\times\theta_{22})\times \text{out}_{a21}\times(1-\text{out}_{a21})\times 1$$

$$\frac{\partial L(\theta)}{\partial \theta_{12}}=((\text{out1}-y1)*\text{out1}*(1-\text{out1})*\theta_{23}+(\text{out2}-y2)*\text{out2}*(1-\text{out2})*\theta_{24})\times \text{out}_{a22}\times(1-\text{out}_{a22})\times x_1$$

$$\frac{\partial L(\theta)}{\partial \theta_{14}}=((\text{out1}-y1)*\text{out1}*(1-\text{out1})*\theta_{23}+(\text{out2}-y2)*\text{out2}*(1-\text{out2})*\theta_{24})\times \text{out}_{a22}\times(1-\text{out}_{a22})\times x_2$$

$$\frac{\partial L(\theta)}{\partial \theta_{16}}=((\text{out1}-y1)*\text{out1}*(1-\text{out1})*\theta_{23}+(\text{out2}-y2)*\text{out2}*(1-\text{out2})*\theta_{24})\times \text{out}_{a22}\times(1-\text{out}_{a22})\times 1$$

可以令 $\delta_i=\hat{y}_i-y_i$，则隐藏层的梯度可以写成

$$\theta_{\delta_i}^2=\delta_i\times\nabla g(\text{out}_i^3)\times \text{out}_i^2$$

则输入层的梯度可以写成

$$\theta_{\delta_i}^1=\left(\sum_{i=1}^{n}\delta_i*\nabla g(\text{out}_i^3)*\theta_i^2\right)\times\nabla g(\text{out}_i^2)\times \text{out}_i^1$$

θ_i^2 表示第 2 层所有的参数，其他以此类推。从上面的推导可知，反向传播求梯度会将输出层的误差 δ_i 传递到隐藏层，然后隐藏层再次将误差 δ_i 传递到输入层，输入层的梯度受上一层所有神经元的影响（相加）。

反向传播过程实际上就是复合函数链式求导的过程，网络的隐藏层越多，梯度连乘也就越多，求梯度也就难一些。好消息是使用深度学习编程框架提供的自动求导功能，可以简化程序的编写，这部分内容将在第 3 章进行介绍。

根据推导公式，使用 NumPy 来完成反向传播和梯度下降的代码如下：

```
#第 2 章/machineLearn/神经网络反向传播.py
import numpy as np

#Sigmoid()函数
def sigmoid(z):
    return 1 / (1 + np.exp(-z))

if __name__ == "__main__":
    #输入 x 的值
    ga = np.array([[2.], [3.]])
    #初始权重值
    init_w = {
        'w1': np.array([[0.03, 0.02, 0.01], [0.013, 0.012, 0.001]]),
        'w2': np.array([[0.03, 0.02, 0.01], [0.013, 0.012, 0.001]])
    }
    #真实值
    y = np.array([[1.], [0.]])
    #学习率
    alpha = 0.001
    #存储每次输出+偏置项
    l_ga = []
    #存储每次权重值
    l_w = []
    #存储每次非线性输出
    l_out = []
    #进行两次前向传播
    for i in range(1, 3):
        #增加偏置项
        ga = np.append(ga, np.array([[1]]), axis=0)
        l_ga.append(ga)
        #获取权重
        w = init_w['w%s' % i]
        l_w.append(w)
        #线性计算
        a = w @ga
        #非线性计算
        out = sigmoid(a)
        ga = out
        l_out.append(out)
    #求梯度,因为有一个隐藏层和一个输出层,所以需要求两次
    for i in range(-1, -3, -1):
        #定义误差
        d = l_out[-1] - y
        if i == -1:
            #先求隐藏层的梯度
            w2 = d * l_out[-1] * (1 - l_out[-1]) * l_ga[-1].T
        else:
            #求输入层的梯度
            #l_w[-1][:, :2]:
             #反向传播求导 out_a21 和 out_a22 时,只跟 θ_21 和 θ_23 有关
             d2 = d * l_out[-1] * (1 - l_out[-1]) * l_w[-1][:, :2]
             #因为两个隐藏神经元影响输出,所以求导时需要相加
```

```
        d_sum = np.reshape(np.sum(d2, axis=0), [2, 1])
        w1 = d_sum * l_out[-2] * (1 - l_out[-2]) * l_ga[-2].T
print('w2 的梯度：', w2)
print('w1 的梯度：', w1)
#梯度下降
init_w['w1'] = init_w['w1'] - alpha * w1
init_w['w2'] = init_w['w2'] - alpha * w2
print('梯度下降后的梯度：', init_w)
```

运行结果如下：

```
w2 的梯度：[[-0.06532779 -0.06327767 -0.12269182]
 [ 0.06702303  0.06491971  0.12587565]]
w1 的梯度：[[-0.00101788 -0.00152682 -0.00050894]
 [-0.0004712  -0.00070679 -0.0002356 ]]
梯度下降后的梯度：{'w1': array([[0.03000102, 0.02000153, 0.01000051],
      [0.01300047, 0.01200071, 0.00100024]]), 'w2': array([[0.03006533,
0.02006328, 0.01012269],
      [0.01293298, 0.01193508, 0.00087412]])}
```

代码中先进行了两次前向传播，然后根据误差进行两次反向传播求梯度，并根据梯度进行了梯度下降，其计算的详细过程可以如图 2-58 和图 2-59 所示。

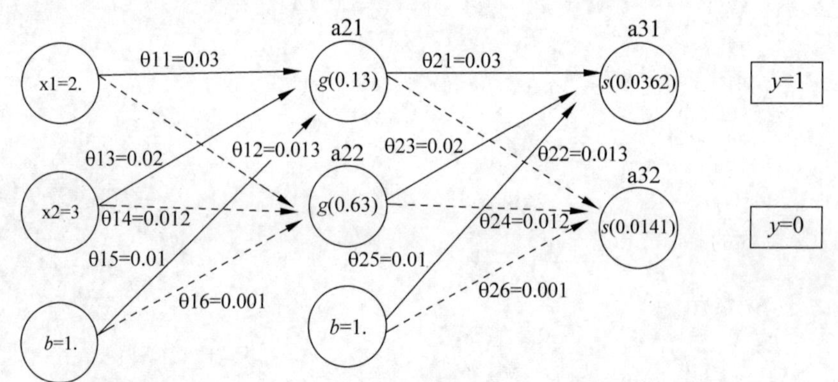

图 2-58　前向传播数值计算

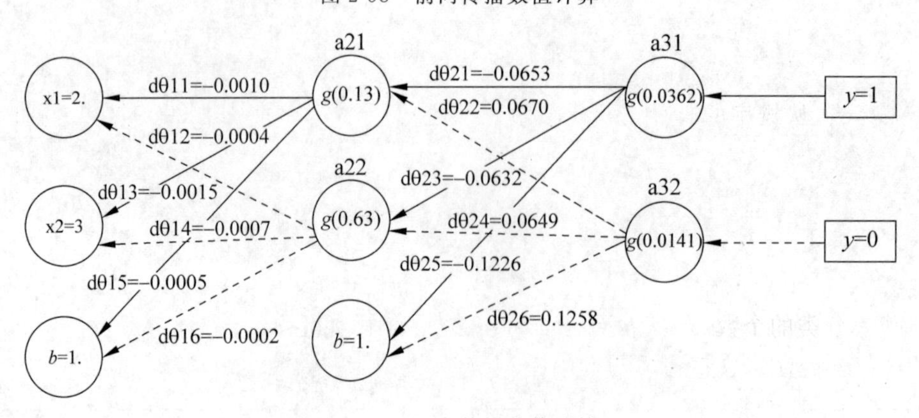

图 2-59　反向传播数值计算

代入数值，前向传播的计算过程如下：

$a21=2\times0.03+3\times0.02+1\times0.01=0.13$

$a22=2\times0.013+3\times0.012+1\times0.001=0.063$

$g(a21)=g(0.13)=0.5324$

$g(a22)=g(0.063)=0.5157$

$a31=g(0.13)\times0.03+g(0.063)\times0.02+1\times0.01=0.0362$

$a32=g(0.13)\times0.013+g(0.063)\times0.012+1\times0.001=0.0141$

$g(a31)=g(0.0362)=0.5090$

$g(a32)=g(0.0141)=0.5035$

代入数值，根据误差反向传播求隐藏层梯度的计算如下：

$$\begin{aligned}\frac{\partial L(\theta)}{\partial\theta_{21}}&=(g(0.0362)-1)\times g(0.0362)\times(1-g(0.0362))\times g(a21)\\&=(0.5090-1)\times0.5090\times(1-0.5090)\times0.5324\\&=-0.0653\end{aligned}$$

再求输入层的梯度：

$$\begin{aligned}\frac{\partial L(\theta)}{\partial\theta_{11}}=&((g(0.0362)-1)\times g(0.0362)\times(1-g(0.0362))\times0.03+(g(0.0141)-0)\times\\&g(0.0141)\times(1-g(0.0141))\times0.013)\times g(0.13)\times(1-g(0.13))\times2\\=&((0.5090-1)\times0.5090\times(1-0.5090)\times0.03+0.5035\times0.5035\times\\&(1-0.5035)\times0.013)\times0.5324\times(1-0.5324)\times2\\=&-0.0010\end{aligned}$$

其他权重参数的梯度与此类似，感兴趣的读者可以计算一下。

2.6.3 Softmax 反向传播

在“2.6.2 节反向传播算法”的示例中，将输出层概率值相加会大于 1，即

$$\begin{aligned}g(a31)&=g(0.0362)=0.5090\\g(a32)&=g(0.0141)=0.5035\\g(a31)+g(a32)&=0.5090+0.5035=1.0125\end{aligned}$$

作为分类算法，从概率的角度来考虑其输出值之和应该等于 1。Softmax 算法的目的就是如此，其公式为

$$\text{Softmax}_{(\text{out}_i)}=\frac{\mathrm{e}^{\text{out}_i}}{\sum_{c=1}^{n}\mathrm{e}^{\text{out}_i}} \tag{2-46}$$

其中，c 代表分类的个数。out_i 代表第几个神经元节点的输出。为什么不是直接映射，而是采用自然对数底数 e，这是因为指数函数是递增的，其函数的斜率也可以增大，也就是如果 x 的变化很小，则可能导致 y 值较大。

如输出值为[2,3,5]直接映射，其概率值为$\left[\frac{2}{10},\frac{3}{10},\frac{5}{10}\right]$=[0.2,0.3,0.5]，相差[0.1,0.2]，而使用 Softmax 映射，其概率值为$\left[\frac{e^2}{e^2+e^3+e^5},\frac{e^3}{e^2+e^3+e^5},\frac{e^5}{e^2+e^3+e^5}\right]$=[0.042 00,0.1142,0.8438]，[0.072,0.730]与[3,5]之间相差的概率值比直接映射的大，如图 2-60 所示。

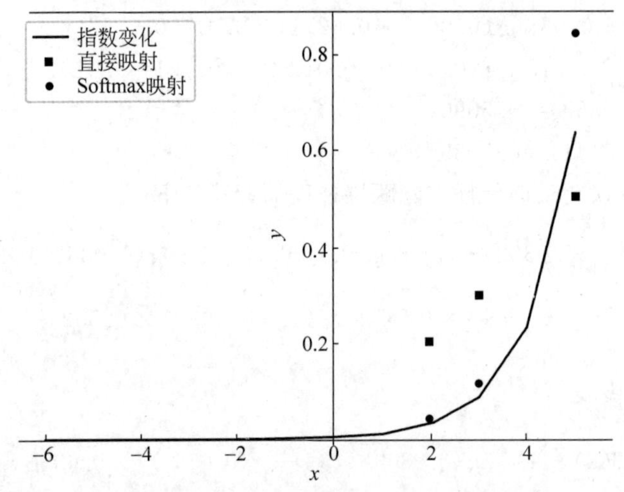

图 2-60　Softmax 值域映射与直接映射概率对比

Softmax 算法一般只用在输出层，此时神经网络的损失函数一般为交叉熵损失，即公式：

$$L(\theta)=-\frac{1}{N}\sum_{i=1}^{N}y_i\times\log\quad\text{Softmax}(\hat{y}_i) \tag{2-47}$$

其中，y_i 为真实值的概率，$\hat{y}_i$ 为预测值的概率。交叉熵描述的是两个概率分布之间的距离，熵值越小两个概率分布越接近，熵值越大两个概率分布越远。

当 $y_i=1$ 时，预测概率越接近 1，损失越小，如图 2-61 所示。

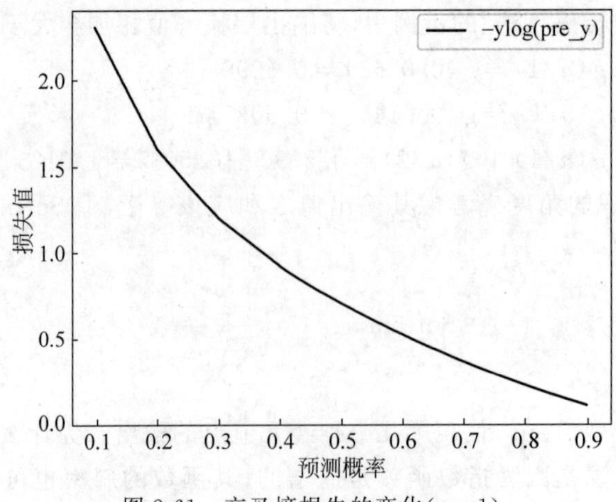

图 2-61　交叉熵损失的变化($y=1$)

当 $y_i=0$ 时，预测概率越接近 0，损失越小，如图 2-62 所示。

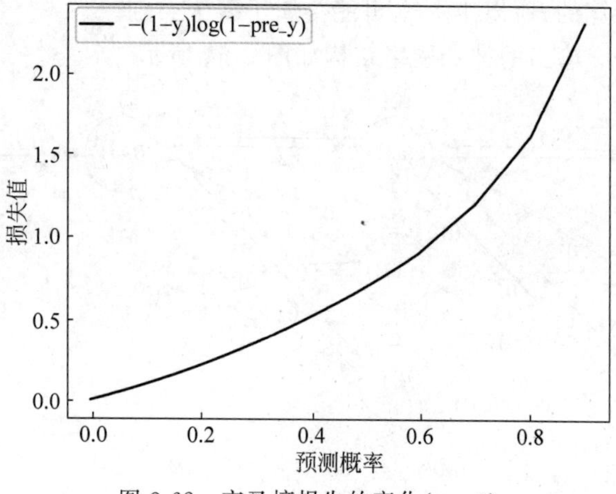

图 2-62　交叉熵损失的变化（$y=0$）

假设 $y_3=[1\quad 0\quad 0]$，预测值 $\hat{y}_3=[0.1\quad 0.1\quad 0.8]$，则交叉熵损失 $L=-1\times\log 0.1=2.302$，如果 $\hat{y}_3=[0.8\quad 0.1\quad 0.1]$，则 $L=-1\times\log 0.8=0.223$。后者的熵值更小，说明两个概率之间分布更接近。

逻辑回归中的损失函数与交叉熵损失函数类似，因此该损失函数又称为二元交叉熵，见式(2-48)：

$$L(\theta)=-\frac{1}{N}\sum_{i=1}^{N}[y_i\times\log\hat{y}_i+(1-y_i)\log(1-\hat{y}_i)] \tag{2-48}$$

在二元交叉熵损失函数中 $y_i\times\log\hat{y}_i$ 与 $(1-y_i)\log(1-\hat{y}_i)$ 必定有一项为 0。当 $y_i=1$ 时，$\hat{y}_i$ 越接近 1，当 $L(\theta)$ 为 0 时，$\hat{y}_i$ 越接近 0，损失值将无穷大；当 $y_i=0$ 时，$\hat{y}_i$ 越接近 0，当 $L(\theta)$ 为 0 时，$\hat{y}_i$ 越接近 1，损失值将无穷大；当预测值接近标签值时损失很小，当预测值远离标签值时损失很大，如图 2-63 所示。

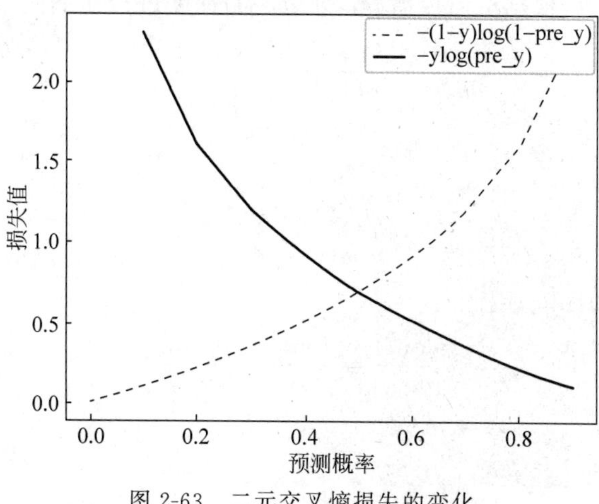

图 2-63　二元交叉熵损失的变化

在多分类时,交叉熵损失由于 Softmax 互斥,那么 1 个 y 只能属于 1 个类别,而二元交叉熵每个 y 是独立计算的,所以 1 个 y 可能会属于多个类别。

假设存在 Softmax 输出的神经网络结构如图 2-64 所示。

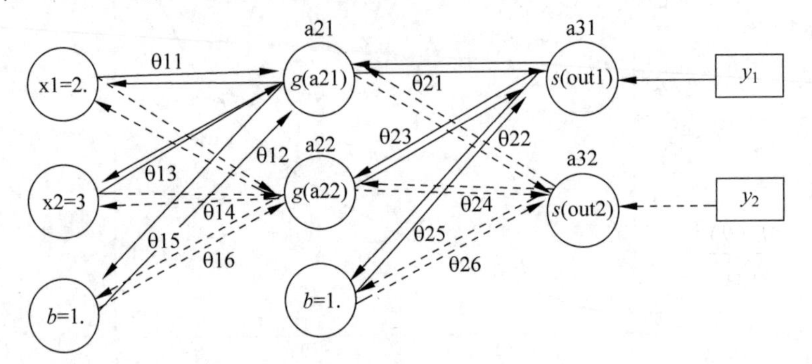

图 2-64 Softmax 神经网络

则前向传播的过程如下:

连接过程 1_1, $a_{21}=\theta_{11}x_1+\theta_{13}x_2+b\times\theta_{15}$

激活函数 1_1, $\text{out}_{a21}=g(a_{21})$

连接过程 1_2, $a_{22}=\theta_{12}x_1+\theta_{14}x_2+b*\theta_{16}$

激活函数 1_2, $\text{out}_{a22}=g(a_{22})$

连接过程 2_1, $a_{31}=\text{out}_{a21}\times\theta_{21}+\text{out}_{a22}\times\theta_{23}+b\times\theta_{25}$

激活函数 2_1, $\text{out1}=\text{Softmax}\left(\frac{e^{a_{31}}}{e^{a_{31}}+e^{a_{32}}}\right)$

连接过程 2_2, $a_{32}=\text{out}_{a21}\times\theta_{22}+\text{out}_{a22}\times\theta_{24}+b*\theta_{26}$

激活函数 2_2, $\text{out2}=\text{Softmax}\left(\frac{e^{a_{32}}}{e^{a_{31}}+e^{a_{32}}}\right)$

损失函数为交叉熵,根据反向传播算法求 θ_{21} 的梯度过程如下:

$$\frac{\partial L(\theta)}{\partial\theta_{21}}=\frac{\partial L(\theta)}{\partial\text{out1}}\times\frac{\partial\text{out1}}{\partial a_{3i}}\times\frac{\partial a_{31}}{\partial\theta_{21}}$$

$\frac{\partial L(\theta)}{\partial\text{out1}}=-y_1*\frac{1}{\hat{y}_1}$,令 $y_1=1$,所以

$$\frac{\partial L(\theta)}{\partial\text{out1}}=-\frac{1}{\hat{y}_1}$$

因 out1 受 Softmax 的影响,因此

$$\frac{\partial\text{out1}}{\partial a_{3i}}=\frac{\partial\text{out1}}{\partial a_{31}}+\frac{\partial\text{out2}}{\partial a_{32}}$$

故

$$\frac{\partial \text{out1}}{\partial a_{31}} = \frac{\partial \dfrac{e^{a_{31}}}{e^{a_{31}} + e^{a_{32}}}}{\partial e^{a_{31}}} = \frac{e^{a_{31}}}{e^{a_{31}} + e^{a_{32}}} \times \left(1 - \frac{e^{a_{31}}}{e^{a_{31}} + e^{a_{32}}}\right) = \hat{y}_1 \times (1 - \hat{y}_1)$$

$$\frac{\partial \text{out1}}{\partial a_{32}} = \frac{\partial \dfrac{e^{a_{31}}}{e^{a_{31}} + e^{a_{32}}}}{\partial e^{a_{32}}} = -\frac{e^{a_{31}}}{e^{a_{31}} + e^{a_{32}}} \times \frac{e^{a_{32}}}{e^{a_{31}} + e^{a_{32}}} = -\hat{y}_1 \times \hat{y}_2$$

代入后

$$\begin{aligned}\frac{\partial L(\theta)}{\partial \theta_{21}} &= -y_1 \times \frac{1}{\hat{y}_1} \times (\hat{y}_1 \times (1 - \hat{y}_1) + (-\hat{y}_1) \times \hat{y}_2) \times \text{out}_{a21} \\ &= -y_1 \times ((1 - \hat{y}_1) - \hat{y}_2) \times \text{out}_{a21} \\ &= -y_1 \times (1 - \hat{y}_1 - \hat{y}_2) \times \text{out}_{a21}\end{aligned}$$

又因为 $y_1=1$，则 $y_2=0$，所以，

$$\frac{\partial L(\theta)}{\partial \theta_{21}} = (-y_1 + \hat{y}_1 + \hat{y}_2 - y_2) \times \text{out}_{a21}$$

$$\frac{\partial L(\theta)}{\partial \theta_{21}} = \sum_{i=1}^{n} (\hat{y}_i - y_i) \times \text{out}_{a21}$$

其他参数的梯度与此类似，在此就不展开了。

总结

反向传播就是在求损失函数的导数，而多层感知机构建了一个较复杂的复合函数，所以反向传播算法是一个链式求导的过程。

练习

使用矩阵推导反向传播算法的过程，并实现 NumPy 版本的反向传播算法的梯度下降代码。

2.7 欠拟合与过拟合

先了解一个决策边界的概念，所谓决策边界就是根据不同特征对样本数据进行分类，不同类型间的分界就是算法模型针对该数据集的决策边界。用于分类问题中，通过决策边界可以更好地可视化分类结果。

例如逻辑回归的决策边界是一条直线，如图 2-65 所示。

神经网络决策边界由隐藏层的数量来确定，如果没有隐藏层，则它只能学习线性函数，如果它有隐藏层，则可以学习到基于样本数据的非线性函数，可以具有任意的决策边界，如图 2-66 所示。

所谓欠拟合即决策边界无论是训练集还是测试集都不能很好地对样本进行区分，算法模型学习到的样本特征太少，如图 2-67 所示。

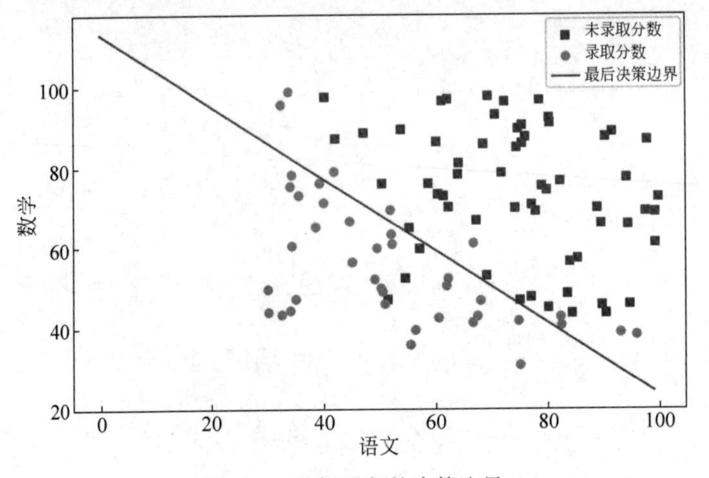

图 2-65　逻辑回归的决策边界

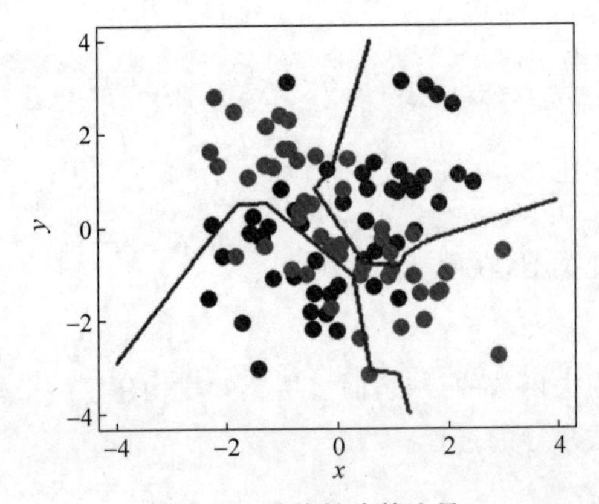

图 2-66　非线性决策边界

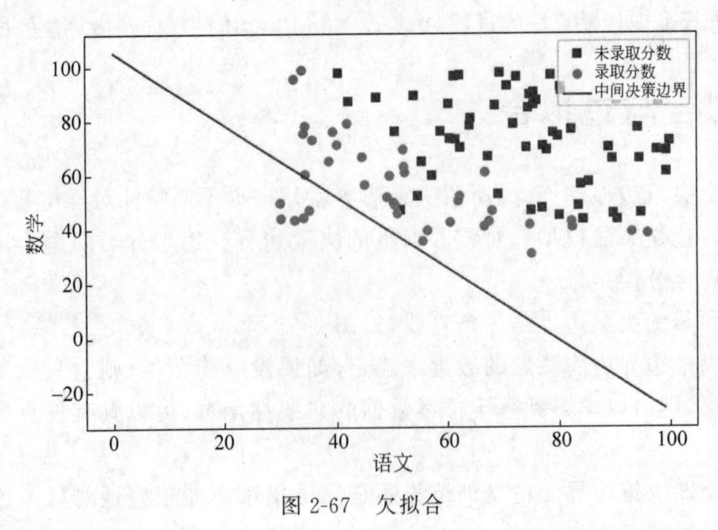

图 2-67　欠拟合

欠拟合的原因主要有以下几点。

(1) 样本有效特征较少。

(2) 模型复杂度较低。

(3) 训练时间不够。

欠拟合的解决方法有以下几种。

(1) 检查数据的分布。可以使用散点图等查看训练样本数据的分布,查看噪声数据的分布,可以通过聚类算法 K-Means 来检测离群数据,将那些落在簇之外的值视为噪声数据,然后删除。

(2) 继续训练,调整学习率等超参数。欠拟合可能是由于训练时间不够或者学习率过大、初始权重参数不合理等导致的。

(3) 增加训练的数据量。无论怎样,增加数据量通常是一个很好的办法,更多的数据往往意味着更多的特征。

(4) 增加模型的复杂度、增加模型权重参数、增加非线性函数。在神经网络中更多的神经元、更多的非线性输出会使模型的表现能力增强,但同时也可能导致过拟合。

所谓过拟合即对训练集中的样本高度拟合,但是对于验证集或者测试集拟合效果较差,算法模型对噪声数据进行了过度学习,从而忽略了数据的一般规律,如图 2-68 所示。

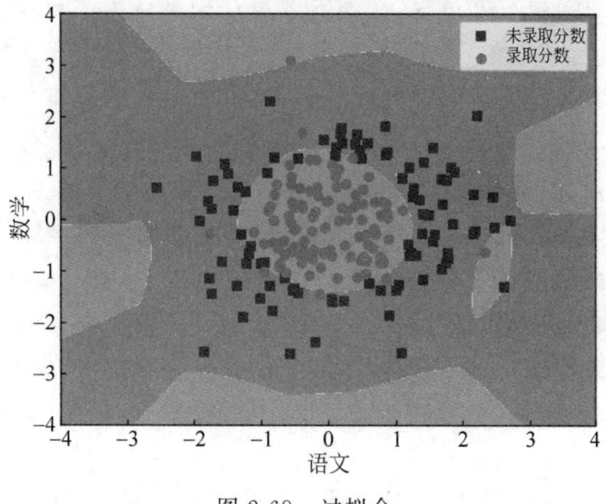

图 2-68 过拟合

图中“录取分数”背景为蓝色,“未录取分数”背景为淡红色,大多异常数据被隔离开。

关于神经网络训练与拟合的过程可以在网站 https://playground.tensorflow.org/动态地进行可视化训练,如图 2-69 所示。

图中训练损失有所下降,但是验证损失并没有下降,说明训练已出现过拟合的情况。过拟合就是训练时把训练样本学得太好了,很可能已经把训练样本自身的一些特点当作了所有潜在样本都会具有的一般性质,从而导致泛化性能下降,在验证集上的效果较差,模型泛化能力较弱。例如,上学时有人把某一张历史试卷中的题目全部记下,以为下一次能考高分,但是下一次考试仍然很差,其根本原因就是在学习时并没有从历史课本中学习到普遍的

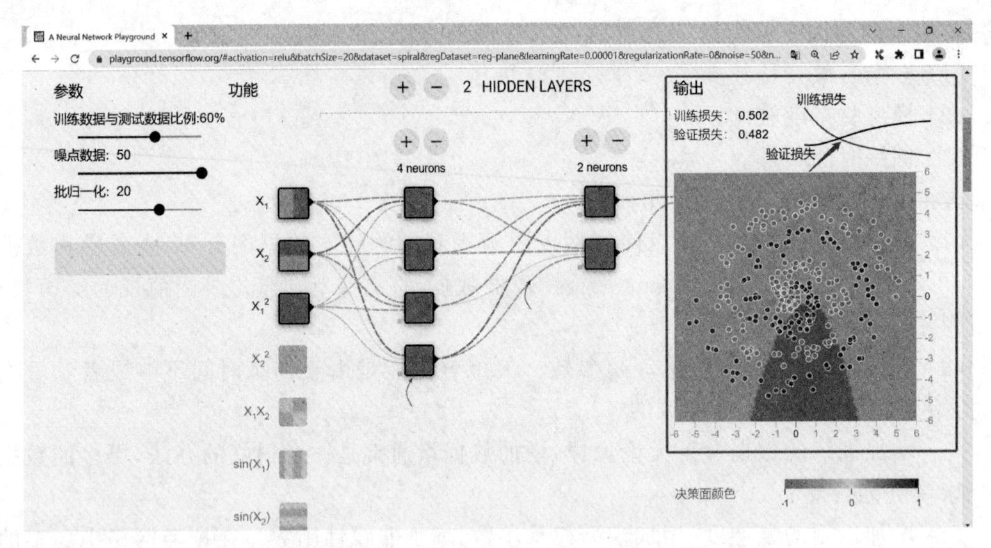

图 2-69 可视化神经网络训练过程过拟合

历史规律知识,从而导致再次考试的能力较差,也就是模型泛化能力较差。

过拟合的主要原因有以下几点。

(1) 训练数据中的噪声数据过多,从而导致算法模型认为部分噪声是特征数据,从而扰乱模型的学习。

(2) 训练数据过少或者选择的数据标签错误,并不能反映样本数据的整体规律。

(3) 模型不合理,权重参数较多,非线性函数较多,模型复杂度较高。

过拟合无法彻底避免,只能缓解,其采用的方法主要有以下几点。

(1) 从源头获取更多的数据。更多的数据更有可能反映样本数据的真实规律及分布,从而使模型可能学习到更好的特征与分布。

(2) 数据增强。通过一定的规则扩充数据,例如图像的平移、缩放等。

(3) 降低模型的复杂度。对于神经网络可以减少神经元,减少非线性输出。减少参数或者减少非线性输出会使拟合曲线减少,从而使模型忽略某些噪声数据,但同时也有可能导致欠拟合。

(4) 使用早停技术,即在图 2-69 训练损失和验证损失交叉时停止学习。

(5) 使用 DropOut 技术。DropOut 即在神经网络前向传播时以一定的概率让神经元随机失活,这样可以使模型不太依赖某些局部特征,从而提高泛化能力。

(6) 使用正则化。正则化通过对损失函数进行修改额外增加约束和惩罚,忽略噪声数据,允许一些错误,从而改善网络模型在测试集上的表现,减少泛化误差、提高模型泛化能力。正则化是缓解过拟合的一个重要方法,将在 2.8 节进行介绍。

总结

欠拟合学习的特征过少,无法反映真实数据的分布;过拟合学习到较多的噪声数据,从而扰乱模型的学习。

练习

总结解决欠拟合、过拟合的方法，并在网站 https://playground.tensorflow.org/观察调整参数后的变化。

2.8 正则化

在 2.4.1 节最大似然估计中介绍了最大似然概率 $L(\theta \mid x)=\prod_{i=1}^{N} P(y_i \mid x_i;\theta)$ 来构建逻辑回归的损失函数，其采用的思想就是从已知样本学习参数 θ，使所有样本出现某种分布的概率最大。

但是想一想，能不能在构建损失函数时把人类社会已知的经验加进去呢？答案是可以的，但是这时不再称为最大似然估计，而是最大后验估计。

假设春节回家某媒婆组织相亲见面，共见了 5 人，其中只有 1 人比较满意，那么根据最大似然概率来计算，相亲成功的概率为 20%，但是相过亲的人都知道，这个概率太高了，是不真实的，通过媒婆介绍的成功率其实只有约 1%，那么通过最大后验概率算出来此媒婆相亲成功的概率值为 2%，也就能更接近真实的情况。

对于最大似然估计需要最大化 $P(D|\theta)$，然而在最大后验估计需要最大化 $P(\theta|D)$，但是 $P(\theta|D)$不好直接计算，此时可以使用贝叶斯公式，将 $P(\theta|D)$转换为 $P(D|\theta)$，贝叶斯的公式为

$$P(\theta;y_i \mid x_i)=\frac{P(y_i \mid x_i;\theta)\times P(\theta)}{P(y_i \mid x_i)} \tag{2-49}$$

式中，$P(y_i|x_i;\theta)$指当前获得数据的似然概率，$P(\theta)$为先验概率，$P(y_i|x_i)$为平均似然值，可以理解为在所有可能的 θ 下，获得该观测数据的期望，是一个常数，可以忽略，因此 $P(\theta;y_i|x_i)$与 $P(y_i|x_i;\theta) * P(\theta)$成比例关系，可得最大后验估计公式：

$$P(\theta;y_i \mid x_i)=\prod_{i=1}^{N} P(y_i \mid x_i;\theta)\times P(\theta) \tag{2-50}$$

假设 $P(\theta)$满足高斯分布，即公式：

$$f(\theta \mid \mu,\sigma)=\frac{1}{\sqrt{2\pi\sigma}}\exp\frac{-(\theta-\mu)^2}{\sigma^2} \tag{2-51}$$

其中，μ 代表均值，σ 代表方差，如果假设均值为 0，方差为 1，则记为 $P(\theta)\sim N(0,1)$。

$$\begin{aligned}
P(\theta;y_i \mid x_i) &= \prod_{i=1}^{N} P(y_i \mid x_i;\theta)\times P(\theta) \\
&= \underset{\theta}{\operatorname{argmax}}\sum_{i=1}^{N} P(y_i \mid x_i;\theta)\times P(\theta) \\
&= \underset{\theta}{\operatorname{argmax}}\sum_{i=1}^{N} P(y_i \mid x_i;\theta)+\log P(\theta)
\end{aligned}$$

$$=\underset{\theta}{\operatorname{argmax}}\sum_{i=1}^{N}P(y_i \mid x_i;\theta)+\log\frac{1}{\sqrt{2\pi\sigma}}\exp\frac{-(\theta-0)^2}{1}$$

$$=\underset{\theta}{\operatorname{argmax}}\sum_{i=1}^{N}P(y_i \mid x_i;\theta)+\log\frac{1}{\sqrt{2\pi\sigma}}\exp\frac{-\theta^2}{1}$$

$$=\underset{\theta}{\operatorname{argmax}}\sum_{i=1}^{N}P(y_i \mid x_i;\theta)-\frac{1}{\sqrt{2\pi\sigma}}\theta^2$$

$$=-\underset{\theta}{\operatorname{argmin}}\sum_{i=1}^{N}P(y_i \mid x_i;\theta)+\frac{1}{\sqrt{2\pi\sigma}}\theta^2$$

对 $\lambda=\frac{1}{\sqrt{2\pi\sigma}}$ 进行代替，那么上式变为

$$P(\theta;y_i \mid x_i)=-\underset{\theta}{\operatorname{argmin}}\sum_{i=1}^{N}\log P(y_i \mid x_i;\theta)+\lambda\theta^2 \tag{2-52}$$

因此最大后验估计可以看成在最大似然估计的基础上增加了先验概率，当我们的样本数据少时，模型算法学不到更好的参数，增加先验概率能够更加反映真实的情况，但是当样本较多时，最大似然概率会越来越准确，先验概率的影响会越来越小。这一特性正好满足样本数据较少时模型函数可能出现过拟合的情况，增加一点先验概率可以更好地反映真实的情况，因此式(2-48)也是正则化损失函数的计算公式。

$$\begin{aligned}l(\theta \mid y_i;x_i)=&-\frac{1}{N}\underset{\theta}{\operatorname{argmin}}\sum_{i=1}^{N}(y_i\log h_\theta(x_i)+\\&(1-y_i)\log(1-h_\theta(x_i)))+\frac{\lambda}{N}\sum_{i=1}^{N}\theta^2\end{aligned} \tag{2-53}$$

如果是均方差损失，则正则化的公式就变为

$$L(\theta)=\frac{1}{2N}\sum_{i=1}^{N}(\hat{y}_i-y_i)^2+\frac{\lambda}{2N}\sum_{i=1}^{N}\theta^2 \tag{2-54}$$

假设 $\hat{y}_i$ 的最佳模型函数为 $\widehat{y_{\text{best}}}=\theta_1\times x_1+\theta_2\times x_2^2+\theta_0$，而此时模型函数为 $\hat{y}=\theta_1\times x_1+\theta_2\times x_2^2+\theta_3\times x_1\times x_2^2+\theta_3\times x_1^2\times x_2^2+\theta_0$，设 $\lambda=1000$，则损失函数变为 $L(\theta)=\frac{1}{2N}\sum_{i=1}^{N}(\theta_1\times x_1+\theta_2\times x_2^2+\theta_3\times x_1\times x_2^2+\theta_4\times x_1^2\times x_2^2+\theta_0-y)^2+\frac{1000}{2N}\times\theta_0^2+\frac{1000}{2N}\times\theta_1^2+\frac{1000}{2N}\times\theta_2^2+\frac{1000}{2N}\times\theta_3^2+\frac{1000}{2N}\times\theta_4^2$。此时 $L(\theta)$ 的损失值会很大，但是模型函数的目的是要 $L(\theta)$ 很小，那么根据梯度下降算法可知，只有当 $[\theta_0、\theta_1、\theta_2、\theta_3、\theta_4]\to 0$ 时 $L(\theta)$ 才可能很小，而此时模型函数可能出现欠拟合。当 λ 设得很小时，参数可能没有起到惩罚的效果，训练结果过拟合，找不到最佳模型函数。只有当 λ 被设为一个恰当的值时，才有可能使 θ_3、$\theta_4\to 0$，从而使模型函数接近 $\widehat{y_{\text{best}}}$。总之 λ 是一个超参数，需要根据算法模型训练的情况进行适当调整。

使用神经网络训练可视化网站，选择 L2 正则化(θ^2)，并设置 $\lambda=0.001$，可以发现相对于没有加正则化(图 2-69)更晚出现过拟合，如图 2-70 所示。

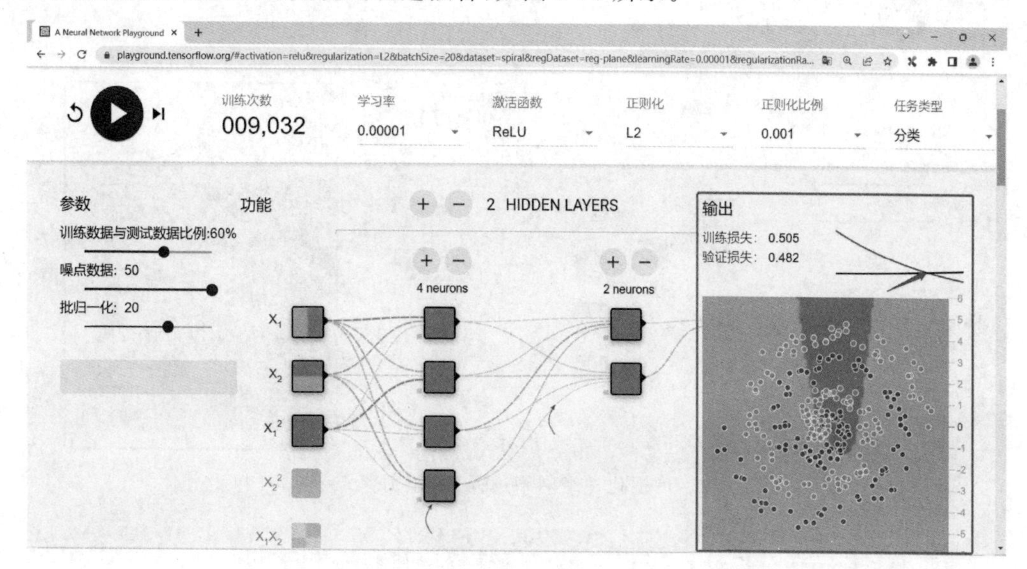

图 2-70 可视化神经网络过程 L2 正则化 $\lambda=0.001$

设置 $\lambda=10$ 相对于图 2-70 拟合效果更好，也更晚出现过拟合的趋势，如图 2-71 所示。

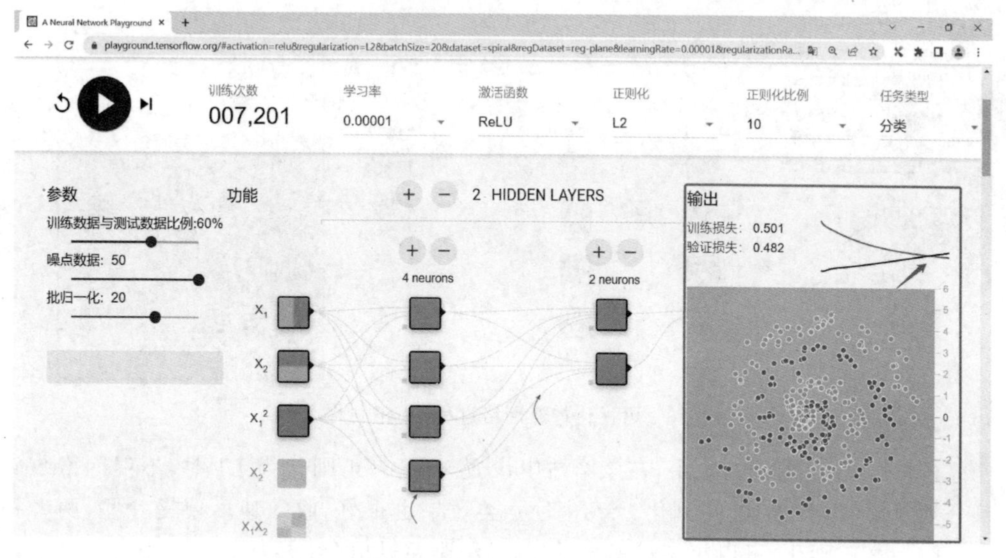

图 2-71 可视化神经网络过程 L2 正则化 $\lambda=10$

当选择 L1 正则化，并设置 $\lambda=0.001$，可以发现神经网络收敛的速度相对于 L2 要快一些，出现过拟合的情况比没有加正则化更晚，如图 2-72 所示。

当选择 L1 正则化，并设置 $\lambda=10$，神经网络收敛的速度最快，而且分类情况比 $\lambda=0.001$ 更好，但是仍然出现了过拟合，如图 2-73 所示。

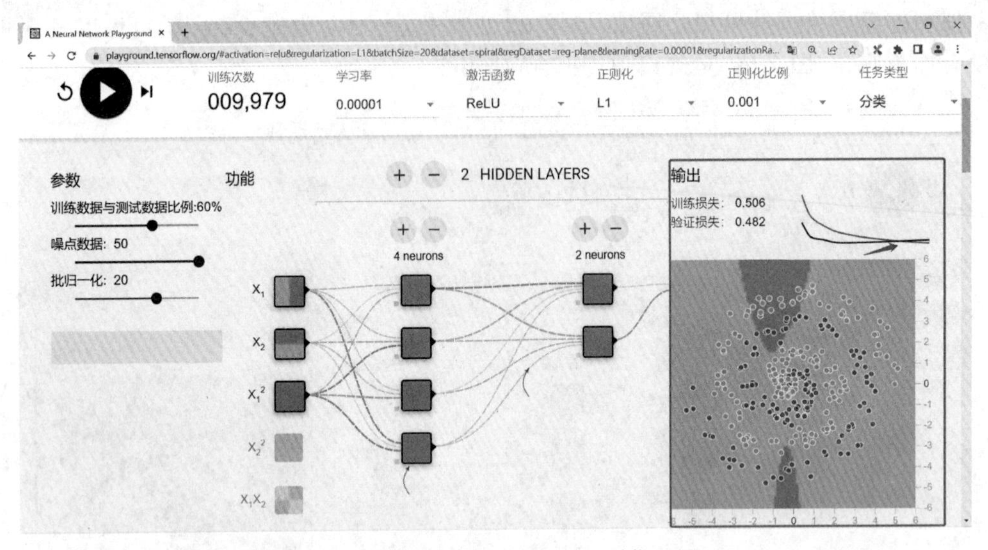

图 2-72 可视化神经网络过程 L1 正则化 $\lambda=0.001$

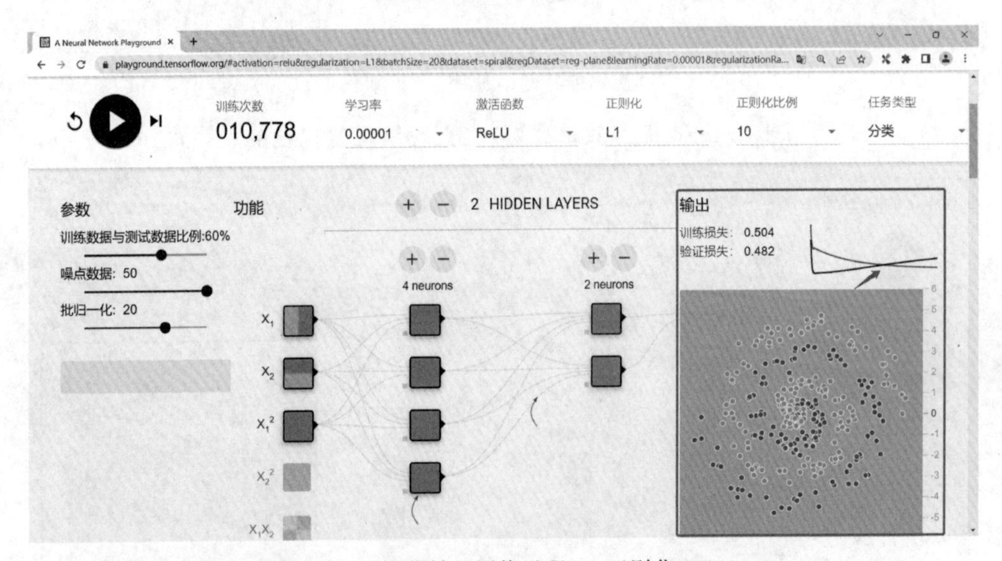

图 2-73 可视化神经网络过程 L1 正则化 $\lambda=10$

综合 L1、L2 正则化的效果，在该网站中设置 $\lambda=10$、正则化为 L1 时，不仅网络收敛快而且分类效果佳，当然 L2 正则化 $\lambda=10$ 时分类效果也很好，收敛速度要慢一些，两者相差不大。L1、L2 正则化，可以缓解过拟合，但是无法避免过拟合。

那么什么是 L1 正则化，为什么 L1 比 L2 正则化收敛更快？

L1 正则化只需在损失函数加上 θ 的绝对值，即公式：

$$L(\theta)=\frac{1}{2N}\sum_{i=1}^{N}(\hat{y}_i-y_i)^2+\frac{\lambda}{2N}\sum_{i=1}^{N}|\theta| \tag{2-55}$$

根据梯度下降的原理 L1、L2 的梯度下降公式可以写成：

$$\theta^i = \theta^i - \alpha * \left[\frac{\partial L(\theta)}{\partial \theta^i} + 1 * \lambda\right]$$

$$\theta^i = \theta^i - \alpha * \left[\frac{\partial L(\theta)}{\partial \theta^i} + 2 \times \lambda \times \theta^i\right] \tag{2-56}$$

图形化展示 L1、L2 及其导数，如图 2-74 所示。

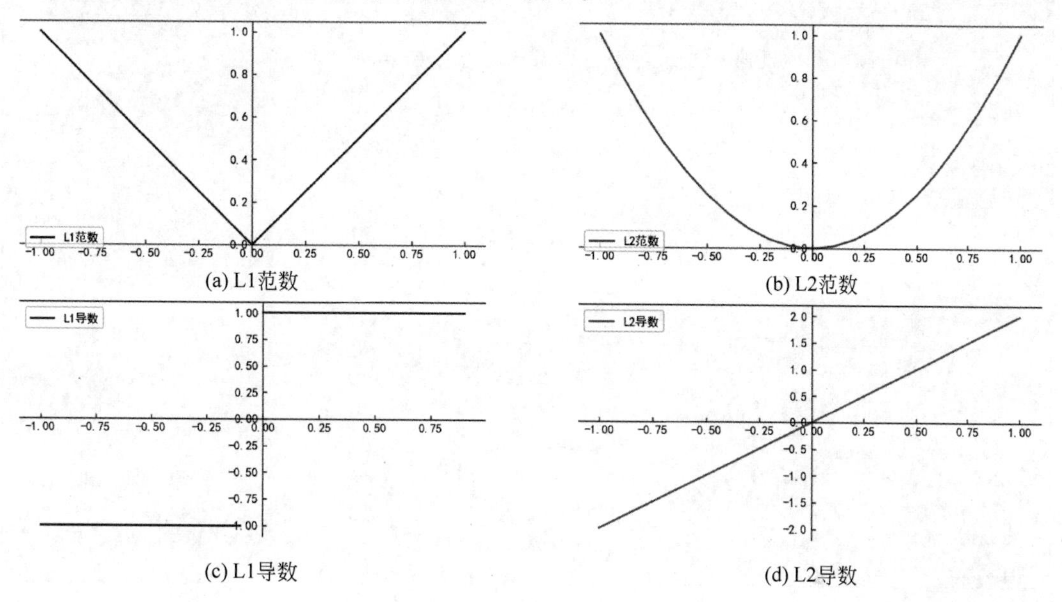

(a) L1范数　(b) L2范数　(c) L1导数　(d) L2导数

图 2-74　L1、L2 及导数的值域

在梯度更新时，不管 L1 是多少(只要不是 0)其梯度都是 1 或 −1，所以每次更新时，它都是稳步向 0 前进的。L2 则需要多次梯度更新才有可能靠近 0。

假设$\frac{\partial L(\theta)}{\partial \theta^1}=0.5, \theta^1=0.1, \alpha=0.001, \lambda=10$，则 L1 的梯度更新为 $\theta^1=0.1-0.001\times[0.5+10\times1]=0.08950$，而 L2 的梯度更新为 $\theta^1=0.1-0.001\times[0.5+2\times10\times0.1]=0.0975$。

将以上计算过程用代码实现，代码如下：

```
#第 2 章/machineLearn/L1 和 L2 图形化展示.py
def when_l1_l2_theta_close0(type='l1'):
    #初始权重
    theta = 0.1
    x = 5
    y = 1
    #惩罚力度
    c = 10
    #学习率
    alpha = 0.0001
    theta_list = []
    for i in range(100):
```

```
        if type == 'l1':
            #abs|theta|的导数,如果 theta 小于 0,则为-1,如果 theta 大于 0,则为 1
            d = -1 if theta < 0 else 1
            theta = theta - alpha * ((y - theta * x) * x + d * c)
        else:
            theta = theta - alpha * ((y - theta * x) * x + 2 * c * theta)
        theta_list.append(theta)
        if theta <= 0:
            print('{}此时 theta 等于{}'.format(type, theta))
            #break
    return len(theta_list), theta_list


def plt_l1_l2_theta_close0():
    for i in range(1, 3):
        plt.subplot(1, 2, i)
        epoch, theta_list = when_l1_l2_theta_close0(type='l{}'.format(i))
        plt.title('L{}时θ递减'.format(i))
        plt.plot(range(epoch), theta_list, label='θ变化', color='red')
        plt.legend(loc='best')
        plt.xlabel(u'学习次数')
        plt.ylabel(u'θ梯度下降值')
    plt.savefig("l1、l2的梯度下降变化")
    plt.show()

if __name__ == "__main__":
    plt_l1_l2_theta_close0()
```

运行后会发现 L1 大概在梯度下降第 75 次 θ 就接近 0,而 L2 在 100 次时仍不为 0,如图 2-75 所示。

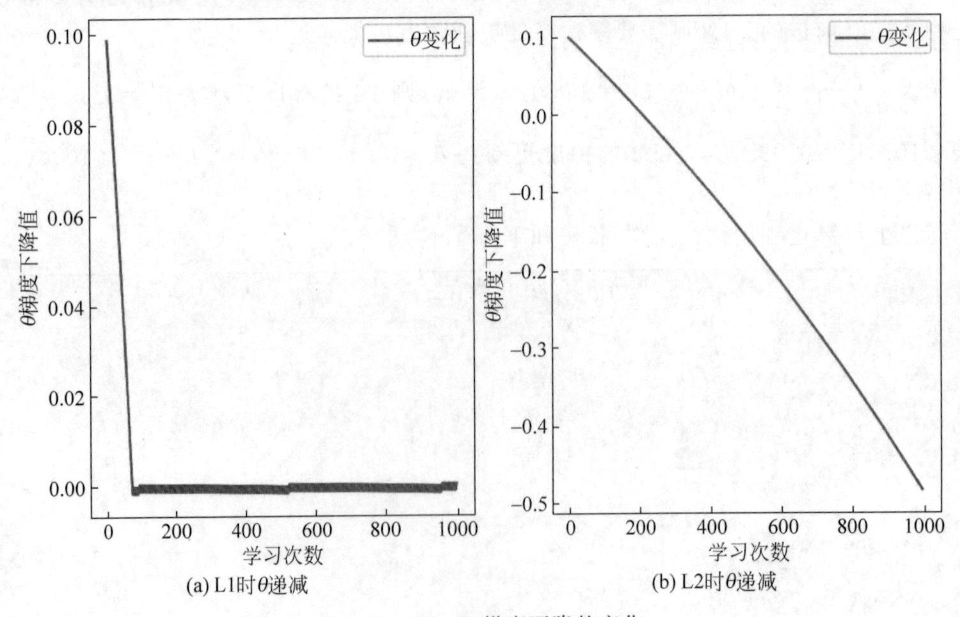

图 2-75　L1、L2 梯度下降的变化

均方差损失函数(是一个凹函数)的最优值取决于0点处的导数,如果损失函数在0点处的导数$\frac{\partial L(\theta)}{\partial \theta^i}$不为0,则加上L2正则化$2\times\lambda\times\theta^i$后导数仍然不为0,说明0这个点不是极值点,最优值不在$\theta=0$处。

而L1正则化时$\theta=0$不可导,不可导点是不是极值点,就要看不可导点左右的单调性。单调性可以通过这个点左、右两侧的导数符号来判断,如果导数符号相同,则不是极值点,如果左侧导数为正、右侧为负,则是极大值,如果左侧导数为负、右侧为正,则是极小值。根据极值点判断原则,$\theta=0$左侧导数$\frac{\partial L(\theta)}{\partial \theta^i}=\frac{\partial L(\theta)}{\partial \theta^i}-\lambda$,只要为正则项系数$\lambda$大于$\frac{\partial L(\theta)}{\partial \theta^i}$,那么左侧导数小于0,右侧导数$\frac{\partial L(\theta)}{\partial \theta^i}+\lambda>0$,所以$\theta=0$就会变成一个极小值点,这样L1经常会把参数变为0,从而产生稀疏性。

神经网络的稀疏性即模型网络中只有少量的神经元被激活,而大多数神经元处于未激活状态,稀疏性可以提高神经网络的表达效果,主要有以下几个原因:

(1) 缓解过拟合。稀疏表达可以减少参数量(此时$\theta=0$),从而缓解过拟合,使网络泛化能力更高。

(2) 提高计算效率。因为只有少量神经元被激活,所以网络的计算量会大大降低。

(3) 提高可解释性。因为只有少量神经元被激活,决策面会更平滑,所以可以更加容易地理解网络的决策过程。

(4) 提高特征的判别能力。因为只有少量神经元被激活,所以模型网络会更加关注那些对于分类任务有用的特征。

L1正则化也相当于假设$P(\theta)$服从拉普拉斯分布,即

$$f(\theta \mid \mu,b)=\frac{1}{2b}\exp\frac{-|\theta-\mu|}{b} \tag{2-57}$$

假设为$P(\theta)\sim N(0,b)$,则

$$\begin{aligned}
P(\theta;y_i \mid x_i) &= \prod_{i=1}^{N} P(y_i \mid x_i;\theta)\times P(\theta) \\
&= \underset{\theta}{\operatorname{argmax}}\sum_{i=1}^{N} P(y_i \mid x_i;\theta)\times P(\theta) \\
&= \underset{\theta}{\operatorname{argmax}}\sum_{i=1}^{N} P(y_i \mid x_i;\theta)+\log P(\theta) \\
&= \underset{\theta}{\operatorname{argmax}}\sum_{i=1}^{N} P(y_i \mid x_i;\theta)+\log\frac{1}{2b}\exp\frac{-|\theta|}{b} \\
&= -\underset{\theta}{\operatorname{argmin}}\sum_{i=1}^{N} P(y_i \mid x_i;\theta)+\frac{1}{b}|\theta|
\end{aligned}$$

$$=-\underset{\theta}{\operatorname{argmin}}\sum_{i=1}^{N}P(y_i \mid x_i;\theta)+\frac{1}{b}\mid\theta\mid$$

对 $\lambda=\frac{1}{b}$ 进行代替，那么上式变为

$$P(\theta;y_i \mid x_i)=-\underset{\theta}{\operatorname{argmin}}\sum_{i=1}^{N}\log P(y_i \mid x_i;\theta)+\lambda\mid\theta\mid \tag{2-58}$$

当然也可以假设 $P(\theta)$ 服从其他概率分布，那么此时得出的正则项也是不一样的，起到的效果也会所差别。

虽然 L1 更加稀疏，但是平时用得更多的是 L2 正则化，主要是由于损失函数中 L1 会对所有 θ 进行相同比例的惩罚，而 L2 对 θ 较大的值有较大力度的惩罚，很小的 θ 给以非常小的惩罚，当 θ 趋近于 0 时几乎不惩罚。这个性质跟 L2 正则化中的 θ^2 有关，θ 越大其平方越大，θ 越小其平方越小，从而导致其惩罚力度不一样，也就更平滑一些。

总结

正则化是缓解过拟合的一种有效手段，给予模型一点点错误，使模型可以避免学到过多的噪声数据；L2 正则化其惩罚力度在变化，通常来说会优先进行选择。

练习

在网站中选择不同的正则化和惩罚力度，观察预测结果的拟合变化。

2.9 梯度消失与梯度爆炸

理论上神经网络有更多的非线性激活函数、更多的隐藏层，模型网络的学习能力会更强，但实际上此时容易出现梯度学习不稳定，从而导致网络出现退化问题。梯度学习不稳定问题，具体表现在梯度消失和梯度爆炸。所谓梯度消失指某个网络层的梯度值接近或者等于 0，梯度爆炸则表现为梯度值为 nan(极大值)、inf(异常值)，从而导致模型网络无法继续学习。

假设拥有 5 个网络层的单个神经元网络如图 2-76 所示。

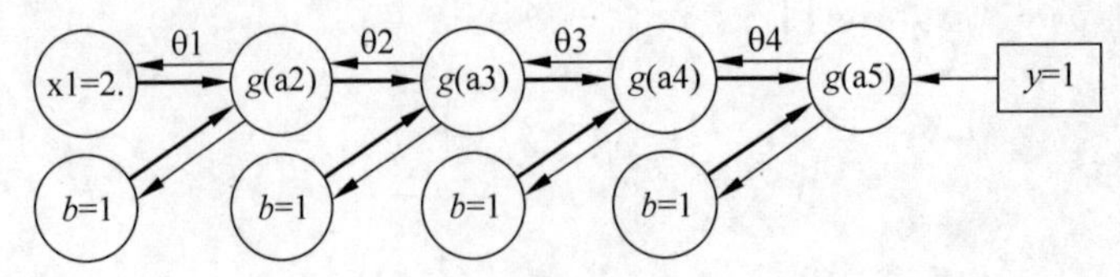

图 2-76　5 层单神经元网络模型

根据反向传播算法 $\theta1$ 的梯度为 $\frac{\partial L}{\partial\theta_1}=\frac{\partial L}{\partial g(a5)}\times\frac{\partial g(a5)}{\partial g(a4)}\times\frac{\partial g(a4)}{\partial g(a3)}\times\frac{\partial g(a3)}{\partial g(a2)}\times\frac{\partial g(a2)}{\partial\theta_1}$，如果网络的激活函数为 Sigmoid 函数，则 $\frac{\partial L}{\partial\theta_1}=(\hat{y}-y)\times g'(a5)\times\theta_4\times g'(a4)\times\theta_3\times g'(a3)\times\theta_2\times g'(a2)\times x_1$，Sigmoid 函数的导数 g' 的值域在[0,0.25]之间，如图 2-77 所示。

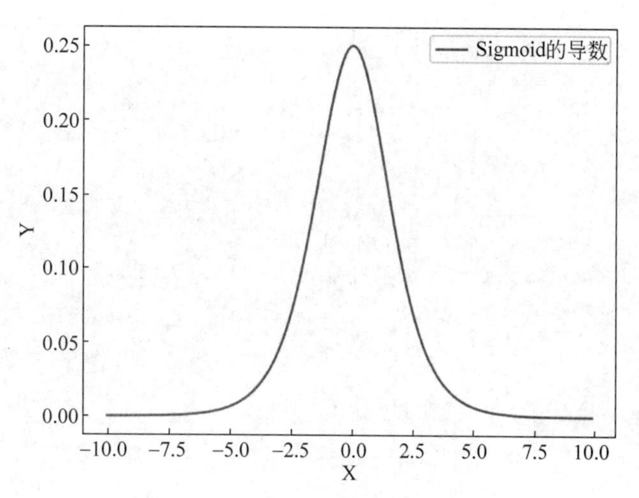

图 2-77 Sigmoid 函数的导数的值域

假设所有 g' 都取 0.25，并且所有 θ 的值都是 0.1，预测 $\hat{y}=0.6$、真值 $y=1$，则 $\frac{\partial L}{\partial \theta_1}=(0.6-1)\times 0.25\times 0.1\times 0.25\times 0.1\times 0.25\times 0.1\times 0.25\times 2.=-0.000\,003\,125$，此时 $\frac{\partial L}{\partial \theta_1}$ 已经非常接近于 0，实际情况可能更糟糕，使用以下代码做一个试验。

```
#第 2 章/machineLearn/梯度消失梯度爆炸.py
import numpy as np
import matplotlib.pyplot as plt
def g(z, type='g'):
    if type == 'g':
        #Sigmoid 激活函数
        return 1 / (1 + np.exp(-z))
    elif type == 'relu':
        #ReLU 激活函数,max(0,x)
        return 0 if z < 0 else z

def grad_g(z, type='g'):
    if type == 'g':
        #Sigmoid 的导数
        return g(z) * (1 - g(z))
    elif type == 'relu':
        #ReLU 的导数
        return 0 if z < 0 else 1

#列表内的值实现连乘
def multiply_list(myList):
    result = 1
    for x in myList:
        result = result * x
    return result
```

```
#将每次的梯度展示出来
def describe_plt_grad(layer_num, grad_list, epoch):
    plt.plot(range(-layer_num, 0, 1), grad_list[::-1, 0], label='θ0')
    plt.plot(range(-layer_num, 0, 1), grad_list[::-1, 1], label='θ1')
    plt.xlabel(u'网络层数')
    plt.ylabel(u'梯度')
    plt.title("第{}次梯度下降前-从隐藏层到输入层的梯度变化".format(epoch))
    plt.savefig('从隐藏层到输入层的梯度变化.jpg')
    plt.legend(loc='upper right')
    plt.legend()
    plt.show()
if __name__ == "__main__":
    #plt_dy_sigmoid()
    #模拟神经元的梯度消失和梯度爆炸
    #网络的层数
    layer = 5
    #初始权重缩放因子
    scale = 1
    #学习率
    alpha = 0.1
    #学习次数
    epoch = 1
    #初始 x 的值
    input_x = 2.
    #初始权重
    theta_init = np.ones([layer, 2]) *scale
    #选择的激活函数
    type = 'g'
    for items in range(epoch):
        #input
        X = np.array([[input_x], ])
        X_forward = [X]
        #前向传播时的权重
        theta_forward = []
        #5 个神经元层
        for i in range(layer):
            #插入偏置项后成为输入 a1 层,变为 3 个神经元
            a1 = np.insert(X, 0, 1, axis=1)
            theta = theta_init[i]
            #记录每次前向传播时的权重
            theta_forward.append(theta)
            a1_g1 = a1 @theta
            X = np.reshape(g(X, type), [1, 1])
            #记录每个层的输出值
            X_forward.append(X)
            #print('前向传播:第{}层的输出{}'.format(i, X))
        #损失函数 loss=0.5 *(y-pred_y)**2
        #反向传播,求导过程是先从损失函数向隐藏层开始
        #误差
        d = X_forward[-1] - 1
        #输出层求导时不会乘上一层的 theta,所以用 1 来代替
        theta = np.array([1, 1])
```

```
        #存储每层的梯度
        grad_theta = np.zeros([layer, 2])
        dy_list = []
        for j in range(-1, -layer - 1, -1):
            dy = d *grad_g(X_forward[j], type) *theta
            #存储每次 bp 时的导数值
            dy_list.append(dy)
            #需要求偏置项的梯度
            X = np.insert(X_forward[j - 1], 0, 1, axis=1)
            #每个隐藏层的梯度。上几层的导数值相乘,以方便求下一层的梯度
            grad = multiply_list(dy_list) *X
            #取历史 theta 中的值
            theta = theta_forward[j]
            print('反向传播时：第{}层的梯度：{}'.format(j, grad))
            grad_theta[j] = grad
        #将反向传播获取的梯度传给绘图软件,查看其趋势
        if items == 0: describe_plt_grad(layer, grad_theta, items + 1)
        #梯度下降
        theta_init = theta_init - alpha *grad_theta
        [print ('梯度下降后：第{}层的权重:{}'.format (index, theta)) for index,
theta in enumerate(theta_init)]
        print('#' *50 + str(items) + "次梯度下降")
```

运行结果如下：

```
反向传播时：第-1 层的梯度：[[-0.0764797   -0.05058703]]
反向传播时：第-2 层的梯度：[[0.00584568 0.00391506]]
反向传播时：第-3 层的梯度：[[-0.00044562 -0.00031505]]
反向传播时：第-4 层的梯度：[[3.35538517e-05 2.95541345e-05]]
反向传播时：第-5 层的梯度：[[-2.36618941e-06 -4.73237883e-06]]
梯度下降后：第 0 层的权重:[1.00000024 1.00000047]
梯度下降后：第 1 层的权重:[0.99999664 0.99999704]
梯度下降后：第 2 层的权重:[1.00004456 1.0000315 ]
梯度下降后：第 3 层的权重:[0.99941543 0.99960849]
梯度下降后：第 4 层的权重:[1.00764797 1.0050587 ]
##################################################0 次梯度下降
```

从代码运行结果可以发现网络在第一4 层时参数 θ 已接近 0,梯度下降后第 0 层基本没有变化(初始 $\theta=0.1$),而第 4 层 θ 下降力度较第 0 层大。同时也可以观察到梯度值,离输出层越近其梯度值越大,越容易学习,离输入层越近,其梯度值越容易消失,如图 2-78 所示。

继续使用代码,但是调节参数 layer＝50,scale＝100,epoch＝10,alpha＝10,input_x＝255,运行结果如下：

```
梯度下降后：第 37 层的权重:[nan nan]
梯度下降后：第 38 层的权重:[nan nan]
梯度下降后：第 39 层的权重:[-8.05887799e+284 -2.64133330e+255]
梯度下降后：第 40 层的权重:[-4.06074124e+145 -1.24980456e+131]
```

其梯度变化如图 2-79 所示。

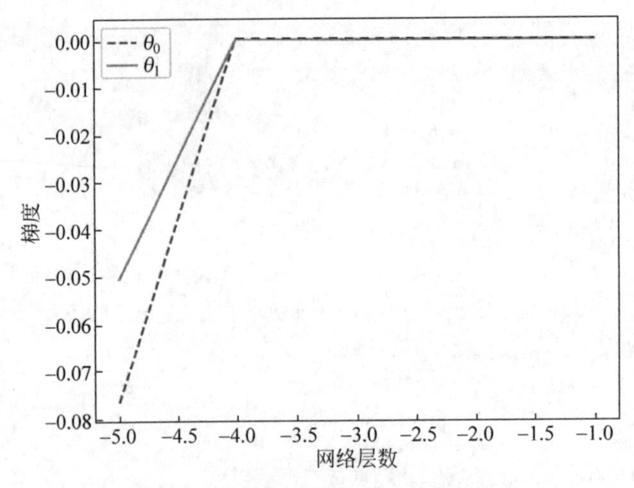

图 2-78 从隐藏层到输入层的梯度消失

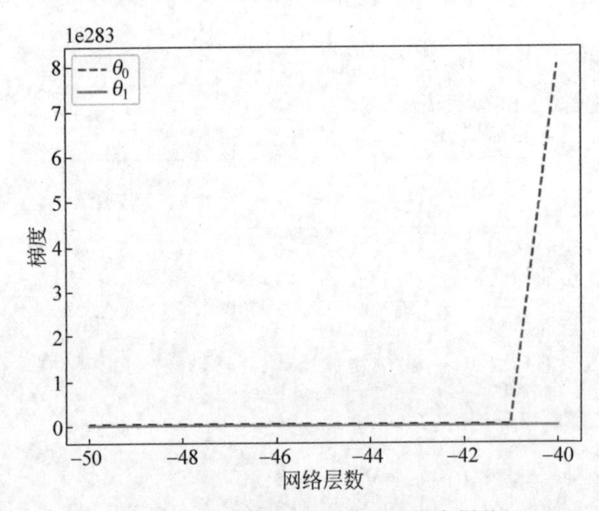

图 2-79 从隐藏层到输入层的梯度爆炸

随着初始的权重参数、学习率、输入值、网络层数的增加，梯度也越来越大，甚至超过了计算机计算的范围，也就是梯度爆炸，其主要原因仍然是由于反向传播连乘时某些值较大，连乘后也就更大。

综上可知，无论是梯度消失还是梯度爆炸都跟神经网络反向传播算法有关，这属于反向传播算法的先天不足，但是这两者也有一些区别，总结如下。

(1) 梯度消失：网络模型层数、激活函数。

(2) 梯度爆炸：网络模型层数、激活函数(某些激活函数可能大于 1)、初始权重、学习率、输入值。

根据产生梯度不稳定的原因，缓解梯度爆炸的主要方法有减少网络层数、将初始权重以高斯分布的形态分布在[0,1]、学习率小一些、将输入值归一化在[0,1]、梯度截断。

改变代码的参数 layer＝50、scale＝0.01、alpha＝0.0001、input_x＝1，运行代码后可以

发现在相同的 50 个网络层中，此时梯度爆炸没有了，但是出现了梯度消失，如图 2-80 所示。

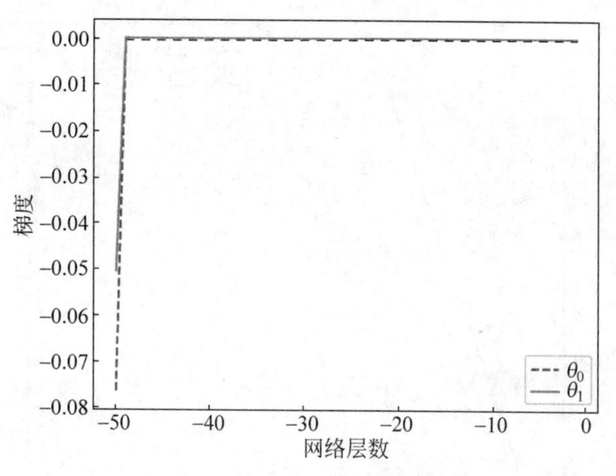

图 2-80 50 层模型网络梯度变化

如果改变代码的参数 layer=2、scale=0.01、alpha=0.0001、input_x=1，则可同时缓解梯度爆炸和梯度消失，如图 2-81 所示。

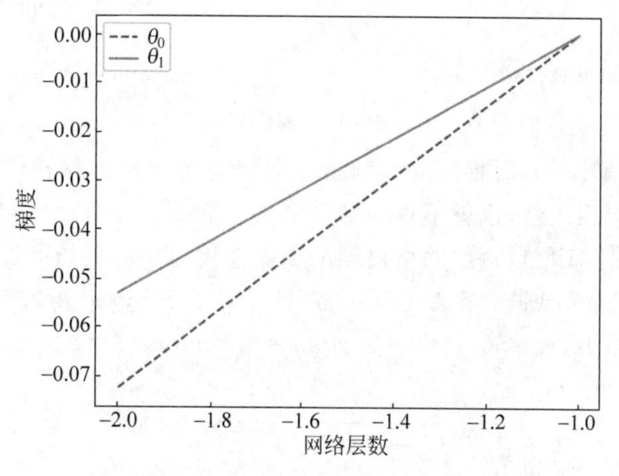

图 2-81 2 层模型网络梯度变化

梯度截断，即设定某个区间[a,b]，如果当前梯度小于 a，则将梯度设为 a，如果当前梯度大于 b，则将梯度设为 b，增加如下代码：

```
#第 2 章/machineLearn/梯度消失梯度爆炸.py
#截断区间
grad[grad[:, :] > -0.000001] = -0.000001
grad[grad[:, :] > 100] = 100
```

缓解梯度消失的方法有减少网络层数、梯度截断、更换激活函数、数据归一化、正则化、使用残差、权重滑动平均等。

更换激活函数，例如使用 ReLU 函数 OutPut $=\max(0,x)$，其函数图像和导数图像如

图 2-82 所示。

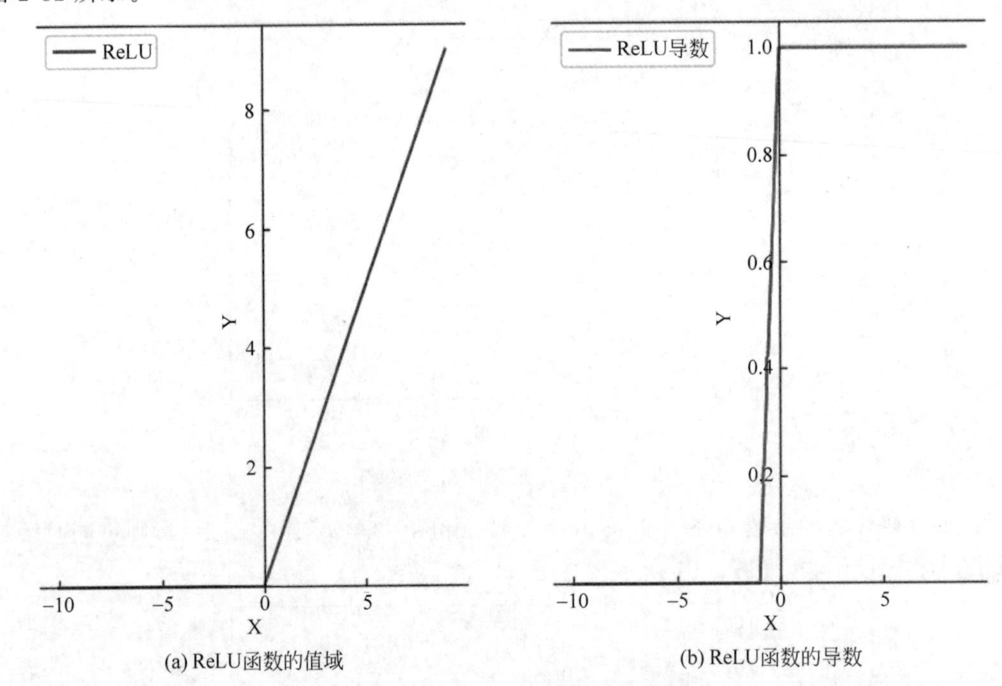

(a) ReLU函数的值域　　(b) ReLU函数的导数

图 2-82　ReLU 函数的值域和导数

当 ReLU 函数值小于 0 时梯度为 0，那么该神经元也就为 0，该神经元就不会被激活，从而使模型网络更具有稀疏性，也就不存在梯度消失的问题。当 ReLU 函数值大于 0 时梯度为 1，那么反向传播时都乘 1，每层网络得到的更新速度是一样的，因此可以缓解梯度爆炸。另一个优点是，ReLU 求导非常容易，其收敛速度相对于 Sigmoid 函数更快。

改变代码的参数 layer＝50、scale＝1、alpha＝0.1、input_x＝2、type＝'relu'、epoch＝2，运行结果如下：

```
梯度下降后：第 17 层的权重：[0.819 0.672]
梯度下降后：第 18 层的权重：[0.81 0.64]
梯度下降后：第 19 层的权重：[0.8 0.6]
##################################################1 次梯度下降
```

其隐藏层梯度变化如图 2-83 所示。

可见 ReLU 函数在一定程度上缓解了梯度不稳定的现象，使模型网络能够做得更深，但是当 layer＝50、其他参数的值不变时，其梯度已趋近小数点后 6 位，梯度下降后的值并没有变化。

当参数 layer＝50，scale＝100，epoch＝10，alpha＝10，input_x＝255、type＝'relu'时，仍然出现了梯度爆炸的情况。

正则化会对 θ 进行惩罚，当 θ 大时就惩罚得大一些，当 θ 小时就惩罚得小一些，其原理已在前文讨论过，也能起到缓解梯度不稳定的情况。

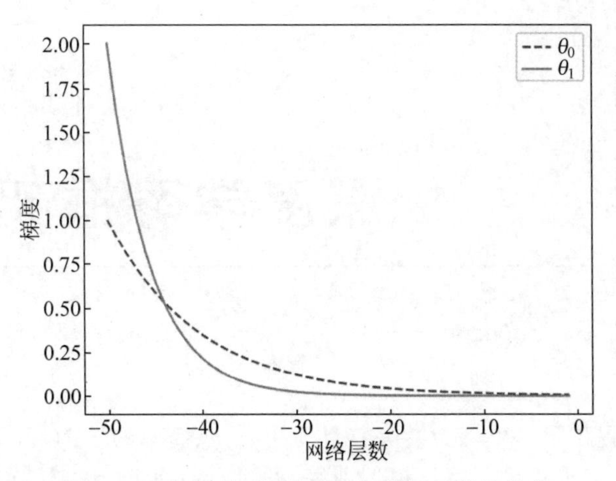

图 2-83 20 层模型网络 ReLU 激活函数的梯度

读者可以自行调节本书代码中的参数，观察梯度的变化，从而对于梯度消失和梯度爆炸有更直观的理解。

总结

由于反向传播算法基于链式求导法则，所以当其中每几层中的权重值过大或过小，则会出现梯度爆炸和梯度消失的情况。

练习

使用本节代码，调整不同的网络层、不同的学习次数，以及不同的输入值，观察梯度爆炸和梯度消失的表现。

第3章 深度学习框架

深度学习框架提供了一整套封装完善且使用方便、简单的API,可以供人们快速地实现各种算法模型,同时它也提供了算法模型训练、评价、调优的方法,是深度学习领域工作、研究的神兵利器。

目前市场占有率最高的框架主要有TensorFlow和PyTorch,它们的核心都是基于张量和计算图进行操作的。TensorFlow各个版本的API管理相对烦琐,学习较困难,而PyTorch要容易得多,但是作为一个深度学习工程师,无论TensorFlow还是PyTorch都需要掌握,因为两者都可能会在工作中被用到。尤其TensorFlow在工业界中占有非常重要的地位,这是无法避开的,因此本章将重点梳理TensorFlow的API,同时也将介绍PyTorch,通过阅读本章可以实现TensorFlow和PyTorch相关API的无缝衔接,帮助读者开启实战深度学习之旅!

3.1 基本概念

在深度学习框架中有4种数据类型,分别如下:

(1) 标量,也就是常量。

(2) 向量,一维数据。

(3) 矩阵,二维数组。

(4) 张量,三维及以上的数据。

但是在广义上它们都被称为张量,又被称为Tensor。无论是TensorFlow还是PyTorch或者其他框架的核心都是基于张量的数据操作。

另一个比较重要的概念是计算图,即将算法模型的计算过程图形化地展示出来,可以方便地查看各个变量之间的关系及张量的流向,如图3-1所示。

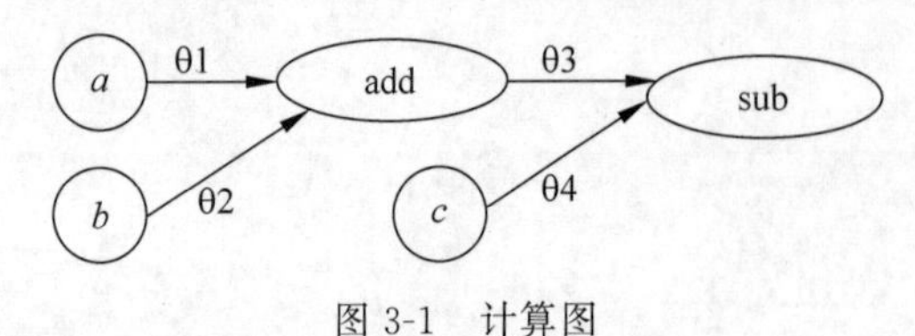

图3-1 计算图

a、b表示数据的输入,add表示数据的运算方法,又称为算子,其实现的计算为 $y=(a\times\theta1+b\times\theta2)\times\theta3-c\times\theta4$。深度学习框架会自动生成计算图,如图3-2所示。

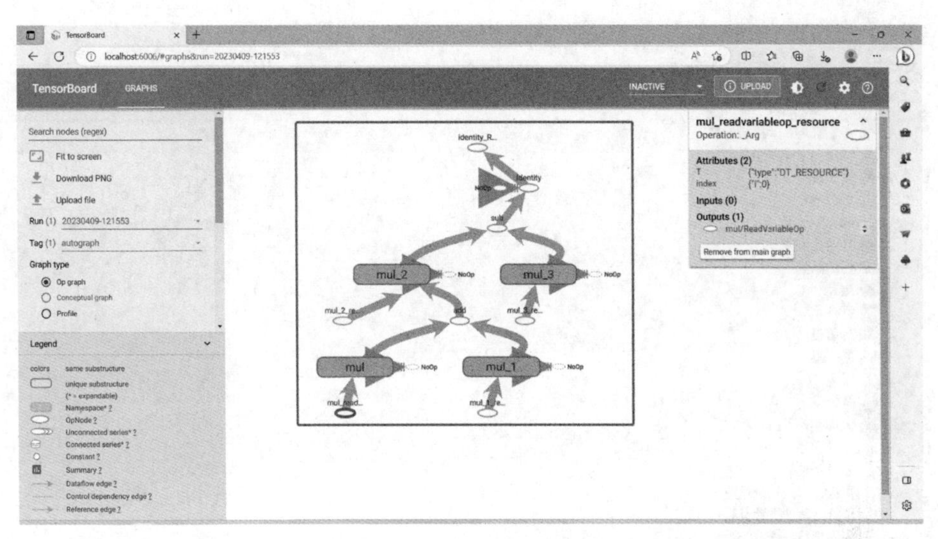

图 3-2 自动生成计算图

TensorFlow 2.x 中提供了动态图和静态图，而 PyTorch 提供的是动态图。所谓静态图是指先构建计算图，然后进行运算，虽然运行效率高，但是调试不方便，也不够灵活。动态图则是运算与计算图的构建同时进行，相对灵活且容易调试，但是运行效率有所下降。

例如静态图中需要先定义规则 $y=(a\times\theta1+b\times\theta2)\times\theta3-c\times\theta4$，然后将结构存储起来，然后把 $\theta1=1,\theta2=2,\theta3=3,\theta4=4$，放到计算图中计算。动态图则是 $y=(a\times1+b\times2)\times3-c\times4$，同时构建计算规则并填充数据，其区别可以类比为，建造房子时先搭建整个房子的框架，再砌砖，这就是静态图。边搭框架边砌砖，这就是动态图。

总结

TensorFlow 2.x 中提供了动态图和静态图，而 PyTorch 提供的是动态图。

练习

运行代码“第 3 章/TensorFlowAPI/计算图的生成.py”在 TensorBoard 中查看计算图的结构。

3.2 环境搭建

深度学习框架的运行环境分为 CPU 和 GPU(显卡)版本，安装 GPU 的 TensorFlow 运行环境，需要满足以下条件：

(1) 显卡支持 CUDA 架构。CUDA 是一种由 NVIDIA 推出的通用并行计算架构，该架构使 GPU 能够解决复杂的计算问题。

本机是否支持 CUDA 架构，可以在设备管理器中查看显卡的型号，如图 3-3 所示。

CUDA 支持的显卡型号及算力，可以在网站 https://developer.nvidia.com/zh-cn/cuda-gpus#compute 中可看，如图 3-4 所示。

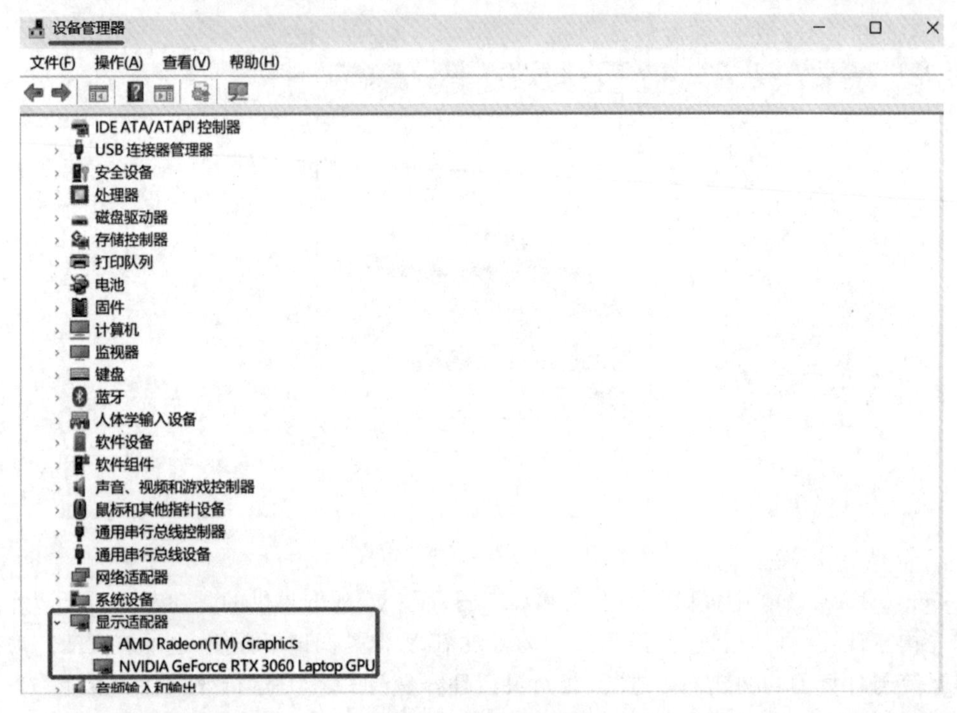

图 3-3 设备管理器

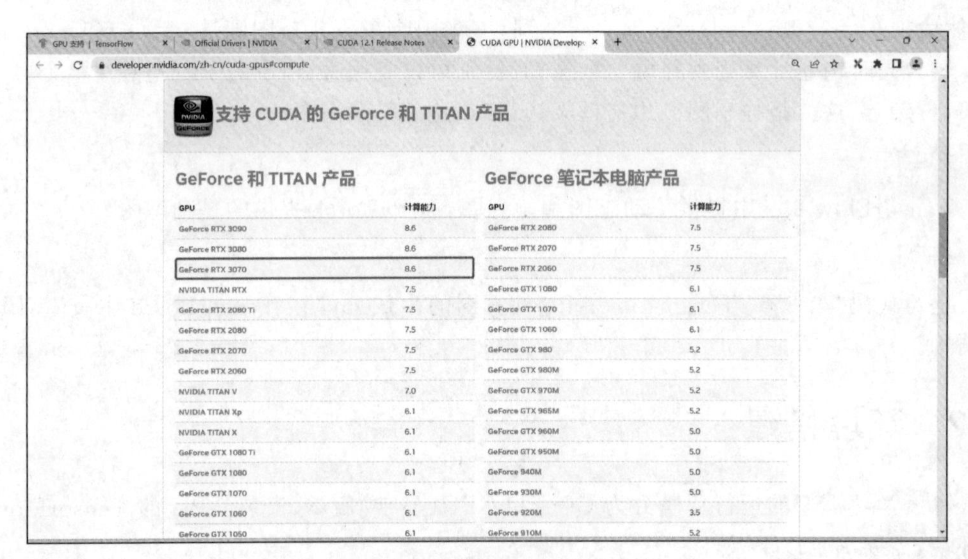

支持 CUDA 的 GeForce 和 TITAN 产品

GeForce 和 TITAN 产品

GPU	计算能力
GeForce RTX 3090	8.6
GeForce RTX 3080	8.6
GeForce RTX 3070	8.6
NVIDIA TITAN RTX	7.5
GeForce RTX 2080 Ti	7.5
GeForce RTX 2080	7.5
GeForce RTX 2070	7.5
GeForce RTX 2060	7.5
NVIDIA TITAN V	7.0
NVIDIA TITAN Xp	6.1
NVIDIA TITAN X	6.1
GeForce GTX 1080 Ti	6.1
GeForce GTX 1080	6.1
GeForce GTX 1070	6.1
GeForce GTX 1060	6.1
GeForce GTX 1050	6.1

GeForce 笔记本电脑产品

GPU	计算能力
GeForce RTX 2080	7.5
GeForce RTX 2070	7.5
GeForce RTX 2060	7.5
GeForce GTX 1080	6.1
GeForce GTX 1070	6.1
GeForce GTX 1060	6.1
GeForce GTX 980	5.2
GeForce GTX 980M	5.2
GeForce GTX 970M	5.2
GeForce GTX 965M	5.2
GeForce GTX 960M	5.0
GeForce GTX 950M	5.0
GeForce 940M	5.0
GeForce 930M	5.0
GeForce 920M	3.5
GeForce 910M	5.2

图 3-4 CUDA 支持显卡型号

一般来说集成显卡都不支持，独立显卡最少要在 1GB 以上才支持 GPU 的深度学习框架的运行。

(2) CUDA Toolkit 的版本与本机计算机的显卡驱动应保持一致。CUDA Toolkit 是 CUDA 开发工具包，主要包含了 CUDA-C、CUDA-C++的编译器、科学库、示例程序等。

通常在安装显卡驱动时会默认安装 CUDA Driver，但是不会安装 CUDA Toolkit。因为只安装 CUDA Driver 就可以正常办公了，而 CUDA Toolkit 通常是为了开发工作的需要才会被安装。

检查 CUDA Toolkit 与显卡驱动版本是否匹配，可以在网站 https://docs.nvidia.com/cuda/cuda-toolkit-release-notes/index.html#title-resolved-issues 中进行，如图 3-5 所示。

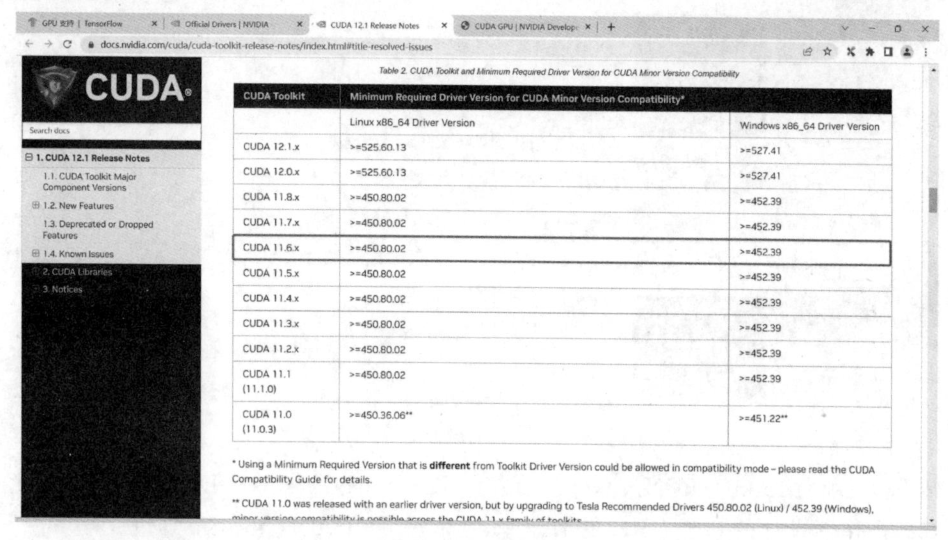

Table 2. CUDA Toolkit and Minimum Required Driver Version for CUDA Minor Version Compatibility

CUDA Toolkit	Minimum Required Driver Version for CUDA Minor Version Compatibility*	
	Linux x86_64 Driver Version	Windows x86_64 Driver Version
CUDA 12.1.x	>=525.60.13	>=527.41
CUDA 12.0.x	>=525.60.13	>=527.41
CUDA 11.8.x	>=450.80.02	>=452.39
CUDA 11.7.x	>=450.80.02	>=452.39
CUDA 11.6.x	>=450.80.02	>=452.39
CUDA 11.5.x	>=450.80.02	>=452.39
CUDA 11.4.x	>=450.80.02	>=452.39
CUDA 11.3.x	>=450.80.02	>=452.39
CUDA 11.2.x	>=450.80.02	>=452.39
CUDA 11.1 (11.1.0)	>=450.80.02	>=452.39
CUDA 11.0 (11.0.3)	>=450.36.06**	>=451.22**

* Using a Minimum Required Version that is **different** from Toolkit Driver Version could be allowed in compatibility mode – please read the CUDA Compatibility Guide for details.

** CUDA 11.0 was released with an earlier driver version, but by upgrading to Tesla Recommended Drivers 450.80.02 (Linux) / 452.39 (Windows),

图 3-5　显卡驱动与 CUDA Toolkit 版本对应

在计算机中打开 NVIDIA 控制面板可以查看当前显卡驱动版本，如图 3-6 所示。

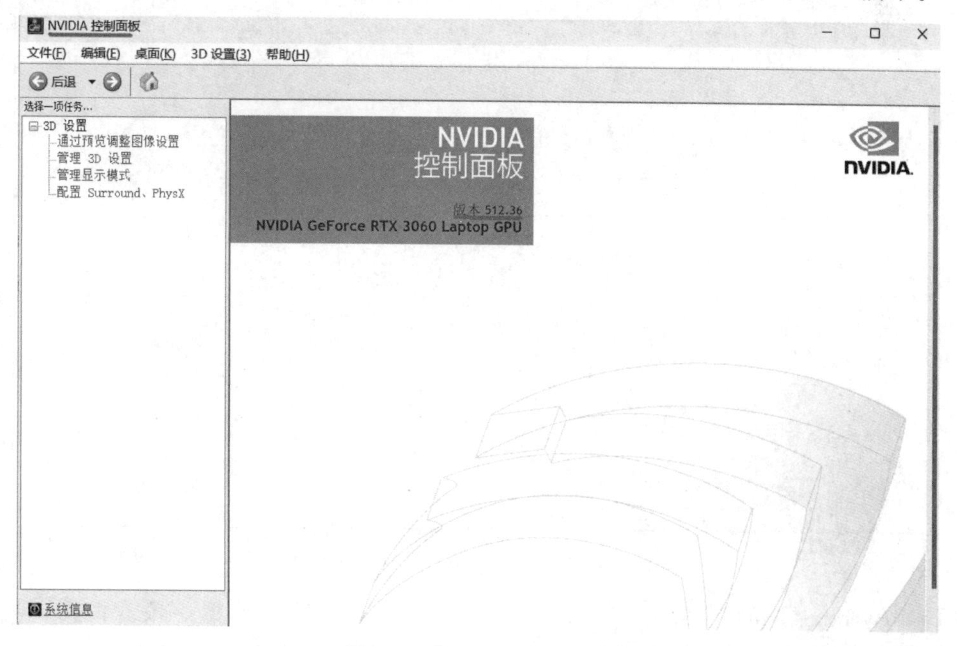

图 3-6　本机显卡驱动版本

如果驱动版本太低,则可以在网站 https://www.nvidia.com/download/index.aspx?lang=en-us 中下载驱动并进行升级,如图 3-7 所示。

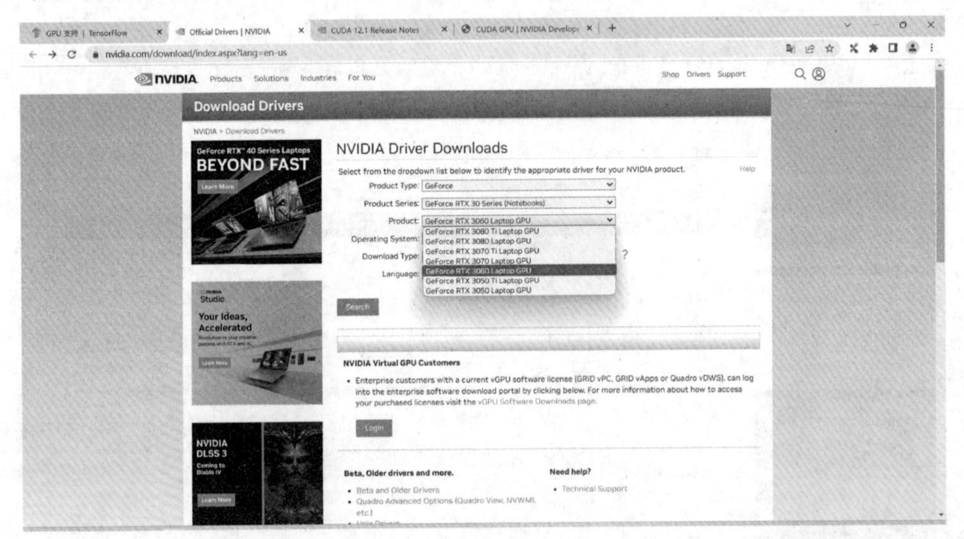

图 3-7 选择显卡型号并下载驱动

(3) CUDA Toolkit、cuDNN 版本应保持一致。cuDNN 是 NVIDIA 针对深度神经网络中的基础操作而设计的基于 GPU 的加速库,其依赖 CUDA Toolkit 工具包。

其依赖关系可以查看网站 https://developer.nvidia.com/rdp/cudnn-archive,如图 3-8 所示。

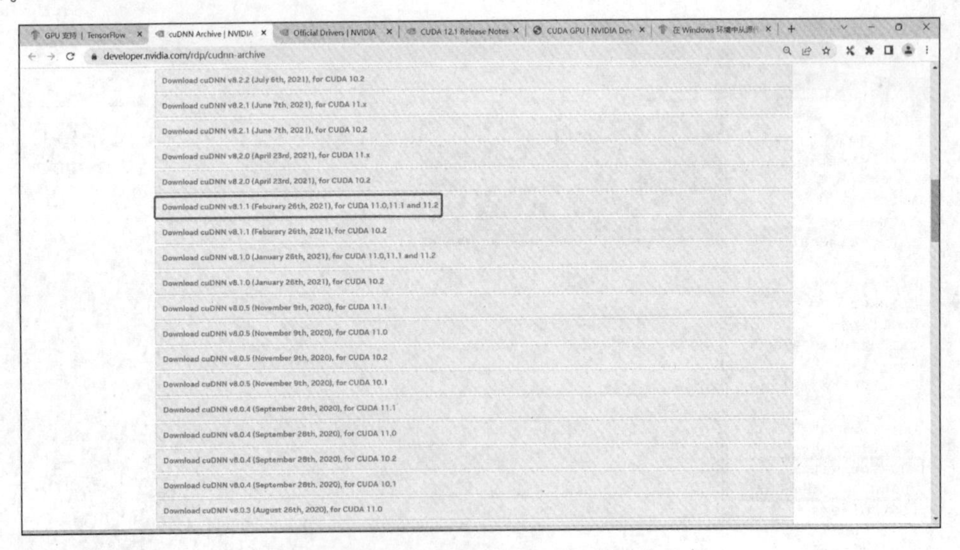

图 3-8 cuDNN 与 CUDA Toolkit 的对应关系

(4) Python、TensorFlow、CUDA Toolkit、cuDNN 版本应保持一致。根据 cuDNN、CUDA 的版本就可以在网站中选择 Python、TensorFlow 的版本进行安装,网站的网址为

https://tensorflow. google. cn/install/source_windows?hl=zh-cn#gpu，如图 3-9 所示。

GPU

版本	Python 版本	编译器	构建工具	cuDNN	CUDA
tensorflow_gpu-2.6.0	3.6-3.9	MSVC 2019	Bazel 3.7.2	8.1	11.2
tensorflow_gpu-2.5.0	3.6-3.9	MSVC 2019	Bazel 3.7.2	8.1	11.2
tensorflow_gpu-2.4.0	3.6-3.8	MSVC 2019	Bazel 3.1.0	8.0	11.0
tensorflow_gpu-2.3.0	3.5-3.8	MSVC 2019	Bazel 3.1.0	7.6	10.1
tensorflow_gpu-2.2.0	3.5-3.8	MSVC 2019	Bazel 2.0.0	7.6	10.1
tensorflow_gpu-2.1.0	3.5-3.7	MSVC 2019	Bazel 0.27.1-0.29.1	7.6	10.1
tensorflow_gpu-2.0.0	3.5-3.7	MSVC 2017	Bazel 0.26.1	7.4	10
tensorflow_gpu-1.15.0	3.5-3.7	MSVC 2017	Bazel 0.26.1	7.4	10
tensorflow_gpu-1.14.0	3.5-3.7	MSVC 2017	Bazel 0.24.1-0.25.2	7.4	10
tensorflow_gpu-1.13.0	3.5-3.7	MSVC 2015 update 3	Bazel 0.19.0-0.21.0	7.4	10
tensorflow_gpu-1.12.0	3.5-3.6	MSVC 2015 update 3	Bazel 0.15.0	7.2	9.0
tensorflow_gpu-1.11.0	3.5-3.6	MSVC 2015 update 3	Bazel 0.15.0	7	9
tensorflow_gpu-1.10.0	3.5-3.6	MSVC 2015 update 3	Cmake v3.6.3	7	9
tensorflow_gpu-1.9.0	3.5-3.6	MSVC 2015 update 3	Cmake v3.6.3	7	9
tensorflow_gpu-1.8.0	3.5-3.6	MSVC 2015 update 3	Cmake v3.6.3	7	9
tensorflow_gpu-1.7.0	3.5-3.6	MSVC 2015 update 3	Cmake v3.6.3	7	9
tensorflow_gpu-1.6.0	3.5-3.6	MSVC 2015 update 3	Cmake v3.6.3	7	9

图 3-9 选择 TensorFlow、Python 版本进行安装

手动配置 TensorFlow 的 GPU 版本比较复杂，推荐使用 conda 命令进行安装，conda 会根据本机的 GPU 型号和驱动版本自动安装 CUDA Toolkit 和 cuDNN。进入"1. 10 包的管理"节由 conda 命令创建的 StudyDNN 环境后，执行命令 conda install tensorflow-gpu=2. 6. 0，如图 3-10 所示。

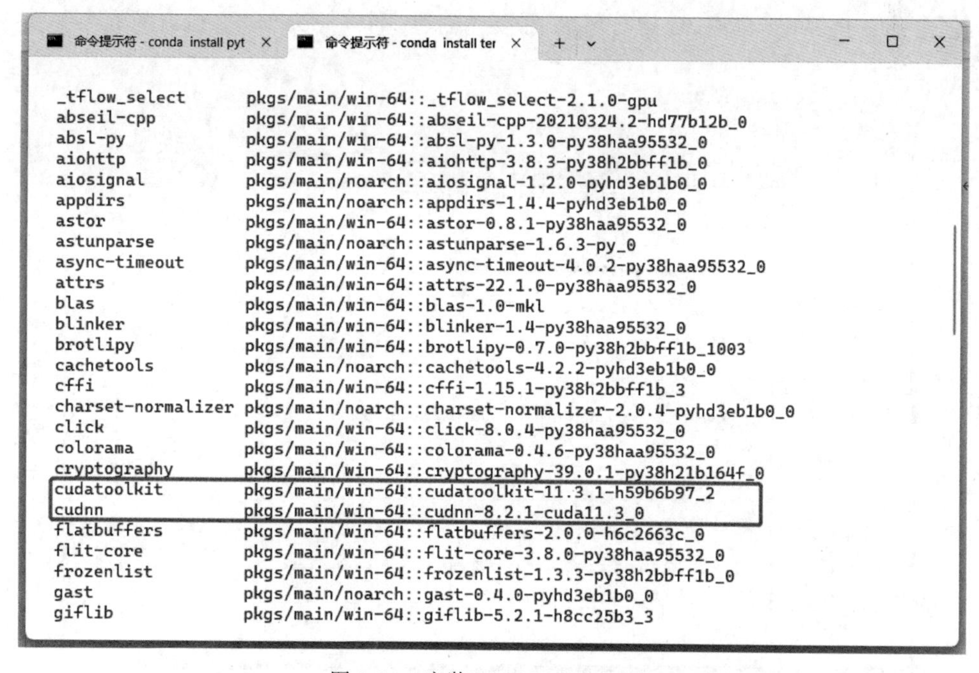

```
_tflow_select        pkgs/main/win-64::_tflow_select-2.1.0-gpu
abseil-cpp           pkgs/main/win-64::abseil-cpp-20210324.2-hd77b12b_0
absl-py              pkgs/main/win-64::absl-py-1.3.0-py38haa95532_0
aiohttp              pkgs/main/win-64::aiohttp-3.8.3-py38h2bbff1b_0
aiosignal            pkgs/main/noarch::aiosignal-1.2.0-pyhd3eb1b0_0
appdirs              pkgs/main/noarch::appdirs-1.4.4-pyhd3eb1b0_0
astor                pkgs/main/win-64::astor-0.8.1-py38haa95532_0
astunparse           pkgs/main/noarch::astunparse-1.6.3-py_0
async-timeout        pkgs/main/win-64::async-timeout-4.0.2-py38haa95532_0
attrs                pkgs/main/win-64::attrs-22.1.0-py38haa95532_0
blas                 pkgs/main/win-64::blas-1.0-mkl
blinker              pkgs/main/win-64::blinker-1.4-py38haa95532_0
brotlipy             pkgs/main/win-64::brotlipy-0.7.0-py38h2bbff1b_1003
cachetools           pkgs/main/noarch::cachetools-4.2.2-pyhd3eb1b0_0
cffi                 pkgs/main/win-64::cffi-1.15.1-py38h2bbff1b_3
charset-normalizer   pkgs/main/noarch::charset-normalizer-2.0.4-pyhd3eb1b0_0
click                pkgs/main/win-64::click-8.0.4-py38haa95532_0
colorama             pkgs/main/win-64::colorama-0.4.6-py38haa95532_0
cryptography         pkgs/main/win-64::cryptography-39.0.1-py38h21b164f_0
cudatoolkit          pkgs/main/win-64::cudatoolkit-11.3.1-h59b6b97_2
cudnn                pkgs/main/win-64::cudnn-8.2.1-cuda11.3_0
flatbuffers          pkgs/main/win-64::flatbuffers-2.0.0-h6c2663c_0
flit-core            pkgs/main/win-64::flit-core-3.8.0-py38haa95532_0
frozenlist           pkgs/main/win-64::frozenlist-1.3.3-py38h2bbff1b_0
gast                 pkgs/main/noarch::gast-0.4.0-pyhd3eb1b0_0
giflib               pkgs/main/win-64::giflib-5.2.1-h8cc25b3_3
```

图 3-10 安装 TensorFlow 2. 6. 0

然后在命令行中执行以下代码测试是否成功地安装了 GPU 版本,如图 3-11 所示。

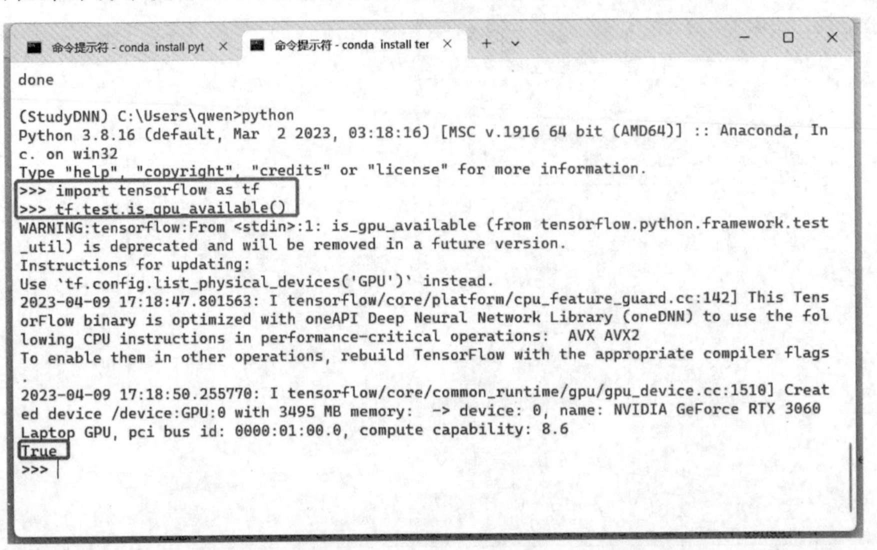

图 3-11　TensorFlow 2.6.0 安装成功 GPU 版本

注意：如果你的计算机是第 1 次安装深度学习框架 GPU 版本，则应先升级显卡驱动；conda 命令可以对相关依赖包进行下载并安装，而 pip 默认下载最新版本，pip 安装的库有可能出现不兼容问题，因此推荐使用 conda 命令安装。

创建一个 PyTorch 的 GPU 运行环境,需要先创建一个 conda 环境,执行的命令如图 3-12 所示。

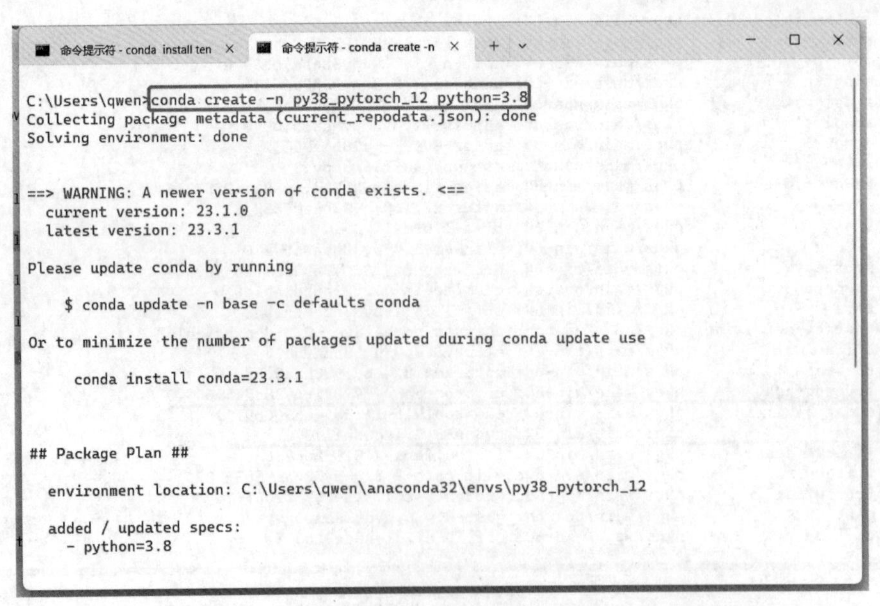

图 3-12　conda 创建环境

然后进入网站 https://pytorch.org/get-started/previous-versions/选择一个版本对应的命令来执行,如图 3-13 所示。

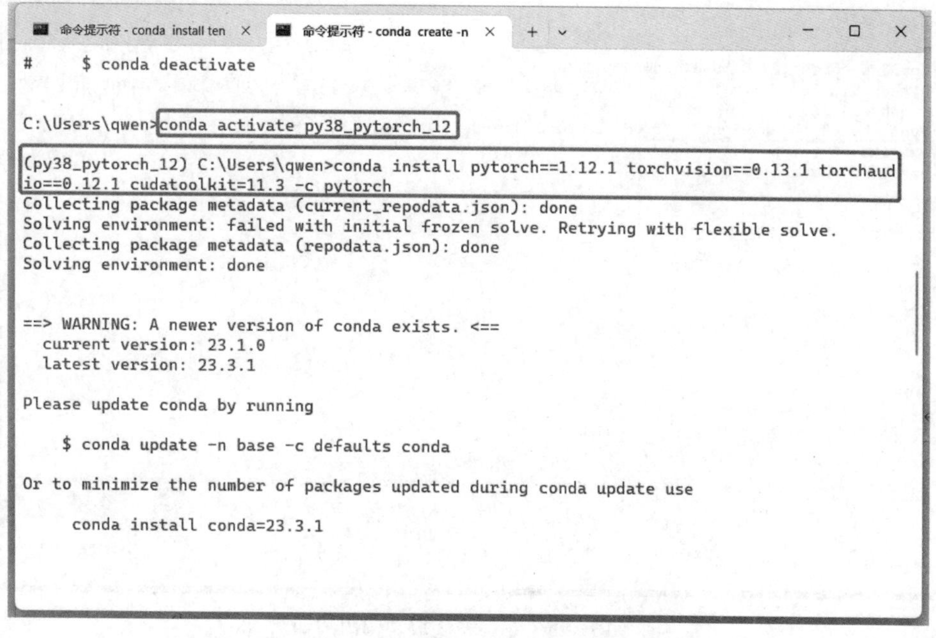

图 3-13　conda 创建 PyTorch 的 GPU 环境

同样 conda 会自动安装 CUDA Toolkit,如图 3-14 所示。

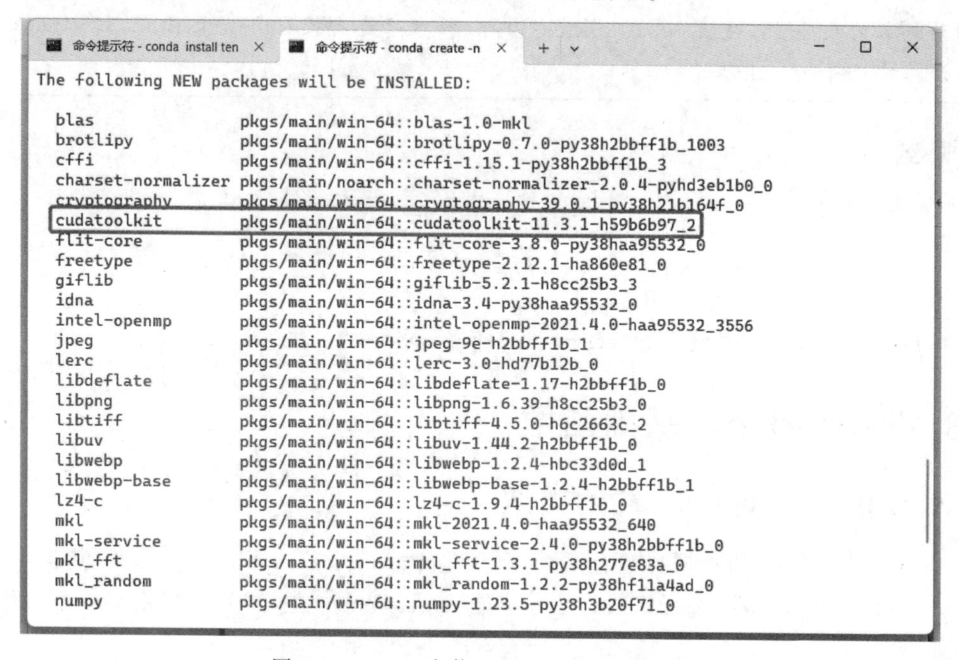

图 3-14　conda 安装 PyTorch 的 CUDA 库

然后在命令行中执行以下代码测试是否成功地安装了 GPU 版本，如图 3-15 所示。

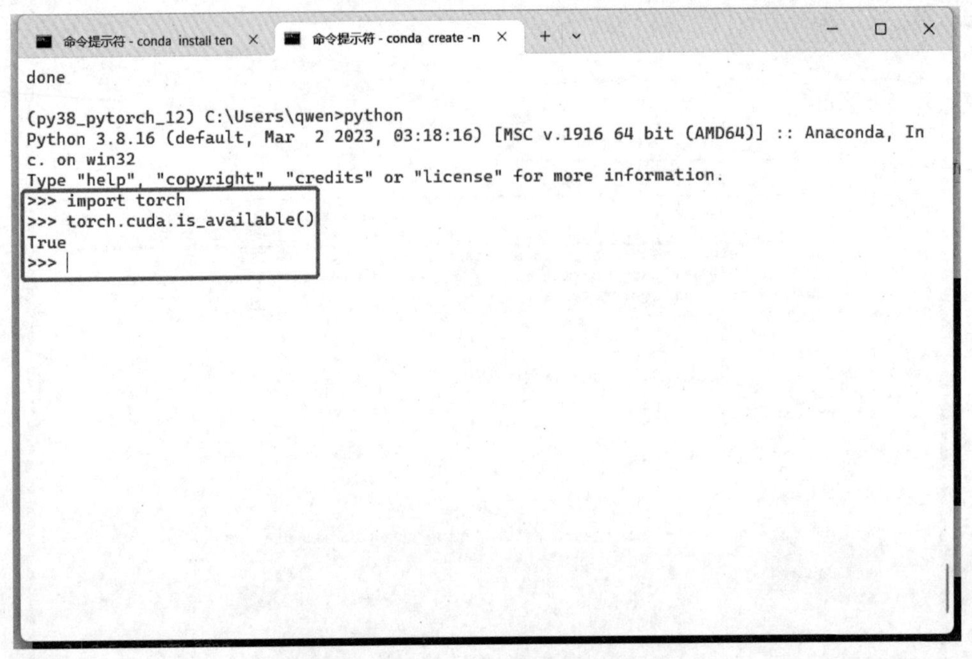

图 3-15 PyTorch 安装成功 GPU 环境

注意：不要将 TensorFlow 和 PyTorch 安装在同一环境中，否则会冲突；笔者当前计算机显卡驱动版本为 512.36，对应 cudatoolkit=11.3，所以只能安装 PyTorch 11.2，读者应根据自己的计算机选择相应版本。

总结

CUDA 环境的安装可使用 conda 命令一键安装。安装的版本需要跟自己的显卡型号相匹配。

练习

在自己的计算机中搭建 TensorFlow 和 PyTorch 运行环境。

3.3 TensorFlow 基础函数

3.3.1 TensorFlow 初始类型

TensorFlow 主要的数据类型有两种，一种是 ResourceVariable 变量类型，其初始后的值可以通过 assign() 函数进行修改；另一种是 EagerTensor 类型，其值域定义后不可修改，代码如下：

```
#第 3 章/TensorFlowAPI/基本函数.py
import tensorflow as tf
import numpy as np

#(1) 创建标量
scalar = tf.constant([[1., 2]], dtype=tf.float32)
print('类型:', type(scalar))
#(2) variable 是一个初始的变量值,随着代码的推进会变化
theta = tf.Variable([[1.], [2.]], dtype=tf.float32)
print('类型:', type(theta))
#(3) 将 NumPy 转换为 Tensor
rnd_numpy = np.random.randn(4, 3)
np_tensor = tf.convert_to_tensor(rnd_numpy)
print("将其他对象转换成 Tensor:", type(np_tensor))
#(4) 将张量转换成其他数据类型
type_conversion = tf.cast(theta, dtype=tf.int8)
print("Tensor 数据类型转换：", type_conversion)

#EagerTensor 对象不能修改,而 ResourceVariable 对象可以修改
theta = theta.assign([[5.], [6.]])
print('ResourceVariable 对象修改后的值:', theta)
#以下语句会报错,因为 EagerTensor 对象不能修改
#EagerTensor' object has no attribute 'assign'
scalar.assign([[11., 22]])
```

运行结果如下：

```
类型: <class 'Tensorflow.python.framework.ops.EagerTensor'>
类型: <class 'Tensorflow.python.ops.resource_variable_ops.ResourceVariable'>

将其他对象转换成 Tensor: <class 'Tensorflow.python.framework.ops.EagerTensor'>
Tensor 数据类型转换: tf.Tensor(
[[1]
[2]], shape=(2, 1), dtype=int8)
ResourceVariable 对象修改后的值: <tf.Variable 'UnreadVariable' shape=(2, 1)
dtype=float32, NumPy=
array([[5.],
       [6.]], dtype=float32)>
Traceback (most recent call last):
  File "D:/DLAI/TensorFlowAPI/基本函数.py", line 24, in <module>
    scalar.assign([[11., 22]])
  File "C:\Users\qwen\anaconda32\envs\py38_tf26\lib\site-packages\TensorFlow\
python\framework\ops.py", line 401, in __getattr__
    self.__getattribute__(name)
AttributeError: 'Tensorflow.python.framework.ops.EagerTensor' object has no
attribute 'assign'
```

tf. Variable()创建的是 ResourceVariable 类型，tf. constant()、tf. convert_to_tensor()、tf. cast()创建的是 EagerTensor 类型，scalar. assign([[11. ,22]])意图对 EagerTensor 类型进行修改，所以会报'Tensorflow. python. framework. ops. EagerTensor' object has no attribute 'assign'错误。

tf.constant()用于创建标量；tf.convert_to_tensor()用于将其他数据类型(例如 NDArray)转换为 EagerTensor；tf.cast()用于将默认的 float32 位数据类型转换为 int 类型。

3.3.2 TensorFlow 指定设备

运行 TensorFlow 可以指定 CPU 或者 GPU 运行(GPU 需要显卡支持)，其语句主要用 tf.device(device_name)来指定，一般来说如果安装的是 TensorFlow-gpu，则会默认用 GPU 来执行，如果 GPU 不存在，则会调 CPU 来运行，代码如下：

```
#第 3 章/TensorFlowAPI/指定运行设备.py
import tensorflow as tf
import time

#在此 cpu 作用域语句下面计算执行时间
with tf.device("cpu"):
    t1 = time.time()
    cpu = tf.constant(1)
    print(time.time() - t1)
#在此 gpu 作用域语句下面计算执行时间
with tf.device('gpu'):
    t2 = time.time()
gpu = tf.constant(1)
print(time.time() - t1)
print(cpu.device, gpu.device, sep='\n')
```

运行结果如下：

```
0.0
0.03126859664916992
/job:localhost/replica:0/task:0/device:CPU:0
/job:localhost/replica:0/task:0/device:GPU:0
```

3.3.3 TensorFlow 数学运算

TensorFlow 已封装好一些基本数学运算函数，例如加、减、乘、除、内积、点乘等，其运算过程与 NumPy 中保持一致，代码如下：

```
#第 3 章/TensorFlowAPI/数学运算.py
import tensorflow as tf

data1 = tf.constant([[2., 2.]], dtype=tf.float32)
data2 = tf.constant([[4., 4.]], dtype=tf.float32)
#加法
print(tf.add(data1, data2))
#减法
print(tf.subtract(data1, data2))
#点乘
print(tf.multiply(data1, data2))
```

```
#除法
print(tf.divide(data1, data2))
#内积
print(tf.matmul(data1, tf.transpose(data2)))
#调用 tf.math 下面的函数
print(tf.math.sqrt(data1))
```

运行结果如下：

```
tf.Tensor([[6. 6.]], shape=(1, 2), dtype=float32)
tf.Tensor([[-2. -2.]], shape=(1, 2), dtype=float32)
tf.Tensor([[8. 8.]], shape=(1, 2), dtype=float32)
tf.Tensor([[0.5 0.5]], shape=(1, 2), dtype=float32)
tf.Tensor([[16.]], shape=(1, 1), dtype=float32)
tf.Tensor([[1.4142135 1.4142135]], shape=(1, 2), dtype=float32)
```

tf. math 类下提供了众多常用的数学函数，如图 3-16 所示。

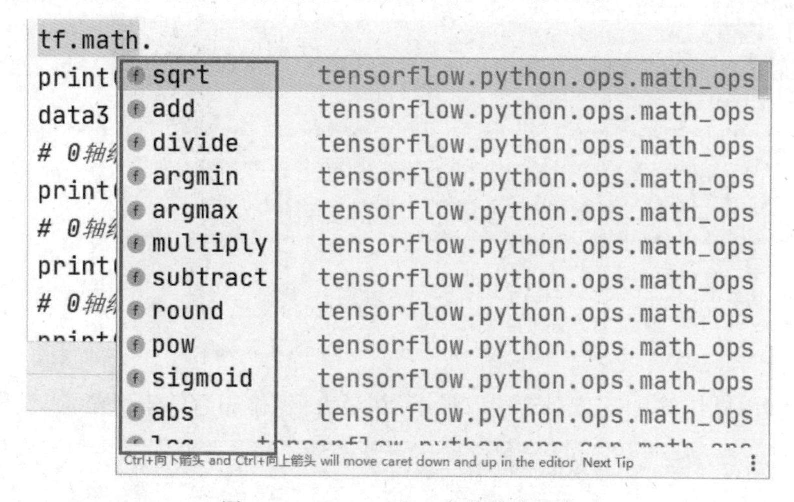

图 3-16 TensorFlow 常见数学函数

根据 axis 轴的位置进行与统计函数相关的操作，代码如下：

```
#第 3 章/TensorFlowAPI/数学运算.py
data3 = tf.convert_to_tensor(
    [[1.1228833, -0.94511896, -1.1491745],
     [0.50430834, -0.1323103, -0.509263]]
)
#1 轴维度最小值
print(tf.reduce_min(data3, axis=1))
#0 轴维度平均值
print(tf.reduce_mean(data3, axis=0))
#0 轴维度求和
print(tf.reduce_sum(data3, axis=0))
#1 轴维度最大
print(tf.reduce_max(data3, axis=1))
```

运行结果如下：

```
tf.Tensor([-1.1491745 -0.509263 ], shape=(2,), dtype=float32)
tf.Tensor([ 0.81359583 -0.53871465 -0.82921875], shape=(3,), dtype=float32)
tf.Tensor([ 1.6271917 -1.0774293 -1.6584375], shape=(3,), dtype=float32)
tf.Tensor([1.1228833  0.50430834], shape=(2,), dtype=float32)
```

代码中计算的维度参照 axis 的位置，由外到内是从 0 轴开始，由于 data3 是二维数据，所以当 axis=0 时竖着算，当 axis=1 时算最内层，其计算顺序如图 3-17 所示。

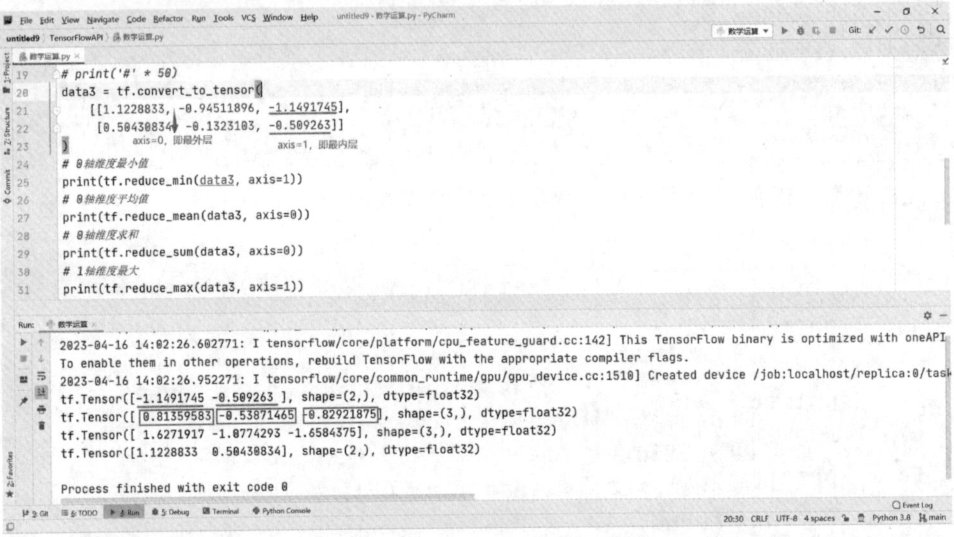

```
# print('#' * 50)
data3 = tf.convert_to_tensor(
    [[1.1228833, -0.94511896, -1.1491745],
     [0.50430834, -0.1323103, -0.509263]]
)
# 0轴维度最小值
print(tf.reduce_min(data3, axis=1))
# 0轴维度平均值
print(tf.reduce_mean(data3, axis=0))
# 0轴维度求和
print(tf.reduce_sum(data3, axis=0))
# 1轴维度最大
print(tf.reduce_max(data3, axis=1))
```

axis=0，即最外层　　axis=1，即最内层

```
tf.Tensor([-1.1491745 -0.509263 ], shape=(2,), dtype=float32)
tf.Tensor([ 0.81359583 -0.53871465 -0.82921875], shape=(3,), dtype=float32)
tf.Tensor([ 1.6271917 -1.0774293 -1.6584375], shape=(3,), dtype=float32)
tf.Tensor([1.1228833  0.50430834], shape=(2,), dtype=float32)

Process finished with exit code 0
```

图 3-17　TensorFlow 中 axis 的计算顺序

另一个常用操作就是根据 axis 获取最小值、最大值、排序后的索引位置，代码如下：

```
#第 3 章/TensorFlowAPI/数学运算.py
data3 = tf.convert_to_tensor(
    [[1.1228833, -0.94511896, -1.1491745],
     [0.50430834, -0.1323103, -0.509263]]
)
#最大值的下标
print(tf.argmax(data3, axis=0))
#最小值的下标
print(tf.argmin(data3, axis=0))
#排序的下标
print(tf.argsort(data3, axis=1))
```

运行结果如下：

```
tf.Tensor([0 1 1], shape=(3,), dtype=int64)
tf.Tensor([1 0 0], shape=(3,), dtype=int64)
tf.Tensor(
[[2 1 0]
 [2 1 0]], shape=(2, 3), dtype=int32)
```

tf.argmax 得到的是索引位置，如果 axis 不同，则其索引值也不同，其计算过程如图 3-18 所示。

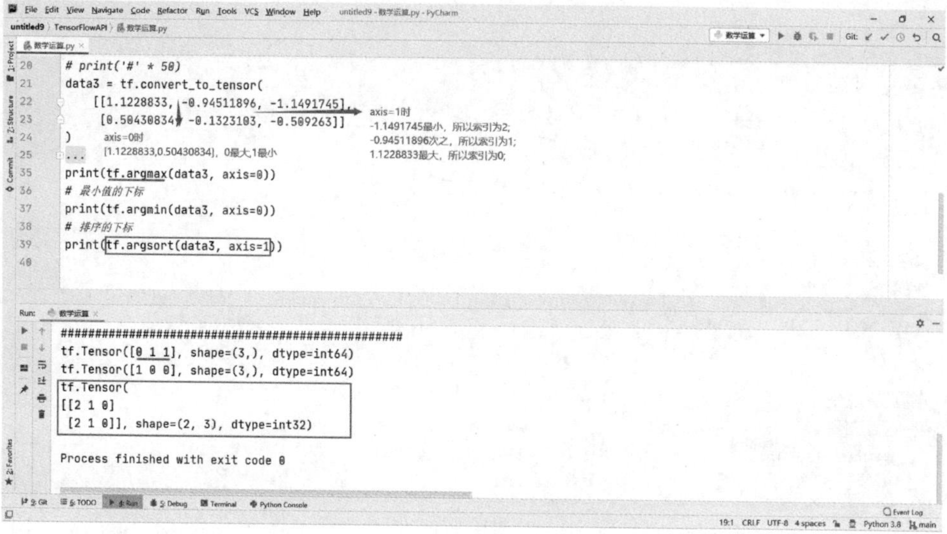

```
# print('#' * 50)
data3 = tf.convert_to_tensor(
    [[1.1228833, -0.94511896, -1.1491745],
     [0.50430834, -0.1323103, -0.509263]]
)
...
print(tf.argmax(data3, axis=0))
# 最小值的下标
print(tf.argmin(data3, axis=0))
# 排序的下标
print(tf.argsort(data3, axis=1))
```

```
##################################################
tf.Tensor([0 1 1], shape=(3,), dtype=int64)
tf.Tensor([1 0 0], shape=(3,), dtype=int64)
tf.Tensor(
[[2 1 0]
 [2 1 0]], shape=(2, 3), dtype=int32)

Process finished with exit code 0
```

图 3-18 TensorFlow 中根据 axis 得到索引值

3.3.4 TensorFlow 维度变化

维度的操作主要包括维度转换、升维、降维、维度交换等，主要使用 tf.reshape()、tf.expand_dims()、tf.squeeze()、tf.transpose()等函数，无论张量的维度如何变化，其 size（总数）应该保持不变，代码如下：

```
#第 3 章/TensorFlowAPI/维度变换.py
import tensorflow as tf

#随机生成 4 维数据
rnd_shape4 = tf.random.normal([4, 32, 32, 3])
print("获取维度:", rnd_shape4.shape)
#数据维度升维、降维
#在 0 轴升一维,变成(1,4, 32, 32, 3)
shape5 = tf.expand_dims(rnd_shape4, axis=0)
print('升维:', shape5.shape)
#在 0 轴降一维,变成(4, 32, 32, 3)
shape5_to4 = tf.squeeze(shape5, axis=0)
print('降维:', shape5_to4.shape)
#在 1 轴升一维,变成(4, 1, 32, 32, 3)
shape6 = tf.expand_dims(rnd_shape4, axis=1)
print('升维:', shape6.shape)
#在 0 轴降一维,变成(4, 32, 32, 3)
shape6_1 = tf.squeeze(shape6, axis=1)
print('降维:', shape6_1.shape)
#改变维度
```

```
to_shape = tf.reshape(shape6_1, [1, 1, 4, 32, 32, 3])
print("改变维度:", to_shape.shape)

transpose_num = tf.random.normal([32, 28, 3])
#按指定的顺序转置,得到(28, 32, 3)
print("按指定的顺序交换：", tf.transpose(transpose_num, perm=[1, 0, 2]).shape)
#以下语句会报错,因为 tf.squeeze 只能对 axis=0 位置是 1 的数组进行操作
error_shape = tf.squeeze(rnd_shape4, axis=0)
```

运行结果如下：

```
获取维度：(4, 32, 32, 3)
升维：(1, 4, 32, 32, 3)
降维：(4, 32, 32, 3)
升维：(4, 1, 32, 32, 3)
降维：(4, 32, 32, 3)
改变维度：(1, 1, 4, 32, 32, 3)
按指定的顺序交换：(28, 32, 3)
Tensorflow. python. framework. errors _ impl. InvalidArgumentError: Can not
squeeze dim[0], expected a dimension of 1, got 4 [Op:Squeeze]
```

属性 shape 为多维数组的形状，在进行张量操作时，需要特别注意张量的 shape 变化。代码中 tf. random. normal([4,32,32,3])初始了一个 shape 为 4×32×32×3 的张量，tf. expand_dims(rnd_shape4,axis=0)即在最外层插入 1 个维度就变成 shape 为 1×4×32×32×3 的张量，其 size 总数不变；tf. squeeze(shape5,axis=0)将最外层的维度去除，就变成了 4×32×32×3 的张量，其 size 总数也不变。tf. expand_dims(rnd_shape4,axis=1)即在 1 轴插入 1 个维度，所以变成了 4×1×32×32×3，总结规律可以发现，axis=?就在该位置插入 1 个维度，而 tf. squeeze 就减少 1 个维度。

tf. reshape()可以实现任意维度变化的操作，可以代替 tf. expand_dims()、tf. squeeze()实现维度的变化。tf. transpose()按指定的 axis 进行交换，交换后 size 保持不变，但是 shape 发生了变化。

注意：tf. squeeze 只能对 axis 所在位置维度为 1 的数组进行降维，所以 error_shape 变量时会报 InvalidArgumentError 错误。

3.3.5 TensorFlow 切片取值

TensorFlow 支持如 NumPy 一样的下标取值方法，代码如下：

```
#第 3 章/TensorFlowAPI/下标、切片操作.py
#初始张量
a = tf.ones([1, 5, 5, 3])
#先取 axis=0 的内容,得到 5*5*3;从 5*5*3 中再得到 axis=0 下标为 2 的内容,即 5*3;从 5*3
#中再得到 axis=0 下标为 2 的内容,即 shape=(3,)
print(a[0][2][2].shape)
```

运行结果如下：

```
(3,)
```

代码 a[0]，即从 1 个 5×5×3 中得到第 0 个内容，所以 shape 为 5×5×3；a[0][2]即从 5 个 5×3 中得到下标为 2 的内容，即 5×3；a[0][2][2]即从 5 个(3,)中得到下标为 2 的内容，即 shape=(3,)。

上面的代码 a[0][2][2]可以被修改成 a[0,2,2]的表达方式，代码如下：

```
#第 3 章/TensorFlowAPI/下标、切片操作.py
import tensorflow as tf

#初始张量
a = tf.ones([1, 5, 5, 3])
print(a[0].shape)                      #[5,5,3] 由外向内
print(a[0, 2].shape)                   #[5,3] 第 0 轴,第 1 轴里的第 3 个
print(a[0, 2, 2].shape)                #[3,]
```

运行结果如下：

```
(5, 5, 3)
(5, 3)
(3,)
```

根据结果可知，a[0][2][2]和 a[0,2,2]的内容是一样的。从 a[0,2,2]的表达式中可知每个 axis 可以用逗号隔开，有多少个逗号就有多少个维度。

多维度的切片与 NumPy 保持一致，代码如下：

```
#第 3 章/TensorFlowAPI/下标、切片操作.py
import tensorflow as tf

#初始张量
a = tf.ones([1, 5, 5, 3])
print("多维度的切片:", a[0, :, :, :].shape)                    #得到 5*5*3
print("多维度的切片:", a[:, :, :, 2].shape)                    #得到 1*5*5
print("多维度间隔取值:", a[:, 0:4:2, 0:4:2, :].shape)          #得到 1*2*2*3
print("...省略:", a[..., 2].shape)                             #得到 1*5*5
```

运行结果如下：

```
多维度的切片: (5, 5, 3)
多维度的切片: (1, 5, 5)
多维度间隔取值: (1, 2, 2, 3)
...省略: (1, 5, 5)
```

在每个 axis 中都可以使用切片的语法，a[0,:,:,:]表示 axis=0 中下标为 0 的所有的内容；a[:,:,:,2]表示 axis=3 中下标为 2 的内容，前面 axis 的内容保留，所以其 shape 为 1×5×5；a[:,0:4:2,0:4:2,:]表示 axis=0 中所有的内容，axis=1 每 2 取 1 个，所以 axis=1 时取下标 0、2 的内容，axis=2 也一样，axis=3 中所有的内容，所以其 shape 为 1×2×2×3；

a[…,0]中的…表示前面 axis=0,1,2 的内容全部保留,与 a[:,:,:,2]相同。

TensorFlow 中下标或者切片只能取值而不能进行修改,但是 NumPy 都可以,代码如下:

```
#第 3 章/TensorFlowAPI/下标、切片操作.py
import tensorflow as tf
import numpy as np

NumPy_a = np.ones([1, 5, 5, 3])
tensor_a = tf.ones([1, 5, 5, 3])
#NumPy 修改值
NumPy_a[..., 0] = 0
#TensorFlow 中不支持下标修改
tensor_a[..., 0] = 0
```

运行结果如下:

```
tensor_a[..., 0] = 0
TypeError: 'Tensorflow.python.framework.ops.EagerTensor' object does not
support item assignment
```

3.3.6 TensorFlow 中 gather 取值

函数 gather(params,indices,axis=None)提供了更灵活、强大的取值方法。params 要求传入张量,indices 可以传入多种形式的索引号的值,axis 为指定的轴位置,代码如下:

```
#第 3 章/TensorFlowAPI/gather 取值.py
import tensorflow as tf
params = tf.constant([10., 11., 12., 13., 14., 15.])
#取 params 索引下标为[2, 0, 2, 5]的内容
print(tf.gather(params, indices=[2, 0, 2, 5]))
#先取 params 索引下标[2, 0]的值,组成 1 个维度,然后取索引下标[2, 5]的值,再组成 1 个维度
#然后将两个维度合并成一个新的维度,变成 2*2 的张量
print(tf.gather(params, [[2, 0], [2, 5]]))
```

运行结果如下:

```
tf.Tensor([12. 10. 12. 15.], shape=(4,), dtype=float32)
tf.Tensor(
[[12. 10.]
 [12. 15.]], shape=(2, 2), dtype=float32)
```

代码中 params 只有一个维度,所以 axis=0。将 params 的内容变为 4×3,其 gather 取值的代码如下:

```
#第 3 章/TensorFlowAPI/gather 取值.py
import tensorflow as tf

params = tf.constant([[0, 1.0, 2.0],
                      [10.0, 11.0, 12.0],
                      [20.0, 21.0, 22.0],
                      [30.0, 31.0, 32.0]])
```

```
#没有指定 axis,默认为 0 轴,即下标索引为 3 和 1 的内容
print(tf.gather(params, indices=[3, 1]))
#axis=1,则取 axis=1 中下标索引为 2 和 1 的内容
print(tf.gather(params, indices=[2, 1], axis=1))
```

运行结果如下：

```
tf.Tensor(
[[30. 31. 32.]
 [10. 11. 12.]], shape=(2, 3), dtype=float32)
tf.Tensor(
[[ 2.  1.]
 [12. 11.]
 [22. 21.]
 [32. 31.]], shape=(4, 2), dtype=float32)
```

代码中 axis=1 时会将索引下标为[2,1]的内容全部取走，所以得到的 shape=(4,2)，其取值过程如图 3-19 所示。

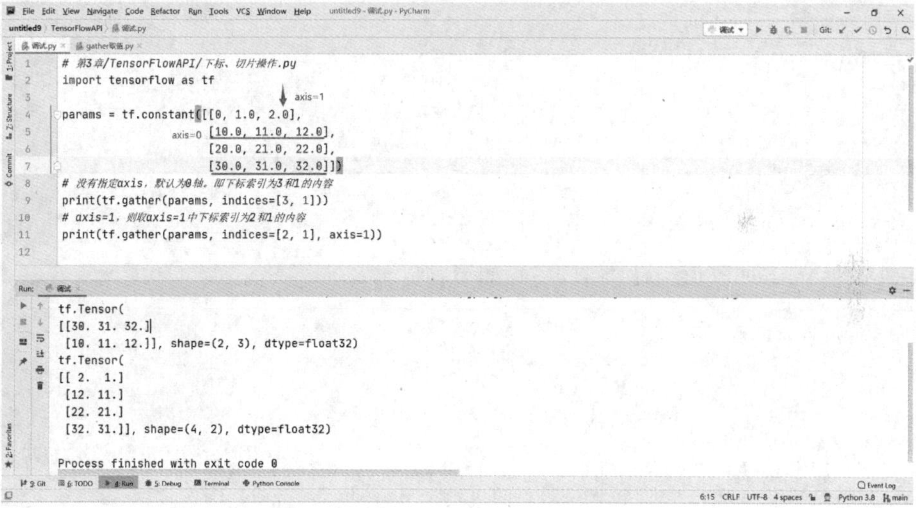

图 3-19　TensorFlow 中 gather 取值

同样 indices 可以为多维张量，代码如下：

```
#第 3 章/TensorFlowAPI/gather 取值.py
import tensorflow as tf

params = tf.constant([[0, 1.0, 2.0],
                      [10.0, 11.0, 12.0],
                      [20.0, 21.0, 22.0],
                      [30.0, 31.0, 32.0]])
#indices 是一个多维张量
indices = tf.constant([
    [2, 4],
    [0, 4]])
```

```
#当 indices 中的 4 超过 params 的索引时会自动补 0
#因为 axis=0,所以 0 轴补 3 个 0
print(tf.gather(params, indices, axis=0))
print('#'*20)
#因为 axis=1,所以 1 轴补 1 个 0
print(tf.gather(params, indices, axis=1))
```

运行结果如下：

```
tf.Tensor(
[[[20. 21. 22.]
  [ 0.  0.  0.]]

 [[ 0.  1.  2.]
  [ 0.  0.  0.]]], shape=(2, 2, 3), dtype=float32)
####################
tf.Tensor(
[[[ 2.  0.]
  [ 0.  0.]]

 [[12.  0.]
  [10.  0.]]

 [[22.  0.]
  [20.  0.]]

 [[32.  0.]
  [30.  0.]]], shape=(4, 2, 2), dtype=float32)
```

多维张量 gather 取值与此类似，代码如下：

```
#第 3 章/TensorFlowAPI/gather 取值.py
import tensorflow as tf

params = tf.random.normal([4, 35, 8])
#axis=0 就是第 1 个维度的变化
#[7, 9, 16]的下标不存在,会补 0,所以是[5,35,8]
result0 = tf.gather(params, axis=0, indices=[2, 3, 7, 9, 16])
#axis=1 就是第 2 个维度的变化,所以是[4,5,8]
result1 = tf.gather(params, axis=1, indices=[2, 3, 7, 9, 16])
#axis=2 就是最里面的维度,所以是[4,35,3]
result2 = tf.gather(params, axis=2, indices=[2, 3, 7])
#axis=2 就是第 3 个维度的变化,所以[4,35]会保留,另外[[2, 3, 7], [0, 1, 2]]即对[8]取了
#两次,所以应该变成[2,3],结合后变成[4,35,2,3]
result3 = tf.gather(params, axis=2, indices=[[2, 3, 7], [0, 1, 2]])
print(result0.shape, result1.shape, result2.shape, result3.shape)
```

运行结果如下：

```
(5, 35, 8) (4, 5, 8) (4, 35, 3) (4, 35, 2, 3)
```

代码 tf.gather(params, axis=0, indices=[2, 3, 7, 9, 16])中 indices 的数量超过了 4，TensorFlow 会自动补 0，如图 3-20 所示。

运行结果如下：

```
误差：tf.Tensor(
[[-0.4909289]
 [ 0.5035277]], shape=(2, 1), dtype=float32)
梯度：[<tf.Tensor: shape=(2, 3), dtype=float32, NumPy=
array([[-0.00101757, -0.00152636, -0.00050879],
       [-0.00047112, -0.00070667, -0.00023556]], dtype=float32)>, <tf.Tensor:
shape=(2, 3), dtype=float32, NumPy=
array([[-0.06529391, -0.06325722, -0.12268066],
       [ 0.0669831 ,  0.06489372,  0.12585449]], dtype=float32)>]
```

自动求梯度只需在 with tf. GradientTape()as tape 作用域下，并使用 tape. gradient(loss,lw)就可以根据 loss 求梯度。相对于“2.6.2 反向传播算法”节中实现的梯度计算，使用自动求梯度功能非常方便。

总结

TensorFlow 的基本 API 主要有维度变换、切片取值、自动求梯度等功能，其中自动求梯度功能对于网络的学习提供了极大的便利。

练习

调试本节所有代码，观察多维数组状态下张量的操作功能；本节相关 API 在 TensorFlow 模型搭建中有重要作用。

3.4　TensorFlow 中的 Keras 模型搭建

3.4.1　tf. keras 简介

Keras 是一个广泛受欢迎的神经网络框架，用 Python 编写而成，能够在 TensorFlow、PyTorch 中运行，其特点是支持快速神经网络模型搭建，同时网络模型具有很强的层次性，可以在 CPU 和 GPU 上运行。

TensorFlow 2. x 版本中的 Keras 主要在 tensorflow. keras 模块下，包含 Models、Layers、Optimizers 等子模块，每个子模块又封装了子方法，如图 3-22 所示。

使用 tensorflow. keras 可以方便、快速地构建神经网络模型，是深度学习代码实战中的重要组成部分。

3.4.2　基于 tf. keras. Sequential 模型搭建

Sequential 的参数是一个列表，列表里面存放着 layers 层下的各个算子，将各个算子按线型结构组装起来，模型就搭建完毕了。这个过程可以形象地类比成美食“串串”，Sequential 就是那一根“竹签”，将各种“美食”(算子)串联起来，就可以“涮串串”了，代码如下：

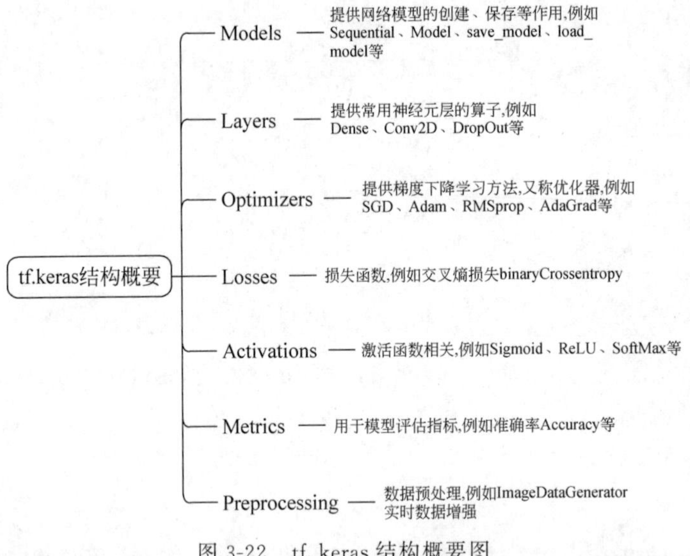

图 3-22　tf.keras 结构概要图

```
#第 3 章/TensorFlowAPI/基于 tf.keras.Sequential 模型搭建.py
import tensorflow as tf
from tensorflow.keras import Sequential, layers, Input

#输入数据的 shape
input_shape = (28, 28, 1)
model = Sequential(
    [
        Input(shape=input_shape),                     #输入数据的维度需要指定
        layers.Flatten(),                             #将多维数据打平成一维数据
        layers.Dense(784, activation='sigmoid'),      #784 个神经元
        layers.Dense(128, activation='sigmoid'),
        layers.Dense(10, activation='softmax')        #10 个类别
    ]
)
#描述模型的结构
model.summary()
```

运行结果如下：

```
Model: "sequential"
_________________________________________________________________
Layer (type)                 Output Shape              Param #
=================================================================
flatten (Flatten)            (None, 784)               0
_________________________________________________________________
dense (Dense)                (None, 784)               615440
_________________________________________________________________
dense_1 (Dense)              (None, 128)               100480
_________________________________________________________________
dense_2 (Dense)              (None, 10)                1290
=================================================================
```

```
Total params: 717,210
Trainable params: 717,210
Non-trainable params: 0
_________________________________________________________________
```

Sequential 提供了网络模型组装方法，layers 模块下提供了各个算子(例如 Dense 类)设置神经元的个数(Dense 又称全连接层)，其主要参数如下：

```
def __init__(self,
             units,                              #神经元的个数
             activation=None,                    #激活函数
             use_bias=True,                      #是否支持偏置项,即 wx+b 中的 b
             kernel_initializer='glorot_uniform',
             **kwargs):
```

model.summary()会将各个网络层的 shape 及 param 参数自动计算出来。Input(shape=input_shape)用来表示输入数据的 shape。

在 model.summary()处断点，查看 model 对象的构成会发现 input、output 分别用于描述网络层的输入与输出，layers 属性用来描述神经网络层中的结构，trainable_variables、trainable_weights 属性会自动化初始神经网络权重中的 θ 值，如图 3-23 所示。

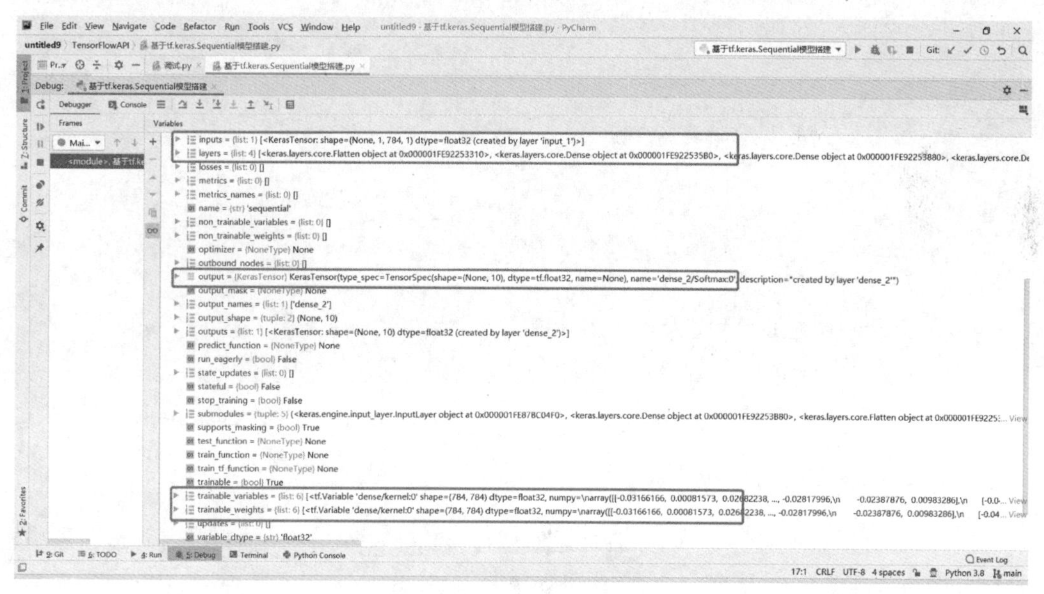

图 3-23 model 对象结构

3.4.3 继承 tf.keras.Model 类模型搭建

创建神经网络模型也可以通过自定义类来继承 tf.keras.Model 类，然后在__init__()构造方法中初始各个网络层，并在 call()构造函数中实现前向传播的逻辑，代码如下：

```python
#第 3 章/TensorFlowAPI/继承 tf.keras.Modle 类模型搭建.py
import tensorflow as tf
from tensorflow.keras import layers


class MyModle(tf.keras.Model):
    def __init__(self):
        #继承 Model 类中的属性
        super(MyModel, self).__init__()
        #全连接
        self.flat = layers.Flatten()
        self.dense1 = layers.Dense(784, activation='sigmoid')
        self.dense2 = layers.Dense(128, activation='sigmoid')
        #输出
        self.out = layers.Dense(10, activation='softmax')

    #call()为类的实例方法,可以使类的实例对象成为可调用对象
    def call(self, inputs):
        #在 call()中实现前向传播
        x = self.flat(inputs)
        x = self.dense1(x)
        x = self.dense2(x)
        #返回预测结果
        return self.out(x)

    def get_config(self):
        #重载 get_config 方法
        base_config = super(MyModel, self).get_config()
        return dict(list(base_config.items()))


if __name__ == "__main__":
    model = MyModel()
    #继承 Model 类,需要通过 build 编译一下
    model.build(input_shape=(28, 28, 1))
    model.summary()
```

继承 Model 的方法,需要调用 model. build(input_shape)指明输入的 shape。

3.4.4 函数式模型搭建

另一种方法是直接描述输入,然后将返回值再传给下一个算子的输入,直到最后一个输出,然后将输入与输出放入 Model 类,代码如下:

```python
#第 3 章/TensorFlowAPI/函数式模型搭建.py
from tensorflow.keras import Input, layers, Model
from tensorflow.keras.activations import sigmoid, softmax

#描述输入
inputs = Input(shape=(28, 28, 1))
#描述中间过程
```

```
x = layers.Flatten()(inputs)
x = layers.Dense(784, activation=sigmoid)(x)
x = layers.Dense(128, activation=sigmoid)(x)
outputs = layers.Dense(10, activation=softmax)(x)
#把输入和输出装载在 Model 这个类里面
#inputs 和 outpus 可以为列表,用来表示多输入或者多输出
model = Model(inputs=inputs, outputs=outputs)
model.summary()
```

激活函数不仅可以传字符串,也可以使用 tensorflow.keras.activations 模块下定义好的方法。只要描述了 inputs 和 outputs 模型 model 便会自动地找到相关层。

总结

TensorFlow 搭建模型可以使用 Sequential,也可以继承 Model 类,同时还支持函数式搭建。

练习

分别使用 3 种搭建模型的方式完成拥有 3 个隐藏层的模型搭建,并调试代码以观察每行变量中对象的变化。

3.5 TensorFlow 中模型的训练方法

3.5.1 使用 model.fit 训练模型

在 TensorFlow 中训练模型可以使用 model.fit()方法,其参数如下:

```
model.fit(
    x=None,                             #训练 x 的数据
    y=None,                             #训练 y 的数据
    batch_size=None,                    #每次训练的样本数
    epochs=1,                           #迭代次数
    verbose=1,                          #日志等级;0:为不在标准输出流输出日志信息;
                                        #1:显示进度条;2:每个 epoch 输出一行记录
    callbacks=None,                     #回调函数,即在训练过程中将被调用执行的函数
    validation_split=0.0,               #从 x 和 y 中设置验证集的比例
    validation_data=None,               #验证集的数据,validation_data 设置后会覆盖
                                        #validation_split
    shuffle=True,                       #是否打乱数据,通常为 True
    class_weight=None,                  #不重要,一般默认
    sample_weight=None,                 #不重要,一般默认
    initial_epoch=0,                    #开始的 epoch 次数设置
    steps_per_epoch=None,               #不重要,一般默认
    validation_steps=None,              #不重要,一般默认
    validation_freq=1,                  #不重要,一般默认
    max_queue_size=10,                  #不重要,一般默认
    workers=1,                          #进程数,Linux 可以设置多个,Windows 为 1
    use_multiprocessing=False           #是否多进程
)
```

一般设置只需传入 x、y、batch_size、epochs、callbacks 参数，代码如下：

```
#第 3 章/TensorFlowAPI/使用 model.fit 训练模型.py
import tensorflow as tf
from tensorflow.keras import Input, layers, Model
from tensorflow.keras.activations import sigmoid, softmax
import numpy as np

#第 1 步：读取数据 ########################
num_classes = 10                        #类别数
input_shape = (28, 28, 1)
#自动下载手写数字识别的数据集
(x_train, y_train), (x_test, y_test) = tf.keras.datasets.mnist.load_data()
#归一化处理，将值放在 0~1
x_train = x_train.astype('float32') / 255.
x_test = x_test.astype('float32') / 255.
#增维，变成[batch,height,width,channel]
x_train = np.expand_dims(x_train, -1)
x_test = np.expand_dims(x_test, -1)
#将 y 转换为 one_hot 编码
y_train = tf.keras.utils.to_categorical(y_train, num_classes)
y_test = tf.keras.utils.to_categorical(y_test, num_classes)

#第 2 步：构建模型 ########################
#描述输入
inputs = Input(shape=input_shape)
#描述中间过程
x = layers.Flatten()(inputs)
x = layers.Dense(784, activation=sigmoid)(x)
x = layers.Dense(128, activation=sigmoid)(x)
outputs = layers.Dense(10, activation=softmax)(x)
#把输入和输出装载在 Model 这个类里面
#inputs 和 outpus 可以为列表，用来表示多输入或者多输出
model = Model(inputs=inputs, outputs=outputs)
model.summary()
#第 3 步：设置回调函数、学习率和损失函数
callbacks = [
    tf.keras.callbacks.ModelCheckpoint(
        filepath='mnist_weights',
        save_best_only=True,
        monitor='val_loss',
    ),
]
#设置学习率和损失函数
model.compile(
    optimizer=tf.keras.optimizers.Adam(learning_rate=0.001),
    loss=tf.keras.losses.CategoricalCrossentropy(from_logits=False),
    metrics=['accuracy'],          #评价指定，此处为准确率
)
#第 4 步：训练 ########################
model.fit(
    x=x_train,
    y=y_train,
```

```
        batch_size=2,
        epochs=10,
        validation_split=0.1,
        callbacks=callbacks
    )
```

执行后就会开始进行训练，控制台会展示每个 epoch 训练后的损失、准确率的结果，如图 3-24 所示。

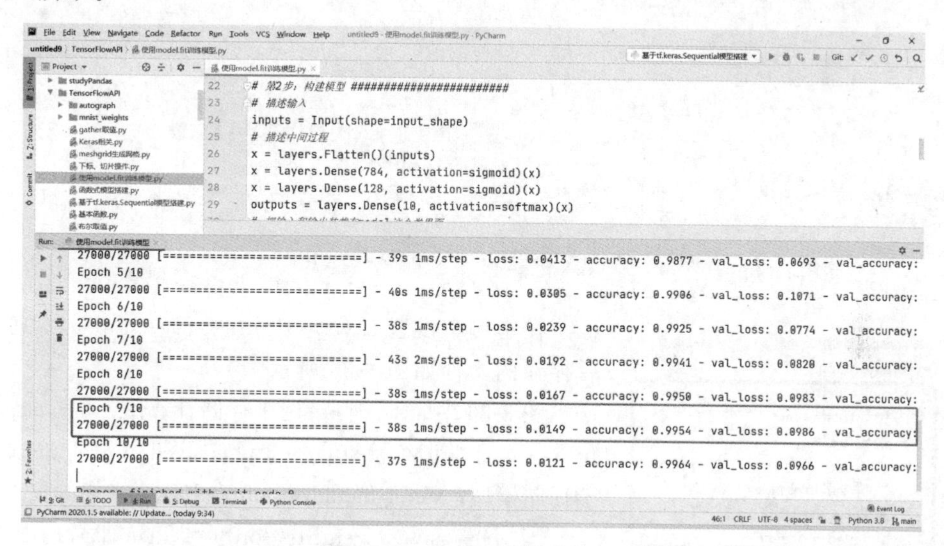

图 3-24　model. fit 训练结果

训练数据用的是手写数字识别数据集，其主要任务是能够识别手写[0,9]的数字，共计 10 个分类，都是灰度图，其尺寸为 28×28。tf. keras. datasets. mnist. load_data()会自动进行下载，共计 6 万张灰度图，如图 3-25 所示。

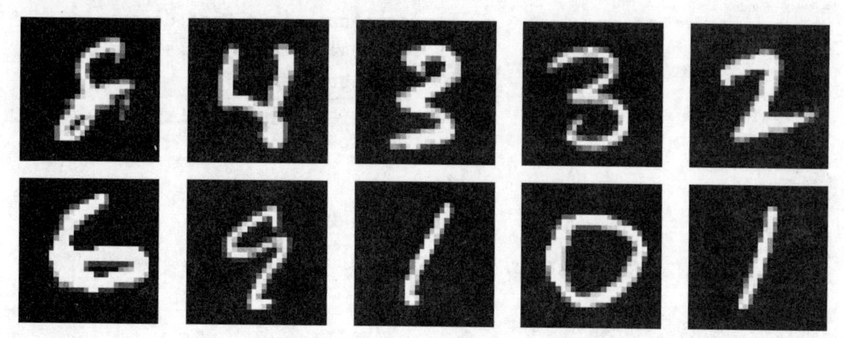

图 3-25　手写数字识别中的数据

因为一张图片，其值域在[0,255]，这里通过 x_train. astype('float32')/255 将其值域限定在[0,1]，可以在一定程度上缓解梯度爆炸。

由于 mnist. load_data()的数据 shape 是 60000×28×28，但是图片应该还有一个通道维度，所以需要通过 np. expand_dims(x_train,－1)变成 60000×28×28×1。

输出值 y 通过 tf.keras.utils.to_categorical(y_test,num_classes)变成[0. 0. 0. 0. 1. 0. 0. 0. 0. 0.]的独热编码，这里表示数字 4。共计 10 个分类，当前图片为哪个分类，那么其位置中的值为 1，其他为 0。

回调函数 callbacks 是一个列表，tf.keras.callbacks.ModelCheckpoint()用来设置模型和权重参数保存等信息，主要参数如下：

```
keras.callbacks.ModelCheckpoint(
    filepath,                   #保存权重文件的位置
    monitor='val_loss',         #需要监视的值,val_accuracy、val_loss 或者 accuracy
    verbose=0,                  #信息展示方式：0 或者 1
    save_best_only=True,        #True 表示当前 epoch 的 monitor 比上一次好时保存
    save_weights_only=False,    #True 只存储权重,下次想用还得搭建模型。
                                #False 会将权重和模型一起保存
    mode='auto',                #save_best_only=True、monitor=val_accuracy,
                                #mode=max;val_loss 时为 min;auto 自动推断
    period=1                    #隔几个 epoch 进行保存
)
```

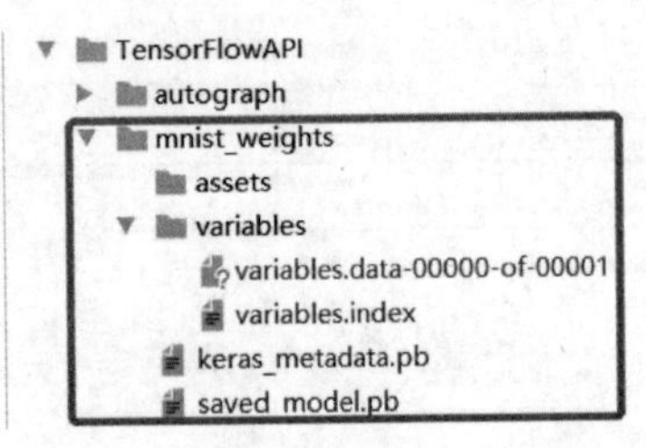

图 3-26　模型和权重文件的目录

在本例中设置 save_weights_only=False，训练完成后存在的文件如图 3-26 所示。

图 3-26 中 assets 目录用于存储计算图；variables 目录包含一个标准训练检查点的变量；saved_model.pb 文件用于存储模型；keras_metadata.pb 为 Keras 格式的权重。

当设置 save_weights_only = True 时，callbacks.ModelCheckpoint()中的 filepath 文件后缀名要为.h5 文件，再次训练得到如图 3-27 所示的内容。

```
56  # 第3步：设置回调函数、学习率和损失函数
57  callbacks = [
58      tf.keras.callbacks.ModelCheckpoint(
59          filepath='mnist_weights/手写.h5',
60          save_best_only=True,
61          save_weights_only=True,
62          monitor='val_loss',
63      ),
```

图 3-27　只存储权重参数文件

可在 model.compile()函数中设置学习率和损失函数，其主要参数如下：

```
def compile(self,
            optimizer='rmsprop',     #梯度下降学习方法,俗称优化器,在
                                     #tf.keras.optimizers 模块下
            loss=None,               #损失函数设置,tf.keras.losses 模块下有多种
                                     #损失函数的封装。当然也可以自定义损失函数
            metrics=None,            #评价标准,自带 accuracy 和 mse
            run_eagerly=None,        #是否支持调试模型
)
```

学习率在 tf.keras.optimizers.Adam()中的 learning_rate 设置,CategoricalCrossentropy()为交叉熵损失。最后调用 model.fit()函数 TensorFlow 会自动根据反向传播算法进行训练、学习。

神经网络的学习训练的过程如下:

(1) 读取并处理数据。

(2) 构建神经网络模型。

(3) 回调函数、学习率、损失函数的设置。

(4) 调用 model.fit()进行训练。

注意: model.fit()中的 epochs 的大小与训练计算机的显存有关系,如果设置得太大,则会报 OOM 错误;调用 model.fit()后会自动进行反向传播算法的学习,但是这个过程在代码中是不可见的。

3.5.2 使用 model.train_on_batch 训练模型

训练方式使用 model.fit()虽然非常方便,但是不能在训练过程中调试代码,也不能在训练过程中进行自定义控制,因此 TensorFlow 还提供了 model.train_on_batch 的方法进行训练,代码如下:

```
#第 3 章/TensorFlowAPI/使用 model_train_on_batch 训练模型.py
#第 4 步: 训练 #########################
BATCH_SIZE = 6
epochs = 10
#按某个 batch_size 切割数据,并进行缓存
#训练集
ds_train = tf.data.Dataset.from_tensor_slices(
    (x_train, y_train)
).shuffle(buffer_size=1000).batch(BATCH_SIZE).prefetch(
    tf.data.experimental.AUTOTUNE
).cache()
#验证集
ds_val = tf.data.Dataset.from_tensor_slices(
    (x_test, y_test)
).shuffle(buffer_size=1000).batch(BATCH_SIZE).prefetch(
    tf.data.experimental.AUTOTUNE
).cache()

for epoch in tf.range(1, epochs + 1):
    #将模型的评价内容置为初始状态
    model.reset_metrics()
    #获取当前学习率
    new_lr = model.optimizer.lr
    #获取数据以进行训练
    for x, y in ds_train:
```

```
        #训练
        train_result = model.train_on_batch(x, y)
    #获取验证数据进行训练
    for val_x, val_y in ds_val:
         valid_result = model.train_on_batch(val_x, val_y)
    #操作每 5 个 epoch 学习率变为 2xnew_lr
    if epoch % 5 == 0:
        model.optimizer.lr.assign(new_lr *2)
        tf.print("当前学习率:", new_lr.NumPy())
    #每隔 10 个 epoch 保存一次权重文件
    if epoch % 10 == 0:
        model.save(f'train_on_batch{epoch}.h5')
    #打印训练中的结果
    tf.print("train:", dict(zip(model.metrics_names, train_result)))
    tf.print("valid:", dict(zip(model.metrics_names, valid_result)))
```

在 3.4.5 节的基础上更改了训练的方式，通过自定义 for epoch in tf. range(1,epochs+1)来控制学习率每 5 个 epoch 变为 2xnew_lr，每 10 个 epoch 保存一次权重文件。model. train_on_batch(x,y)每次循环的单个数据，同样该函数也自动进行反向传播并进行训练、学习，学习的过程并没有用代码来体现。

运行后可以看到目录中存在 train_on_batch10. h5 这个权重文件。观察控制台的输出会发现当前学习率已从 0.001 变为 0.004，如图 3-28 所示。

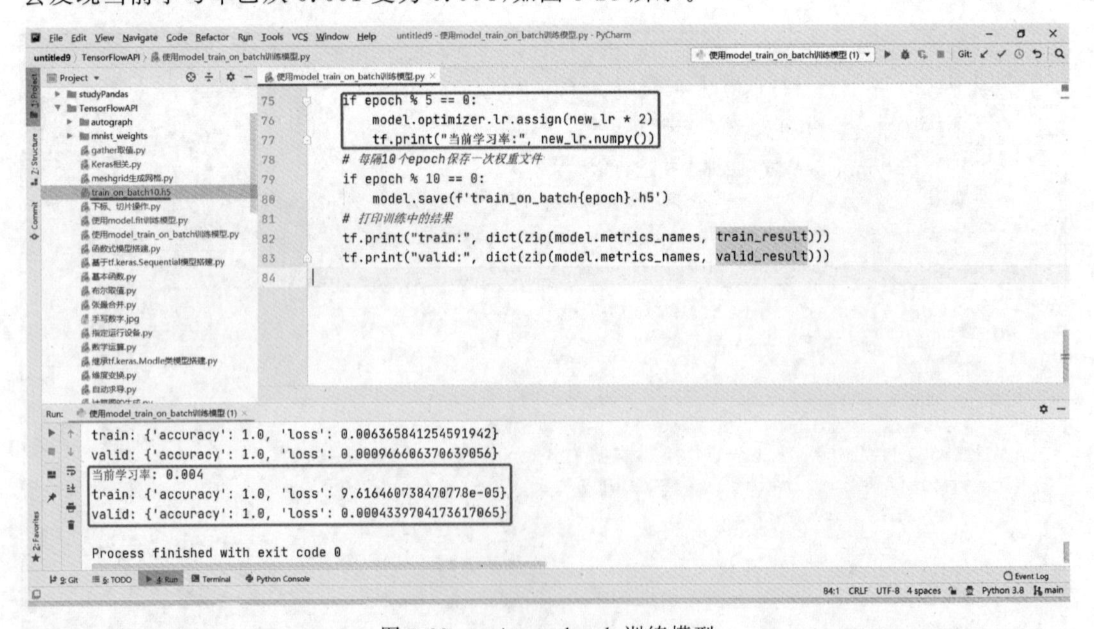

图 3-28 train_on_batch 训练模型

如果 model. save()不指定后缀名，则默认保存为 TensorFlow 的格式，如图 3-29 所示。

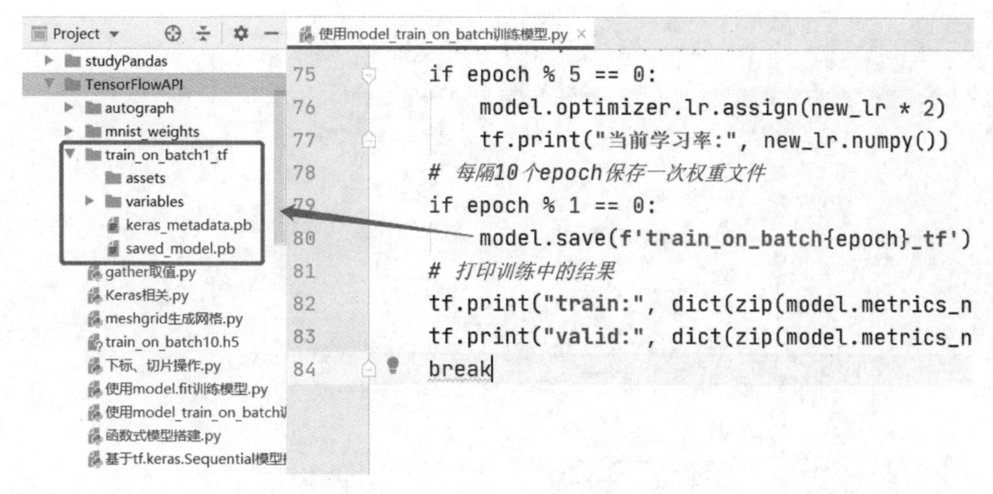

图 3-29　save 保存为 tf 格式的模型和参数

3.5.3　自定义模型训练

训练方式 model.fit()完全是一个黑箱操作，而 model.train_on_batch 虽然能够在一定程度上对训练过程进行控制，但是与反向传播学习的原理并不对应，同时其灵活程度有时不够，此时就可以使用完全自定义模型进行训练、学习，代码如下：

```
#第 3 章/TensorFlowAPI/完全自定义模型训练.py
#第 4 步：训练 ##########################
def train_step(model, x, y):
    #model:模型
    #features:训练集 x
    #labels:训练集 y
    #tf.GradientTape()自动求梯度的作用域语句
    with tf.GradientTape() as tape:
        #前向传播
        predictions = model(x, training=True)
        #使用损失函数
        loss = loss_func(y, predictions)
    #根据损失函数自动求梯度
    gradients = tape.gradient(loss, model.trainable_variables)
    #将梯度更新到 model.trainable_variables 属性中，然后由 optimizers 进行指定优化器
    #的梯度下降
    optimizer.apply_gradients(zip(gradients, model.trainable_variables))
    #更新评价指标
    train_loss.update_state(loss)
    train_metric.update_state(y, predictions)


#验证集
def valid_step(model, features, labels):
    #验证集不进行梯度下降更新学习
    predictions = model(features)
```

```
        batch_loss = loss_func(labels, predictions)
        valid_loss.update_state(batch_loss)
        valid_metric.update_state(labels, predictions)


BATCH_SIZE = 6
epochs = 10
#按某个 batch_size 切割数据，并进行缓存
#训练集
ds_train = tf.data.Dataset.from_tensor_slices(
    (x_train, y_train)
).shuffle(buffer_size=1000).batch(BATCH_SIZE).prefetch(
    tf.data.experimental.AUTOTUNE
).cache()
#验证集
ds_val = tf.data.Dataset.from_tensor_slices(
    (x_test, y_test)
).shuffle(buffer_size=1000).batch(BATCH_SIZE).prefetch(
    tf.data.experimental.AUTOTUNE
).cache()

for epoch in tf.range(1, epochs + 1):
    #训练集
    for x, y in ds_train:
        train_step(model, x, y)
    #验证集
    for val_x, val_y in ds_val:
        valid_step(model, val_x, val_y)
    logs = f'Epoch={epoch},' \
           f'Loss:{train_loss.result()},' \
           f'Accuracy:{train_metric.result()},' \
           f'Valid Loss:{valid_loss.result()},' \
           f'Valid Accuracy:{valid_metric.result()}'
    #每隔 10 个 epoch 保存一次权重文件
    if epoch % 5 == 0:
        model.save(f'完全自定义训练{epoch}')
    tf.print(logs)
    #将评价指标重置为 0
    train_loss.reset_states()
    valid_loss.reset_states()
    train_metric.reset_states()
    valid_metric.reset_states()
```

运行结果如图 3-30 所示。

在函数 train_step(model,x,y)中使用 tf.GradientTape()来控制 model(x,training=True)进行前向传播，并定义损失函数 loss_func(y,predictions)，然后根据损失 loss 对象结合 tape.gradient(loss,model.trainable_variables)求梯度。

然后根据优化器的选择 optimizer.apply_gradients(zip(gradients,model.trainable_variables))更新梯度下降的权重参数。

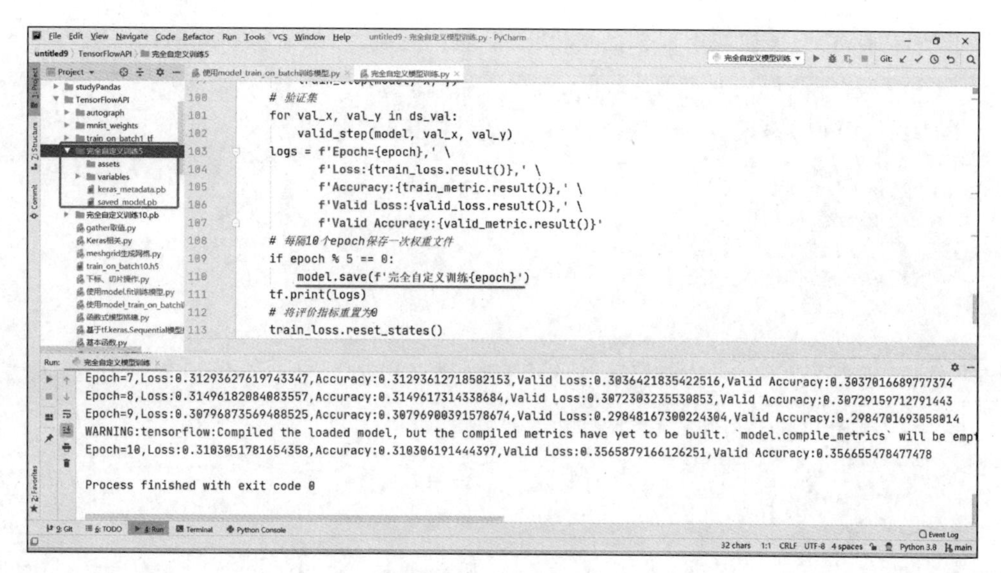

图 3-30 自定义训练结果

函数 valid_step(model,features,labels)是验证集,并不需要进行梯度下降的更新。整个训练、学习过程与神经网络反向传播算法的原理相对应,相对 model.fit()更容易理解,但构造过程稍复杂一些。

总结

在 TensorFlow 中搭建模型可使用 model.fit、model.train_on_batch 和自定义模型训练的方式,其形式灵活多变。

练习

分别使用 3 种搭建模型的方式及 3 种不同的训练方式进行组合,完成手写数字识别的训练。

3.6 TensorFlow 中 Metrics 评价指标

3.6.1 准确率

准确率是分类问题中最为原始的评价指标,准确率的定义是预测正确的结果占总样本的百分比,其公式如下:

$$\text{Accuracy}=\frac{\text{TP}+\text{TN}}{\text{TP}+\text{TN}+\text{FP}+\text{FN}} \tag{3-1}$$

公式中各评价指标的含义如下。

(1) TP(True Positive):真正例,被模型预测为正的正样本。

(2) FP(False Positive):假正例,被模型预测为正的负样本。

(3) FN(False Negative):假负例,被模型预测为负的正样本。

(4) TN(True Negative):真负例,被模型预测为负的负样本。

在 kearas. metrics 模块下框架已自带正例、负例的统计，只需将训练的代码修改如下：

```
#第 3 章/TensorFlowAPI/神经网络的评价指标.py
#设置学习率和损失函数
METRICS = [
    tf.keras.metrics.TruePositives(name='TP'),
    tf.keras.metrics.FalsePositives(name='FP'),
    tf.keras.metrics.TrueNegatives(name='TN'),
    tf.keras.metrics.FalseNegatives(name='FN'),
]
model.compile(
    optimizer=tf.keras.optimizers.Adam(learning_rate=0.001),
    loss=tf.keras.losses.CategoricalCrossentropy(from_logits=False),
    metrics=METRICS, #评价指定,此处为准确率
    run_eagerly=True
)


def plt_accuracy(history):
    h = history.history
    #history.history 得到的是字典
    #但是 v 不是一个 list,转换成 NDarray 格式以方便计算
    h = {k: np.array(v) for k, v in h.items()}
    train_accuracy = (h['TP'] + h['FP']) / (h['TP'] + h['FP'] + h['TN'] + h['FN'])
    val_accuracy = (h['val_TP'] + h['val_FP']) / (h['val_TP'] + h['val_FP'] +
h['val_TN'] + h['val_FN'])
    plt.plot(history.epoch, train_accuracy, label='训练准确率')
    plt.plot(history.epoch, val_accuracy, label='验证准确率')
    plt.xlabel('迭代次数')
plt.ylabel('准确率')
plt.title('迭代次数与准确率的变化')
    plt.legend(loc='best')
    plt.show()


#第 4 步：训练 ########################
history = model.fit(
    x=x_train,
    y=y_train,
    batch_size=6,
    epochs=3,
    validation_split=0.1,
    callbacks=callbacks
)
#对训练结果进行绘图
plt_accuracy(history)
```

训练结束 history 中将自带 METRICS 的结果，plt_accuracy()通过解析 history 中的内容，通过准确率公式计算后得到结果，如图 3-31 所示。

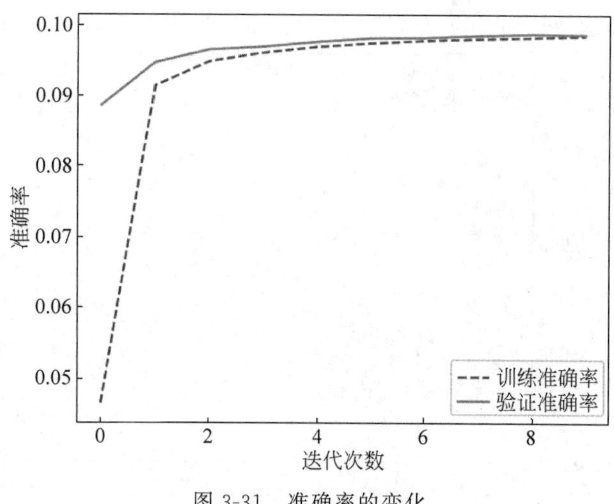

图 3-31　准确率的变化

准确率在类别数据量不平衡时，并不能客观地评价算法的优劣，例如测试集中有 100 个样本，其中 99 个是负样本，1 个是正样本，假设算法模型都能正确地预测出负样本，那么模型的准确率就是 99%，从数值来看效果不错，但事实上这个算法没有任何预测能力（因为正样本没有预测成功，模型偏向于预测负样本）。

3.6.2　精确率

精确率又叫查准率，在所有被预测为正的样本中实际为正样本的概率，其公式如下：

$$\text{Precision}=\frac{\text{TP}}{\text{TP}+\text{FP}} \tag{3-2}$$

精确率和准确率看上去有些类似，但它们是两个不同的概念。精确率代表对正样本结果的预测准确程度，而准确率代表整体的预测准确程度，包括正样本和负样本，代码如下：

```
#第 3 章/TensorFlowAPI/神经网络的评价指标.py
#精准率
def plt_precision(history):
    h = history.history
    #history.history 得到的是个字典
    #但是 v 不是一个 list,转换成 NDarray 格式以方便计算
    h = {k: np.array(v) for k, v in h.items()}
    train_accuracy = (h['TP']) / (h['TP'] + h['FP'])
    val_accuracy = (h['val_TP']) / (h['val_TP'] + h['val_FP'])
    plt.plot(history.epoch, train_accuracy, label='训练精准率')
    plt.plot(history.epoch, val_accuracy, label='验证精准率')
    plt.xlabel('迭代次数')
    plt.ylabel('精准率')
    plt.legend(loc='best')
    plt.title('迭代次数与精准率的变化')
    plt.show()
```

调用执行后的结果如图 3-32 所示。

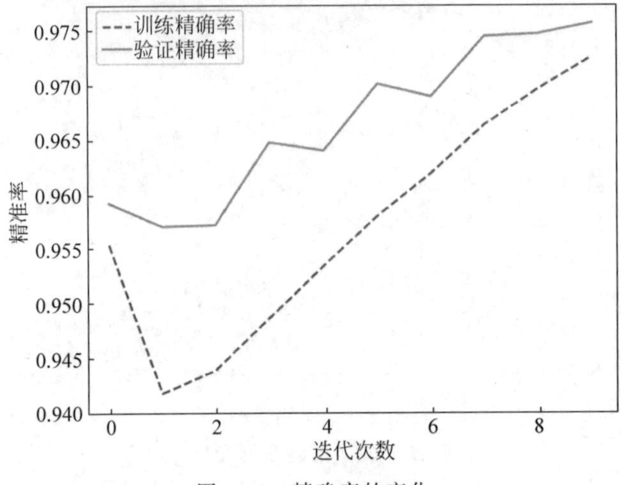

图 3-32　精确率的变化

3.6.3　召回率

召回率又叫查全率，它的含义是在实际为正的样本中被预测为正样本的概率，公式如下：

$$\text{Recall}=\frac{\text{TP}}{\text{TP}+\text{FN}} \tag{3-3}$$

代码如下：

```
#第 3 章/TensorFlowAPI/神经网络的评价指标.py
#召回率
def plt_recall(history):
    h = history.history
    #history.history 得到的是个字典
    #但是 v 不是一个 list，转换成 NDarray 格式以方便计算
    h = {k: np.array(v) for k, v in h.items()}
    train_accuracy = (h['TP']) / (h['TP'] + h['FN'])
    val_accuracy = (h['val_TP']) / (h['val_TP'] + h['val_FN'])
    plt.plot(history.epoch, train_accuracy, label='训练召回率')
    plt.plot(history.epoch, val_accuracy, label='验证召回率')
    plt.xlabel('迭代次数')
    plt.ylabel('召回率')
    plt.legend(loc='best')
    plt.title('迭代次数与召回率的变化')
    plt.show()
```

调用执行后得到的结果如图 3-33 所示。

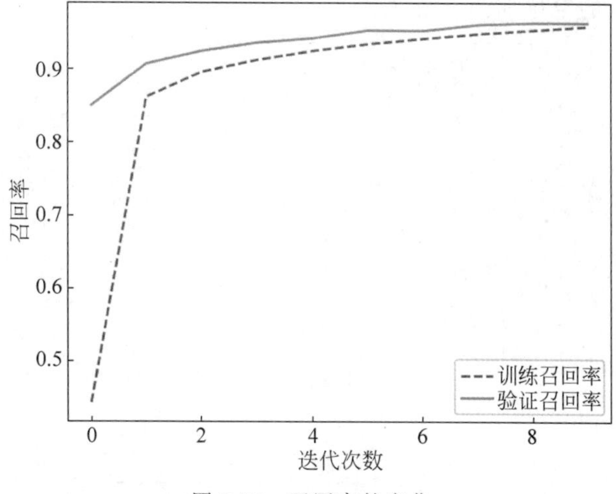

图 3-33 召回率的变化

3.6.4 P-R 曲线

在不同的应用场景中关注的指标不同。例如，在预测股票时更关心精准率，人们更关心那些升值的股票里真正升值的有多少，而在医疗病患中更关心的是召回率，即真的病患里我们预测错的情况应该越小越好。

精准率和召回率是一对此消彼长的关系。例如，在推荐系统中，我们想推送的内容应尽可能地让用户感兴趣，那只能推送我们把握较高的内容，这样就漏掉了一些用户感兴趣的内容，召回率就低了；如果想让用户感兴趣的内容都被推送，就只能推送全部内容，但是这样精准率就低了。在实际工程中，往往需要结合两个指标的结果，从而寻找到一个平衡点，使算法的性能达到最大化，这就是 P-R 曲线，代码如下：

```
#第 3 章/TensorFlowAPI/神经网络的评价指标.py
def plt_PR(history):
    h = history.history
    #history.history 得到的是个字典
    #但是 v 不是一个 list,转换成 NDarray 格式以方便计算
    h = {k: np.array(v) for k, v in h.items()}
    #精准率
    train_precision = (h['TP']) / (h['TP'] + h['FP'])
    val_precision = (h['val_TP']) / (h['val_TP'] + h['val_FP'])
    #召回率
    train_recall = (h['TP']) / (h['TP'] + h['FN'])
    val_recall = (h['val_TP']) / (h['val_TP'] + h['val_FN'])
    plt.plot(train_recall, train_precision, label='训练 P-R 曲线')
    plt.plot(val_recall, val_precision, label='验证 P-R 曲线')
    plt.xlabel('ReCall')
    plt.ylabel('Precision')
    plt.legend(loc='best')
    plt.title('P-R 曲线')
    plt.show()
```

调用执行后得到的结果如图 3-34 所示。

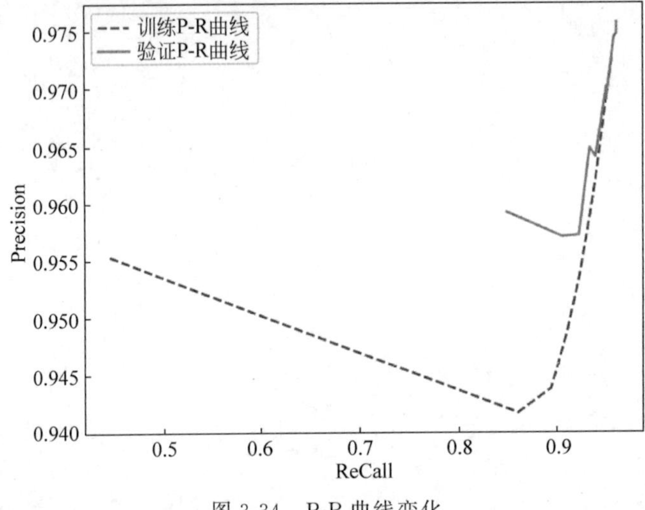

图 3-34　P-R 曲线变化

模型的精准率和召回率互相制约，P-R 曲线越向右上凸出，表示模型的性能越好；由于精准率和召回率更关注正样本的情况，当负样本比较多时 P-R 曲线的效果一般，此时使用 ROC 曲线更合适。

3.6.5　F1-Score

平衡精准率和召回率的另一个指标是 F1-Score，其公式如下：

$$F1=\frac{2\times P\times R}{P+R} \tag{3-4}$$

F1 值越高，代表模型的性能越好，代码如下：

```
#第 3 章/TensorFlowAPI/神经网络的评价指标.py
def plt_F1_Score(history):
    h = history.history
    #history.history 得到的是个字典
    #但是 v 不是一个 list，转换成 NDarray 格式以方便计算
    h = {k: np.array(v) for k, v in h.items()}
    #精准率
    train_precision = (h['TP']) / (h['TP'] + h['FP'])
    val_precision = (h['val_TP']) / (h['val_TP'] + h['val_FP'])
    #召回率
    train_recall = (h['TP']) / (h['TP'] + h['FN'])
    val_recall = (h['val_TP']) / (h['val_TP'] + h['val_FN'])

    f1_train = (2 * train_precision * train_recall) / (train_precision + train_
recall)
    f1_val = (2 * val_precision * val_recall) / (val_precision + val_recall)

    plt.plot(history.epoch, f1_train, label='训练集 F1-Score')
    plt.plot(history.epoch, f1_val, label='验证集 F1-Score')
```

```
plt.xlabel('迭代次数')
plt.ylabel('F1-Score')
plt.legend(loc='best')
plt.title('迭代次数 F1-Score 的变化')
plt.show()
```

调用执行后得到的结果如图 3-35 所示。

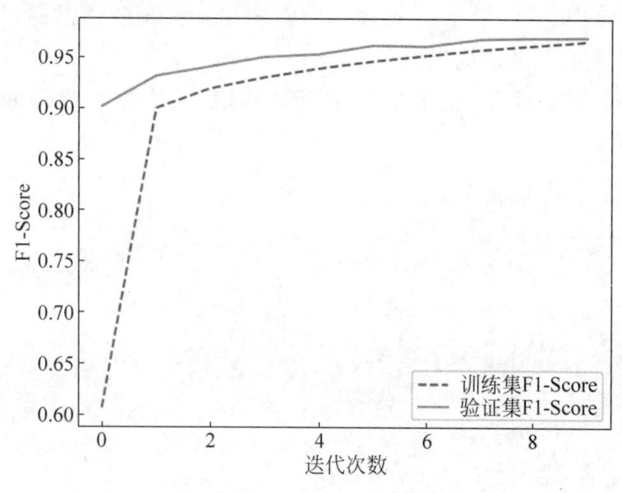

图 3-35　F1-Score 曲线变化

3.6.6　ROC 曲线

在实际数据集中经常会出现类别不平衡现象，即负样本比正样本多(或者相反)，而测试数据中的正负样本的分布也可能随着时间的变化而变化，ROC 曲线和 AUC 曲线可以很好地消除样本类别不平衡对评估指标结果的影响。

ROC 曲线可以无视样本的不平衡，主要取决于灵敏度(sensitivity)和特异度(specificity)，又称为真正率(TPR)和假正率(FPR)。

(1) 真正率(True Positive Rate，TPR)，又称为灵敏度。

(2) 假负率(False Negative Rate，FNR)。

(3) 假正率(False Positive Rate，FPR)。

(4) 真负率(True Negative Rate，TNR)，又称为特异度。

对应公式为

$$
\begin{aligned}
\text{TPR} &= \frac{\text{TP}}{\text{TP}+\text{FN}} \\
\text{FNR} &= \frac{\text{FN}}{\text{TP}+\text{FN}} \\
\text{FPR} &= \frac{\text{FP}}{\text{TN}+\text{FP}} \\
\text{TNR} &= \frac{\text{TN}}{\text{TN}+\text{FP}}
\end{aligned}
\tag{3-5}
$$

从式(3-5)可知，灵敏度 TPR 与召回率一样，是正样本的召回率，特异度 TNR 是负样本的召回率，而假负率 FNR＝1－TPR，假正率 FPR＝1－TNR，这 4 个维度都是针对单一类别的预测结果而言的，所以对整体样本是否均衡并不敏感。例如，假设总样本中，90％是正样本，10％是负样本，在这种情况下如果使用准确率进行评价，则是不科学的，但是 TPR 和 TNR 是可以的，因为 TPR 只关注 90％正样本中有多少是被预测正确的，而与那 10％的负样本没有关系，同样 FPR 只注意 10％负样本中有多少是被预测错误的，也与 90％的正样本无关，这样就避免了样本不平衡的问题。

ROC 曲线主要由灵敏度 TPR 和假正率 FPR 构成，其中横坐标为假正率 FPR、纵坐标为灵敏度 TPR，代码如下：

```
#第 3 章/TensorFlowAPI/神经网络的评价指标.py
def plt_ROC(history):
    h = history.history
    #history.history 得到的是个字典
    #但是 v 不是一个 list,转换成 NDarray 格式以方便计算
    h = {k: np.array(v) for k, v in h.items()}
    train_FPR = (h['FP']) / (h['TN'] + h['FP'])
    train_TPR = (h['TP']) / (h['TP'] + h['FN'])

    val_FPR = (h['val_FP']) / (h['val_TN'] + h['val_FP'])
    val_TPR = (h['val_TP']) / (h['val_TP'] + h['val_FN'])

    plt.plot(train_FPR, train_TPR, label='训练集 ROC')
    plt.plot(val_FPR, val_TPR, label='验证集 ROC')
    plt.xlabel('False Positive Rate')
    plt.ylabel('True Positive Rate')
    plt.legend(loc='best')
    plt.title('ROC 曲线')
    plt.show()
```

调用执行后得到的结果如图 3-36 所示。

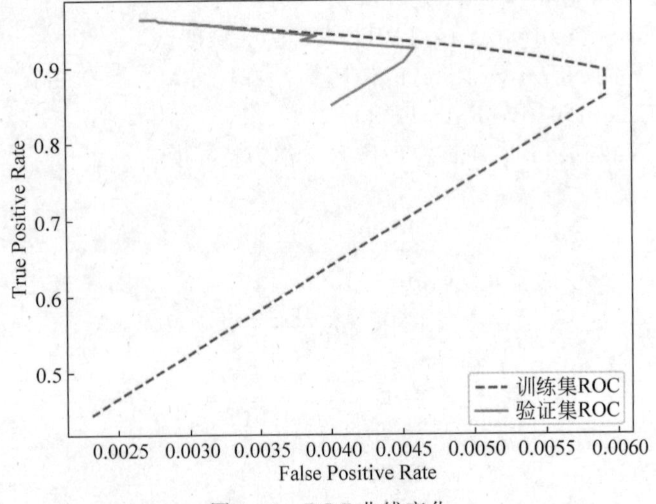

图 3-36　ROC 曲线变化

ROC 曲线中如果 FPR 越低 TPR 越高，则代表性能越高。

3.6.7　AUC 曲线

AUC(Area Under Curve)又称曲线下面积，是处于 ROC 曲线下方的面积的大小，ROC 曲线下方面积越大说明模型性能越好，因此 AUC 同理。如果是完美模型，则它的 AUC=1，说明所有正例排在负例的前面，模型达到最优状态，代码如下：

```
#第 3 章/TensorFlowAPI/神经网络的评价指标.py
#增加评价指标
METRICS = [
    tf.keras.metrics.AUC(name='AUC')
]
def plt_AUC(history):
    h = history.history
    #history.history 得到的是个字典
    #但是 v 不是一个 list，转换成 NDArray 格式以方便计算
    h = {k: np.array(v) for k, v in h.items()}
    train_auc = h['AUC']
    val_auc = h['val_AUC']
    plt.plot(history.epoch, train_auc, label='训练集 AUC')
    plt.plot(history.epoch, val_auc, label='验证集 AUC')
    plt.xlabel('Epoch')
    plt.ylabel('AUC')
    plt.legend(loc='best')
    plt.title('Epoch 与 AUC 的变化')
    plt.show()
```

调用执行后得到的结果如图 3-37 所示。

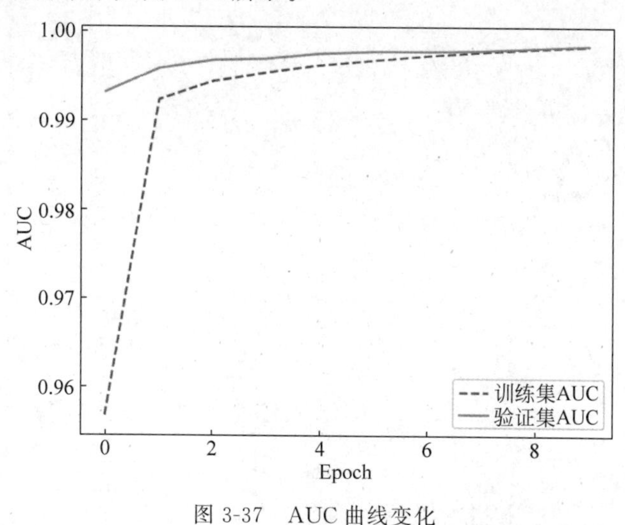

图 3-37　AUC 曲线变化

3.6.8　混淆矩阵

混淆矩阵(Confusion Matrix)又被称为错误矩阵，通过它可以直观地观察到算法的分

类效果，它的每列表示真值样本的分类概率，每行表示预测样本的分类概率，它反映了分类结果的混淆程度，代码如下：

```
#第 3 章/TensorFlowAPI/神经网络的评价指标.py
from collections import deque

#自己实现的混淆矩阵函数，只保存最后 1 个 epoch 的结果
val_ll = deque([], maxlen=1)
def confusion_matrix_acc(y_true, y_pred):
    #取预测值下标的最大值的位置
    pre_label = tf.argmax(y_pred, axis=-1)
    #取真实值下标的最大值的位置
    true_lable = tf.argmax(y_true, axis=-1)
    #调用混淆矩阵函数
    matrix_mat = tf.math.confusion_matrix(true_lable, pre_label)
    #只保存最后一次预测与真实值的矩阵值
    val_ll.append(matrix_mat)
    #这个返回值没有使用，语法要求要有返回值，metrics 默认求平均
    return matrix_mat

#在 compile 中的 metrics 增加自定义评价函数 confusion_matrix_acc
model.compile(
    optimizer=tf.keras.optimizers.Adam(learning_rate=0.001),
    loss=tf.keras.losses.CategoricalCrossentropy(from_logits=False),
    metrics=[METRICS, confusion_matrix_acc],      #评价指定，此处为准确率
    run_eagerly=True
)

#将矩阵中的数据转换成浮点数
def getMatrix(matrix):
    #求和
    per_sum = matrix.sum(axis=1)
    #得到每个单元格的概率
    ll = [matrix[i, :] / per_sum[i] for i in range(10)]
    #保留两位小数
    matrix = np.around(np.array(ll), 3)
    #如果有异常值，则赋为 0
    matrix[np.isnan(matrix)] = 0
    return matrix


#绘制混淆矩阵
def plt_confusion_matrix(class_num=10):
    #设置类别名称
    kind = [f"数字_{x}" for x in range(class_num)]
    #将分类概率转换为百分比
    data = getMatrix(val_ll[0].NumPy())
    plt.imshow(data, cmap=plt.cm.Blues)
    plt.title("手写数字识别混淆矩阵")                #title
    plt.xlabel("预测值")
    plt.ylabel("真实值")
    plt.yticks(range(class_num), kind)            #y 轴标签
```

```
    plt.xticks(range(class_num), kind, rotation=45)      #x 轴标签
    #数值处理
    for x in range(class_num):
        for y in range(class_num):
            value = float(data[y, x])
            color = 'red' if value else 'green'
            plt.text(x, y, value, verticalalignment='center',
horizontalalignment='center', color=color)
    #自动调整子图参数,使之填充整个图像区域
    plt.tight_layout()
    plt.colorbar()
    plt.savefig("手写数字识别的混淆矩阵.jpg")
    plt.show()
```

调用执行后得到的结果如图 3-38 所示。

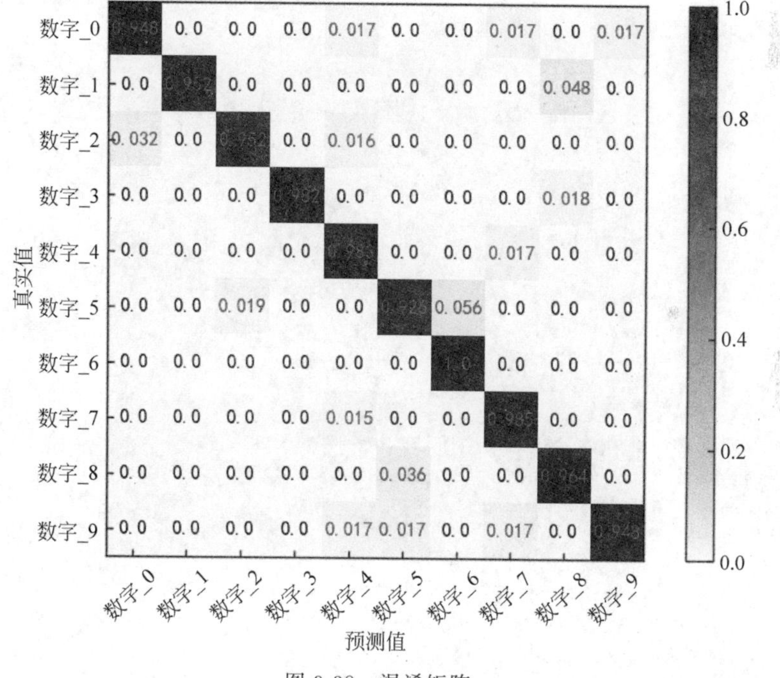

图 3-38 混淆矩阵

如图 3-38 中,当预测"数字_0"时,真实值"数字_0"的概率为 0.948,预测值为"数字_3"的概率是 0.02,也就是说有 0.052 的概率当前模型将"数字_0"预测为其他数字。代码中主要使用了 tf.math.confusion_matrix(true_lable,pre_label)来统计验证集中预测为正样本的数量,因为其返回的是一个整数,所以 getMatrix(matrix)将其转换成了概率。deque([],maxlen=1)是一个队列,因为 model.fit()在训练时,metrics 不能从 confusion_matrix_acc()函数得到返回值 matrix_mat(metrics 本身没有混淆矩阵这个指标),所以这里使用队列来更新 matrix_mat 的值。

总结

评估模型的指标有准确率、精确率、召回率、P-R 曲线、F1-Score、混淆矩阵等指标。

练习

调试并重写评价指标的代码，观察模型评估指标与代码实现的关系。

3.7 TensorFlow 中的推理预测

推理预测调用 load_model()函数加载保存的权重文件，然后使用 cv2.imread()读入灰度图，并进行维度的变化，最后由 tf.argmax 取最大下标对应的类别，得到推理结果，代码如下：

```
#第 3 章/TensorFlowAPI/预测推理.py
import tensorflow as tf
import cv2
from tensorflow.keras.models import load_model

if __name__ == "__main__":
    model = load_model('完全自定义训练 400')
height,width = 28, 28
    #读预测图片
    img = cv2.imread('test.png', cv2.IMREAD_GRAYSCALE)
    #转换为模型输入的 shape
    img = cv2.resize(img, [height,width])
    #转换为 torch.tensor()类型
    img_tensor = tf.cast(img, dtype=tf.float32)
    #因为模型输入要为[1,28,28,1],所以升维
    img_tensor = tf.reshape(img_tensor, [1, height,width, 1])
    output = model.predict(img_tensor)
    #取预测结果最大值的下标
    index = tf.argmax(output, -1).NumPy()
    #预测类型
    kind = [f"数字_{x}" for x in range(10)]
    print(f"预测={kind[int(index+1)]},概率={round(float(output[:, int(index)]),
 2)}")
```

运行结果如图 3-39 所示。

总结

模型搭建、模型训练、模型保存、模型推理预测是神经网络的 4 大阶段，推理阶段也是实际软件应用的主要内容。

练习

完成手写数字识别模型的推理代码。

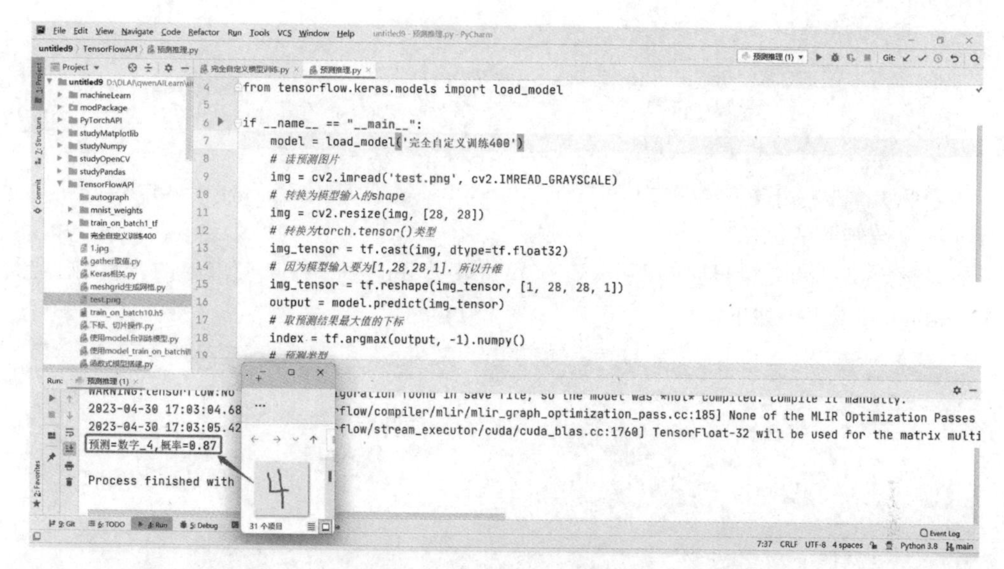

图 3-39 TensorFlow 预测结果

3.8 PyTorch 搭建神经网络

3.8.1 PyTorch 中将数据转换为张量

PyTorch 中将数据转换为张量的函数常见的有 torch.tensor()、torch.from_numpy()、torch.as_tenor()等,代码如下:

```
#第 3 章/PyTorchAPI/转换为张量.py
import torch
import numpy as np
from torch.autograd import Variable

#(1) 创建标量
scalar = torch.tensor([[1, 2], [3, 4]], dtype=torch.float32)
scalar[0, :] = 0
print(scalar)
#(2) 从 NumPy 转换为张量
np_tensor = torch.from_numpy(np.random.rand(2, 2))
print(np_tensor)
#(3) torch 随机生成指定维度的数据
rnd = torch.randn([2, 2])
print(rnd)
print('#'*30)
```

运行结果如下:

```
tensor([[0., 0.],
        [3., 4.]])
```

```
tensor([[0.5540, 0.9942],
        [0.3511, 0.1088]], dtype=torch.float64)
tensor([[ 1.6299, -2.0948],
        [-0.1387, -0.4610]])
```

代码 scalar[0，:]=0 会运行成功，而 TensorFlow 则无法运行。PyTorch 中的张量可以通过索引操作改变值的内容。

如果想定义一个变量，即专门用来存储权重参数并能够进行自动求导的张量，则需要使用 Variable()类，代码如下：

```
#第 3 章/PyTorchAPI/转换为张量.py
#(4) 创建权重参数变量
var_tensor = Variable(scalar, requires_grad=True)
#(5) 损失函数
v_out = torch.mean(var_tensor *var_tensor)
#(6) 反向传播
v_out.backward()
#(7) 获取梯度值
print(var_tensor.grad)
```

运行结果如下：

```
tensor([[0.0000, 0.0000],
        [1.5000, 2.0000]])
```

3.8.2 PyTorch 指定设备

指定设备使用 torch.device()传入 cpu 或者 cuda 可指定硬件来运行，代码如下：

```
#第 3 章/PyTorchAPI/指定设备.py
import torch

tensor1 = torch.tensor([[1, 2], [3, 4]])
tensor2 = torch.tensor([[3, 3], [5, 5]])
device = torch.device("cuda" if torch.cuda.is_available() else "cpu")
if device:
   print(tensor1 + tensor2)
```

运行结果如下：

```
tensor([[4, 5],
        [8, 9]])
```

3.8.3 PyTorch 数学运算

常见的数学操作已被 PyTorch 封装，代码如下：

```
#第 3 章/PyTorchAPI/数学运算.py
import torch
```

```
tensor1 = torch.tensor([[1., 2.], [3., 4.]])
tensor2 = torch.tensor([[3., 3.], [5., 5.]])

#矩阵内积
print(torch.matmul(tensor1, tensor2))
#矩阵点乘
print(torch.mul(tensor1, tensor2))
#矩阵相加
print(torch.add(tensor1, tensor2))
#矩阵相减
print(torch.sub(tensor1, tensor2))
#矩阵平方根
print(torch.sqrt(tensor1))
#矩阵求平均
print(torch.mean(tensor1))
```

运行结果如下：

```
tensor([[13., 13.],
        [29., 29.]])
tensor([[ 3., 6.],
        [15., 20.]])
tensor([[4., 5.],
        [8., 9.]])
tensor([[-2., -1.],
        [-2., -1.]])
tensor([[1.0000, 1.4142],
        [1.7321, 2.0000]])
tensor(2.5000)
```

3.8.4 PyTorch 维度变化

维度的变化主要使用 torch. reshape()、torch. unsqueeze()、torch. squeeze()、torch. transpose()函数，代码如下：

```
#第 3 章/PyTorchAPI/维度变化.py
import torch

tensor1 = torch.tensor([[1., 2.], [3., 4.]])
#由原来的 2*2,变换为[1,2,2]
print(torch.reshape(tensor1, [1, 2, 2]).shape)
#在原来的 2*2 的 axis=0 处增加 1 这个维度
new_tensor = torch.unsqueeze(tensor1, 0)
print(new_tensor.shape)
#在原来的 1*2*2 的 axis=0 处,减去 1 这个维度
print(torch.squeeze(new_tensor, 0).shape)
#使用 transpose 变换维度,按 axis 进行变换,由原来的 1*2*2 变成 2*1*2
print(torch.transpose(new_tensor, 1, 0).shape)
```

运行结果如下：

```
torch.Size([1, 2, 2])
torch.Size([1, 2, 2])
torch.Size([2, 2])
torch.Size([2, 1, 2])
```

3.8.5 PyTorch 切片取值

其切片取值的语法与 TensorFlow 保持一致，需要注意的是 PyTorch 中通过切片取值可以进行修改，代码如下：

```
#第 3 章/PyTorchAPI/切片取值.py
import torch

a = torch.randn([4, 28, 28, 3])
print(a[0, ...].shape)              #第 1 个[28,28,3]
print(a[0, 1, :, :].shape)          #[28,3]
print(a[:, :, :, 2].shape)          #[4,28,28],因为它取的是前面所有的值,而最里层的是
                                    #第 0 个,所以是[4,28,28]
print(a[..., 2].shape)              #[4,28,28]
print(a[:, 0, :, :].shape)          #[4,28,3]
```

运行结果如下：

```
torch.Size([28, 28, 3])
torch.Size([28, 3])
torch.Size([4, 28, 28])
torch.Size([4, 28, 28])
torch.Size([4, 28, 3])
```

3.8.6 PyTorch 中 gather 取值

PyTorch 中的 gather 取值与 TensorFlow 类似，代码如下：

```
#第 3 章/PyTorchAPI/gather 取值.py
import torch

params = torch.tensor([[1, 2], [3, 4]])
#沿着 1 轴取值:[0,0]表示第 1 行取下标为 0,0 的值;[3,4]表示第 2 行取下标为 1,1 的值
print(torch.gather(params, 1, index=torch.tensor([[0, 0], [1, 1]])))
#沿着 0 轴取值
print(torch.gather(params, 0, index=torch.tensor([[1, 1], [0, 0]])))
```

运行结果如下：

```
tensor([[1, 1],
        [4, 4]])
tensor([[3, 4],
        [1, 2]])
```

3.8.7 PyTorch 中布尔取值

在 PyTorch 中通常使用切片的语法来过滤数据，当然也可以使用 torch.masked_select()函数，代码如下：

```
#第 3 章/PyTorchAPI/布尔取值.py
import torch

data = torch.tensor([[1, 2], [3, 4], [5, 6]])
#获取 data 中>2 的数
mask = data > 2
print('mask:', mask)
#通常直接使用 data[data > 2]来表示
print("输出满足条件的数：", data[mask], data[data > 2])
#按布尔掩码选择元素
print("输出满足条件的数：", torch.masked_select(data, mask, out=None))
```

运行结果如下：

```
mask: tensor([[False, False],
        [ True, True],
        [ True, True]])
输出满足条件的数：tensor([3, 4, 5, 6]) tensor([3, 4, 5, 6])
输出满足条件的数：tensor([3, 4, 5, 6])
```

3.8.8 PyTorch 张量合并

张量合并使用 torch.cat()函数来完成，代码如下：

```
#第 3 章/TensorFlowAPI/张量合并.py
import torch

t1 = torch.randn([4, 28, 28])
t2 = torch.randn([2, 28, 28])
#按 axis=0 进行合并
print(torch.cat([t1, t2], dim=0).shape)
```

运行结果如下：

```
torch.Size([6, 28, 28])
```

3.8.9 PyTorch 模型搭建

PyTorch 搭建神经网络模型主要集中在 torch.optim 模块和 torch.nn 模块下，其模块下的概要结构如图 3-40 所示。

PyTorch 模型创建继承自 torch.nn.Module 类，可以在__init__()初始构造函数中描述属性，然后在 forward()方法中实现网络的前向传播，其描述方法与 3.4.3 节类似，代码如下：

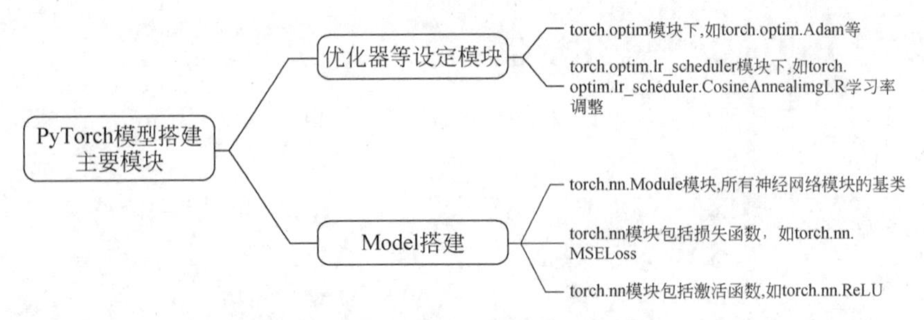

图 3-40 PyTorch 模型搭建主要模块

```
#第 3 章/PyTorchAPI/手写数字识别.py
import torch
#pip install thop
from thop import profile

#搭建模型
class MyModel(torch.nn.Module):
    def __init__(self):
        #继承 Model 类中的属性
        super(MyModel, self).__init__()
        #全连接
        self.flat = torch.nn.Flatten()
        #Linear(input_channel,output_channel)
        self.fc = torch.nn.Sequential(
            torch.nn.Linear(784, 784),
            torch.nn.Linear(784, 128)
        )
        #输出
        self.out = torch.nn.Linear(128, 10)

    def forward(self, input_x):
        x = self.flat(input_x)
        #第 1 个网络层的计算和输出
        x = torch.sigmoid(self.fc(x))
        return torch.sigmoid(self.out(x))

if __name__ == "__main__":
    model = MyModel()
    print("输出网络结构：", model)
    #PyTorch 输入的维度是 batch_size,channel,height,width
    input = torch.randn(1, 1, 28, 28)
    flops, params = profile(model, inputs=(input,))
    print('浮点计算量：', flops)
    print('参数量：', params)
```

运行结果如下：

```
输出网络结构：MyModel(
  (flat): Flatten(start_dim=1, end_dim=-1)
  (fc): Sequential(
```

```
    (0): Linear(in_features=784, out_features=784, bias=True)
    (1): Linear(in_features=784, out_features=128, bias=True)
  )
  (out): Linear(in_features=128, out_features=10, bias=True)
)
[INFO] Register count_linear() for <class 'torch.nn.modules.linear.Linear'>.
[INFO] Register zero_ops() for <class 'torch.nn.modules.container.Sequential'>.
浮点计算量: 716288.0
参数量: 717210.0
```

torch.nn.Sequential()与 tf.keras.Sequential()一样,将神经网络各层"串起来"。获取模型的参数量使用了 thop 这个库。

注意:PyTorch 的输入维度是[batch_size,channel,height,width],而 TensorFlow 的输入维度是[batch_size,height,width,channel]。

3.8.10 PyTorch 模型自定义训练

PyTorch 的自定义训练与 3.5.3 节类似,主要步骤如下:

(1) 读取训练数据和验证数据。

(2) 设置优化器、学习率等。

(3) 设置损失函数。

(4) 对训练集进行学习。

(5) 对验证集进行学习。

(6) 保存模型和参数。

使用手写数字识别模型 MyModel 进行训练,代码如下:

```
#第 3 章/PyTorchAPI/手写数字识别自定义训练.py
if __name__ == "__main__":
    model = MyModel()
    print("输出网络结构: ", model)
    #训练过程
    #(1)加载数据集
    batch_size = 60
    learning_rate = 1e-5
    epochs = 10
    #手写数字识别的数据集会被自动下载到当前./data 目录
    #训练集
    train_loader = torch.utils.data.DataLoader(
        torchvision.datasets.MNIST(
            './data/',
            train=True,
            transform=torchvision.transforms.ToTensor(),
            download=True
        ),
        batch_size=batch_size,
```

```
        shuffle=True,
    )
    #验证集
    val_loader = torch.utils.data.DataLoader(
        torchvision.datasets.MNIST(
            './data/',
            train=False,
            download=True,
            transform=torchvision.transforms.ToTensor(),
        ),
        batch_size=batch_size, shuffle=True)
    #(2)设置优化器、学习率
    optimizer = torch.optim.Adam(model.parameters(), lr=learning_rate)
    #(3)设置损失函数
    criterion = torch.nn.CrossEntropyLoss()
    #(4)训练
    def model_train(data, label):
        #清空模型的梯度
        model.zero_grad()
        #前向传播
        outputs = model(data)
        #计算损失
        loss = criterion(outputs, label)
        #计算 acc
        pred = torch.argmax(outputs, dim=1)
        accuracy = torch.sum(pred == label) / label.shape[0]
        #反向传播
        loss.backward()
        #权重更新
        optimizer.step()
        #进度条显示 loss 和 acc
        desc = "train.[{}/{}] loss: {:.4f}, Acc: {:.2f}".format(
            epoch, epochs, loss.item(), accuracy.item()
        )
        return loss, accuracy, desc
    #验证
    @torch.no_grad()
    def model_val(val_data, val_label):
        #前向传播
        val_outputs = model(val_data)
        #计算损失
        loss = criterion(val_outputs, val_label)
        #计算前向传播结果与标签一致的数量
        val_pred = torch.argmax(val_outputs, dim=1)
        num_correct = torch.sum(val_pred == val_label)
        return loss, num_correct

    #根据指定的 epoch 进行学习
    for epoch in range(1, epochs + 1):
        #进度条
        process_bar = tqdm(train_loader, unit='step')
        #开始训练模式
        model.train(True)
```

```
#因为 train_loader 是按 batch_size 划分的,所以都要进行学习
for step, (data, label) in enumerate(process_bar):
    #调用训练函数
    loss, accuracy, desc = model_train(data, label)
    #输出日志
    process_bar.set_description(desc)
    #在训练的最后 1 组数据后面进行验证
    if step == len(process_bar) - 1:
        #用来计算总数
        total_loss, correct = 0, 0
        #根据验证集进行验证
        model.eval()
        for _, (val_data, val_label) in enumerate(val_loader):
            #验证集前向传播
            loss, num_correct = model_val(val_data, val_label)
            total_loss += loss
            correct += num_correct
        #计算总测试的平均 acc
        val_acc = correct / (batch_size *len(val_loader))
        #计算总测试的平均 loss
        val_Loss = total_loss / len(val_loader)
        #验证集的日志
        var_desc = " val.[{}/{}]loss: {:.4f}, Acc: {:.2f}".format(
            epoch, epochs, val_Loss.item(), val_acc.item()
        )
        #显示训练集和验证集的日志
        process_bar.set_description(desc + var_desc)
#进度条结束
process_bar.close()
#保存模型和权重
torch.save(model, './torch_mnist.pt')
```

运行结果如图 3-41 所示。

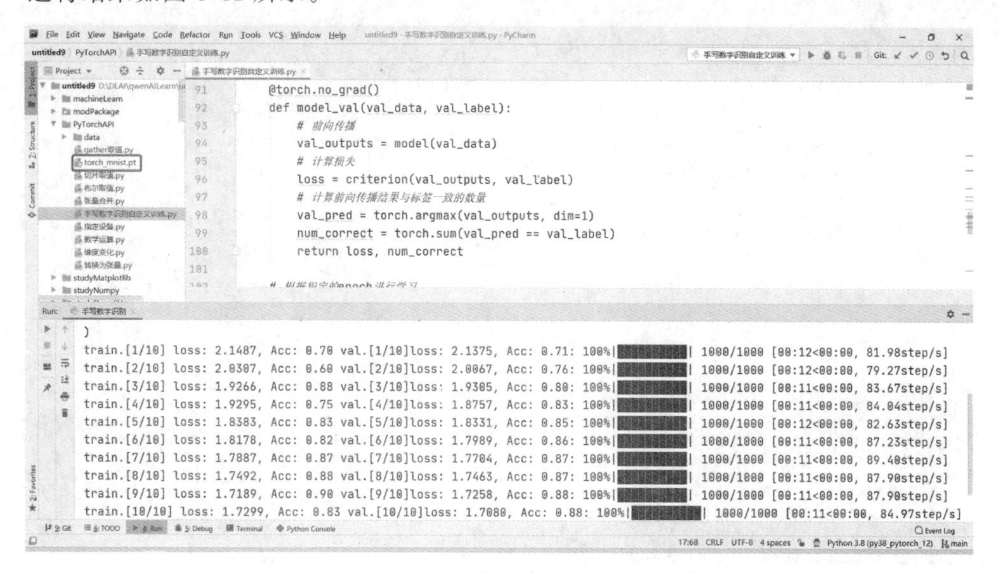

图 3-41　PyTorch 训练手写数字识别结果

仔细观察代码会发现与 TensorFlow 的训练代码类似，例如训练数据处理通过 torch. utils. data. DataLoader()来控制，优化器通过 torch. optim. Adam()设置，损失函数的设置通过 torch. nn. CrossEntropyLoss()实现。在模型训练函数 model_train(data，label)中，由 outputs＝model(data)进行前向传播，由 loss＝criterion(outputs，label)进行损失函数的计算，由 loss. backward()进行反向传播，由 optimizer. step()进行梯度下降后的权重更新，而当训练完毕时调用 torch. save(model，'. /torch_mnist. pt')保存模型和权重。

3.8.11 PyTorch 调用 Keras 训练

PyTorch 也支持调用 Keras 模块进行训练，首先调用 torchkeras 库中的 KerasModel()类，然后使用 model. fit()就能进行训练了，代码如下：

```
#第 3 章/PyTorchAPI/手写数字识别 Keras 风格训练.py
#需要安装的包
#pip install torchkeras,wandb,accelerate
from torchkeras import KerasModel
if __name__ == "__main__":
    #(1)加载数据集
    batch_size = 60
    learning_rate = 1e-5
    epochs = 10
    #手写数字识别的数据集会被自动下载到当前./data 目录
    #训练集
    train_loader = torch.utils.data.DataLoader(
        torchvision.datasets.MNIST(
            './data/',
            train=True,
            transform=torchvision.transforms.ToTensor(),
            download=True
        ),
        batch_size=batch_size,
        shuffle=True,
    )
    #验证集
    val_loader = torch.utils.data.DataLoader(
        torchvision.datasets.MNIST(
            './data/',
            train=False,
            download=True,
            transform=torchvision.transforms.ToTensor(),
        ),
        batch_size=batch_size, shuffle=True)

    #(2) 调用集成函数，设置优化器、学习率等
    model = MyModel()
    model = KerasModel(
        net=model,
        loss_fn=torch.nn.CrossEntropyLoss(),
        optimizer=torch.optim.Adam(model.parameters(), lr=1e-5)
```

```
    )
    print(model)
    #使用 model.fit()函数进行训练
    history = model.fit(
        train_data=train_loader,
        val_data=val_loader,
        epochs=10,
        patience=3, #每隔 3 个 epoch 保存权重
        ckpt_path='./torch_mnist_keras.pt',
    )
```

运行结果如图 3-42 所示。

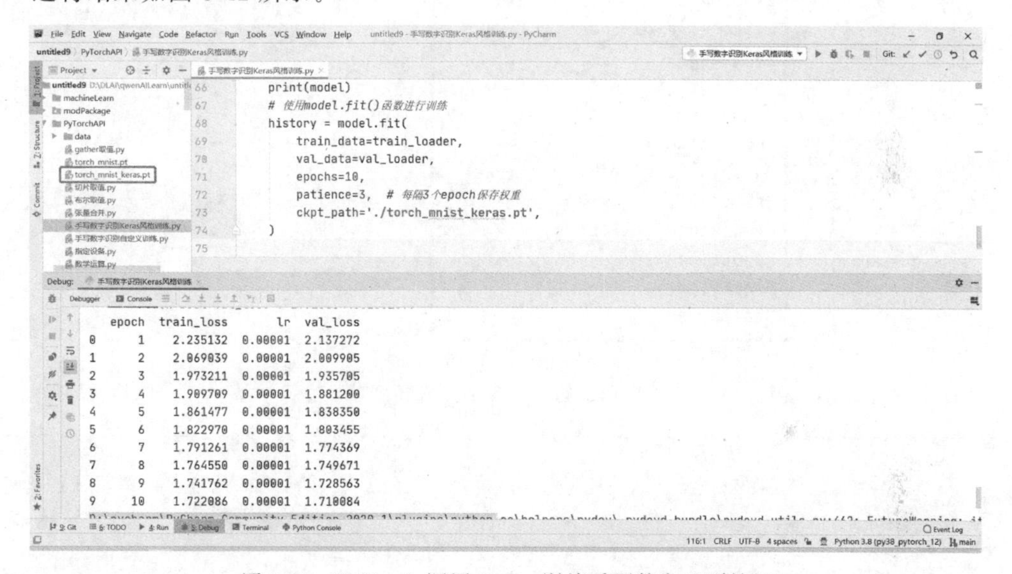

图 3-42　PyTorch 调用 Keras 训练手写数字识别结果

3.8.12　PyTorch 调用 TorchMetrics 评价指标

评价指标库 TorchMetrics 在分类模型中支持 20 种指标，如图 3-43 所示。

使用方法，代码如下：

```
#第 3 章/PyTorchAPI/手写数字识别评价指标.py
from torchmetrics import MetricCollection, Accuracy, Precision, Recall,
ConfusionMatrix
#绘制混淆矩阵
def plt_confusion_matrix(data, class_num=10):
    #设置类别名称
    kind = [f"数字_{x}" for x in range(class_num)]
    data = np.around(data.NumPy(), 2)
    plt.imshow(data, cmap=plt.cm.Blues)
    plt.title("手写数字识别混淆矩阵") #title
    plt.xlabel("预测值")
```

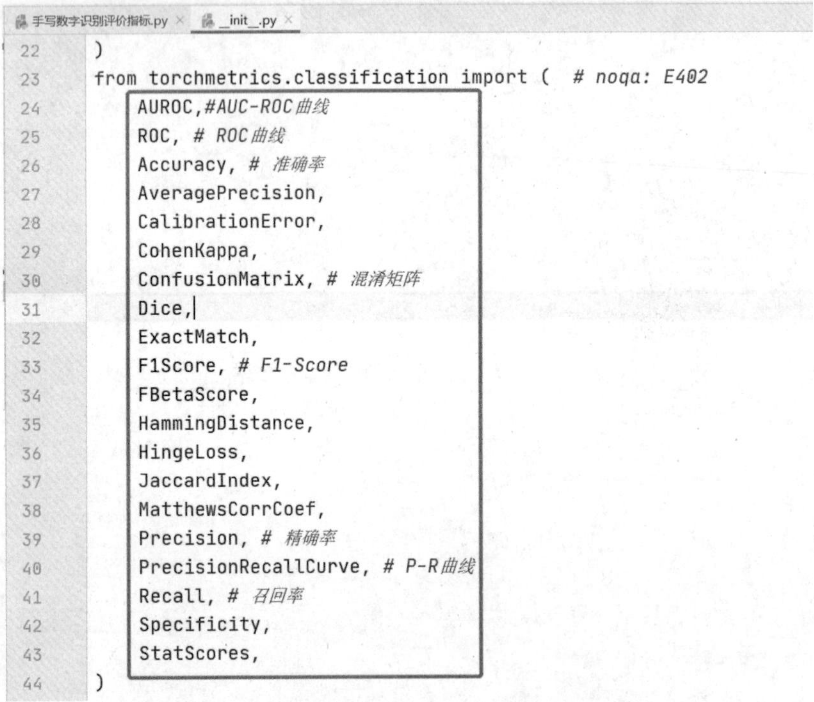

```
手写数字识别评价指标.py × | __init__.py ×
22    )
23    from torchmetrics.classification import (  # noqa: E402
24        AUROC,#AUC-ROC曲线
25        ROC, # ROC曲线
26        Accuracy, # 准确率
27        AveragePrecision,
28        CalibrationError,
29        CohenKappa,
30        ConfusionMatrix, # 混淆矩阵
31        Dice,
32        ExactMatch,
33        F1Score, # F1-Score
34        FBetaScore,
35        HammingDistance,
36        HingeLoss,
37        JaccardIndex,
38        MatthewsCorrCoef,
39        Precision, # 精确率
40        PrecisionRecallCurve, # P-R曲线
41        Recall, # 召回率
42        Specificity,
43        StatScores,
44    )
```

图 3-43　TorchMetrics 评价指标

```
    plt.ylabel("真实值")
    plt.yticks(range(class_num), kind)                          #y 轴标签
    plt.xticks(range(class_num), kind, rotation=45)             #x 轴标签
    #数值处理
    for x in range(class_num):
        for y in range(class_num):
            value = data[y, x]
            color = 'red' if value else None
            plt.text(x, y, value, verticalalignment='center',
horizontalalignment='center', color=color)
    #自动调整子图参数,使之填充整个图像区域
    plt.tight_layout()
    plt.colorbar()
    plt.savefig("手写数字识别的混淆矩阵.jpg")
    plt.show()


if __name__ == "__main__":
    model = MyModel()
    #训练过程
    #(1)加载数据集
    batch_size = 60
    learning_rate = 1e-5
    epochs = 10
```

```
    #训练集
    train_loader = torch.utils.data.DataLoader(
        torchvision.datasets.MNIST(
            './data/',
            train=True,
            transform=torchvision.transforms.ToTensor(),
            download=True
        ),
        batch_size=batch_size,
        shuffle=True,
    )
    #验证集
    val_loader = torch.utils.data.DataLoader(
        torchvision.datasets.MNIST(
              './data/',
            train=False,
            download=True,
            transform=torchvision.transforms.ToTensor(),
        ),
        batch_size=batch_size, shuffle=True)
    #(2)设置优化器、学习率
    optimizer = torch.optim.Adam(model.parameters(), lr=learning_rate)
    #(3)设置损失函数
    criterion = torch.nn.CrossEntropyLoss()
    #定义评价指标
    metrics = {
        'acc': Accuracy(num_classes=10, task='multiclass'),
        'prec': Precision(num_classes=10, task='multiclass',
average='macro'),
        'recall': Recall(num_classes=10, task='multiclass',
average='macro'),
        'matrix': ConfusionMatrix(num_classes=10, task='multiclass',
normalize='true')
    }
    train_collection = MetricCollection(metrics)
    val_collection = MetricCollection(metrics)
    #(4)训练
    def model_train(data, label):
        #清空模型的梯度
        model.zero_grad()
        #前向传播
        outputs = model(data)
        #计算损失
        loss = criterion(outputs, label)
        train_collection.forward(outputs, label)
        #反向传播
        loss.backward()
        #权重更新
        optimizer.step()
        return loss
```

```
#验证
@torch.no_grad()
def model_val(val_data, val_label):
    #前向传播
    val_outputs = model(val_data)
    #计算损失
    loss = criterion(val_outputs, val_label)
    val_collection.forward(val_outputs, val_label)
    return loss


#根据指定的 epoch 进行学习
for epoch in range(1, epochs + 1):
    #进度条
    process_bar = tqdm(train_loader, unit='step')
    #开始训练模式
    model.train(True)
    #因为 train_loader 是按 batch_size 划分的,所以都要进行学习
    for step, (data, label) in enumerate(process_bar):
        #调用训练函数
        loss = model_train(data, label)
        #在训练的最后 1 组数据后面进行验证
        if step == len(process_bar) - 1:
            #根据验证集进行验证
            model.eval()
            for _, (val_data, val_label) in enumerate(val_loader):
                #验证集前向传播
                loss = model_val(val_data, val_label)
    train_metrics = train_collection.compute()
    val_metrics = val_collection.compute()
    train_desc = f"train: acc={train_metrics['acc']}, precision={train_
metrics['prec']},recall=={train_metrics['recall']}"
    val_desc = f" val:acc={val_metrics['acc']},precision={val_metrics['prec']},
recall=={val_metrics['recall']}"
    process_bar.close()
    #保存模型和权重
    torch.save(model, './torch_mnist.pt')
#展示最后 1 次 epoch 的混淆矩阵图
plt_confusion_matrix(val_metrics['matrix'])
```

运行结果如图 3-44 所示。

代码中通过 from torchmetrics import MetricCollection, Accuracy 来引入模型评价指标,然后由 train_collection = MetricCollection(metrics)来收集每次训练和验证中的指标值,而在 model_train(data,label)训练函数中只需增加 train_collection. forward(outputs, label)就可以完成指标的收集工作,训练完成后由 train_collection. compute()完成所有数据的计算并得到相关指标的值,然后就可以根据需要对数据进行可视化操作,例如代码中混淆矩阵的可视化。

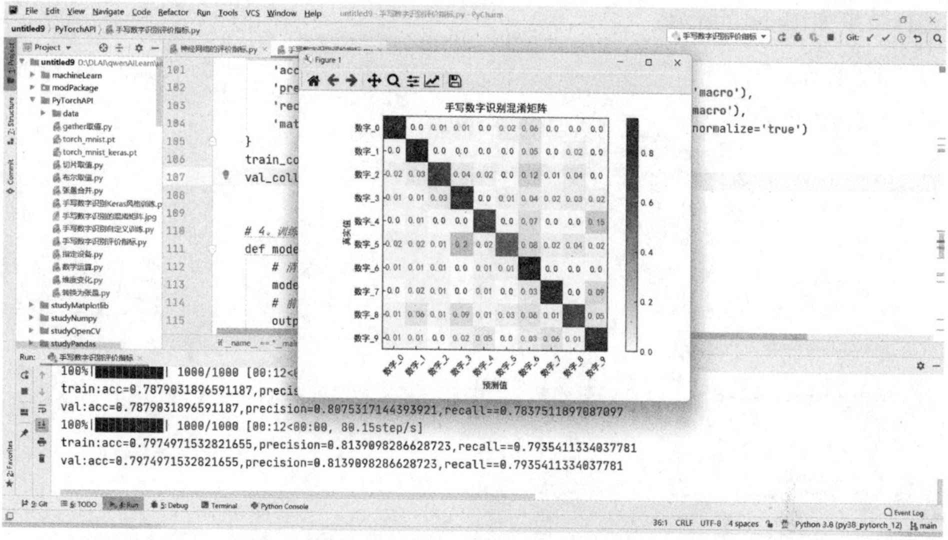

图 3-44　增加 TorchMetrics 评价指标运行结果

3.8.13　PyTorch 中推理预测

预测推理代码只需由 torch.load()导入模型权重文件，然后由 cv2.imread()读取图片，转换为模型后输入 shape，因为模型的输入维度为[batch_size，channel，height，width]，所以需要由 img_tensor.unsqueeze(0)进行升维，然后由 model(img_tensor)进行前向传播，再根据输出的概率通过 torch.max(output，1)排序，得到概率最大值，代码如下：

```
#第 3 章/PyTorchAPI/预测推理.py
import torch
import cv2

if __name__ == "__main__":
    #加载模型文件
    model = torch.load("torch_mnist.pt")
    #读预测图片
    img = cv2.imread('test.png', cv2.IMREAD_GRAYSCALE)
    #转换为模型输入的 shape
    img = cv2.resize(img, [28, 28])
    #转换为 torch.tensor()类型
    img_tensor = torch.tensor(img, dtype=torch.float32)
    #因为模型输入为[1,28,28,1],所以要升维
    img_tensor = img_tensor.unsqueeze(0)
    #前向传播
    output = model(img_tensor)
    #取预测结果最大值的下标
    max_value, pre = torch.max(output, 1)
    #预测类型
    kind = [f"数字_{x}" for x in range(10)]
    print(f"预测={kind[pre+1]},概率={round(max_value.item(), 2)}")
```

运行结果如图 3-45 所示。

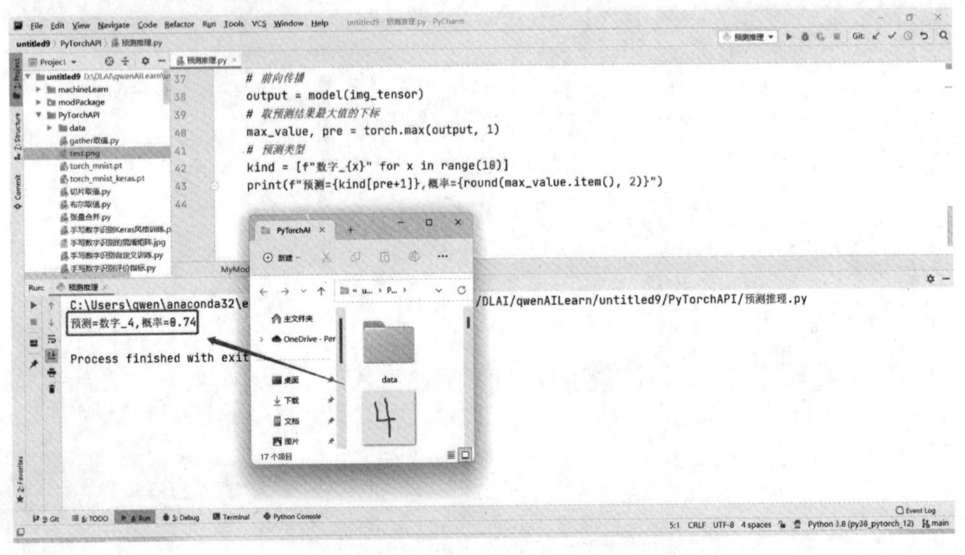

图 3-45 PyTorch 手写数字推理结果

总结

PyTorch 的主要 API 也由张量操作、模型搭建、模型训练、TorchMetrics 模型评估、模型推理构成，相对 TensorFlow 来讲 API 管理更加规范，调试起来较为方便。

练习

完成由 PyTorch 构建手写数字识别的模型，并进行训练、推理工作。

第4章 卷积神经网络

卷积神经网络(Convolutional Neural Networks,CNN)是一类包含卷积计算且具有网络深度的神经网络模型,其权重参数的学习采用反向传播和梯度下降算法。通过卷积神经网络的构建可以实现图像分类、目标检测、图像分割、图像生成等众多任务,是计算机视觉技术的一个重要组成部分。

本章将详细介绍卷积神经网络中的一些结构,并结合深度学习框架代码实战经典卷积神经网络中的图像分类模型,例如 LeNet5、AlexNet、VGG-16、GoogLeNet、MobiLeNet、ShuffLeNet、RepVGG 等。通过阅读本章内容,读者可以掌握图像分类任务的理论和代码的实现,进而为目标检测等任务奠定基础。

4.1 卷积

4.1.1 为什么用卷积

在 2.6 节中已知神经网络都采用矩阵内积来建立输入与输出之间的关系,假设有 M 个输入和 N 个输出,在训练过程中就需要 $M \times N$ 个参数去描述神经网络之间的关系,如果有多层神经网络,则这个参数量将会变得更大。例如一张图片为 1024×768,以此作为输入,下一层输出 512×384,则神经网络参数将达到千亿级,这是很难支撑的,而使用卷积神经网络则可以缓解此问题,因为卷积神经网络具有参数共享、局部感知、平移不变性等特点。

4.1.2 单通道卷积计算

所谓参数共享是指使用相同的卷积核参数在整个输入数据上进行滑动计算,从而减少模型参数量,降低过拟合的风险,并提高计算效率。

将输入图像按卷积核的大小进行划分,输入图像中的每个像素与卷积核对应的参数进行相乘后相加,最后对得到的张量进行累加就可以得到本次卷积后的输出。卷积核在输入图像中的滑动步长是人为控制的,假设步长为 1,则卷积的计算过程如图 4-1 所示。

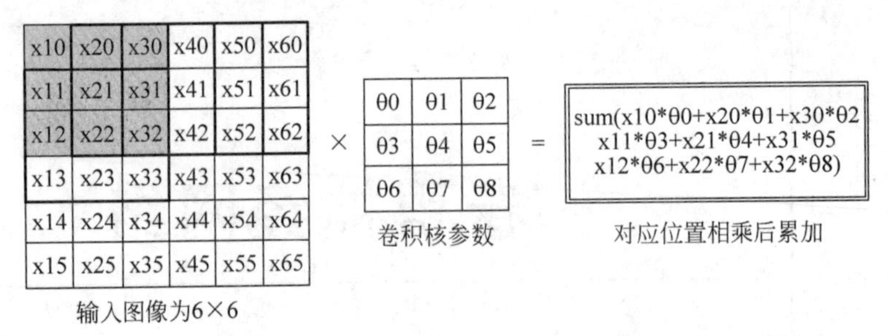

图 4-1　卷积计算($s=1$)(见彩插)

将输入图像按卷积核大小进行划分,从 x10 开始,因为步长 $s=1$,所以第 1 次计算为浅绿色背景区域与卷积核进行计算得到 1 像素值;第 2 次计算则由蓝色边框区域进行计算,以此类推,第 3 次由绿色边框区域进行计算,第 4 次由黑色边框区域进行计算,横向计算完成后再计算纵向,从红色边框区域开始重复计算,最后得到一个 4×4 的图像。

假设步长 $s=3$,则得到 2×2 的图像,其计算过程如图 4-2 所示。

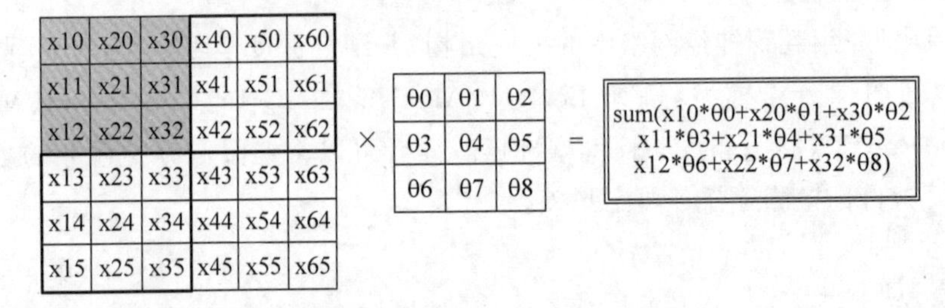

图 4-2　卷积计算($s=3$)(见彩插)

将上述计算过程数值化,假设存在一个 6×6 的输入图像,并且将卷积核参数设置为垂直方向,其卷积计算过程如图 4-3 所示。

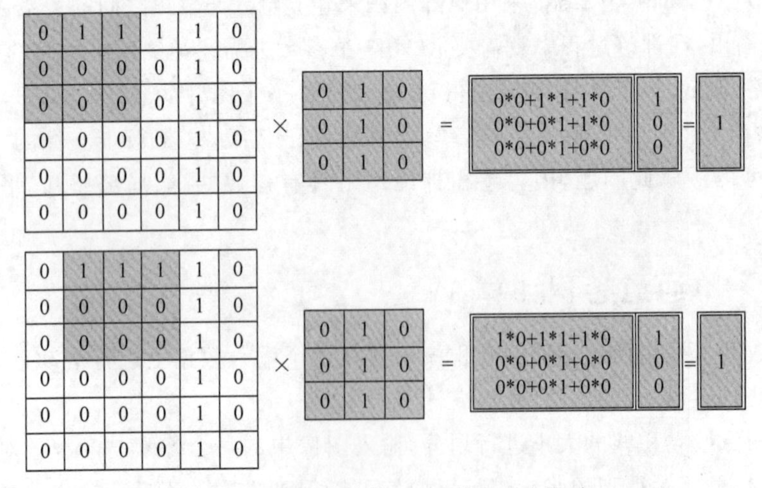

图 4-3　卷积数值计算($s=1$ 从左向右滑动)

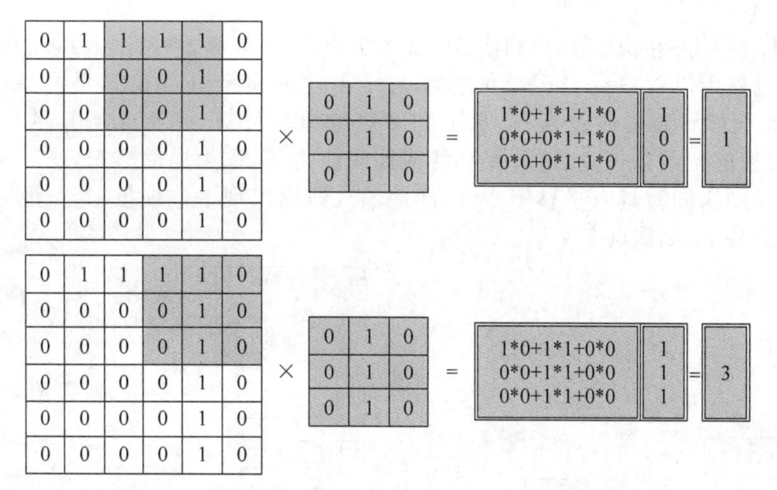

图 4-3(续)

从上向下滑动,计算过程如图 4-4 所示。

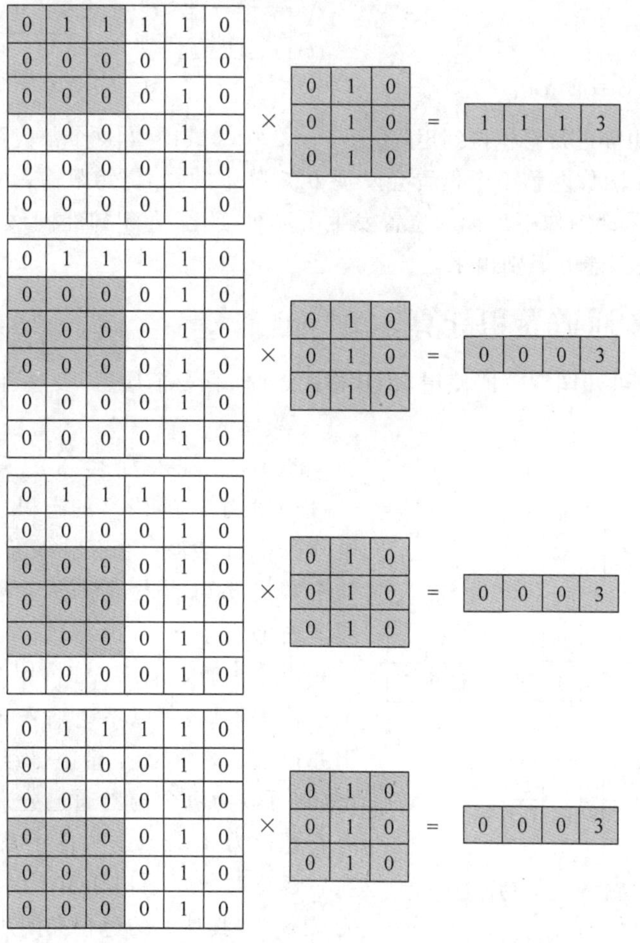

图 4-4 卷积数值计算($s=1$ 从上向下滑动)

通过步长 $s=1$ 的卷积计算，得到输出图像大小为 4×4，如果 $s=3$，则为 2×2，如图 4-5 所示。

从计算过程可以观察到卷积核的参数对于图像的每个局部信息是共享的，一共只有 9 个参数。卷积计算通过从左到右、从上到下进行滑动计算，捕捉到数据的局部特征和结构，能够有效地保留图像信息，这种局部特征捕获能力，又称为局部感知。

卷积神经网络同时具有平移不变性，即使输入数据出现平移现象，卷积神经网络也能够识别出相同的特征，如图 4-6 所示。

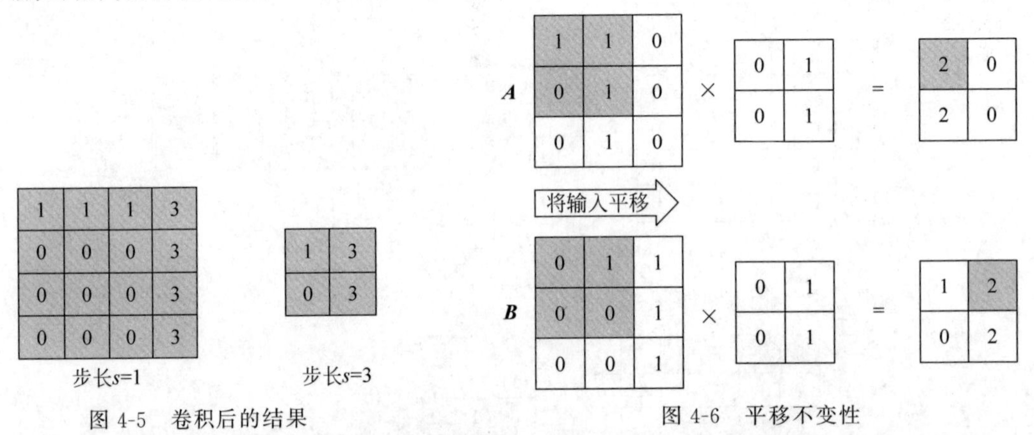

图 4-5　卷积后的结果

图 4-6　平移不变性

图 4-6 中 ***A*** 矩阵进行卷积得到矩阵[[2,0],[2,0]]，其中[2]这个数值跟输入中的[[1,1],[0,1]]相关，将 A 图像左移 1 个位置会发现 ***B*** 矩阵中的[2]仍然跟输入中的[[1,1],[0,1]]相关。换句话说无论图像中目标的位置发生了哪种变化，卷积神经网络都能识别出相同的特征，即神经网络都能够识别出来。

4.1.3　多通道卷积计算

在 1.14 节中可知图像一般使用 RGB 格式，即拥有 3 个通道，多通道输入的卷积计算过程如图 4-7 所示。

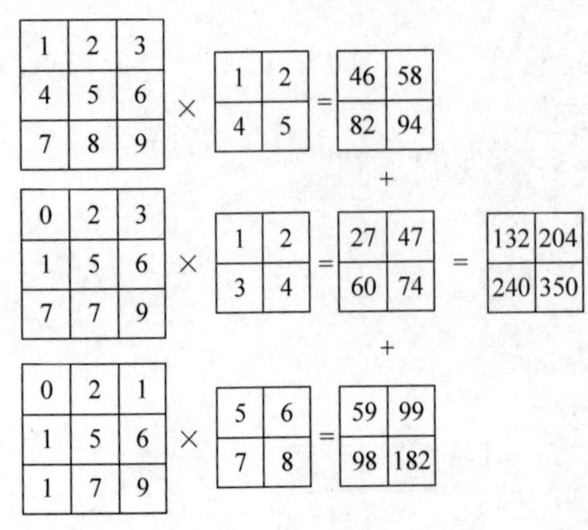

图 4-7　多通道输入卷积的计算

图 4-7 中输入为 3×3 的三通道图像，以步长 $s=1$ 分别与 3 个 2×2 的卷积进行计算，得到 3 个矩阵后相加即为多通道输入卷积的计算。卷积核的个数需要与输入通道数相等，并且输出为各个通道输出相加。

当输入通道有多个时，因为对各个通道的结果进行了累加，所以不论输入通道数是多少个，输出通道数总是 1，如果希望得到含有多个通道的输出，则可以为每个输出通道分别创建数组，假设原有卷积核为 3×2×2，并且此时变成 n×3×2×2，在做运算时每个输出通道上的结果由原来的 1 个变成 n 个，其计算的过程如图 4-8 所示。

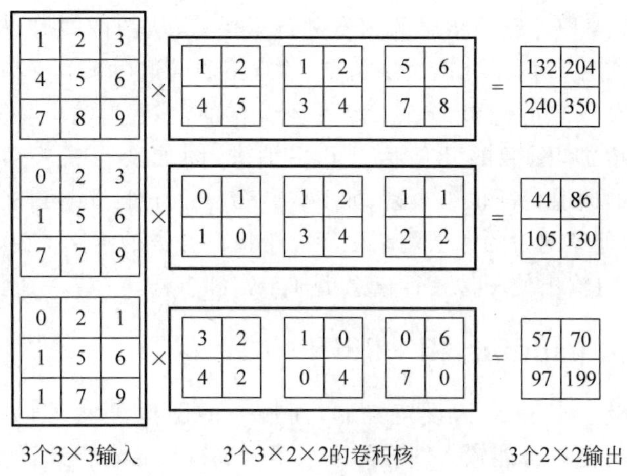

图 4-8 多通道输入与多通道输出(见彩插)

图 4-8 中输入的图像为 3×3 的三通道，将图像分别为红、黄、绿色边框的 3 个 3×2×2 的卷积核进行计算，得到 3 个 2×2 的多通道输出。因为卷积核的参数不同，以及局部感知等原因，所以多通道输出将会捕捉到不同方向、尺度、颜色、边缘、角点、纹理等更多的特征信息，使神经网络模型能够学到更丰富的特征层次和功能组合，从而提高预测能力。卷积后得到的输出又称为 Feature Map。

如果对图片进行卷积神经网络分类模型输出，然后随机取多通道中的像素数值进行图像输出，则可以得到如图 4-9 所示的图像。

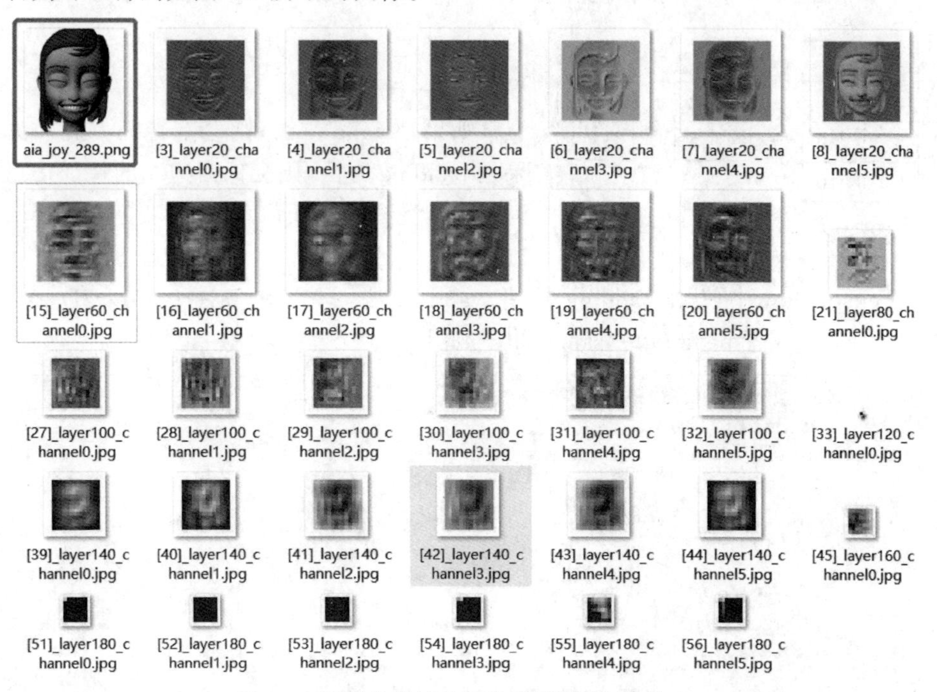

图 4-9 不同卷积神经网络不同通道图像输出

图 4-9 对 201 个卷积神经网络层的图像进行了保存，从图像中可以观察到神经网络浅层（如 layer20）得到的更多的是几何边缘信息，并且不同通道（如 layer20_channel0、layer20_channel1）得到的几何信息的色彩、饱和度等是不一样的，多通道输出特征信息非常丰富。

随着网络层的增加，图像的几何信息几乎消失，图像变得模糊且倾向于单像素，如 layer180，深层网络的通道特征也很丰富，如 layer180_channel4、layer180_channel5。

综上可知，卷积神经网络的深层网络更倾向于提供分类的语义信息，浅层网络得到更多的是几何信息，多通道输出使图像特征信息更加丰富和多样。

4.1.4 卷积 padding 和 valid

卷积核的大小及步长由人为设定控制，而输入的图像虽然可以通过 OpenCV 中的 Resize()函数统一，但是有可能输入图像的大小不能进行充足卷积计算。假设步长 $s=2$，则[3,6]不足以进行 1 次横向卷积计算，此外[7,8]也不足以进行 1 次纵向卷积计算，如图 4-10 所示。

如果遇到图 4-10 中的情况，则通常的处理方法有两种，即丢弃（valid）或者补 0（padding）。如果是 valid 方法，则将不足计算的输入信息全部丢弃，计算过程如图 4-11 所示。

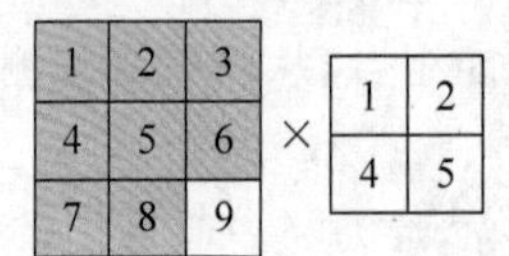

图 4-10 卷积计算不能进行 2 次

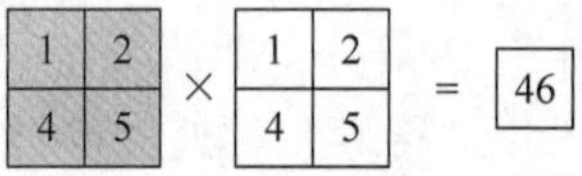

图 4-11 卷积计算 valid

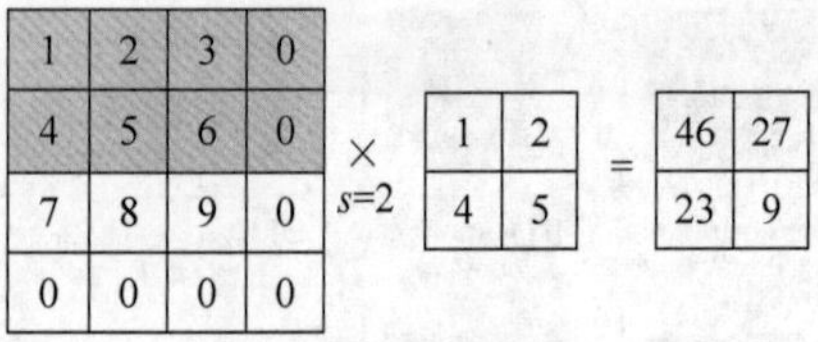

图 4-12 卷积计算 padding

如果是 padding 方法，则在不足处补 0，计算过程如图 4-12 所示。

如果是 valid 方法，则会有特征信息丢失，而如果是 padding 方法，则会保留边缘的特征信息，通常为了保留更多的特征信息，所以我们会选择 padding 方法。

在 TensorFlow 中使用 padding 或者 valid 方法，只需在 layers. Conv2D()中设置 padding='valid'或者 padding='same'，代码如下：

```
#第 4 章/CNNStudy/TensorFlow 卷积补 0.py
import tensorflow as tf
from tensorflow.keras import layers

input = tf.convert_to_tensor([
    [1, 2, 3, 4, 5],
    [4, 5, 6, 4, 5],
    [7, 8, 9, 4, 5],
    [1, 2, 3, 4, 5],
    [1, 2, 3, 4, 5]]
)
```

```
#因为 TensorFlow 输入维度要求为[batch_size,height,width,channel]
input = tf.reshape(input, [1, 5, 5, 1])
#filters 指多通道卷积输出的个数
net1 = layers.Conv2D(filters=1, kernel_size=(3, 3), strides=(3, 3), padding='valid')
result1 = net1(input)
print("valid 结果：", result1.shape)
print('#' *50)
net2 = layers.Conv2D(filters=1, kernel_size=(3, 3), strides=(3, 3), padding='same')
#前向传播输出
result1 = net2(input)
print('same 结果：', result2.shape)
```

运行结果如下：

```
valid 结果：(1, 1, 1, 1)
same 结果：(1, 2, 2, 1)
```

代码中 net1 为 padding，输出为 2×2，net2 为 valid，输出为 1×1。

如果是 PyTorch，则根据 padding 的个数来决定，代码如下：

```
#第 4 章/CNNStudy/Pytorch 卷积补 0.py
import torch as tt
input = tt.tensor([
    [1, 2, 3, 4, 5],
    [4, 5, 6, 4, 5],
    [7, 8, 9., 4, 5],
    [1, 2, 3, 4, 5],
    [1, 2, 3, 4, 5]
])
#因为 PyTorch 输入维度要求为[batch_size,channel,height,width]
input = input.view([1, 1, 5, 5])
#in_channels 为输入的通道数
#out_channels 为输出时的多通道数
net1 = tt.nn.Conv2d(in_channels=1, out_channels=1, kernel_size=3, stride=3,
padding=0)
result1 = net1(input)
print("valid 结果：", result1.shape)
print('#' *50)
net2 = tt.nn.Conv2d(in_channels=1, out_channels=1, kernel_size=3, stride=3,
padding=1)
result2 = net2(input)
print("padding=1 结果：", result2.shape)
print('#' *50)
net3 = tt.nn.Conv2d(in_channels=1, out_channels=1, kernel_size=3, stride=3,
padding=2)
result3 = net3(input)
print("padding=2 结果：", result3.shape)
```

运行结果如下：

```
valid 结果：torch.Size([1, 1, 1, 1])
padding=1 结果：torch.Size([1, 1, 2, 2])
padding=2 结果：torch.Size([1, 1, 3, 3])
```

从代码中可以发现如果在 PyTorch 中丢弃，则设置 padding=0，如果是 padding 方法，则需要计算 padding 的个数，当 padding=1 时输出为 2×2，而当 padding=2 时输出为 3×3。

需要注意 TensorFlow 与 PyTorch 卷积后图像输出的大小略有不同，TensorFlow 的计算公式为

$$\text{如果补 0，则 } N=\left\lceil\frac{W}{S}\right\rceil \tag{4-1}$$

$$\text{如果丢弃，则 } N=\left\lceil\frac{W-F+1}{S}\right\rceil$$

其中，W 指输入图像的大小，S 为步长，F 为卷积核，其结果向上取整输出图像的大小 N。假设 $W=5$、$S=3$、$F=3$，如果是 valid 方法，则 $N=\left\lceil\frac{5-3+1}{3}\right\rceil=1$；如果是 padding 方法，则 $N=\left\lceil\frac{5}{3}\right\rceil=2$。

PyTorch 的计算公式如下：

$$N=\left\lfloor\frac{W-F+2P}{S}\right\rfloor+1 \tag{4-2}$$

P 为 padding 的个数，此处为向下取整。假设 $W=5$、$S=3$、$F=3$，如果 $P=0$，则 $N=\left\lfloor\frac{5-3+2\times0}{3}\right\rfloor+1=1$，如果 $P=1$，则 $N=\left\lfloor\frac{5-3+2\times1}{3}\right\rfloor+1=2$；如果 $P=2$，则 $N=\left\lfloor\frac{5-3+2\times2}{3}\right\rfloor+1=3$。

当然输入图像或者卷积核不一定都是正方形，当宽和高不等时只需分别计算，假设输入为 4×3，卷积核为 2×3，步长 $S=2$、$P=1$，利用式(4-2)计算，则 $N(h)=\left\lfloor\frac{4-2+2\times1}{2}\right\rfloor+1=3$，$N(w)=\left\lfloor\frac{3-3+2\times1}{2}\right\rfloor+1=2$。

4.1.5 感受野

特征图 Feature Map 在输入 Input 图上卷积的区域映射又称为感受野，即特征图上的点是由输入图像中感受野区域计算得到的，如图 4-13 所示。

特征图 1 中的红色背景 9 是相对于输入图像红色边框的映射，绿色背景 9 则是绿色边框区域的映射，特征图 2 中的 36 则是相对于特征图 1 所有区域的映射。对于特征图 2 来讲，其感受野是 5×5，也就是接受了整个输入图的映射区域。

感受野大小的计算公式如下：

$$\mathrm{RF}_{i+1}=(K-1)\times\prod_{i=1}^{i}\mathrm{Stride}_i+\mathrm{RF}_i \tag{4-3}$$

其中，RF_{i+1} 表示当前层的感受野，RF_i 表示上一层的感受野，K 表示卷积核的大小，

$\prod_{i=1}^{i}$ Stride$_i$ 表示之前所有层的步长的乘积(不包括当前层)。设初始 $RF_1=1$,特征图 1 的感受野 $RF_2=(3-1)\times1+1=3$,特征图 2 的感受野 $RF_3=(2-1)\times2\times1+3=5$。

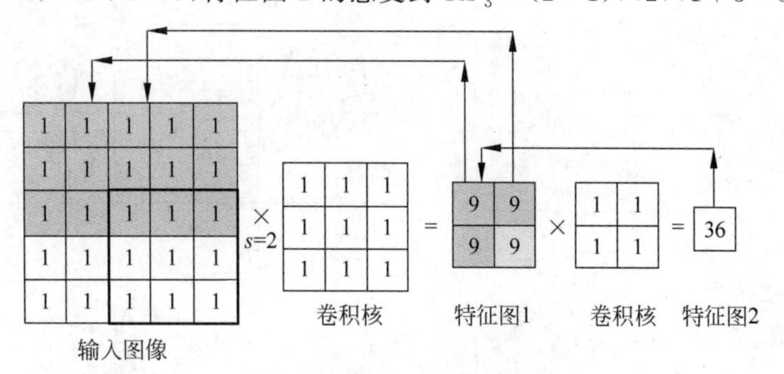

图 4-13 感受野的映射(见彩插)

卷积层越多,感受野越大,获取的特征信息就越多,忽略重要特征信息的可能性就越低,对于分类任务、语义分割任务来讲,越大的感受野其网络的识别性能可能越好。对于目标检测任务,如果感受野过小,则只能观察到局部的特征,而如果感受野过大,则会引入无效信息,对于目标检测任务要注意锚框 Anchor 与感受野对齐,在论文 *Single Shot Scale-invariant Face Detector* 中指出锚框 Anchor 的大小应该小于感受野。

关于感受野还需要了解以下几点:

(1) 经过多分支结构卷积,按照感受野最大路径计算,所以残差结构不会改变感受野。

(2) 空洞卷积,引入空洞率 a,经过空洞卷积,Kernel Size 由原来的 k 变为 $a\times(k-1)+1$。

(3) ReLU、BN、DropOut 等元素级操作不会影响感受野。

(4) 步长 $s=1$ 的 1 如图 4-10 中 1 卷积不会改变感受野。

(5) 经过全连接其感受野为整个输入图像。

(6) 步长 $s=1$ 的卷积层线性增加感受野,深度网络可以通过堆叠多层卷积增加感受野。

(7) 步长 $s=2$ 的卷积层乘性增加感受野,但受限于输入分辨率不能随意增加。

(8) 步长 $s=1$ 的卷积层加在网络后面位置会比加在前面位置增加更多感受野。

(9) 卷积神经网络输出层的感受野通常大于输入图像分辨率,例如 ResNet-101 的输出层感受野是 843,而输入图像大小为 227×227。

(10) 在 *Understanding the Effective Receptive Field in Deep Convolutional Neural Networks* 论文中指出,实际感受野要小于理论感受野,因为输入层中边缘点的使用次数明显比中间点要少,因此贡献程度不同,经过多层卷积堆叠后,输入层对于特征图点做出的贡献分布呈高斯分布形状。

计算卷积神经网络感受野可以使用开源工具 pytorch-receptive-field 和 TensorFlow 版本的工具 receptive_field。

注意:影响卷积神经网络预测结果感受野只是其一,其他因素也非常重要。

4.1.6 卷积程序计算过程

4.1.2 节介绍了理论上卷积的计算过程，用程序实现其过程如图 4-14 所示。

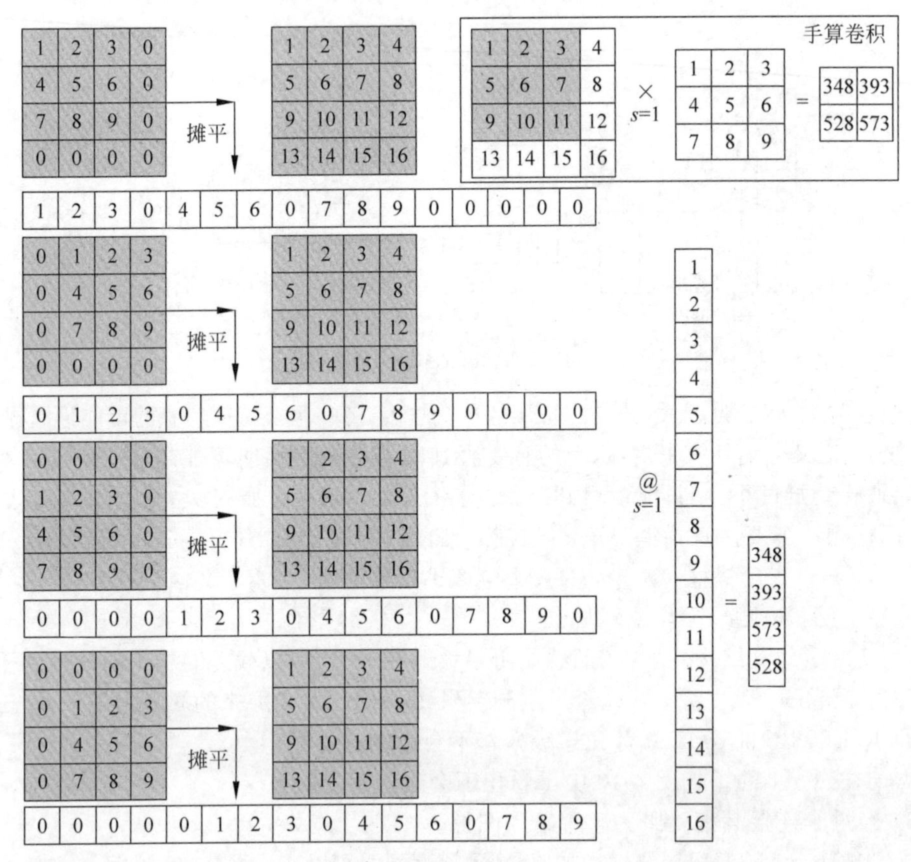

图 4-14 计算机中卷积的计算（$s=1$）（见彩插）

图 4-14 中输入为 4×4 的图像、卷积核为 3×3、步长为 1，根据公式可知输出图像为 2×2。图中根据计算的位置在卷积核中填充 0（图中红色区域），摊平后卷积核变为 4×16 的矩阵，将输入也摊平为 16 维向量，通过内积可得 4 维向量，Reshape 后可得结果与手动计算矩阵相同。

如果卷积核为 2×2，步长为 2，则输入图像 3×3 不足计算，此时通过 padding 变成 4×4 的输入，卷积核每次移动补 0 的数量与步长相同，摊平后变成 4×16 的矩阵与输入摊平做内积得到 1 个 4 维向量，如图 4-15 所示。

将上述计算过程展开，可以发现刚好是前向传播 $Y=G(\Theta_i @ X_i)$ 的计算，此时可以通过反向传播算法求出最优卷积核参数。卷积的前向传播计算与全连接的不同点在于每次卷积时卷积核的参数值相同，所处位置不同（按步长滑动），并且有效连接（不为 0 的参数）仅与卷积核参数有关，如图 4-16 所示。

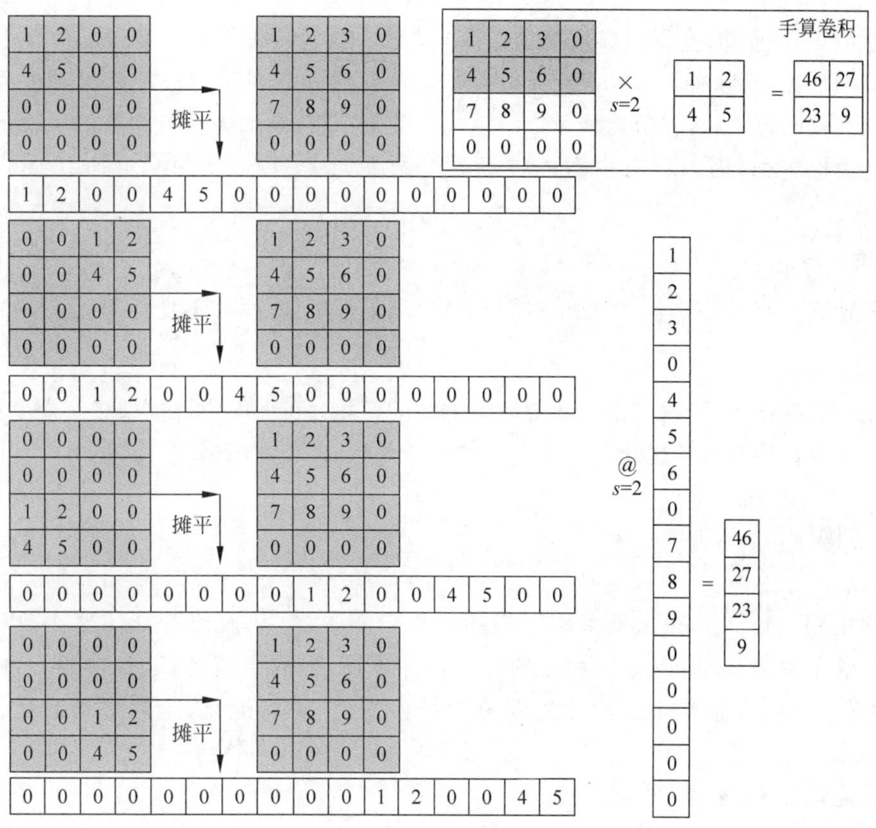

图 4-15　计算机中卷积的计算($s=2$)

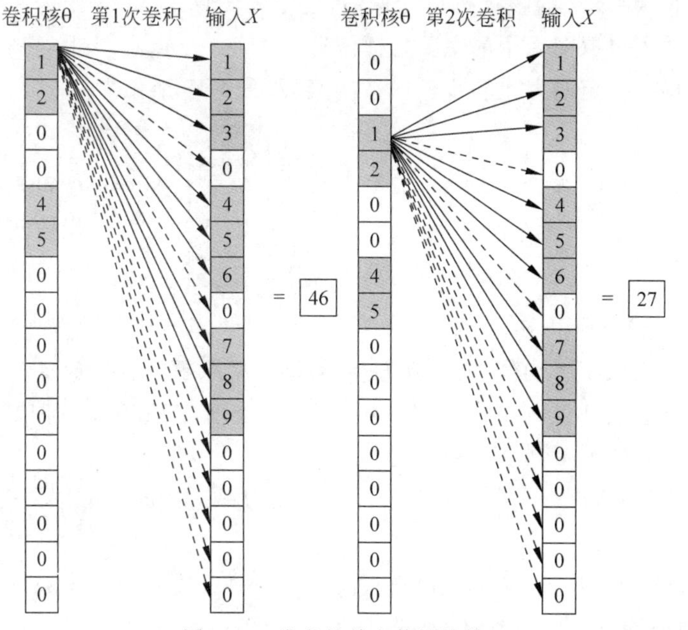

图 4-16　卷积的前向传播计算

观察图 4-16 可知，卷积核参数按全连接的方法 1 次需要计算 16×16 次，但是由于虚线处均为 0 的计算，实际有效计算只有 4×9 次，大量参数为 0 表明卷积计算相对于全连接是一种稀疏连接的方式，而稀疏连接可以降低模型的复杂度，从而降低过拟合的风险。稀疏连接泛化能力也更强，因为只有少数的神经元对结果产生影响，使网络更加专注于重要的特征。

总结

卷积计算具有参数共享、稀疏连接、滑动窗口、局部感知、平移不变性等优点。

练习

手动计算卷积 padding 和 valid 之后特征图的变化；在深度学习框架中实现卷积，并根据原理调节参数以观察其变化。

4.2 池化

池化主要是仿照人的视觉系统对特征图进行降采样，从而使特征图变小，以便简化模型的计算复杂度，减少下一层网络的参数和计算量，提取主要特征，保持卷积的平移不变性，并且能够增大感受野。池化主要分为最大池化、平均池化、全局平均池化、全局最大池化。

最大池化计算过程如图 4-17 所示。

最大池化同卷积一样，按指定的核大小和步长滑动，以便取最大值。图 4-17 中使用大小为 2×2 和步长为 2×2 的核进行滑动取值，使输入图像由 4×4 变为 2×2。最大池化由于得到最大值，故能提取图像中的主要特征信息。

平均池化，即按指定核大小和步长取平均值，如图 4-18 所示。

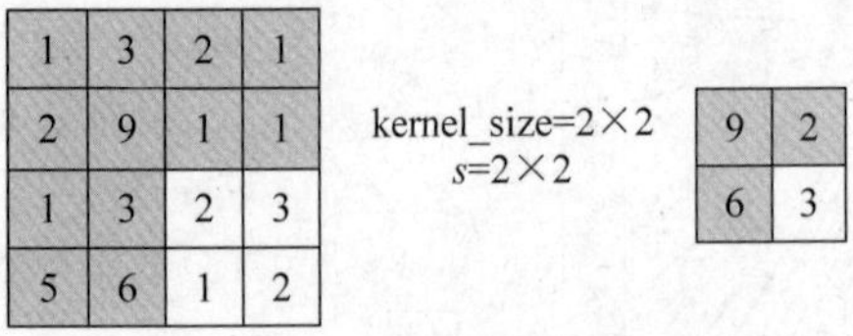

1	3	2	1
2	9	1	1
1	3	2	3
5	6	1	2

kernel_size=2×2
s=2×2

9	2
6	3

图 4-17 最大池化

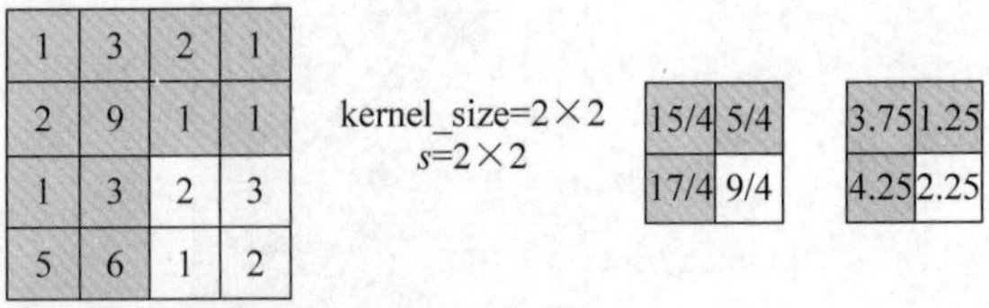

1	3	2	1
2	9	1	1
1	3	2	3
5	6	1	2

kernel_size=2×2
s=2×2

15/4	5/4
17/4	9/4

3.75	1.25
4.25	2.25

图 4-18 平均池化(1)

平均池化，因为求了平均值，所以其方差小，能够提取更多背景信息。

如果输入特征大小不够进行池化，则与卷积处理方法相同，可以对输入特征进行 padding 或者 valid，如图 4-19 所示。

图 4-19 中按 kernel_size＝3×3，步长 s＝2×2，并采用 padding 方法进行最大池化，同时部分区域会有重叠。

如果输入是多通道特征，则池化操作只需分别在每个输入通道中进行池化操作，对输入 2 通道特征分别进行 2 次平均池化，如图 4-20 所示。

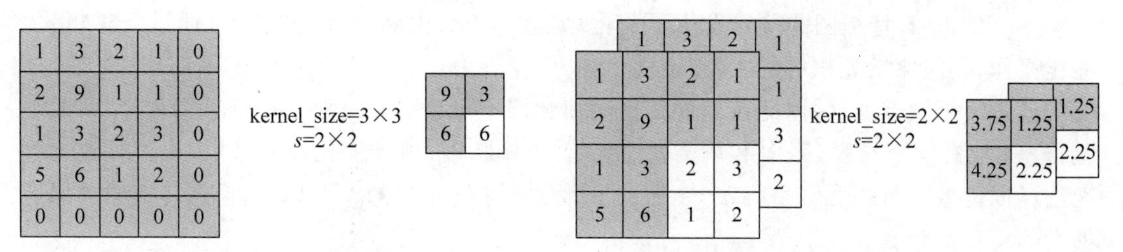

图 4-19 平均池化(2)(见彩插)　　图 4-20 平均池化(3)

全局平均池化,对于输入的每个通道的特征计算 1 个平均值,如图 4-21 所示。

全局平均池化,没有核大小和步长,它针对每个通道求 1 个平均值。通常全局平均池化用于网络的最后一层,以此来代替全连接,因为池化没有参数,所以可以显著地降低模型的参数量,防止过拟合的风险。

全局最大池化与全局平均池化类似,即每个通道取 1 个最大值,如图 4-22 所示。

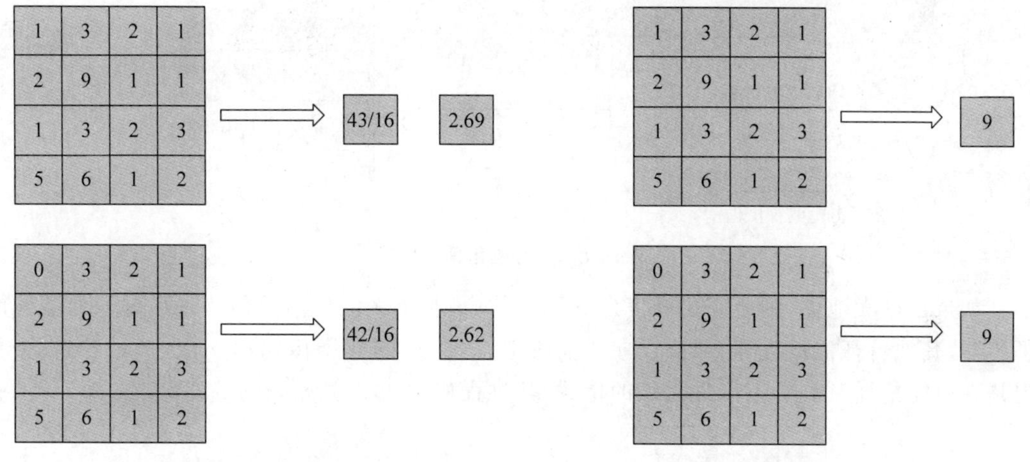

图 4-21 全局平均池化　　图 4-22 全局最大池化

最大池化能够保留卷积的平移不变性,如图 4-23 所示。

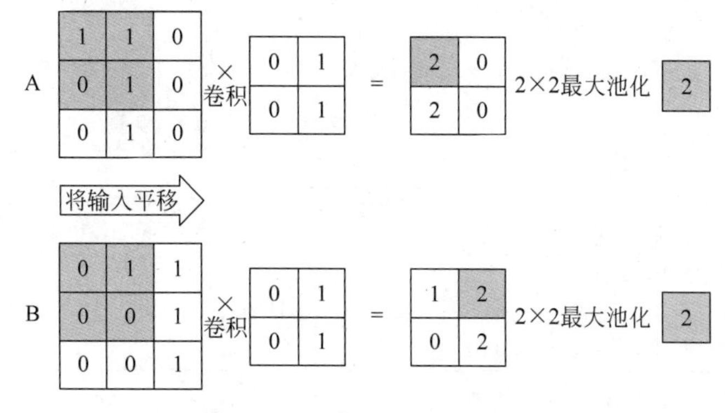

图 4-23 最大池化平移不变性

图 4-23 中 B 特征图的位置发生了变化，通过卷积后得到 2×2 的特征，通过 2×2 最大池化操作，仍能保留卷积后的主要特征。

池化由于没有参数，因此在反向传播时池化层不可导。假设对 k 层特征图做 2×2 的池化，$k+1$ 层有 4 个梯度，那么第 k 层就应该有 16 个梯度，由于池化没有参数，也就导致梯度无法反向传播。解决思路是将 $k+1$ 层的 1 梯度传递给 k 层的 4 个梯度，但是传递时要保证梯度的总和不变。

如果是平均池化，则反向传播只需把某个元素的梯度按池化的 kernel_size 进行等分，然后分配给前一层，这样就保证了池化前后梯度之和不变，如图 4-24 所示。

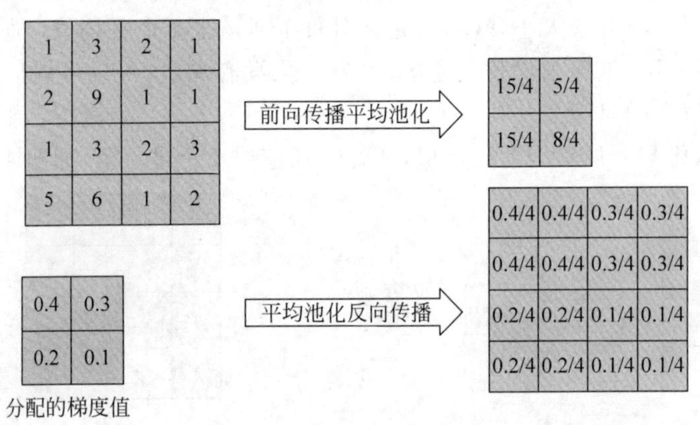

图 4-24 平均池化梯度处理

如果是最大池化，则要满足梯度之和不变的原则，最大池化前向传播时只对最大值进行了传递，其他特征值被丢弃，那么反向传播时把 $k+1$ 层的梯度传递给 k 层，其他梯度为 0，但是在前向传播时需要记住最大值的位置，其过程如图 4-25 所示。

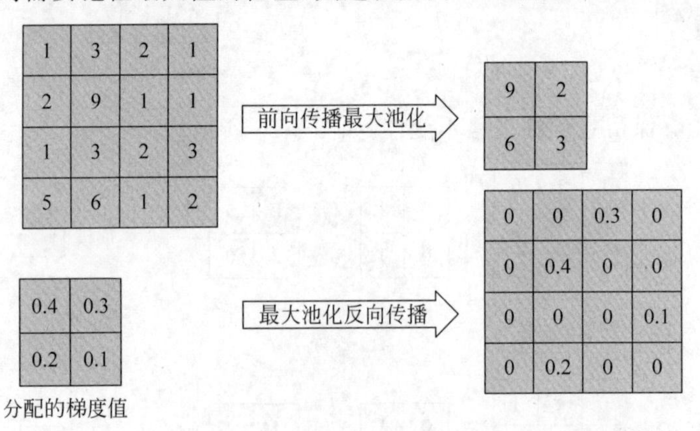

图 4-25 最大池化梯度处理

注意：池化操作没有学习参数，虽然降低了参数量，但是有可能会丢失某些特征信息。

总结

池化的主要作用是将特征图缩小，池化包括最大池化、平均池化、全局平均池化、全局最大池化。

练习

在深度学习框架中实现池化，并根据原理调节参数以观察其变化。

4.3　卷积神经网络的组成要素

一个典型的卷积神经网络由输入层、隐藏层、输出层构成，而隐藏层又由多通道卷积、激活函数、池化、全连接层等元素组成。不同的网络模型其卷积和池化层的通道数不同、卷积核大小不同、步长不同、全连接个数不同、隐藏层的个数也不相同，不同任务的网络结构都有可能不同，因此诞生了很多经典卷积神经网络，例如 LeNet5、AlexNet、VGGNet、GoogLeNet、ResNet、MobiLeNet、ShuffLeNet、RepVGGNet 等。卷积神经网络的主要组成如图 4-26 所示。

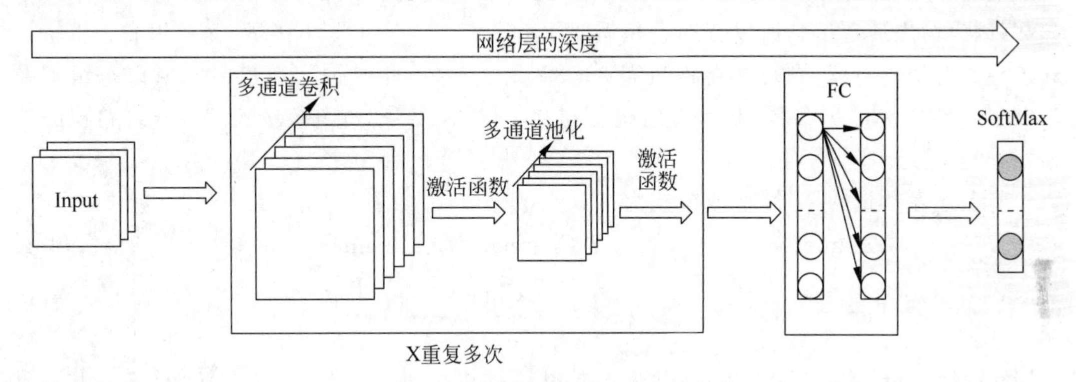

图 4-26　卷积神经网络的主要组成结构

卷积神经网络中多通道的数量又称为网络的宽度，而网络层的个数又称为网络的深度。

总结

卷积神经网络由输入层、隐藏层、输出层构成，不同任务的网络模型其组成结构不同，但满足神经网络构成的模式。

练习

观察 LeNet5 的网络结构是否满足图 4-26 所示的构成。

4.4　常见卷积分类

4.4.1　分组卷积

原始卷积操作中每个输出通道都与输入的每个通道相连接，通道之间是以稠密方式进行连接的，而分组卷积(Group Convolution)中输入和输出的通道会被划分为多个组，每个组的输出通道只和对应组内的输入通道相连接，而与其他组的通道无关，如图 4-27 所示。

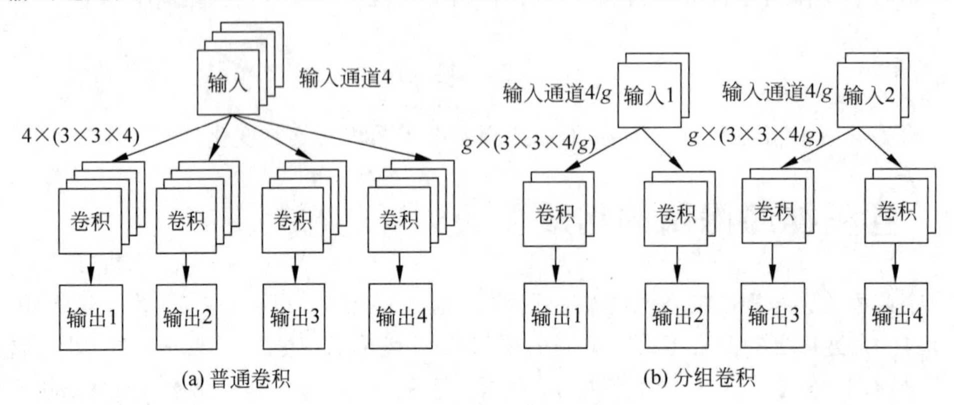

图 4-27　普通卷积与分组卷积

图 4-27 中普通卷积，输入通道数=4，假设卷积核为 3×3，则每个卷积需要 3×3×4 个参数，因为输出通道=4，所以普通卷积需要 4×(3×3×4)=144 个参数，而分组卷积将输入通道=2，卷积核不变，则每个卷积只需 3×3×2 个参数，此时输出通道=2，所以每组需要 2×(3×3×2)=36 个参数，共计 2 组进行拼接，一共只需 72 个参数。显然，分组卷积的一大好处是参数量减少了。

将上述参数量的推导，换成公式为

$$\text{Parameters}_{\text{conv}} = (k \times k \times \text{Channel}_{\text{in}}) \times \text{Channel}_{\text{out}} \tag{4-4}$$

$$\text{Parameters}_{\text{group\_conv}} = \left(k \times k \times \frac{\text{Channel}_{\text{in}}}{g}\right) \times \frac{\text{Channel}_{\text{out}}}{g} \times g$$

其中，k 为 kernel_size，$\text{Channel}_{\text{in}}$ 为输入通道数，$\text{Channel}_{\text{out}}$ 为输出通道数，g 为分组的数量，化简后可知分组卷积的参数量是普通卷积的 $1/g$。

通过将卷积按通道划分为多个路径，可以利用分布式计算的能力，有利于提高大规模卷积神经网络的训练能力。同时分组卷积将不同的输入通道分组，在每组的内部进行卷积操作，这样可以使模型更具可解释性，更容易理解每个特征起到的作用。

4.4.2　逐点卷积

逐点卷积(Pointwise Convolution，PW)又称为 1×1 卷积，计算过程如图 4-28 所示。

图 4-28 中因为输入是 3×3×1，卷积核为 1×1×1，所以输出特征图的大小没有变化，仍然是 3×3，因为只有一个通道，所以输出是 3×3×1。如果从计算方式来看，则 1×1 卷积与普通卷积没有区别，但是如果输入为 3×3×64，卷积核为 1×1×1，则其输出为 3×3×1，此时 1×1 卷积实现了通道输出上的减少，以及多通道输入信息的融合，如图 4-29 所示。

如果 1×1 卷积有多个通道，则输出可以根据通道个数的设置来调节输出通道的增加或者减少(有文章将其称为升维、降维)。如果有多组 1×1 卷积的加入，则会使模型网络的通道信息从多到少(或从小到多),也就能够实现跨通道信息的交换(例如从 64 个通道变为 128 个通道),如图 4-30 所示。

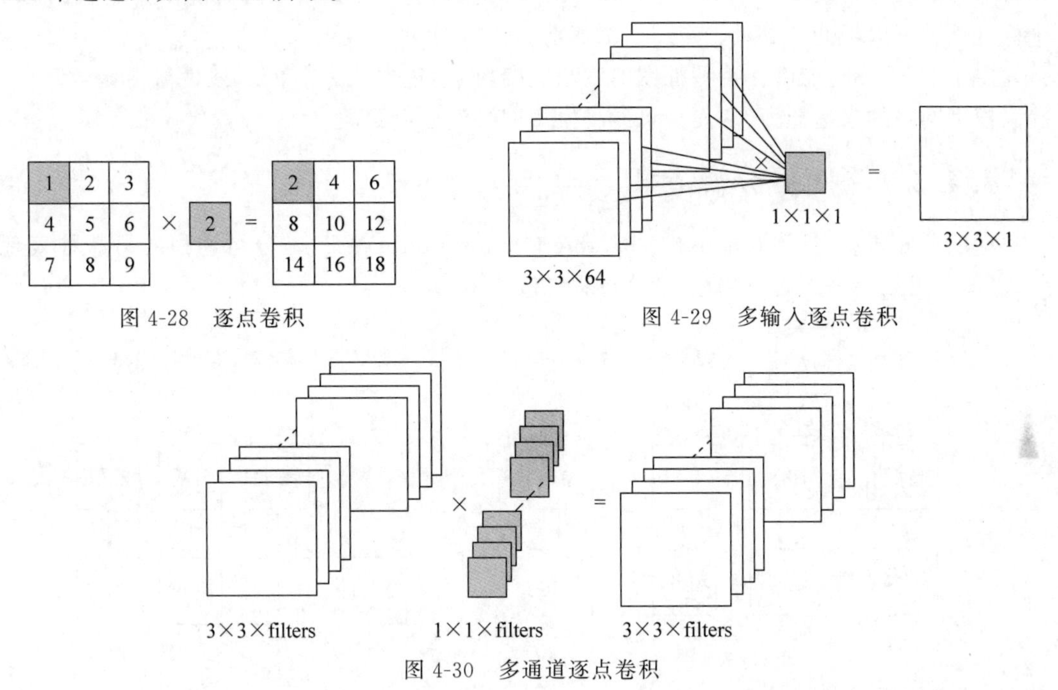

图 4-28 逐点卷积

图 4-29 多输入逐点卷积

图 4-30 多通道逐点卷积

图 4-30 中假设 filters=32,则由输入 3×3×64 通过 1×1 卷积计算,降维成 3×3×32 的输出;如果 filters=256,则由输入 3×3×64 通过 1×1 卷积,升维成 3×3×256。

1×1 卷积的另 1 个好处是通过降维后再升维,可以减少计算量、参数量,如图 4-31 所示。

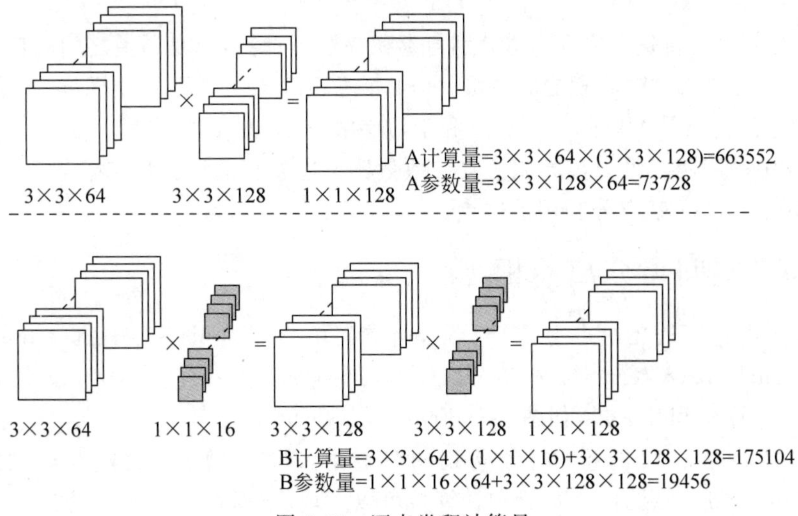

图 4-31 逐点卷积计算量

图 4-31 中 3×3×64 的输入与 3×3×128 进行卷积，其计算量比 3×3×64 与 1×1×16 进行通道降维后再跟 3×3×128 进行通道升维的计算量和参数都多，因此 1×1 卷积的另 1 个好处就是可以降低计算量、参数量，加速前向传播的计算。

由于 1×1 卷积能够保持特征图的大小不变，而通过卷积之后的非线性激活函数，可以提高模型网络的表达能力，从而有可能将模型网络做得更深。

4.4.3 深度可分离卷积

深度可分离卷积(Depthwise Separable Convolution，DWC)是分组卷积的另一种表现形式，即分组数量等于通道数量，计算过程如图 4-32 所示。

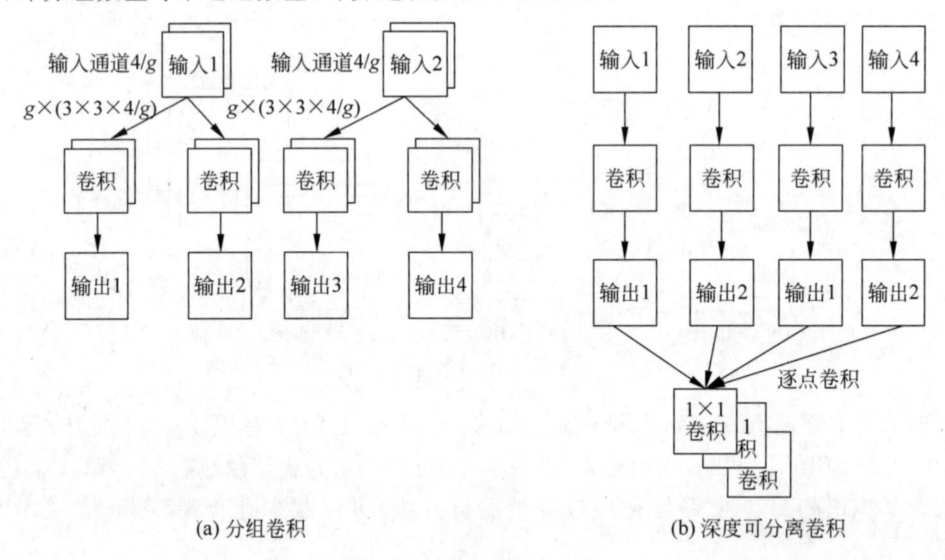

图 4-32 深度可分离卷积

深度可分离卷积将每个输入通道分别与卷积核进行计算，但是通道之间的信息并没有实现交互，所以一般说来 DW 卷积之后会跟一个 PW 卷积(1×1 卷积)，以实现跨通道信息的整合。

深度卷积的优点是可以降低计算量和学习参数量，如图 4-33 所示。

图 4-33 中利用深度可分离卷积使其参数量比普通卷积的计算量和参数量少了很多，轻量级网络 MobiLeNet 就充分利用了这点。

4.4.4 空间可分离卷积

深度可分离卷积是在通道上进行卷积，而空间可分离卷积(Separable Convolution)是将原卷积 kernel_size×kernel_size 拆分成 1×kernel_size 和 kernel_size×1 的卷积进行计算，从而降低计算量和参数量，如图 4-34 所示。

图 4-34 中将原 3×3 的卷积拆分成 1×3 的卷积，通过计算后再经过 3×1 的卷积，其输出结果仍然为 1×1。从图中参数量和计算量可以发现，空间可分离卷积的计算量和参数量更少，网络推理速度更快。

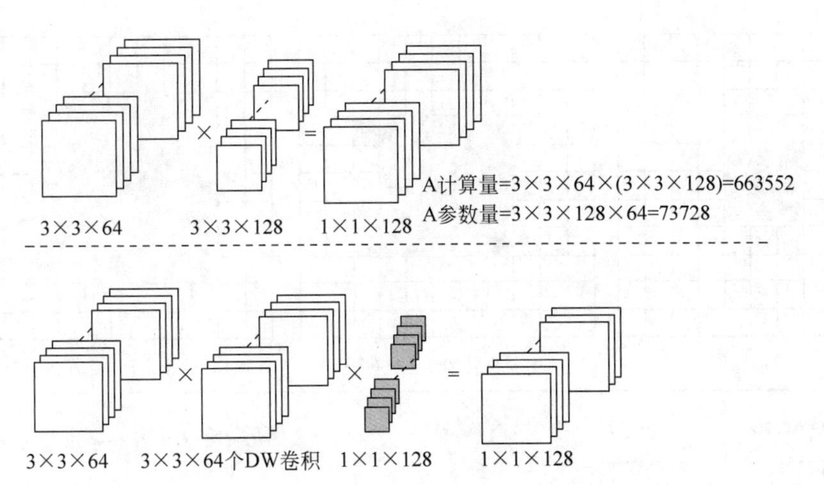

图 4-33　深度可分离卷积参数量

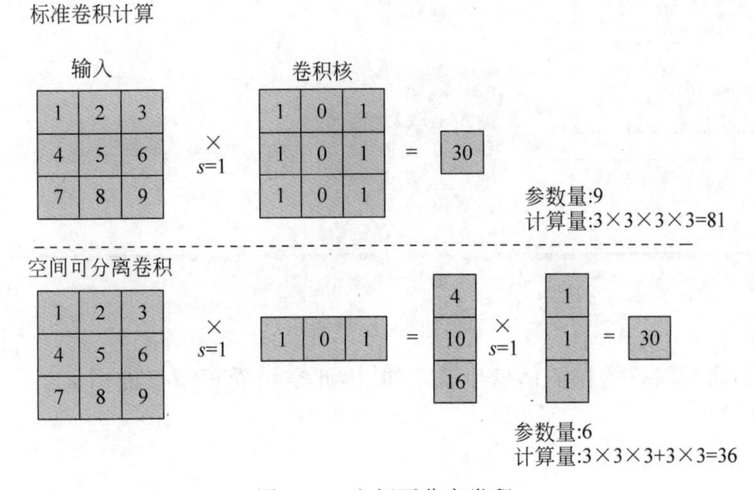

图 4-34　空间可分离卷积

4.4.5　空洞卷积

空洞卷积(Atrous Convolution)又称为膨胀卷积或扩展卷积，通过调节扩张率(Dilation Rate)的超参数，在不增加卷积核参数的情况下能够扩大感受野，其计算过程如图 4-35 所示。

图 4-35 中普通卷积是空洞卷积扩张率＝1 的特殊情况，其在原图计算的范围为 3×3，感受野＝3；当扩张率＝2 时，空洞卷积在四周扩张 1 行 1 列，变成 5×5，然后将原值每 2 赋值，其感受野＝5；当扩张率＝3 时，空洞卷积在四周扩张 2 行 2 列，变成 7×7，然后将原值每 3 赋值，感受野＝7；空洞卷积扩张率越大，其感受野也就越大，能够捕获的特征信息也就

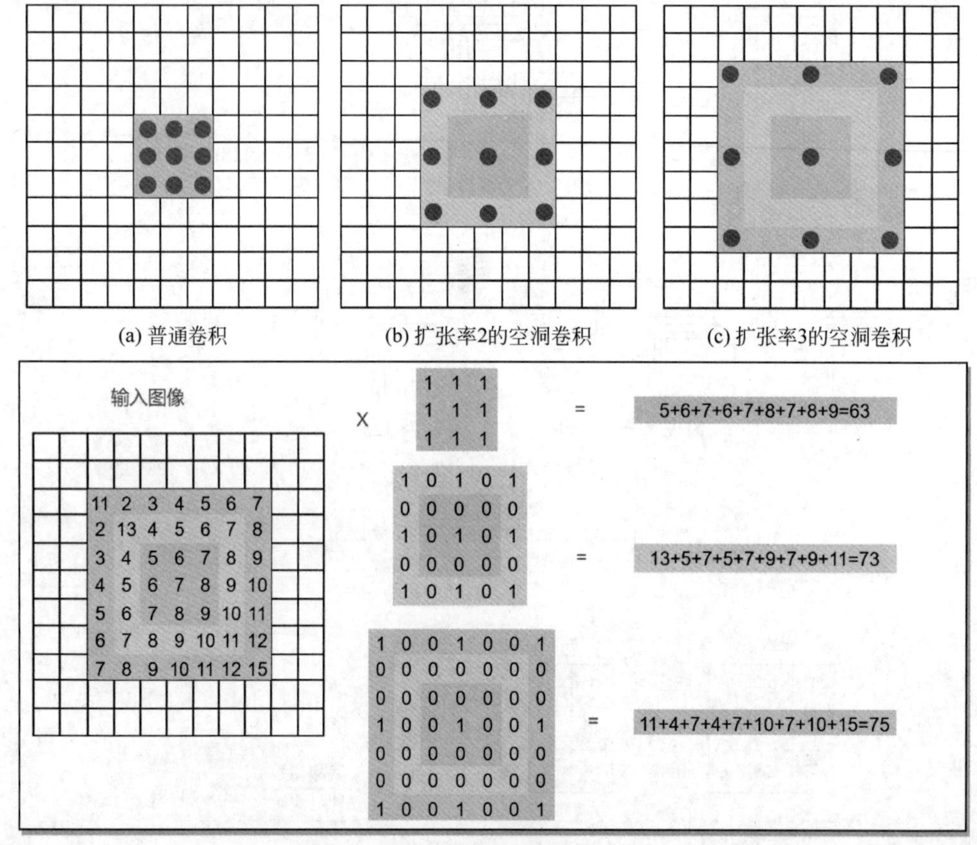

图 4-35　空洞卷积

越大，扩张率不同所捕获的特征信息也就不同，因此空洞卷积不仅能增大感受野，也能够获取多尺度信息。

空洞卷积感受野的计算公式如下：

$$RF = K + (K-1)(R-1) \tag{4-5}$$

其中，RF 指感受野，K 为原始卷积核步长，R 为扩张率。假设 $K=3$、$R=3$，则 $RF=3+(3-1)\times(3-1)=7$。

另外，因为空洞卷积会扩张卷积核的大小，而进行卷积时原输入的特征图可能需要 padding 或者 valid 方法才能保留原普通卷积输出特征图的大小。

空洞卷积并不会对所有输入的特征进行计算，如图 4-35 中扩张率=3，实际上参与运算的只有 7/49，因此空洞卷积会丢失特征的连续性。如果多次叠加具有相同扩张率的卷积核，则会造成输入特征中有一些像素自始至终都不会参与运算，这对像素集密集预测任务并不友好。

4.4.6 转置卷积

转置卷积(Transpose Convolution)又被称为反卷积,是一种用来对特征图进行上采样的方法之一。上采样也就是将低分辨率特征转换为高分辨率特征。

普通卷积的任务是将多个输入映射为 1 个输出,而卷积的反方向操作就是将 1 个输出逆向为多个输入,如图 4-36 所示。

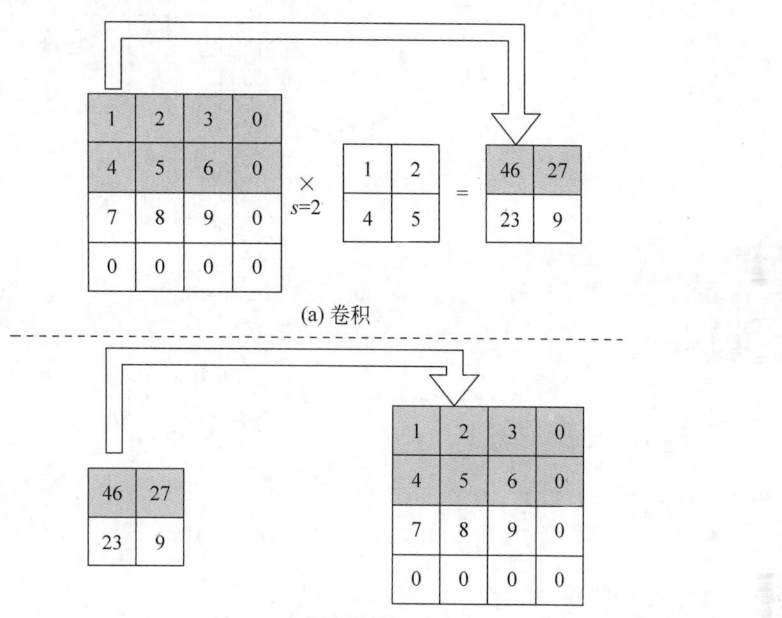

图 4-36　卷积与转置卷积

图 4-36 中希望将 2×2 的输出矩阵,通过转置卷积上采样成 4×4 的矩阵,那么就需要实现 1 变 4 的映射。

在 4.1.6 节中已知卷积的计算可以将卷积核摊平为 16 维的向量,因为要进行 4 次卷积,所以卷积核可变为 4×16 的矩阵,然后将输入转置为 16×1,通过矩阵内积运算可得 4×1 的输出,转置之后变成 2×2。

转置卷积输入是 2×2 的矩阵,转置后变成 4×1,因为要上采样成 4×4,所以转置卷积核需要 16×4 个参数(宽和高各 2 倍上采样),进行矩阵内积后输出为 16×1,转置之后变成 4×4,也就完成了上采样操作。具体过程如图 4-37 所示。

注意:卷积核、转置卷积核的参数是通过学习得来的,并不是一个自定义值。转置卷积描述的是卷积输入与输出的反向处理,而不是指卷积输出与卷积核参数计算出原输入的特征值。

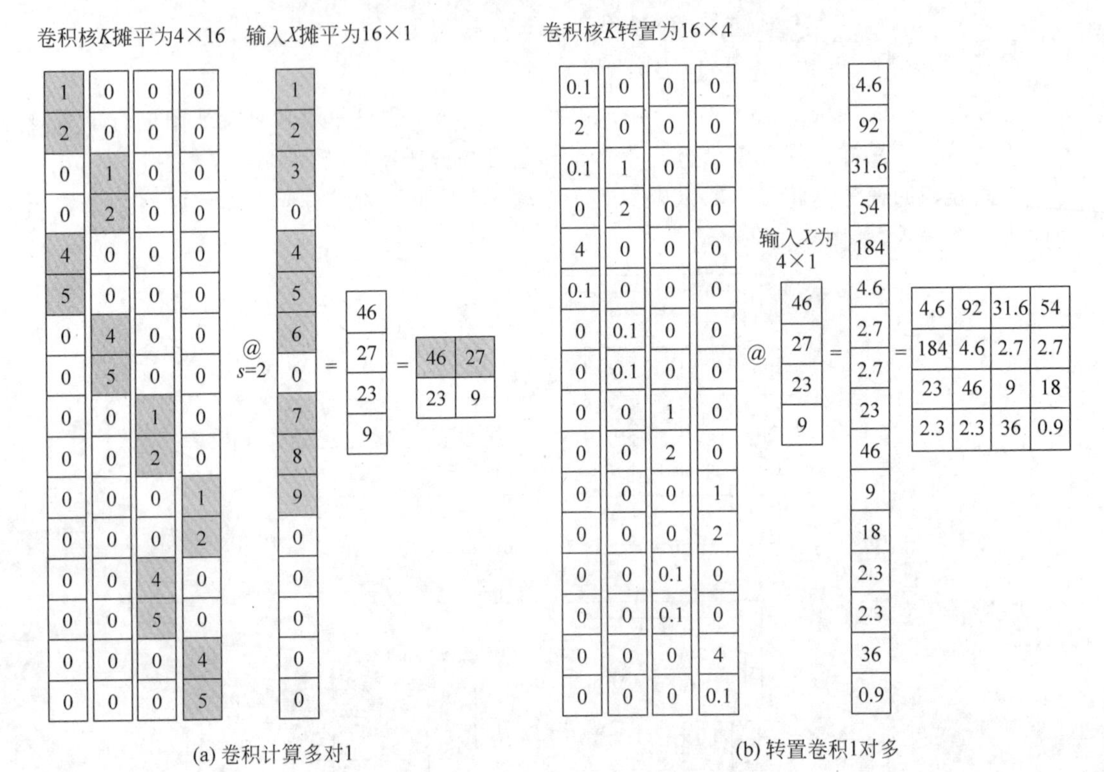

(a) 卷积计算多对1　　(b) 转置卷积1对多

图 4-37　转置卷积的计算

4.4.7　可变形卷积

标准卷积操作是将特征图分成多个与卷积核大小相同的区域，然后进行卷积操作，每个卷积后的值在特征图上的感受野相同。常规卷积的卷积核为固定的大小与形状，对于形状规则的物体会有更好的效果，但是对于形变比较复杂的物体其泛化能力就可能会下降，为解决此问题就出现了可变卷积(Deformable Convolution)。

可变形卷积在卷积核每个参数的基础上额外增加了一个方向参数，通过对神经网络的训练，卷积核就能在训练过程中扩展到更大的范围，如图 4-38 所示。

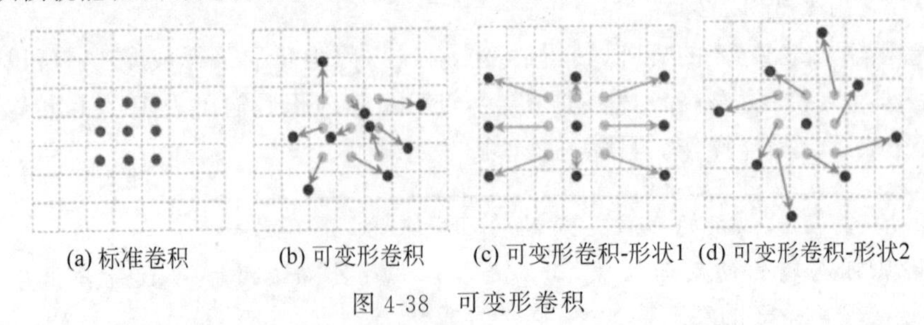
(a) 标准卷积　(b) 可变形卷积　(c) 可变形卷积-形状1　(d) 可变形卷积-形状2

图 4-38　可变形卷积

在图 4-38(a)的基础上(b)、(c)、(d)增加了方向参数，使卷积核可变为任意形状。

可变形卷积的实现步骤如图 4-39 所示。

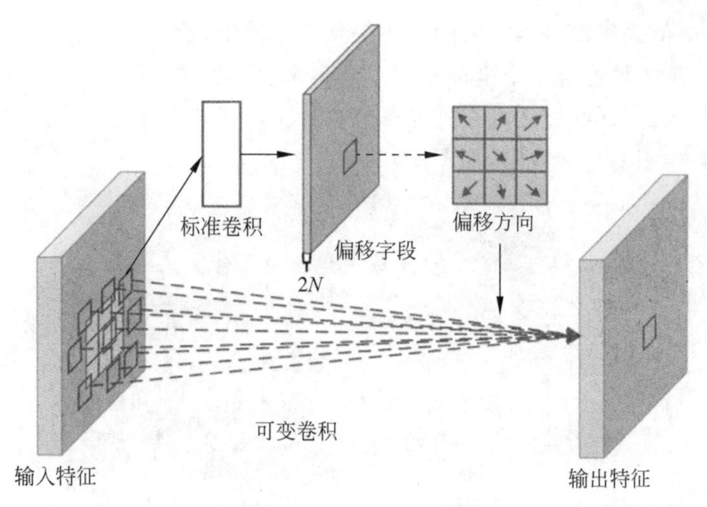

图 4-39 可变形卷积的计算过程

图 4-39 展示了可变形卷积的学习过程，首先通过一个标准卷积得到偏差，输出的偏差特征需要与输入的特征图保持一致，其通道数为 2×kernel_size×kernel_size（宽和高方向），然后将偏差值与输入特征值的坐标进行相加，得到坐标值，最后通过双线性插值的方法得到输出特征。将图 4-39 的过程实例化，其计算过程如图 4-40 所示。

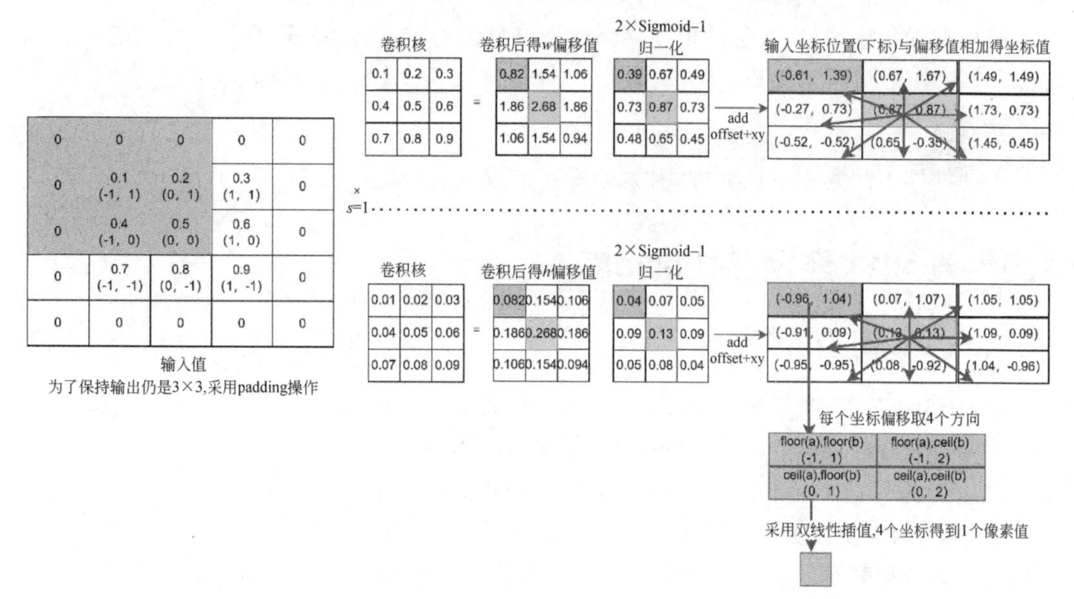

图 4-40 可变形卷积的实例化计算过程

图 4-40 中输入为 4×4×1，因为需要保持输出与输入的高和宽相等，所以采用了 padding 方法，然后通过 2×3×3 的卷积得到高、宽的偏移值，通过 2×Sigmoid()−1 归一化，对偏移值与输入值的坐标进行相加，得到输出坐标值，而此时从图中观察可知已从中心点向各个方向发生偏移。因为输出特征要为整数，所以对每个坐标通过 floor(a)向下取整，通过 ceil(b)向上

取整,然后将 4 个值采用双线性插值的方法得到 1 个输出,所以最后输出为 2×(3×3×1)。

双线性插值是一种通过 4 个坐标预测 1 个新像素的算法,拥有可学习参数,可保障可变形卷积的反向传播,其详细算法将在其他章节讲解。

关于可变形卷积的直观效果,如图 4-41 所示。

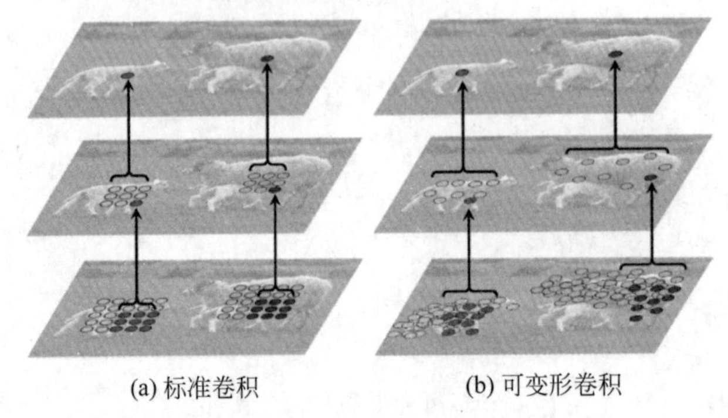

(a) 标准卷积　　(b) 可变形卷积

图 4-41　可变形卷积与标准卷积可视化效果

图 4-41 中可变形卷积识别动物的形状比标准卷积更立体、感受野范围更大,但是噪声数据也有可能更多。

总结

除标准卷积模块,在大量的模型网络中还有分组卷积、逐点卷积、深度可分离卷积、空洞卷积、转置卷积、可变形卷积等结构。

练习

手动推导不同卷积的计算方式,并分析其参数量的变化。

4.5 卷积神经网络 LeNet5

4.5.1 模型介绍

LeNet5 是在 1998 年由 Y. LeCun 等发表的论文 *Gradient-based Learning Applied to Document Recognition* 中提出。该算法是卷积神经网络的开山之作,在手写数字识别任务中可以达到 98%以上的准确率,这在当时是最先进的技术水平。

LeNet5 的成功证明了深度学习的潜力,吸引了更多研究者加入深度学习的研究中,同时也为后来更加复杂的卷积神经网络奠定了基础,例如 ResNet、DarkNet 等。虽然 LeNet5 在当今深度学习的发展中已经不再是最先进的技术,但它的经典结构和训练方法仍然对深度学习的发展和应用有重要意义。

其主要网络结构如图 4-42 所示。

输入是 32×32×1 的灰度图像,经过 5×5 卷积后得到 28×28×6,使用平均池化得到 14×14×6,再经过 5×5 卷积后得到 10×10×16,平均池化得到 5×5×16,然后摊平进入全

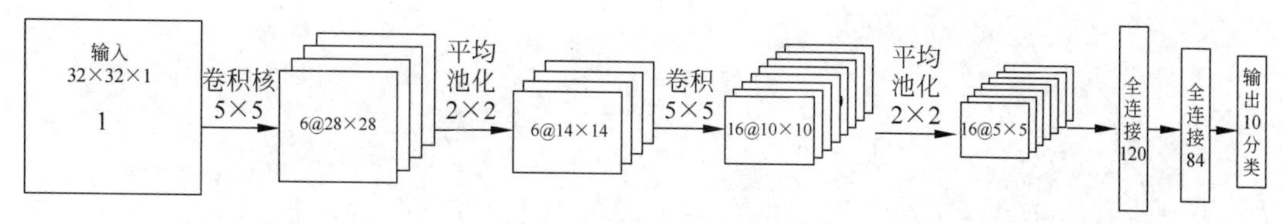

图 4-42 LeNet5 网络结构

连接层，120、84 是作者经过尝试后的最佳值，最后经过全连接输出 10 个数字的类别。因为浅层网络的特征信息相对丰富，而深层网络的特征信息可能丢失，所以随着网络深度的增加，卷积层的通道数也在增加。

4.5.2 代码实战

LeNet5 无论是使用 TensorFlow 或者 PyTorch 实现，都可以划分为模型构建部分、模型训练部分、模型推理部分。分类模型网络搭建的代码统一使用 TensorFlow 实现，并且模型采用 3.4.3 节搭建的方式，训练采用 3.5.1 节使用 model.fit 训练模型的方式。

为了使后面更大规模模型的搭建更方便，这里定义了 operator_matching()函数，用来管理各个算子，代码如下：

```
#第 4 章/ClassModleStudy/common.py

from tensorflow.keras.layers import Dense, Flatten, Conv2D, DropOut, DepthwiseConv2D
from tensorflow.keras.layers import AvgPool2D, MaxPool2D, GlobalAveragePooling2D


def operator_matching(operator_type, unit, kernel_size, strides, padding,
activation):
    """
    根据在配置文件中传入的算子类型返回算子
    :param operator_type: 算子类型
    :param unit: 神经元或者通道数
    :param kernel_size: 卷积、池化，尺寸
    :param strides: 步长
    :param padding: 补 0 或者丢弃
    :param activation: 激活函数
    :return:
    """
    if operator_type == 'Conv':
        #标准卷积
        operator = Conv2D(filters=unit,
                          kernel_size=kernel_size,
                          strides=strides,
                          padding=padding,
                          activation=activation)
    elif operator_type == 'DConv':
        #深度可分离卷积
        operator = DepthwiseConv2D(kernel_size=kernel_size,
```

```
                                    strides=strides,
                                    padding=padding,
                                    activation=activation)
    elif operator_type == 'DropOut':
        operator = DropOut(unit)
    elif operator_type == 'AvgPool':
        #平均池化
        operator = AvgPool2D(pool_size=kernel_size,
                             strides=strides,
                             padding=padding)
    elif operator_type == 'MaxPool':
        #最大池化
        operator = MaxPool2D(pool_size=kernel_size,
                             strides=strides,
                             padding=padding)
    elif operator_type == 'GlobalAvgPool':
        #全局平均池化
        operator = GlobalAveragePooling2D()
    elif operator_type == 'Flat':
        operator = Flatten()
    elif operator_type == 'FC':
        operator = Dense(unit, activation=activation)
    elif operator_type == 'OutPut':
        operator = Dense(unit, activation=activation)
    return operator
```

模型搭建的代码如下：

```
#第 4 章/ClassModleStudy/LeNet5/model.py
import tensorflow as tf
from tensorflow.keras import Model, Sequential
from tensorflow.keras.activations import sigmoid, softmax
#自定义模块,用来处理配置文件
from ClassModleStudy.common import operator_matching


class LeNet5(Model):
    def __init__(self):
        super(LeNet5, self).__init__()
        #格式为[算子类型,通道数/神经元,卷积核大小,步长大小,补 0 或丢弃,激活函数类型]
        #0 代表不设置,其主要目的是格式对齐
        config = [
            ['Conv', 6, [5, 5], [1, 1], 'same', sigmoid],
            ['AvgPool', 0, [2, 2], [1, 1], 'same', sigmoid],
            ['Conv', 16, [5, 5], [1, 1], 'same', sigmoid],
            ['AvgPool', 0, [2, 2], [1, 1], 'same', sigmoid],
            ['Flat', 0, 0, 0, 0, 0],
            ['FC', 120, 0, 0, 0, sigmoid],
            ['FC', 84, 0, 0, 0, sigmoid],
            ['OutPut', 10, 0, 0, 0, softmax],
        ]
        #根据配置文件调用 operator_matching()返回各类算子,然后加入 Sequential()中
```

```
        operator_list = []
        for operator_type, unit, kernel_size, \
            strides, padding, activation in config:
            operator = operator_matching(
                operator_type, unit,
                kernel_size, strides,
                padding, activation
            )
            operator_list.append(operator)
        self.LeNet5_Forward = Sequential(operator_list)

    def call(self, input, **kwargs):
        return self.LeNet5_Forward(input)
```

代码中将 LeNet5 的模型结构赋给了 config 变量，然后使用一个循环按顺序读取并传递给 operator_matching()函数，将每层结构追加到 operator_list 列表中，最后使用 Sequential()实现模型的构建。

LeNet5 训练的代码如下：

```
#第 4 章/ClassModleStudy/LeNet5/train.py
import tensorflow as tf
import numpy as np
#导入 LeNet5 的模块
from ClassModleStudy.LeNet5.model import LeNet5
from ClassModleStudy.common import plt_loss

#第 1 步：读取数据
num_classes = 10                                #类别数
input_shape = (28, 28, 1)
epochs = 50
batch_size = 300
#自动下载手写数字识别的数据集
(x_train, y_train), (x_test, y_test) = tf.keras.datasets.mnist.load_data()

#归一化处理，将值放在 0~1
x_train = x_train.astype('float32') / 255.
#增维，变成[batch,height,width,channel]
x_train = np.expand_dims(x_train, -1)
#将 y 转换为 one_hot 编码
y_train = tf.keras.utils.to_categorical(y_train, num_classes)

#第 2 步：构建模型
net = LeNet5()
net.build([None, 28,28, 1])
net.summary()
#第 3 步：设置回调函数、学习率和损失函数
callbacks = [
    tf.keras.callbacks.ModelCheckpoint(
        filepath='LeNet5_mnist_weights',
        save_best_only=True,
        save_weights_only=False,
```

```
        monitor='val_loss',
    ),
]
#设置学习率和损失函数
net.compile(
    optimizer=tf.keras.optimizers.Adam(learning_rate=0.0001),
    loss=tf.keras.losses.CategoricalCrossentropy(from_logits=False),
    metrics=['accuracy'],        #评价指定,此处为准确率
    run_eagerly=True
)
#第 4 步: 训练
his = net.fit(
    x=x_train,
    y=y_train,
    batch_size=batch_size,
    epochs=epochs,
    validation_split=0.1,
    callbacks=callbacks
)
plt_loss(his, epochs, 'LeNet5 手写数字识别训练结果')
```

代码中基于 3.5.1 节的模板,只更改了 net＝LeNet5()的引用,其他代码并没有变化。50 轮训练的结果如图 4-43 所示。

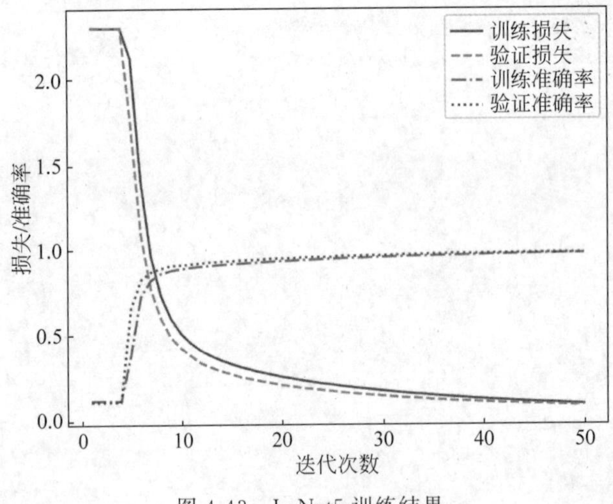

图 4-43　LeNet5 训练结果

从图 4-43 中观察可知,随着迭代次数的增加,损失在下降,而准确率却在增长。

推理代码使用 3.7 节 TensorFlow 中的推理预测中的模板内容,导入相关图片即可。

总结

LeNet5 是卷积神经网络的开山之作,由卷积、池化、全连接构成。

练习

完成 LeNet5 模型的搭建、训练、推理代码。

4.6 深度卷积神经网络 AlexNet

4.6.1 模型介绍

AlexNet 是在 2012 年论文 *ImageNet Classification with Deep Convolutional Neural Networks* 中提出的，由 A. Krizhevsky，I. Sutskever，G. Hinton 发表的。AlexNet 将网络层数扩展到 8 层，证明了应用大规模卷积神经网络的可能性。

AlexNet 在 ImageNet LSVRC-2010(ImageNet 数据集有 120 万张图像，1000 个不同的类别)比赛中取得了 37.5%和 17.0%的 top-1 和 top-5 的错误率，这比以前的先进水平要好得多。

受限于当时 GTX 580 GPU 硬件的限制，原始模型使用两块显卡并行训练，随着硬件性能的提高，本书将其合并为 1 个 GPU 实现训练，故其结构如图 4-44 所示。

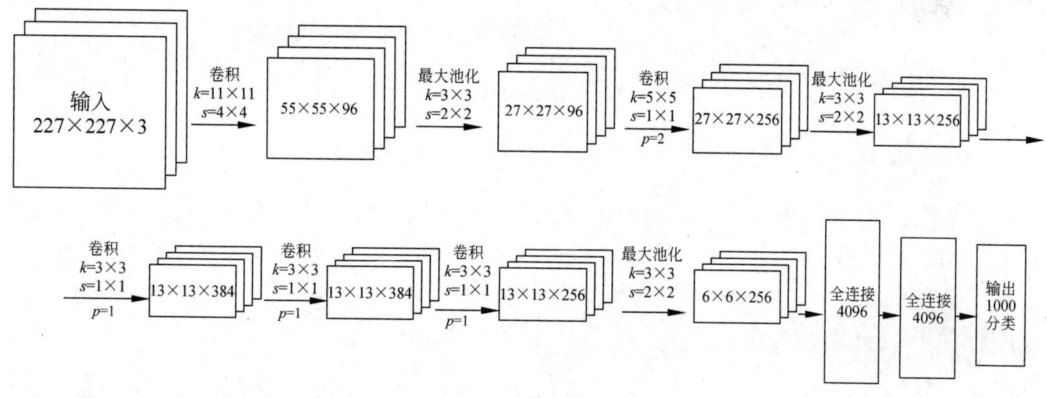

图 4-44　AlexNet 网络结构

AlexNet 的第 1 个卷积核为 11×11，这是因为浅层网络特征信息丰富，较大的卷积核依然能够提取主要特征，但可能会忽略一些次要特征。

使用 3×3 卷积的好处是刚好能够捕捉输入特征图左、右、上、下、中间点的信息。连续 3 个 3×3 步长为 1 的卷积，相当于经过 1 个 7×7 卷积，但是其参数量减少了 256×7×7×256−(256×3×3×384+384×3×3×384+384×3×3×256)=114 688 个。进入全连接后其参数量快速增长至 37 748 736，为了减少过拟合，AlexNet 使用了 DropOut 技术。

另外采用 padding=same 的方式，第 2 组卷积池化后输出特征图的尺寸不会发生变化，同时 3 个 3×3 卷积也行了 padding，其尺寸也不会变化。AlexNet 激活函数使用 ReLU 使网络收敛更快(导数为 1 或者 0)，更易训练。

4.6.2 代码实战

代码实战部分采用 FERG_DB 数据集，该数据集由 7 个卡通表情类别构成，训练集共计 44 499 张图片，验证集共计 11 122 张图片，如图 4-45 所示。

图 4-45　FERG_DB 数据集

其训练、验证数据集构建的形式如图 4-46 所示。

名称	修改日期	类型	大小
anger	2023/6/17 16:53	文件夹	
disgust	2023/6/17 16:53	文件夹	
fear	2023/6/17 16:53	文件夹	
joy	2023/6/17 16:54	文件夹	
neutral	2023/6/17 16:54	文件夹	
sadness	2023/6/17 16:54	文件夹	
surprise	2023/6/17 16:54	文件夹	

图 4-46　FERG_DB 训练集的存放格式

在模型搭建之前，创建 ConfNet()类，传入一个配置文件后进行读取，并根据配置文件创建一个模型，代码如下：

```
#第 4 章/ClassModleStudy/common.py
def get_operator_conf(config):
    #从配置文件中解析算子
    #格式为[算子类型,通道数/神经元,卷积核大小,步长大小,补 0 或丢弃,激活函数类型]
    #0 代表不设置,其主要目的是格式对齐
    operator_list = []
    for operator_type, unit, kernel_size, \
        strides, padding, activation in config:
        operator = operator_matching(
            operator_type, unit,
            kernel_size, strides,
            padding, activation
        )
        operator_list.append(operator)
    return operator_list


class ConfNet(Model):
    def __init__(self, config, index_output=[]):
        super(ConfNet, self).__init__()
        #config 根据传入的配置文件进行模型的生成
        #index_output 如果有多个输出,则指定哪一层进行输出即可
        self.operator_list = []
        self.index_out = index_output
```

```
        self.operator_list = get_operator_conf(config)

    def call(self, input, **kwargs):
        #对配置的算子进行前向传播
        x = input
        x_result = []
        for operator_forward in self.operator_list:
            x = operator_forward(x)
            x_result.append(x)
        #如果没有多个输出,则默认返回最后 1 个结果
        if not len(self.index_out):
            return x_result[-1]
        #如果有多个输出,则返回指定的输出
        return [x_result[i] for i in self.index_out]
```

ConfNet()类可根据配置文件解析成某个模型。因为传入的 config 是列表,所以又定义了函数 get_operator_conf(config),用于解析成各个算子。在 call()方法中,如果 index_output=[]存在内容,则可以返回多个输出。

同时对训练过程进行封装,代码如下:

```
#第 4 章/ClassModleStudy/common.py
#训练 FERG_DB 数据集
def train_FERG_DB(net_class, img_height=227, img_width=227, epochs=50, batch_
size=32, lr=0.0001):
    train_dir = "../FERG_DB/train"
    var_dir = "../FERG_DB/val"
    #读取数据集对象
    train_gen = ImageDataGenerator(rescale=1. / 255)
    val_gen = ImageDataGenerator(rescale=1. / 255)
    #从路径读取数据集
    train_data = train_gen.flow_from_directory(
        directory=train_dir,
        batch_size=batch_size,
        shuffle=True,
        target_size=(img_height, img_width),
        class_mode='categorical'
    )
    #训练数据集的数量
    total_train = train_data.n
    val_data = val_gen.flow_from_directory(
        directory=var_dir,
        batch_size=batch_size,
        shuffle=False,
        target_size=(img_height, img_width),
        class_mode='categorical'
    )
    #验证数据集的数量
    total_val = val_data.n
    #将文件夹的名称生成 key,value 是编号
    class_indices = train_data.class_indices
    inverse_dict = {v: k for k, v in class_indices.items()}
```

```
    print("训练类别顺序：", inverse_dict)
    #第 2 步：构建模型
    net = net_class
    net.build([None, img_height, img_width, 3])
    net.summary()
    #第 3 步：设置回调函数、学习率和损失函数
    callbacks = [
        tf.keras.callbacks.ModelCheckpoint(
            filepath='AlexNet_weights',
            save_best_only=True,
            save_weights_only=False,
            monitor='val_loss',
        ),
    ]
    #设置学习率和损失函数
    net.compile(
        optimizer=tf.keras.optimizers.Adam(learning_rate=lr),
        loss=tf.keras.losses.CategoricalCrossentropy(from_logits=False),
        metrics=['accuracy'],      #评价指定，此处为准确率
        run_eagerly=True
    )
    #第 4 步：训练
    his = net.fit(
        x=train_data,
        steps_per_epoch=total_train //batch_size,
        epochs=epochs,
        validation_data=val_data,
        validation_steps=total_val //batch_size,
        callbacks=callbacks
    )
    return his
```

在 common.py 文件中定义了函数 train_FERG_DB()，用来训练 FERG_DB 的 7 个类别，传入模型为 net_class，输入尺寸根据需要传播 RGB 通道。ImageDataGenerator()的功能是将图片数据归一化到 0～1，并创建生成器对象，然后由 train_gen.flow_from_directory()实现从文件目录转换为训练数据集，此时文件夹名称即为分类的名称，并按 batch_size 进行读取、训练、验证。最后仍然调用 net.fit()方法实现按 batch_size 进行训练，并返回 his 对象。

然后新建 train.py 文件，配置 alex_config，并调用 train_FERG_DB()实现训练，代码如下：

```
#第 4 章/ClassModleStudy/AlexNet/train.py
from ClassModleStudy.common import plt_loss, train_FERG_DB
from ClassModleStudy.common import ConfNet
from tensorflow.keras.activations import relu, softmax

alex_config = [
    ['Conv', 96, [11, 11], [4, 4], 'same', relu],
    ['MaxPool', 0, [3, 3], [2, 2], 'same', 0],
    ['Conv', 256, [5, 5], [1, 1], 'same', relu],
```

```
    ['MaxPool', 0, [3, 3], [2, 2], 'same', 0],
    ['Conv', 384, [3, 3], [1, 1], 'same', relu],
    ['Conv', 384, [3, 3], [1, 1], 'same', relu],
    ['Conv', 256, [3, 3], [1, 1], 'same', relu],
    ['MaxPool', 0, [3, 3], [2, 2], 'same', 0],
    ['Flat', 0, 0, 0, 0, 0],
    ['FC', 4096, 0, 0, 0, relu],
    ['DropOut', 0.5, 0, 0, 0, 0],
    ['FC', 4096, 0, 0, 0, relu],
    ['DropOut', 0.5, 0, 0, 0, 0],
    ['OutPut', 7, 0, 0, 0, softmax],
]

#实例化自定义模型类,总计参数量: 87 670 151
alex_net = ConfNet(alex_config)
#调用训练代码模板
his = train_FERG_DB(alex_net, img_height=227, img_width=227, epochs=5, batch_size=32, lr=0.0001)
#以图的形式展现训练结果
plt_loss(his, his.params['epochs'], 'AlexNet 卡通表情识别训练结果')
```

代码中在全连接层进行了 DropOut(0.5),因为数据集只有 7 个类别,所以 OutPut 为 7。训练结果如图 4-47 所示。

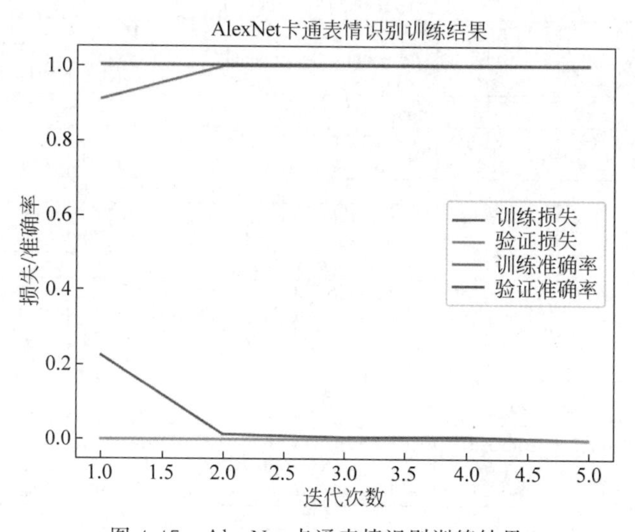

图 4-47　AlexNet 卡通表情识别训练结果

从图 4-47 中可知,训练结果不错。训练结果与训练集有关,因为本数据集相对简单,所以训练结果较好。

总结

AlexNet 将网络层数扩展到 8 层,证明了应用大规模卷积神经网络的可能性,并使用 3 个 3×3 的小卷积来提取特征。

练习

完成 AlexNet 模型的搭建、训练、推理代码。

4.7 使用重复元素的网络 VGG

4.7.1 模型介绍

VGG 的论文 *Very Deep Convolutional Networks for Large-Scale Image Recognition* 由 Karen Simonyan 和 Andrew Zisserman 发表于 2015 年，是 ILSVRC 2014 目标检测任务的冠军，分类任务的亚军，其主要特点是将网络的深度扩展到 16 层，并全部使用 3×3 的卷积。反复堆叠 3×3 的卷积可以减少参数量，并使决策函数更具有判别力。两个 3×3 卷积相当于一个 5×5 的卷积，3 个 3×3 卷积相当于一个 7×7 的卷积，VGG 由 3×3 的卷积和 2×2 的池化层构成，其结构具有对称性，具有更大的感受野，使网络有更强的表达力。

VGG 提供了 6 个尺度的深度网络，最多达到 19 层。在 C 网络中使用了 1×1 的卷积，虽然通道没有发生变化，但是卷积后经过 ReLU 操作增加了网络非线性表达能力，如图 4-48 所示。

ConvNet Configuration					
A 11 weight layers	A-LRN 11 weight layers	B 13 weight layers	C 16 weight layers	D 16 weight layers	E 19 weight layers
input(224×224 RGB image)					
Conv3-64	Conv3-64 **LRN**	Conv3-64 **Conv3-64**	Conv3-64 Conv3-64	Conv3-64 Conv3-64	Conv3-64 Conv3-64
maxpool					
Conv3-128	Conv3-128	Conv3-128 **Conv3-128**	Conv3-128 Conv3-128	Conv3-128 Conv3-128	Conv3-128 Conv3-128
maxpool					
Conv3-256 Conv3-256	Conv3-256 Conv3-256	Conv3-256 Conv3-256	Conv3-256 Conv3-256 **Conv1-256**	Conv3-256 Conv3-256 **Conv3-256**	Conv3-256 Conv3-256 Conv3-256 **Conv3-256**
maxpool					
Conv3-512 Conv3-512	Conv3-512 Conv3-512	Conv3-512 Conv3-512	Conv3-512 Conv3-512 **Conv1-512**	Conv3-512 Conv3-512 **Conv3-512**	Conv3-512 Conv3-512 Conv3-512 **Conv3-512**
maxpool					
Conv3-512 Conv3-512	Conv3-512 Conv3-512	Conv3-512 Conv3-512	Conv3-512 Conv3-512 **Conv1-512**	Conv3-512 Conv3-512 **Conv3-512**	Conv3-512 Conv3-512 Conv3-512 **Conv3-512**
maxpool					
FC-4096					
FC-4096					
FC-1000					
SoftMax					

图 4-48 VGG 网络结构

在训练方法上，先训练 A 网络，初始网络权重参数满足均值为 0 且方差为 0.01 的正态分布随机数。偏置初始为 0，然后在训练更深网络（例如 D）时，将 A 学习到的第 1 层卷积参数和最后 3 层全连接的参数（中间层权重参数随机初始化）迁移过去。因为 A 网络已完成学习，当迁移到 D 时，可加快 D 网络的学习效率，使 D 网络有一个更好的初始权重参数。

VGG 的默认输入尺寸为 224×224，原则上可设为任何不小于 224 的值，以 D 网络为例，最后 1 个 maxpool 的感受野是 212，感受野为整个图像的大部分区域，如果大于 224，则感受野的区域会减小。

虽然如此，但考虑到现实情况，图像尺寸往往是不统一的，所以作者在训练时分别采用两种方法来训练多尺度输入，第 1 种先将尺寸固定为 256 进行训练，然后将 256 尺寸的权重迁移到 384 再进行微调训练；另一种，将每幅图像的尺寸随机上采样到[256,512]，模型训练后就拥有多尺度图像的表达能力，能够提高模型网络的泛化能力。作者将 D 网络的尺寸固定为 256 时，top-1 的错误率为 27.0%，将尺寸固定为 384 时错误率为 26.8%，[256,512]错误率为 25.6%。

将图像随机上采样到[256,512]，这种方法也可看作一种数据增强的方法（裁剪、扩增），数据增强是减轻过拟合，增强网络表达能力的方法之一。

4.7.2　代码实战

以 D 模型为例，使用 ConfNet()类和配置文件快速搭建模型并进行训练，代码如下：

```
#第 4 章/ClassModleStudy/VGG_D16/train.py
from ClassModleStudy.common import plt_loss, train_FERG_DB
from ClassModleStudy.common import ConfNet
from tensorflow.keras.activations import relu, softmax

vgg_config = [
    ['Conv', 64, [3, 3], [1, 1], 'same', relu],          #224×224×64
    ['Conv', 64, [3, 3], [1, 1], 'same', relu],
    ['MaxPool', 0, [2, 2], [2, 2], 'same', 0],           #112×112×64

    ['Conv', 128, [3, 3], [1, 1], 'same', relu],         #112×112×128
    ['Conv', 128, [3, 3], [1, 1], 'same', relu],
    ['MaxPool', 0, [2, 2], [2, 2], 'same', 0],           #56×56×128

    ['Conv', 256, [3, 3], [1, 1], 'same', relu],         #56×56×256
    ['Conv', 256, [3, 3], [1, 1], 'same', relu],
    ['Conv', 256, [3, 3], [1, 1], 'same', relu],
    ['MaxPool', 0, [2, 2], [2, 2], 'same', 0],           #28×28×256

    ['Conv', 512, [3, 3], [1, 1], 'same', relu],         #28×28×512
    ['Conv', 512, [3, 3], [1, 1], 'same', relu],
    ['Conv', 512, [3, 3], [1, 1], 'same', relu],
    ['MaxPool', 0, [2, 2], [2, 2], 'same', 0],           #14×14×512

    ['Conv', 512, [3, 3], [1, 1], 'same', relu],         #14×14×512
    ['Conv', 512, [3, 3], [1, 1], 'same', relu],
```

```
    ['Conv', 512, [3, 3], [1, 1], 'same', relu],
    ['MaxPool', 0, [2, 2], [2, 2], 'same', 0],          #7×7×512

    ['Flat', 0, 0, 0, 0, 0],
    ['FC', 4096, 0, 0, 0, relu],
    ['DropOut', 0.5, 0, 0, 0, 0],
    ['FC', 4096, 0, 0, 0, relu],
    ['DropOut', 0.5, 0, 0, 0, 0],
    ['OutPut', 7, 0, 0, 0, softmax],
]

#实例化自定义模型类,总计参数量: 134 289 223
vgg_net = ConfNet(vgg_config)
#调用训练代码模板
his = train_FERG_DB(vgg_net, img_height=224, img_width=224, epochs=5, batch_
size=16, lr=0.00001)
#以图的形式展现训练结果
plt_loss(his, his.params['epochs'], 'VGG-16 卡通表情识别训练结果')
```

运行代码可以发现卷积层参数量为 14 714 688,而全连接层的参数量为 119 574 535, VGG 参数量主要集中在全连接层。训练结果如图 4-49 所示。

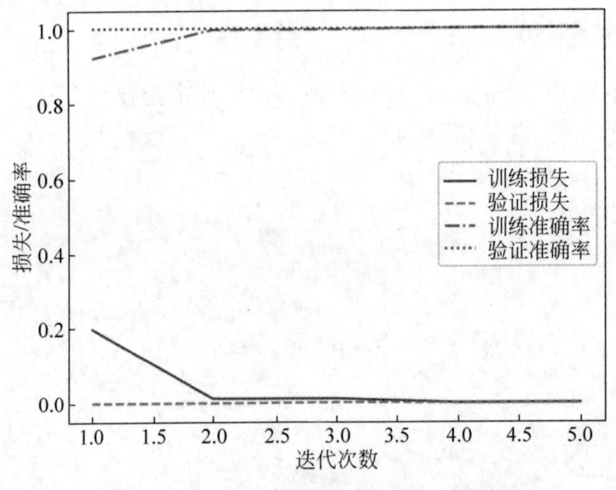

图 4-49 VGG-16 卡通表情识别训练结果

VGG 同样设置了 5 个 Epoch,需要更低的学习率才能达到图像的效果,其参数量是 AlexNet 的 2 倍,训练时间更久。BatchSize 在本机设置为 16,如果设置得过大,则会抛出 OOM when allocating tensor with shape(显存不足)的提示。

GPU 显存的占用与模型宽度、深度有关,模型的参数量、浮点计算量、隐藏层的缓存、设置的 BatchSize 均可影响 GPU 的使用,不同的模型会有不同的 BatchSize 设置。

总结

VGG 反复堆叠 3×3 的卷积可以减少参数量,并使决策函数更具有判别力,并使用 ReLU 激活函数;VGG 是较多目标检测网络特征提取的重要组成部分。

练习

完成 VGG-16 模型的搭建、训练、推理代码。

4.8 合并连接网络 GoogLeNet

4.8.1 模型介绍

GoogLeNet 论文 *Going Deeper with Convolutions* 由 Christian Szegedy 等发表于 2014 年，是 ILSVRC 2014 分类任务的冠军（VGG 为亚军），其主要特点是引入了 Inception 结构，并将网络深度扩展到 22 层，并且参数量约为 VGG 的 1/10，因此在内存或计算资源有限时是比较好的选择，并且性能当时也优于其他模型。

Inception 结构分成 4 路进行不同尺度的卷积、池化，然后将各个分支的结果合并在一起，从而形成一个网络块，如图 4-50 所示。

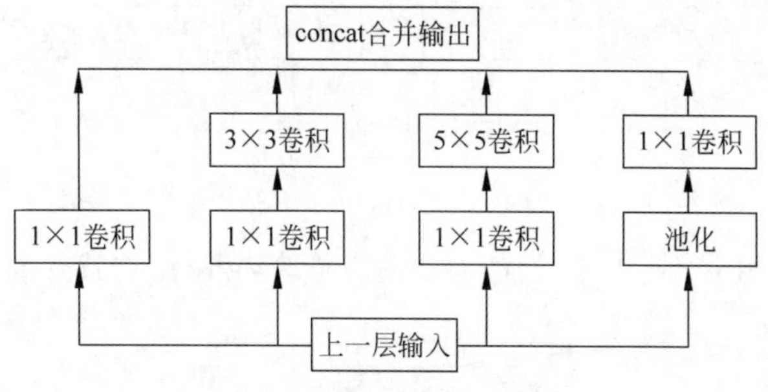

图 4-50 Inception v1 结构

利用 1×1 卷积的特性，对 channel 进行降维，从而减少参数量。不同尺度的卷积，其感受野不同，提取的特征信息也不同，合并操作后其特征信息会更加丰富，从而有利于模型网络的训练与预测。

又由于两个 3×3 卷积相当于 1 个 5×5 卷积，但参数量更少，所以又产生了 Inception v2 的结构，如图 4-51 所示。

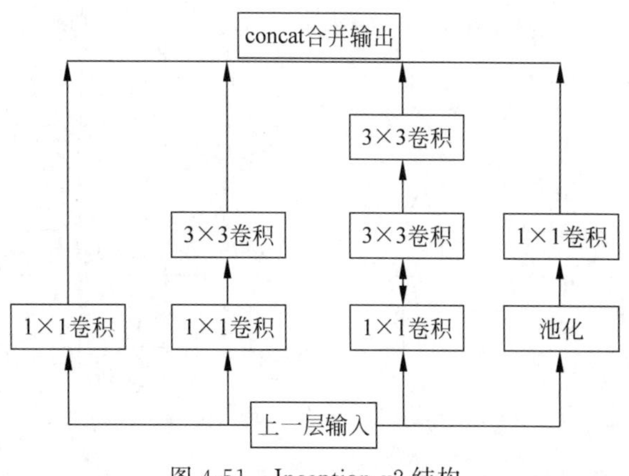

图 4-51 Inception v2 结构

利用 4.4.4 节中的思想，进一步减少参数量，将 3×3 卷积拆分成 1×3 和 3×1 的卷积，又产生了 Inception v3 的结构，如图 4-52 所示。

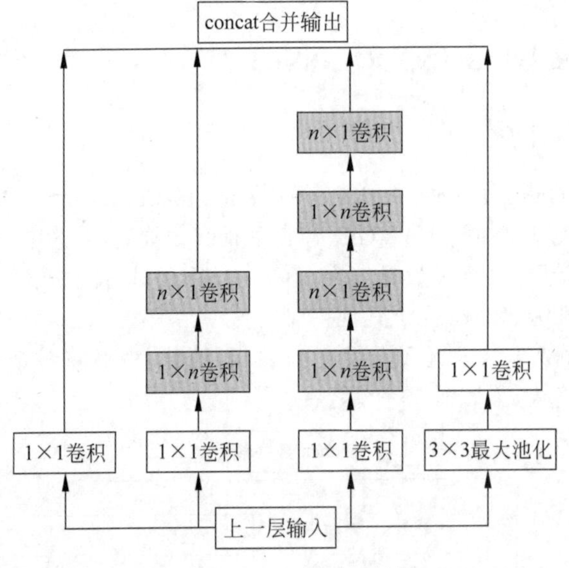

图 4-52　Inception v3 结构

将 v3 中最后一组 $n\times1$ 和 $1\times n$ 的卷积由串行改为并行，这就变成了 Inception v4 的结构，如图 4-53 所示。

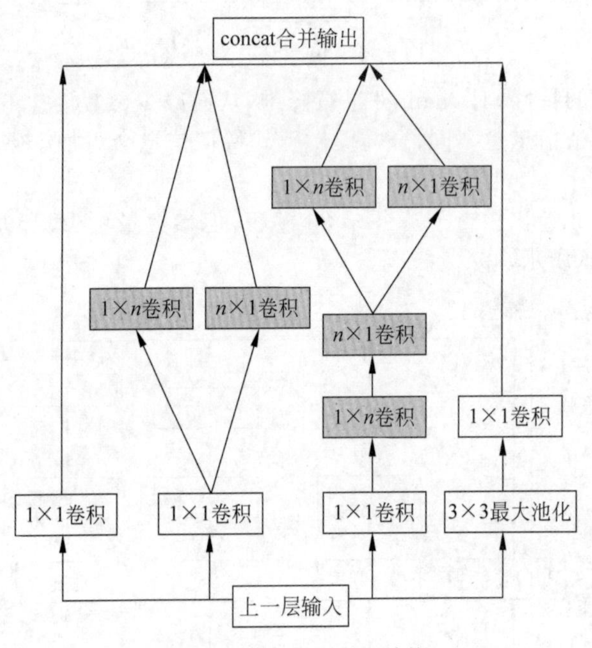

图 4-53　Inception v4 结构

Inception 结构在 ResNet 诞生后又产生了 Inception-ResNet 模块，结构如图 4-54 所示。

(a) Inception-ResNet-1　　(b) Inception-ResNet-2

图 4-54　Inception-ResNet 结构

在 Inception-ResNet 中去掉了池化层，主要是由于残差本身会保留较多特征信息。残差的作用可参考 4.9 节。

以 Inception v1 为主体的 GoogLeNet v1 网络结构如图 4-55 所示。

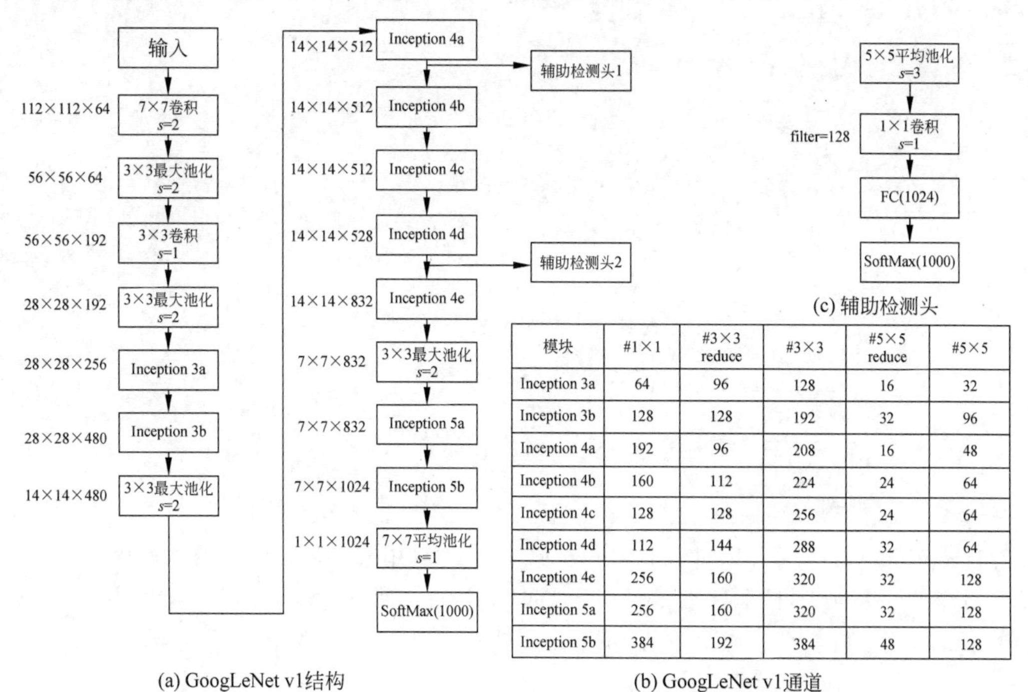

模块	#1×1	#3×3 reduce	#3×3	#5×5 reduce	#5×5
Inception 3a	64	96	128	16	32
Inception 3b	128	128	192	32	96
Inception 4a	192	96	208	16	48
Inception 4b	160	112	224	24	64
Inception 4c	128	128	256	24	64
Inception 4d	112	144	288	32	64
Inception 4e	256	160	320	32	128
Inception 5a	256	160	320	32	128
Inception 5b	384	192	384	48	128

(a) GoogLeNet v1结构　　(b) GoogLeNet v1通道

图 4-55　GoogLeNet v1 网络结构

GoogLeNet 的作者发现网络结构的浅层也有很好的预测能力，因此训练时在网络中增加了两个辅助分类器，并在损失函数中给予 0.3 的权重，但在推理时不使用。原作者还使用了 LocalRespNorm 进行局部归一化，但在本书中没有应用。

4.8.2 代码实战

根据 Inception 结构，使用代码对该结构进行实现，代码如下：

```
#第 4 章/ClassModleStudy/common.py
class Inception_v1(layers.Layer):
   def __init__(self, list_filters, activation):
       #list_filters=[conv1x1, conv3x3_reduce, conv3x3, conv5x5_reduce,
conv5x5, pool_proj]
       #即 list_filters=[64, 96, 128, 16, 32, 32]
       super(Inception_v1, self).__init__()
       self.branch1 = layers.Conv2D(list_filters[0], kernel_size=1,
strides=1, activation=activation)
       #先降维再升维
       self.branch2 = Sequential([
          layers.Conv2D(list_filters[1], kernel_size=1, strides=1,
activation=activation),
          layers.Conv2D(list_filters[2], kernel_size=3, padding='same',
strides=1, activation=activation),
       ])
       self.branch3 = Sequential([
          layers.Conv2D(list_filters[3], kernel_size=1, strides=1,
activation='relu'),
          layers.Conv2D(list_filters[4], kernel_size=5, padding='same',
strides=1, activation=activation),
       ])
       self.branch4 = Sequential([
          layers.MaxPool2D(pool_size=3, strides=1, padding='same'),
          layers.Conv2D(list_filters[5], kernel_size=1, strides=1,
activation=activation),
       ])

   def call(self, inputs, **kwargs):
       outputs = layers.concatenate([
          self.branch1(inputs),
          self.branch2(inputs),
          self.branch3(inputs),
          self.branch4(inputs)
       ])
       return outputs
```

Inception_v1(layers. Layer)继承 layers. Layer 类，用来构建自定义网络模块层，并要求传入对应的一个列表，按顺序表示对应分支的 filter。同时在 common. py 文件中 operator_matching()用于解析算子函数没有 Inception 结构的内容，还需增加以下代码：

```
#第 4 章/ClassModleStudy/common.py
    elif operator_type == 'Inception_v1':
        #此时 unit 为 list_filter,否则为 None
        operator = Inception_v1(unit, activation) if isinstance(unit, list)
else None
```

使用 isinstance()判断是否为 list,如果是,就调用 Inception_v1()来处理,否则返回 None。由于有 3 个检测头,ConfNet()不再适用,所以需要重新搭建 GoogLeNet 类,代码如下:

```
#第 4 章/ClassModleStudy/GoogLeNet/model.py
from ClassModleStudy.common import get_operator_conf
from tensorflow.keras import Model
from tensorflow.keras.activations import relu, softmax

#主干的配置
google_config_ouput = [
    ['Conv', 64, [7, 7], [2, 2], 'same', relu],              #112×112×64
    ['MaxPool', 0, [3, 3], [2, 2], 'same', 0],               #56×56×64

    ['Conv', 192, [3, 3], [1, 1], 'same', relu],             #56×56×192
    ['MaxPool', 0, [3, 3], [2, 2], 'same', 0],               #28×28×192
    #inception 3a 28×28×256
    ['Inception_v1', [64, 96, 128, 16, 32, 32], 0, 0, 0, relu],
    #inception 3b 28×28×480
    ['Inception_v1', [128, 128, 192, 32, 96, 64], 0, 0, 0, relu],
    ['MaxPool', 0, [3, 3], [2, 2], 'same', 0],               #14×14×480
    #inception 4a 14×14×512
    ['Inception_v1', [192, 96, 208, 16, 48, 64], 0, 0, 0, relu],
    #inception 4b 14×14×512 ouput0
    ['Inception_v1', [160, 112, 224, 24, 64, 64], 0, 0, 0, relu],
    #inception 4c 14×14×512
    ['Inception_v1', [128, 128, 256, 24, 64, 64], 0, 0, 0, relu],
    #inception 4d 14×14×528 ouput1
    ['Inception_v1', [112, 144, 288, 32, 64, 64], 0, 0, 0, relu],
    #inception 4e 14×14×832
    ['Inception_v1', [256, 160, 320, 32, 128, 128], 0, 0, 0, relu],
    ['MaxPool', 0, [3, 3], [2, 2], 'same', 0],               #7×7×832
    #inception 5a 7×7×832
    ['Inception_v1', [256, 160, 320, 32, 128, 128], 0, 0, 0, relu],
    #inception 5b 7×7×1024
    ['Inception_v1', [384, 192, 384, 48, 128, 128], 0, 0, 0, relu],
    ['AvgPool', 0, [7, 7], [1, 1], 'valid', 0],              #1×1×1024
    ['Flat', 0, 0, 0, 0, 0],
    ['DropOut', 0.4, 0, 0, 0, 0],
    ['OutPut', 7, 0, 0, 0, softmax],                         #ouput2
]
#辅助分类器的配置
google_config_ouput_x = [
    ['AvgPool', 0, [5, 5], [3, 3], 'valid', 0],
    ['Conv', 128, [1, 1], [1, 1], 'valid', relu],
    ['Flat', 0, 0, 0, 0, 0],
    ['FC', 1024, 0, 0, 0, relu],
```

```
        ['DropOut', 0.7, 0, 0, 0, 0],
        ['OutPut', 7, 0, 0, 0, softmax],
    ]
    google_config_ouput_x_2 = google_config_ouput_x.copy()
    class GoogLeNetV1(Model):
        def __init__(self):
            super(GoogLeNetV1, self).__init__()
            #获取主路径的配置算子
            self.main_net = get_operator_conf(google_config_ouput)
            self.aux = get_operator_conf(google_config_ouput_x)
            self.aux2 = get_operator_conf(google_config_ouput_x_2)

        def aux_forward(self, x, aux_head):
            #辅助头的前向传播
            aux_x = x
            for aux_operator_forward in aux_head:
                aux_x = aux_operator_forward(aux_x)
            return aux_x

        def call(self, input, trainable=True, **kwargs):
            x = input
            x_result = []
            for index, operator_forward in enumerate(self.main_net):
                #解析主路径的前向传播过程
                x = operator_forward(x)
                #如果是,则训练时进行辅助头的训练
                if trainable:
                    #当主路径输出下标满足以下条件时进行辅助头的训练
                    if index == 7:
                        #只对辅助头的结果进行存储
                        x_result.append(self.aux_forward(x, self.aux))
                    elif index == 10:
                        #只对辅助头的结果进行存储
                        x_result.append(self.aux_forward(x, self.aux2))
            #最后1次前向传播的结果
            x_result.append(x)
            return x_result
```

按图4-54的描述配置主干网络并赋给变量google_config_output、将辅助头赋给变量google_config_output_x,然后在GoogLeNetV1()中由get_operator_conf()解析配置文件,再遍历主干网络main_net以实现前向传播。如果是在训练时,则可在下标为7、10时分别再进行google_config_output_x配置的解析,并进行辅助头的前向传播。前向传播的结果,通过x_result.append()进行追加。

训练代码参考3.5.3节进行自定义控制并封装在train_FERG_DB_custom()函数中,实现多损失的反向传播,并在train_step()中将系数设定为0.3,但是在验证valid_step()中设置trainable=False,核心代码如下:

```
#第4章/ClassModleStudy/common.py
def train_step(model, x, y):
```

```
        #model:模型
        #features:训练集 x
        #labels:训练集 y
        #tf.GradientTape()自动求梯度的作用域语句
        with tf.GradientTape() as tape:
            #前向传播
            predictions = model(x, training=True)
            #使用损失函数
            loss0 = loss_func(y, predictions[0])
            loss1 = loss_func(y, predictions[1])
            loss2 = loss_func(y, predictions[2])
            loss = 0.3 *loss0 + 0.3 *loss1 + loss2
        #根据损失函数自动求梯度
        gradients = tape.gradient(loss, model.trainable_variables)
        #将梯度更新到 model.trainable_variables 属性中,然后由 optimizers 进行指定优
        #化器的梯度下降
        optimizer.apply_gradients(zip(gradients, model.trainable_variables))
        #更新评价指标
        train_loss.update_state(loss)
        train_metric.update_state(y, predictions)

    #验证集
    def valid_step(model, features, labels):
        #验证集不进行梯度下降更新学习
        predictions = model(features,trainable=False)
        batch_loss = loss_func(labels, predictions[0])
        valid_loss.update_state(batch_loss)
        valid_metric.update_state(labels, predictions)
```

最后进行训练模型的调用,代码如下:

```
#第 4 章/ClassModleStudy/GoogLeNet/train.py

from ClassModleStudy.common import plt_loss, train_FERG_DB_custom
from ClassModleStudy.GoogLeNet.model import GoogLeNetV1

#实例化自定义模型类 GoogLeNet 主干网络
net = GoogLeNetV1()
#调用训练代码模板
his = train_FERG_DB_custom(net, img_height=224, img_width=224, epochs=5, batch_
size=64, lr=0.0001)
#以图的形式展现训练结果
plt_loss(his, his['epochs'], 'GoogLeNet 卡通表情识别训练结果',is_history=False)
```

训练结果如图 4-56 所示。

总结

Inception 拥有多个版本,其主要目的是更好地实现多尺度特征信息的融合,使网络更具有表达能力。

练习

完成 GoogLeNet v1 模型的搭建、训练、推理代码。

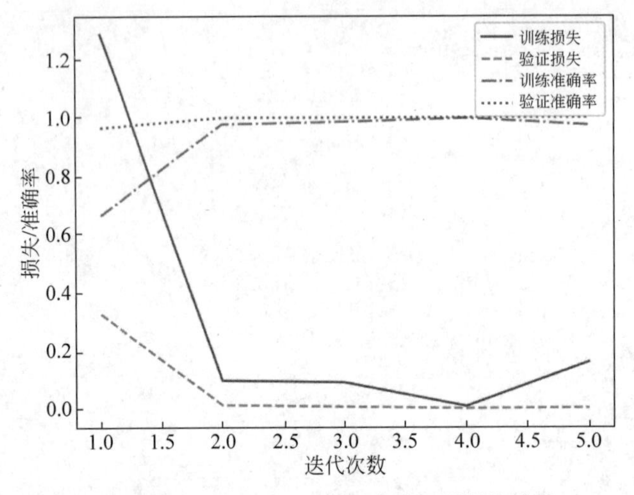

图 4-56　GoogLeNet v1 卡通表情识别训练结果

4.9　残差网络 ResNet

4.9.1　残差块

从 LeNet 的 5 层、AlexNet 的 8 层，VGG 的 16 层，GoogLeNet 的 22 层，随着网络层次的增加，模型的性能也应该越来越强，但是现实情况并不是这样的，随着网络层增加到一定的数量，模型性能反而出现下降，这个问题称为网络退化问题，如图 4-57 所示。

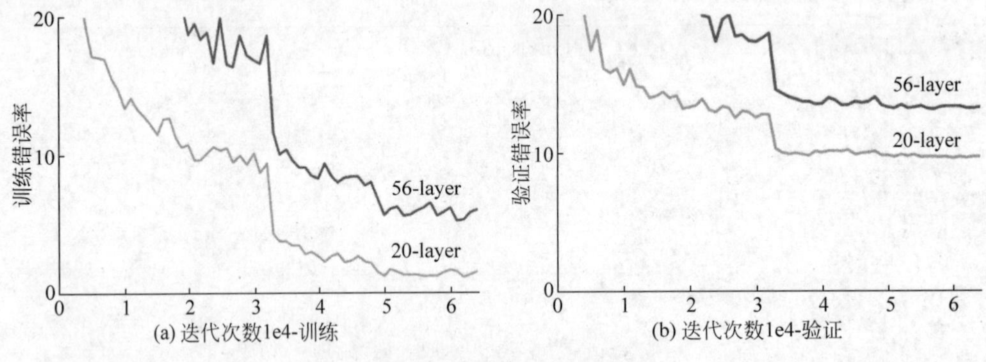

图 4-57　网络退化问题

2015 年 Kaiming He 等发表了论文 *Deep Residual Learning for Image Recognition*，提出使用残差思想解决网络退化问题。ResNet 通过堆叠残差块就可以将网络扩展到上千层，并且模型性能优异，不得不说有了残差，深度学习才开始绚丽多彩。

残差模块的主要思想是恒等映射，如图 4-58 所示。

假设第 3 层模型的性能 ACC 已达到 96%，新增加第 4 层我们期望模型的性能大于 96%，那么就只能在第 3 层输出结果的基础上微调，如果微调失败，则不要第 4 层的结果也

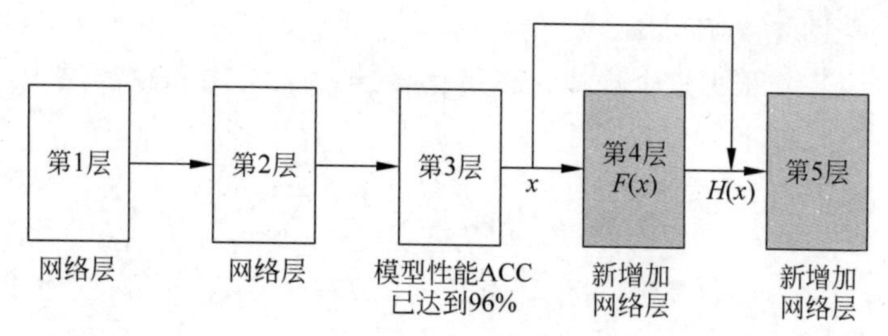

图 4-58 恒等映射

不会影响第 3 层的输出，所以这里可以构建恒等映射公式，公式如下：

$$H(x_{L+1}) = x_L + F(x_L, w_L) \tag{4-6}$$

其中，x_L 代表上一层的输入，$H(x_{L+1})$代表恒等映射后的输出。如果 $F(x)=0$，则 $H(x)=x$；如果 $F(x)!=0$，则 $H(x)$也是在 x 的基础上进行叠加，模型在 x 的基础上进一步学习，模型效果则更有可能更好。

代入数值计算，如图 4-59 所示。

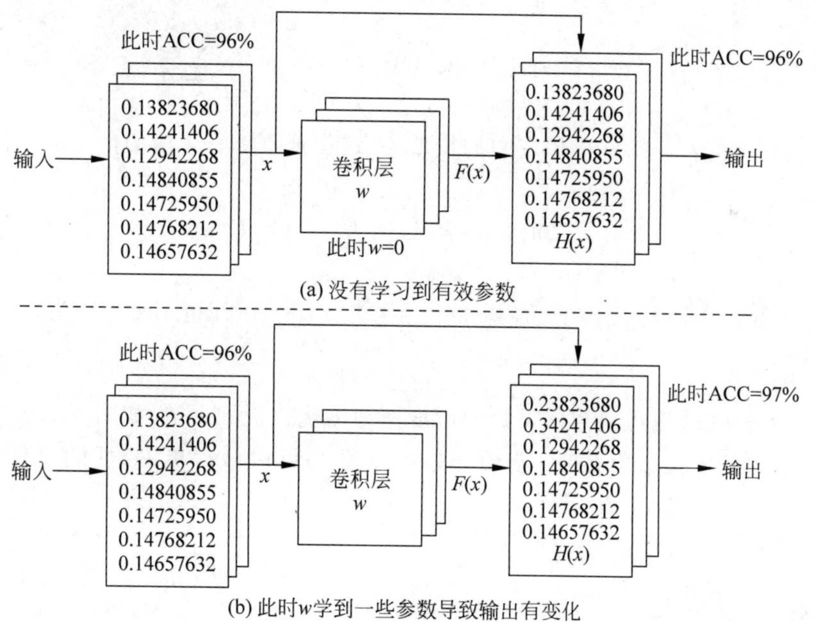

图 4-59 恒等映射数值计算

要实现 $H(x)=x$，则此时需要 $F(x)=0$，通过反向传播学习，则参数 $w=0$，但是 w 不大可能全部为 0，总会有一些参数会进行有效学习，从而导致输出值发生变化，进而影响预测值，所以预测值会通过 $H(x)=x+F(x)$进行稳步提升学习。另外，从图 4-59 也可以观察到残差学习会使网络参数趋近于 0，从而使网络更稀疏，使网络更专注于主要特征信息。

如果恒等映射，中间再多加 1 层，则是远跳连接，将式(4-6)展开，可得 $H(x_{L+2}) = x_{L+1} + F(x_{L+1}, w_{L+1}) = x_L + F(x_L, w_L) + F(x_{L+1}, w_{L+1})$。如果用图形化表示，则如

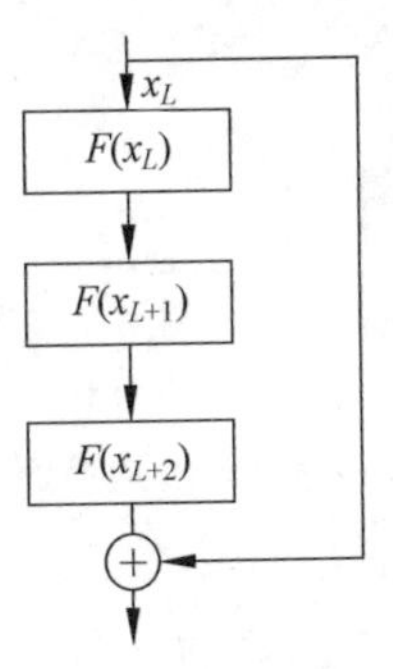

图 4-60　远跳残差

图 4-60 所示。

根据这个规律，任意较深的第 x_{L+1} 层和较浅层 x_L 的映射关系公式如下：

$$H(x_{L+1})=x_L+\sum_{i=L}^{L-1}F(x_i,w_i) \tag{4-7}$$

公式表明，对于任意较深的 $L+1$ 层，特征 x_{L+1} 可以表示为较浅 L 层的特征 x_L 加上 $L+1$ 层与 L 层之间所有残差函数之和 $\sum_{i=L}^{L-1}F(x_i,w_i)$；对于任意 L 层，它的输出特征 $x_L=x_0+\sum_{i=0}^{L-1}F(x_i,w_i)$，即为之前第 1 层残差函数之和加上 x_0。

根据反向传播算法，假设损失函数为 Loss，则求梯度公式如下：

$$\frac{\partial \text{Loss}}{\partial x_L}=\frac{\partial \text{Loss}}{\partial x_{L+1}} * \frac{\partial x_{L+1}}{\partial x_L}=\frac{\partial \text{Loss}}{\partial x_{L+1}}\times\left(1+\frac{\partial}{\partial x_L}\sum_{i=L}^{L-1}F(x_i,w_i)\right) \tag{4-8}$$

公式表明，梯度 $\frac{\partial \text{Loss}}{\partial x_L}$ 被分解成两部分，其中 $\frac{\partial \text{Loss}}{\partial x_{L+1}}\times\frac{\partial}{\partial x_L}\sum_{i=L}^{L-1}F(x_i,w_i)$ 通过权重层传递，而 $\frac{\partial \text{Loss}}{\partial x_{L+1}}$ 不涉及 L 与 $L+1$ 层之间的权重层，这保证了 $L+1$ 层的信息能够传回任意浅层 L；同时 $\frac{\partial \text{Loss}}{\partial x_L}$ 也可以缓解梯度消失的情况，因为 $\frac{\partial}{\partial x_L}\sum_{i=L}^{L-1}F(x_i,w_i)$ 不大可能为 -1（如果激活函数是 ReLU，则为 0 或者 1），所以 $1+\frac{\partial}{\partial x_L}\sum_{i=L}^{L-1}F(x_i,w_i)$ 也不大可能为 0，这样就能缓解梯度消失的情况。

一些实验表明，残差远跳连接可以使深度网络的优化更加平滑，如图 4-61 所示。

从论文给出的实验数据表明，使用残差模块网络可以做得更深，性能更佳，如图 4-62 所示。

(a) 没有使用残差远跳连接

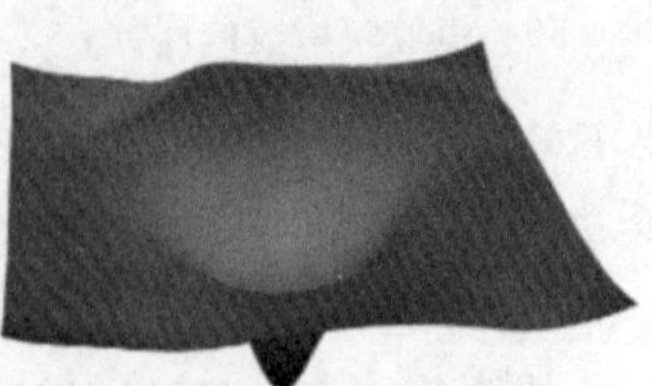
(b) 使用残差远跳连接

图 4-61　不使用和使用残差网络优化可视图

model	top-1 err.	top-5 err.
VGG-16[41]	28.07	9.33
GoogLeNet[44]	—	9.15
PReLU-net[13]	24.27	7.38
plain-34	28.54	10.02
ResNet-34 A	25.03	7.76
ResNet-34 B	24.52	7.46
ResNet-34 C	24.19	7.40
ResNet-50	22.85	6.71
ResNet-101	21.75	6.05
ResNet-152	**21.43**	**5.71**

图 4-62　ResNet 模型性能对比

4.9.2　归一化

数据归一化指将数据的值域映射到某个区间,如[0,1]或[−1,1],本质上是一种线性变换,这种线性变换决定对数据改变后不会造成数据的失效,反而能提高数据的表现,并有利于模型的学习训练。例如在使用梯度下降学习时,归一化的数据可以加快梯度下降的收敛速度,并且能够避免太大的数值引发梯度爆炸等问题。

最常见的归一化为离差标准化(Min-Max Normalization),对原始数据进行线性变换,使结果值在[0,1],其公式为

$$x_{\text{new\_i}}=\frac{x_i-x_{\min}}{x_{\max}-x_{\min}} \tag{4-9}$$

其中,$x_{\max}$ 为样本数据中的最大值,$x_{\min}$ 为样本数据中的最小值。这种归一化的方法比较适用于数值比较集中的情况,如果 $x_{\max}$ 与 $x_{\min}$ 不稳定,则容易使归一化的结果也不稳定。

另一种方法是将数据标准化(Z-Score Normalization),经过处理后其数据分布符合标准正态分布,即均值为0,标准差为1,公式为

$$x_{\text{new\_i}}=\frac{x_i-\mu}{\sigma} \tag{4-10}$$

其中,μ 为本数据的均值,σ 为样本数据的标准差。标准化后数据会保持异常值中的有用信息,使算法对异常值不敏感。

```
#第 4 章/ClassModleStudy/Normalization/max_min.py
import numpy as np

rnd = [474, 797, 252, 492, 938]
for _ in range(2):
    print("随机值: ", rnd)
    print("归一化: ", [round((x - min(rnd)) / (max(rnd) - min(rnd)), 3) for x in
rnd])
    print("标准化: ", [round((x - np.mean(rnd)) / np.std(rnd), 3) for x in rnd])
rnd = [445, 298, 747, 157, 741, 0]
```

运行结果如下:

```
随机值:[474, 797, 252, 492, 938]
归一化:[0.324, 0.794, 0.0, 0.35, 1.0]
标准化:[-0.475, 0.841, -1.379, -0.402, 1.415]
随机值:[445, 298, 747, 157, 741, 0]
归一化:[0.596, 0.399, 1.0, 0.21, 0.992, 0.0]
标准化:[0.168, -0.358, 1.249, -0.863, 1.228, -1.425]
```

从运行结果观察,Min-Max Normalization 归一化将使最小值为0,最大值为1,如果值为0,则该神经元将不再进行梯度学习,所以该方法是不佳的;Z-Score Normalization 标准化后值域满足高斯分布,对极值并不敏感(最小不为0),所以该方法可用。

正因为标准化能够起到加快模型收敛、统一样本数据度量单位、保证数值最小值仍存

活，所以在 AlexNet、GoogLeNet 中对于网络层的输出采用了局部归一化（LocalRespNorm）操作，以加快网络的收敛并预防梯度爆炸，其公式如下：

$$b_{x,y}^{i}=a_{x,y}^{i}\Bigg/\left(k+\alpha\sum_{j=\max\left(0,i-\frac{n}{2}\right)}^{\min\left(N-1,i+\frac{n}{2}\right)}(a_{x,y}^{i})^{2}\right)^{\beta} \tag{4-11}$$

其中，a 表示卷积、池化操作后输出的结果，即[batch，height，width，channel]；$a_{x,y}^{i}$ 表示输出结果中的某个位置[a，b，c，d]，即第 a 张图中的 d 通道下高度为 b 宽度为 c 的点；N 表示通道；a、$n/2$、k、α、β 表示 input、depth_radius、bias、alpha、beta，其中 $n/2$、k、alpha、beta 都是自定义参数，在 TensorFlow 中默认设置 depth_radius＝5、$k=1$、alpha＝1、beta＝0.5；需要注意∑叠加的方向是沿着通道的方向，即每个点值的平方和是沿着 a 中的通道方向进行的，也就是一个点同方向的前面 $n/2$ 个通道和后 $n/2$ 通道的点的平方和。

假设存在 3×3×4 的特征，设 $k=0$、alpha＝1、beta＝1、$N=2$，其中 $N=2$ 表示沿通道方向参考最近的 2/2 个通道，即前 1 个通道点值和后 1 个通道点值的平方和，如图 4-63 所示。

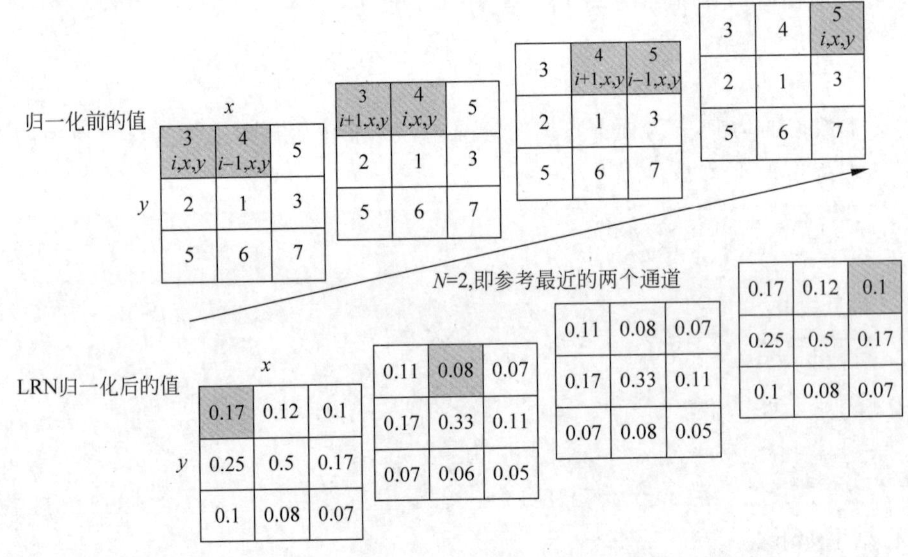

图 4-63　局部响应归一化计算过程（见彩插）

第 1 张特征图中绿色背景的归一化 $b=(a_{x,y}^{i})/(0+1\times((a_{x,y}^{i-1})^{2}+(a_{x,y}^{i})^{2}+(a_{x,y}^{i+1})^{2}))$，由于$(a_{x,y}^{i-1})$不存在，则设$(a_{x,y}^{i-1})=0$，则代入 $b=3/(0+1\times(0^{2}+3^{2}+3^{2}))=0.17$；第 2 张特征图中红色背景的归一化 $b=4/(0+1\times(4^{2}+4^{2}+4^{2}))=0.11$；第 4 张特征图中蓝色背景的归一化 $b=5/(0+1\times(5^{2}+4^{2}+0^{2}))=0.1$，其他以此类推，可使用以下代码进行计算。

```
#第 4 章/ClassModleStudy/Normalization/LocalRespNorm.py

def lpn(up, p, down):
    #up 为上一个通道 x,y 值；p 为当前通道 x,y 值；down 为下一个通道 x,y 值
```

```
    b = p / (0 + 1 * (up ** 2 + p ** 2 + down ** 2))
    return round(b, 2)
```

从上面的计算方式可知，局部响应归一化是按通道方向进行归一化的，并且没有学习参数，在 AlexNet 论文中作者使用此方法，在 ImageNet 数据集中提高约 1.4%。

另一种归一化称为批归一化（Batch Normalization），数学公式为

$$\mathrm{BN}_{\gamma,\beta}(X_i)=\gamma\hat{x}_i+\beta \tag{4-12}$$

其中，$\hat{x}_i=\dfrac{x_i-\mu_\beta}{\sqrt{\delta_\beta^2+\epsilon}}$，即标准化的公式；$\delta_\beta^2=\dfrac{1}{m}\sum_{i=1}^{m}(x_i-\mu_\beta)^2$，即为每个批数据的方差；$\mu_\beta=\dfrac{1}{m}\sum_{i=1}^{m}x_i$，即为每个批数据的平均值；$\epsilon$ 为一个极小值，主要为了防止除数为 0。γ 为缩放因子，β 为偏置项，γ 和 β 通过反向传播算法进行学习。

BN 将数据规整到统一区间，使每层的值域分布在均值为 0 且标准差为 1 的正态分布，从而减少了数据发散程度，降低了网络的学习难度。BN 使用 γ 和 β 作为还原参数，在一定程度上能够保留原数据的分布。

在训练时 μ_β、δ_β^2 取的是当前批中的训练数据，但是在预测时我们只有一个或者很少的样本，此时 μ_β、δ_β^2 的计算一定是有偏估计，因此在训练时，对每个网络中的批训练数据的 $\mu_{\text{batch}\beta}$、$\sigma_{\text{batch}\beta^2}$ 进行存储，然后使用整个样本的统计数据来对预测数据进行归一化，具体来讲使用均值和方差的无偏估计，即公式：

$$\mu_{\text{test}\beta}=E(\mu_{\text{batch}\beta})$$

$$\sigma_{\text{test}\beta^2}=\frac{m}{m-1}E(\sigma_{\text{batch}\beta^2}) \tag{4-13}$$

得到每个特征的均值与方差的无偏估计后，再对预测数据采用同样的归一化方法，即公式：

$$\mathrm{BN}_{\gamma,\beta}(X_{\text{test\_i}})=\gamma\frac{x_{\text{test\_i}}-\mu_{\text{test}\beta}}{\sqrt{\sigma_{\text{test}\beta^2}+\epsilon}}+\beta \tag{4-14}$$

在实际计算的过程中，并不会保留训练过程中的所有历史值再求平均，因为训练早期模型并没有收敛，其值的价值较低，甚至可能会干扰对均值和标准差的估计，因此加入指数平均的算法（参考 2.2.4 节 Momentum 优化算法）将历史批中的均值和方差的作用延续到当前批，但是早期批的影响将会变小，最近批的影响较大。

BN 带来的好处，可以总结如下：

(1) BN 使网络每层输入数据的分布相对稳定，加速模型学习速度。每层的值域输出满足均值为 0 且标准差为 1 的正态分布，使后一层网络不需要去适应上一层输入的变化，从而实现了网络中层与层之间的解耦，允许每层进行独立学习，有利于提高整个神经网络的学习速度。

(2) BN 使模型对网络中的参数不敏感，能够简化调参过程，使网络学习更稳定。在使

用 BN 之后，抑制了参数的微小变化，随着网络层数加深被放大的问题，使网络对参数大小的适应能力更强，因此可以设置较大的学习率，以加速模型的学习。

(3) BN 允许网络使用饱和性激活函数，如 Sigmoid，可以缓解梯度消失。在不使用 BN 层时，由于网络的深度、复杂性，根据反向传播算法，很容易出现底层网络变化累积到上层网络中，进而导致模型的训练进入激活函数的梯度饱和区，出现参数趋 0 的现象。BN 可以让激活函数的输出落在梯度非饱和区，从而缓解梯度消失的现象。

(4) BN 具有一定正则化的效果。在 BN 中，由于使用批数据的均值与方差作为对整体训练样本的均值与方差的估计，尽管每个批中的数据都是从总体样本中抽样得到的，但不同批的均值与方差会有所不同，这就为网络的学习过程中增加了随机噪声，与 DropOut 通过关闭神经元给网络训练带来噪声类似，在一定程度上对模型起到了正则化的效果。

虽然 BN 的优点较多，但是也存在以下缺点：

(1) 显存占用较高。由于是在批的维度进行归一化，BN 要求有较大的 Batch Size 才能有效地工作，而物体检测等任务由于占用内存、显存较高，限制了 Batch Size 的大小，这会限制 BN 有效地进行发挥。

(2) 依赖训练集的均值和标准差。训练的 Batch Size 与测试时不一样，在训练时采用滑动平均来计算均值和标准差，在测试时直接用训练集的平均值来预测，这种方式会导致测试集的数据分布依赖训练集，然而有时训练集与测试集的数据分布可能并不一致。

(3) 对图像的质量会有所破坏。BN 会对数据进行均值为 0 且标准差为 1 的归一化，那么会把图像的对比度进行拉升，破坏图像的色彩分布，影响对比度，从而降低图像的输出质量。

(4) 过小的 BN 效果较差。BN 对 Batch Size 非常敏感，过小的 Batch Size 网络效果较差，因为此时样本很少，均值近似噪声数据。

Layer Normalization 归一化的计算同公式(4-12)，主要区别是 LN 在通道方向，BN 在 Batch 维度上，如图 4-64 所示。

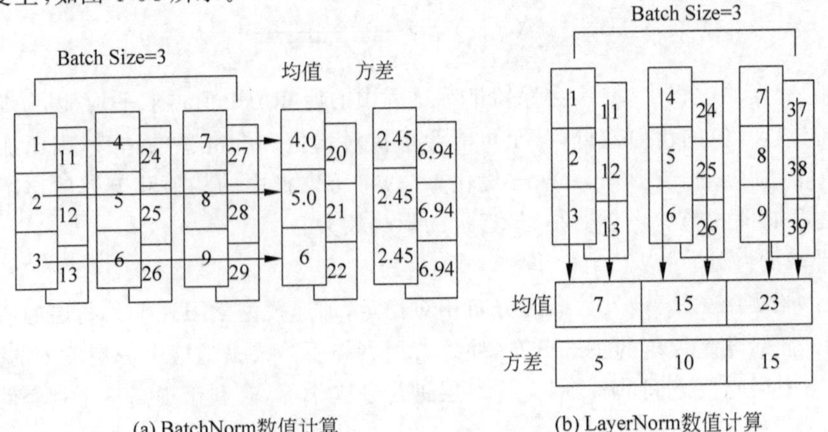

图 4-64　LayerNorm 与 BatchNorm 的计算区别

Layer Norm 是对 Batch 的通道进行归一化，Layer Norm 针对同一样本的不同特征进行归一化，而 Batch Norm 针对不同样本的同一特征进行归一化，因此，Layer Norm 可以不受样本数的限制。具体而言，Batch Norm 在每个通道上统计所有样本，计算均值和方差；Layer Norm 在每个样本上统计所有通道，计算均值和方差。Layer Norm 一般用于 NLP 自然语言处理方向。

Instance Normalization 归一化，在计算归一化统计量时并没有像BN 那样跨样本、单通道，也没有像 LN 那样单样本、跨通道，而是单个样本、单个通道的所有元素，如图 4-65 所示。

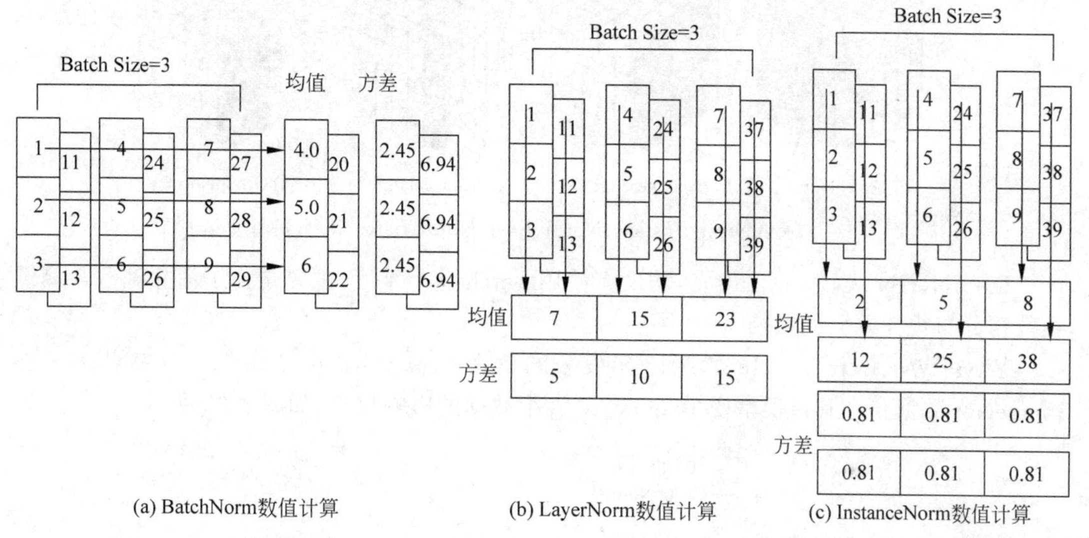

图 4-65　InstanceNorm、LayerNorm、BatchNorm 的计算区别

IN 是基于单一样本的，使用单个通道来求均值和方差。该方法通常应用于生成网络中。

Group Normalization 归一化，按通道方向分组，然后在每组内计算均值和标准差，如图 4-66 所示。

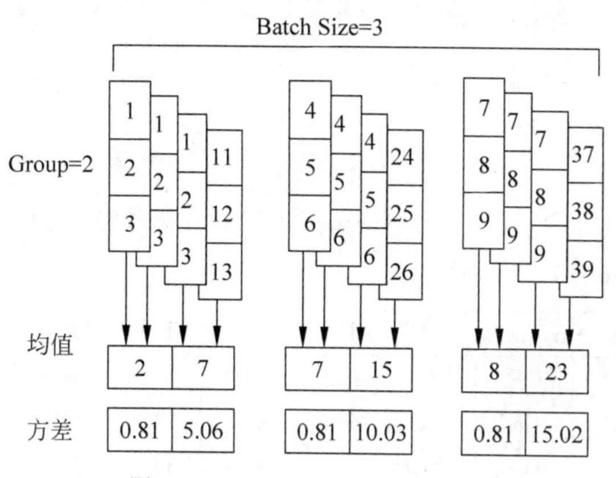

图 4-66　GroupNorm 数值计算

LN和IN分别是GN的特殊情况，当GN中的group＝channel时为LN，当group＝1时为IN。

4种不同归一化方法的区别如图4-67所示，其中N为Batch Size、C为channel、H为height、W为weight。BN按N的方向进行归一化，计算$N \times H \times W$的均值和标准差；LN按C的方向，计算$C \times H \times W$；IN按某1个channel计算$1 \times H \times W$；GN按group计算(C//group)$\times H^{*} \times W$。

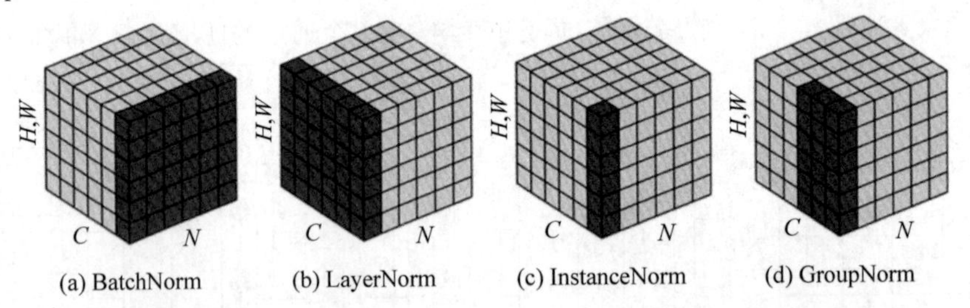

图4-67　GroupNorm、InstanceNorm、LayerNorm、BatchNorm的计算区别

LN、IN、GN从BN算法出发，均没有使用Batch这个维度，主要是为了消除训练与测试集数据分布规律差异的影响。

Yuxin Wu、Kaiming He在2018年发表的论文*Group Normalization*中表明，GN在ImageNet数据集中其错误率受Batch Size大小波动的影响较小，如图4-68所示。

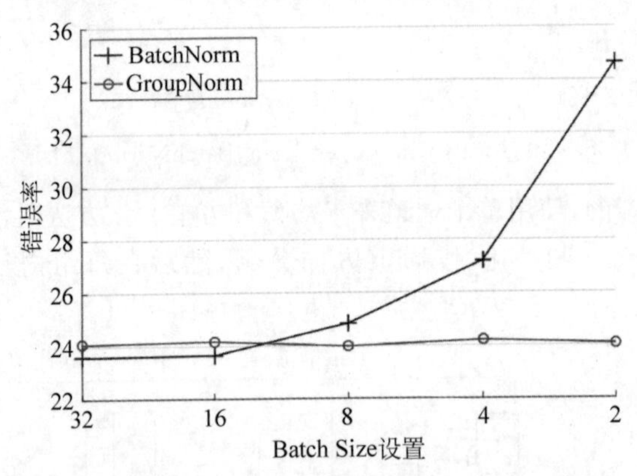

图4-68　GN与BN的错误率比较

作者在论文中给出了使用GN在Mask-R-CNN目标检测与语义分割的指标数据，相对于BN提高了1.4%，如图4-69所示。

backbone	box head	AP^{bbox}	AP^{bbox}_{50}	AP^{bbox}_{75}	AP^{mask}	AP^{mask}_{50}	AP^{mask}_{75}
BN*	—	38.6	59.5	41.9	34.2	56.2	36.1
BN*	GN	39.5	60.0	43.2	34.4	56.4	**36.3**
GN	GN	**40.0**	**61.0**	**43.3**	**34.8**	**57.3**	**36.3**

图4-69　GN与BN的AP性能差异

在 TensorFlow 中使用这 4 种归一化，只需安装 TensorFlow-addons 库，并直接调用，代码如下：

```
#第 4 章/ClassModleStudy/Normalization/TensorFlowNorm.py
from tensorflow.keras import layers
#pip install -q -U TensorFlow-addons
import tensorflow_addons as tfa
#BN
layers.BatchNormalization()
#LN
layers.LayerNormalization()
#GN
tfa.layers.GroupNormalization()
#IN
tfa.layers.InstanceNormalization()
```

4.9.3 模型介绍

ResNet50 以瓶颈残差为核心，通过多次堆叠，在 ImageNet 数据集上取得错误率为 22.85%的成绩(VGG-16 为 28.07%)，模型结构如图 4-70 所示。

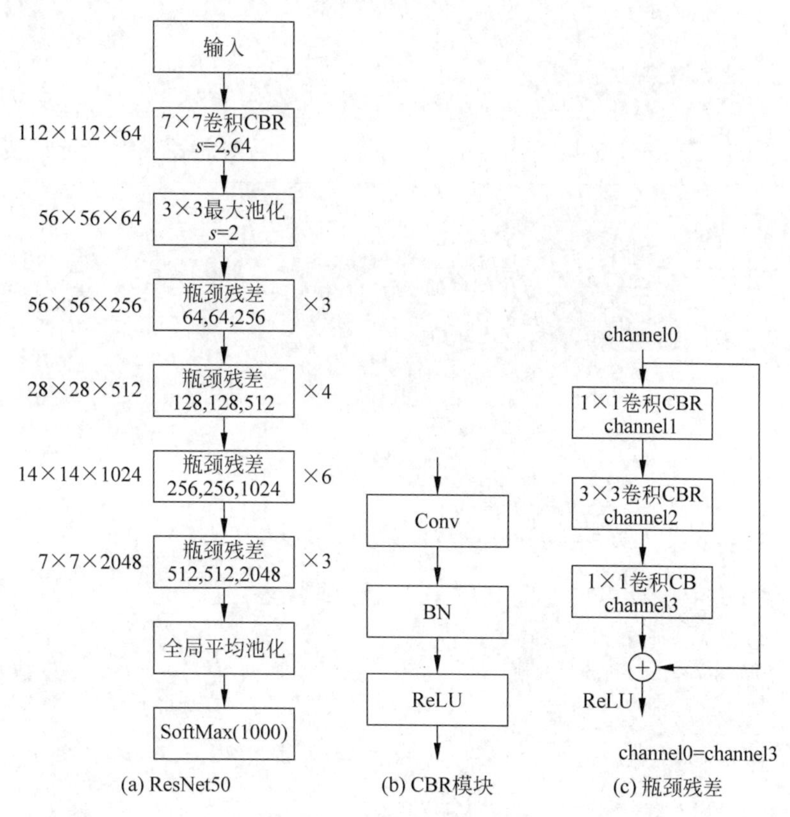

图 4-70 ResNet50 模型结构

瓶颈残差,即 channel 先降维再升维。ResNet50 中有 4 部分的瓶颈残差,其重复的次数分别是 3、4、6、3 次。需要注意,当输入 channel 与输出 channel 不同时,中间有一个 1×1 卷积进行下采样。图 4-70 中的 CB 模块只通过卷积、BN 归一化,不进行 ReLU 激活输出(因为 ReLU 会将小于 0 的值变为 0)。

4.9.4 代码实战

首先,对 CBL 模块进行封装,代码如下:

```
#第 4 章/ClassModleStudy/common.py
class CBR(layers.Layer):
    def __init__(self, filters, kernel_size, strides, padding="same", isRelu=True):
        super(CBR, self).__init__()
        self.conv = layers.Conv2D(filters, kernel_size=kernel_size, strides=
strides, padding=padding)
        self.bn = layers.BatchNormalization(momentum=0.9, epsilon=1e-5)
        self.relu = layers.ReLU()
        self.isRelu = isRelu

    def call(self, inputs, training=True, **kwargs):
        x = self.conv(inputs)
        x = self.bn(x, training=training)
        if self.isRelu:
            x = self.relu(x)
        return x
```

如果传入 isRelu=False,则不进行 relu()的输出,然后对瓶颈残差进行封装,代码如下:

```
#第 4 章/ClassModleStudy/common.py
class BottleNeck(layers.Layer):
    def __init__(self, filters, middle_strides, downsample=None):
        #middle_strides,残差中间层的步长
        #downsample,下采样的算子
        super(BottleNeck, self).__init__()
        self.cbr1 = CBR(filters, 1, 1)
        self.cbr2 = CBR(filters, 3, strides=middle_strides)
        #输出通道刚好是输入通道的 4 倍
        self.cb3 = CBR(filters *4, 1, 1,isRelu=False)
        self.downsample = downsample
        self.relu = layers.ReLU()
        self.add = layers.Add()

    def call(self, inputs, training=True, **kwargs):
        #输入值
        identity = inputs
        #如果需要改变通道数,则对输入 input 进行改变
        if self.downsample:
            identity = self.downsample(inputs)
        x = self.cbr1(inputs, training=training)                    #1×1
        x = self.cbr2(x, training=training)                         #3×3
        x = self.cb3(x)                                             #1*1
        return self.relu(self.add([x, identity]))                   #实现残差
```

cb3 不进行 relu()的输出，而是在 self.relu(self.add([x, identity]))处实现该操作。参数 middle_strides 用来控制 3×3 卷积的步长，用来实现下采样特征图的变换。

然后定义 1 个解析函数，用来组成结构图中的瓶颈残差块，代码如下：

```
#第 4 章/ClassModleStudy/common.py
def parse_resnet_block(BottleNeck, filter, repetitions, strides=1, is_first=
False):
    #BottleNeck:残差对象
    #filter:输入通道数
    #repetitions:残差重复次数
    #strides:步长
    downsample = None
    if strides != 1 or is_first == True:
        #如果步长不是 1,并且也不是第 1 个块,则进行 channel 的变换
        downsample = CBR(filter * 4, 1, strides, isRelu=False)
    #对每个残差块中的第 1 个残差进行下采样
    layer_list = [BottleNeck(filter, strides, downsample)]
    #重复次数-1,并且对其他残差不进行下采样,所以步长=1
    for _ in range(1, repetitions):
        layer_list.append(BottleNeck(filter, middle_strides=1))
    return Sequential(layer_list)
```

对 common.py 文件中的 operator_matching()函数增加以下解析代码：

```
#第 4 章/ClassModleStudy/common.py
elif operator_type == 'CBR':
        operator = CBR(unit, kernel_size, strides, padding)
    elif operator_type == 'ResnetBlock':
        block = BottleNeck
        #unit 按以下格式传值
        #[filter, repetitions, strides, is_first]
        if isinstance(unit, list):
            operator = parse_resnet_block(block, unit[0], unit[1], unit[2], unit[3])
    elif operator_type == 'GlobalAvgPool2D':
        operator = layers.GlobalAvgPool2D()
```

然后根据网络结构图进行配置，训练代码如下：

```
#第 4 章/ClassModleStudy/ResNet50/train.py
from ClassModleStudy.common import plt_loss, train_FERG_DB
from ClassModleStudy.common import ConfNet
from tensorflow.keras.activations import softmax

if __name__ == "__main__":
    resnet50_config = [
        #name,filter,kernel_size,stride,padding,activation
        #算子名称,特征图数量,卷积尺寸,步骤,补 0 数,激活函数
        ['CBR', 64, [7, 7], [2, 2], 'same', 0],
        ['MaxPool', 0, [3, 3], [2, 2], 'same', 0],
        #unit 按以下格式传值
        #[filter, repetitions, strides, is_first]
```

```
        ['ResnetBlock', [64, 3, 1, True], 0, 0, 0, 0],
        ['ResnetBlock', [128, 4, 2, False], 0, 0, 0, 0],
        ['ResnetBlock', [256, 6, 2, False], 0, 0, 0, 0],
        ['ResnetBlock', [512, 3, 2, False], 0, 0, 0, 0],
        ['GlobalAvgPool2D', 0, 0, 0, 0, 0],
        ['OutPut', 7, 0, 0, 0, softmax]
    ]
    resnet50 = ConfNet(resnet50_config)
    #调用训练代码模板
    his = train_FERG_DB(resnet50, img_height=224, img_width=224, epochs=5,
batch_size=32, lr=0.0001)
    #以图的形式展现训练结果
    plt_loss(his, his.params['epochs'], 'ResNet50 卡通表情识别训练结果')
```

运行结果如图 4-71 所示。

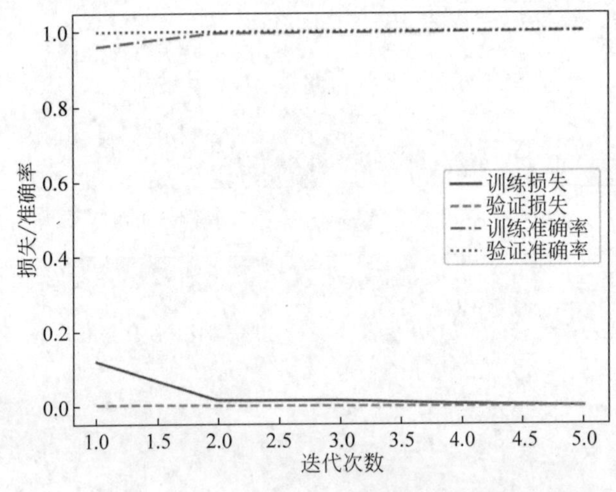

图 4-71　ResNet50 卡通表情识别结果

总结

残差解决了网络退化问题，缓解了梯度消失的现象；归一化避免了梯度爆炸，也能缓解梯度消失。大量的卷积神经网络会使用残差结构及 BN 归一化操作。

练习

完成 ResNet50 模型的搭建、训练、推理代码。

4.10　轻量级网络 MobiLeNet

4.10.1　模型介绍

2017 年谷歌公司提出了一个应用于移动端和嵌入式端的卷积神经网络结构 MobiLeNet v1，其主要使用了深度可分离卷积，极大地减少了参数量。

2018 年 MobiLeNet v2 发布，引入倒残差结构，即 channel 先升维再降维，如图 4-72 所示。

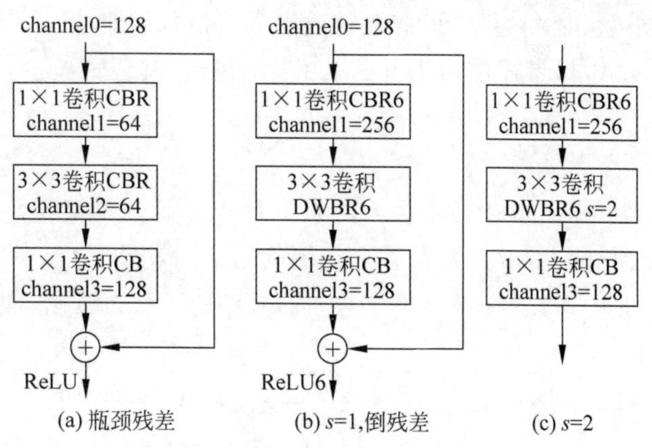

图 4-72 MobiLeNet v2 中的残差结构

图 4-72 中 DWBR6 指经过深度可分享卷积(DW 卷积)，R6 为 ReLU6 激活函数。倒残差(Inverted Residual Block)先进行 channel 的增加，然后经过 DW 卷积之后，再通过 1×1 卷积进行 channel 降维。主要原因是由于 DW 卷中引入了残差，如果按瓶颈残差的方式先压缩特征图，则将会使 DW 输出的特征信息不足，影响模型的性能，所以这里采用先在 channel 上升维再降维的操作。当步长 $s=2$ 时，直接线性输出，主要因为 1×1 卷积已经对特征信息进行了压缩，而 relu()函数对于负值输出为 0，这会造成特征信息的丢失，所以这里进行了线性输出。

ReLU6 的激活函数公式如下：

$$\mathrm{ReLU6}(x)=\min(\max(x,0),6) \tag{4-15}$$

将 ReLU6 函数图形化展示，如图 4-73 所示。

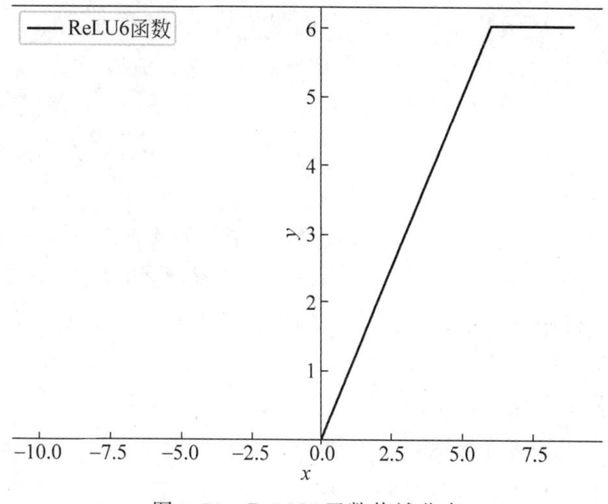

图 4-73 ReLU6 函数值域分布

使用 ReLU6 的主要原因是嵌入式设备其值域一般是在[0,255]之间的 uint 类型，而 ReLU 函数没有限制最大值的边界，可能会超过嵌入式设备的值域。

在 MobiLeNet v3 中作者还使用了 h-swish 激活函数，其公式如下：

$$\text{h-swish}(x) = x \times \text{ReLU6}(x + 3)/6 \tag{4-16}$$

其函数的值域如图 4-74 所示。

从图 4-74 中可知，h-swish 函数不再将所有负数变为 0，这样对于某些神经元可以再次激活，经过作者实验，将 h-swish 函数加在更深层的网络中效果更佳。

在 MobiLeNet v3 中，作者在某些网络层加入了 SE 注意力机制，使网络瓶颈残差变为如图 4-75 所示的结构。

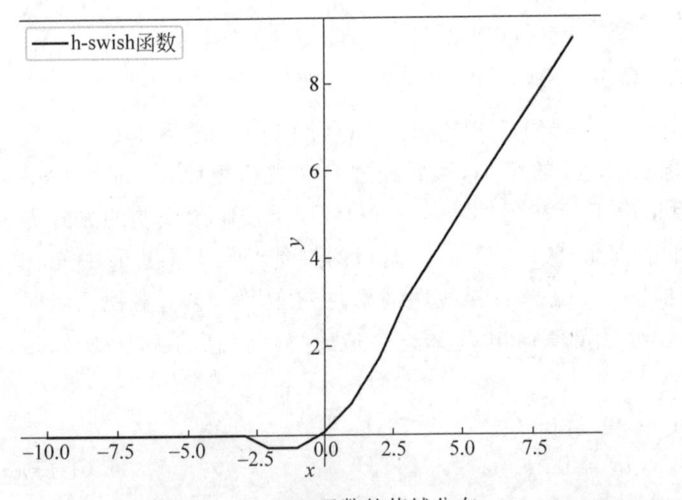

图 4-74 h-swich 函数的值域分布

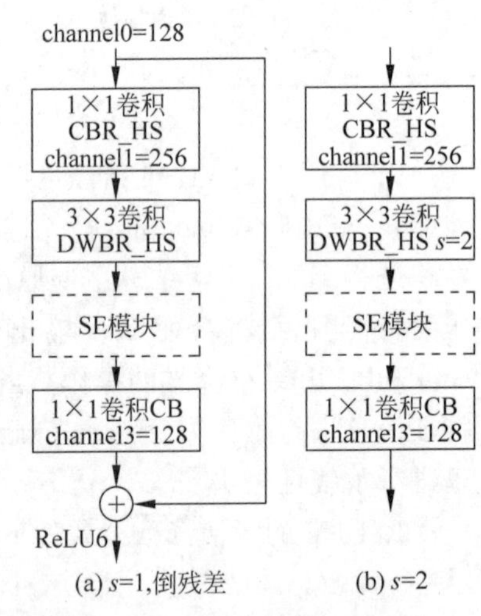

图 4-75 倒残差加入 SE 结构

HS 代表 h-swich 激活函数。SE 注意力机制的加入，使模型更加专注于主要特征，可以提高模型的精度。SE 模块及其他注意力机制将在 4.10.2 节展开。

在输出层，MobiLeNet v3 在全局平均池化后，经过两个 1×1 卷积，再进行预测，经作者实验发现可将推理时间减少 11%。以 MobiLeNet v3-small 为例，其网络结构如图 4-76 所示。

图 4-76 中 k 代表卷积的 kernel_size、s 代表步长，se 代表加入 SE 注意力机制，re 代表 ReLU 激活函数，hs 代表 h-swish 激活函数，exp 代表通道的基数，通过输入 alpha 参数可以调节通道数(exp×alpha)。进入倒残差块时通道数为 exp×alpha，然后输出为 out 通道数。

在论文中作者给出了 MobiLeNet v3 的性能，如图 4-77 所示。

P-1 代表 Pixel-n Phone 手机型号，推理时间，单位为毫秒(ms)。图 4-77 中 V3-Small 模型，Top-1 的准确率为 67.4%，参数量只有 2.5M，在谷歌 Pixel-1 上推理时间为 15.8ms。

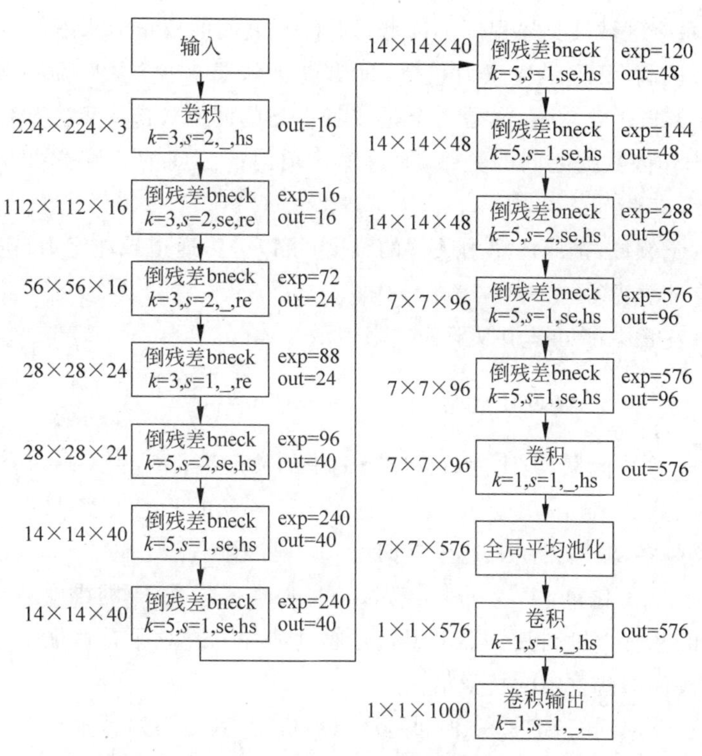

图 4-76　MobiLeNet v3 网络结构

Network	Top-1	MAdds	Params	P-1	P-2	P-3
V3-Large 1.0	**75.2**	**219**	5.4M	**51**	**61**	**44**
V3-Large 0.75	73.3	155	4.0M	39	46	40
MnasNet-A1	75.2	315	3.9M	71	86	61
Proxyless[5]	74.6	320	4.0M	72	84	60
V2 1.0	72.0	300	3.4M	64	76	56
V3-Small 1.0	**67.4**	**56**	2.5M	**15.8**	**19.4**	**14.4**
V3-Small 0.75	65.4	44	2.0M	12.8	15.6	11.7
Mnas-small[43]	64.9	65.1	1.9M	20.3	24.2	17.2
V2 0.35	60.8	59.2	1.6M	16.6	19.6	13.9

图 4-77　MobiLeNet v3 论文性能

4.10.2　注意力机制

注意力机制来源于 20 世纪 90 年代，有认知领域学者发现，人类在处理信息时会过滤掉不太关注的信息，着重于感兴趣的信息，这种处理信息的机制称为注意力机制。

2014 年 Volodymyr 在论文 *Recurrent Models of Visual Attention* 中将其应用于视觉领域。2017 年 Ashish Vaswani 在论文 *Attention is all You Need* 中提出 transformer 结构，注意力机制在计算机视觉、自然语言等领域被广泛应用。

注意力机制主要分为全局注意力、局部注意力及自注意力。全局注意力机制，又称软注意力机制，根据每个区域被关注程度的高低，用[0,1]之间的权重值来缩小、放大注意力的程度；局部注意力机制，又称强注意力机制，即哪些区域是被关注的，哪些区域是不被关注的，不关注区域会被直接舍弃，这是一个是或者不是的问题；自注意力机制对每个输入赋予的权重取决于输入数据之间的关系，通过输入项内部之间的相互博弈决定每个输入项的权重。

在计算机视觉领域，按照注意力关注的维度，可以分为通道域注意力、空间域注意力、混合域注意力和自注意力等。

无论是哪种注意力都可以用以下公式来表示：

$$\text{Attention}(Q,K,V)=\text{Softmax}\left(\frac{\boldsymbol{QK}^{\mathrm{T}}}{\sqrt{d_k}}\right)V \tag{4-17}$$

其中，Q 为 Query 输入信息，类比为数据库查询中的关键字；K 为 Key 内容信息，类比为数据库已存储的内容；Attention(Q,K)表示 Query 和 Key 的匹配程度，Key 越多 Query 查询出来的信息就会越多，但是要注意这里查询出来的是匹配程度，通过 Softmax()归一化与 V 相乘，得到 V 的注意力比重；V 为 value 特征信息本身；d_k 为 K 的维度。

从公式可知，注意力机制其核心思想就是通过某种方式对原有特征信息给予不同权重的关注度，着重关注那些重要的特征信息。

MobiLeNet v3 用到的 SE 是一种通道注意力机制，其结构如图 4-78 所示。

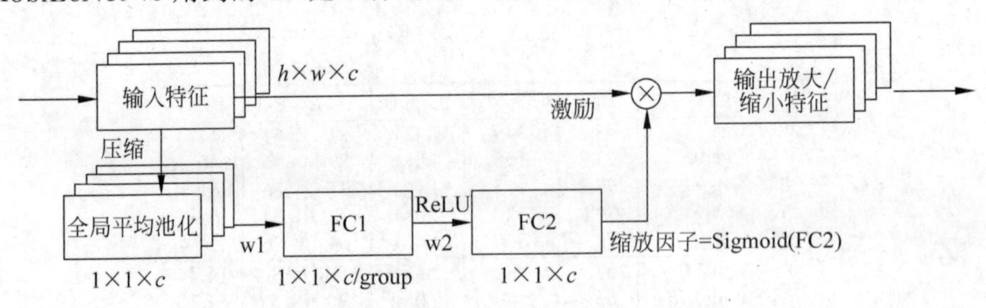

图 4-78　SE 注意力机制

将输入特征信息通过全局平均池化，压缩(Squeeze)成 1×1×channel 的张量，然后通过 FC1 全连接(或者 1×1 卷积)输入 FC2，将 FC2 的输出值通过 Sigmoid()函数映射到[0,1]之间，其中 FC1 的神经元个数是自定义的，在 MobiLeNet v3 网络中设置为 group=4。缩放因子，即注意力激励(excitation)系数，对其值跟输入特征进行点乘，得到不同尺度激励后的特征信息。在 SE 中 FC2 的输出受 w1、w2 权重参数的影响，w1、w2 由反向传播算法学习得到，使输入特征后经过 FC 能够恰当地映射到缩放因子值。套用式(4-17)，则输入特征是 K，FC1、FC2 是 Query，缩放因子是 Attention(Q,K)，即通过 FC 来查询特征的比例关系。

将以上过程使用具体化数值进行计算，如图 4-79 所示。

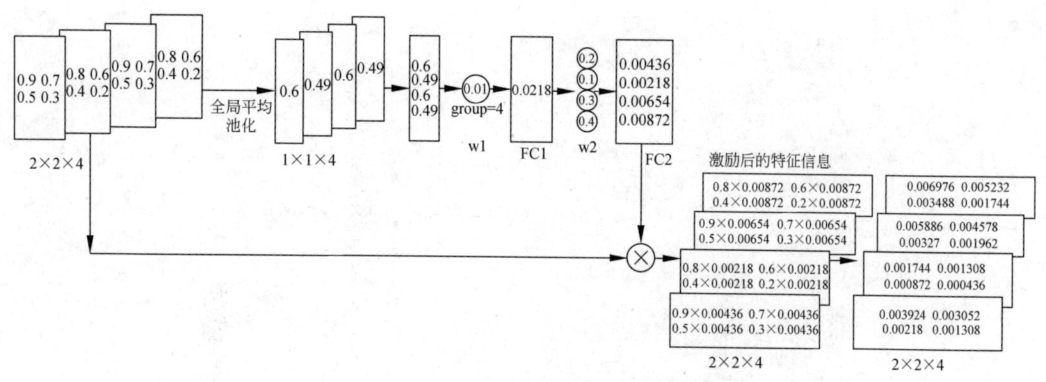

图 4-79 SE注意力机制具体数值化

从图 4-79 观察可知，第 1 个通道的特征给予 0.00436 的关注，第 2 个通道特征给予 0.00218 的关注，不同特征给予的关注度是不同的，所以 SE 注意力机制称为通道域注意力机制。虽然使用 SE 理论上能够更好地提取通道特征信息，但是如果训练不够，或者所处网络层位置的噪声数据较多，则其效果有可能更差。

SE 模块的实现代码如下：

```
#第 4 章/ClassModleStudy/attention.py
from tensorflow.keras.layers import Conv2D, Multiply
from tensorflow.keras.layers import GlobalAveragePooling2D, ReLU
from tensorflow.keras import layers


def make_divisible(channel, divisor=8, min_ch=None):
    """确保通道数是 8 倍的整数倍"""
    min_ch = divisor if min_ch is None else min_ch
    new_ch = max(min_ch, int(channel + divisor / 2) //divisor *divisor)
    if new_ch < 0.9 *channel: new_ch += divisor
    return new_ch


class HardSigmoid(layers.Layer):
    """Hard Sigmoid 激活函数"""

    def __init__(self, **kwargs):
        super(HardSigmoid, self).__init__(**kwargs)
        self.relu6 = ReLU(max_value=6.)

    def call(self, inputs, *args, **kwargs):
        x = self.relu6(inputs + 3) *(1. / 6)
        return x


class SE(layers.Layer):
    def __init__(self, filters, se_ratio=1 / 4.):
        super(SE, self).__init__()
        #通道全局平均池化
```

```
        self.squeeze = GlobalAveragePooling2D()
        #1*1*C/4 卷积
        self.conv1 = Conv2D(make_divisible(filters *se_ratio), 1, padding='same',
use_bias=False)
        self.relu = ReLU()
        #1*1*C 卷积
        self.conv2 = Conv2D(filters, 1, padding='same' ,use_bias=False)
        self.sigmoid = HardSigmoid()
        self.mult = Multiply()

    def call(self, inputs, **kwargs):
        #SE 注意力机制的实现
        #没有进行 BN 层,因为 BN 层强行将均值为 0,方差为 1 会有一些特征信息的数据丢失
        x = self.squeeze(inputs)
        x = layers.Reshape([1, 1, -1])(x)
        x = self.conv1(x)
        x = self.relu(x)
        x = self.conv2(x)
        x = self.sigmoid(x)
        #Multiply()执行输入数组的逐元素乘法,得到新的权重数据,使网络更加专注主要特征
        return self.mult([inputs, x])
```

SE()类按图 4-77 进行了实现，make_divisible()函数确保输入的 filters 是 8 的整数倍。SE 针对的是通道 channel，而忽略了特征图中的 height、width，如图 4-80 所示。

空间域注意力机制，则针对 height、width 的每个特征信息给予不同的关注度，空间域注意力机制主要包括自注意力机制(Self Attention)及非局部注意力机制(Non-local Attention)。

自注意力机制的核心是捕获向量之间的相似性，分别计算 x1 与 x1、x2、x3 之间的相似性，相似程度值越大给予的注意力程度就越高，如图 4-81 所示。

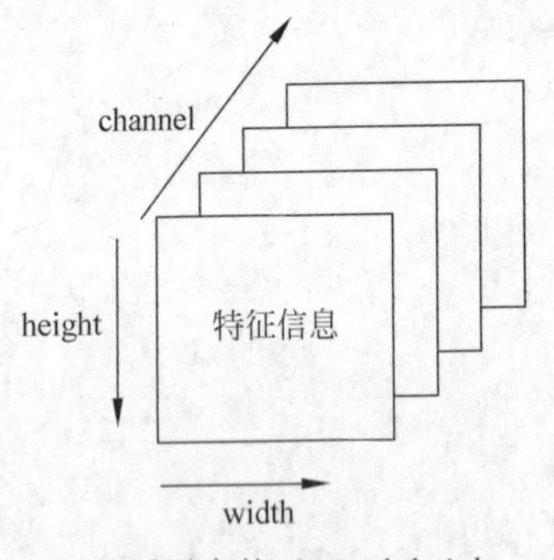

图 4-80 特征图中的 channel、height、width

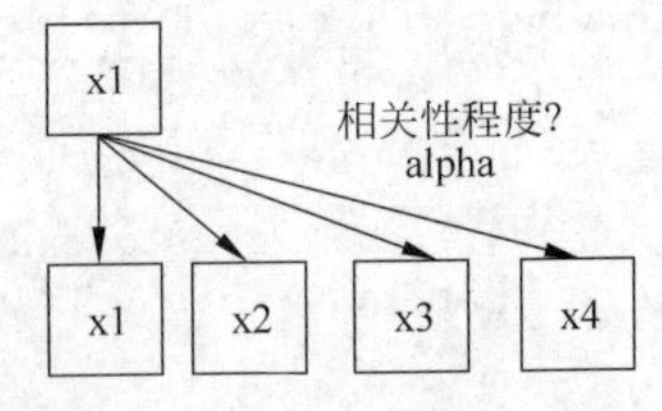

图 4-81 x1 与向量的相似度

向量的相似度的计算可以采用向量点乘相似度、余弦相似度等。向量 $\boldsymbol{x}$ 在向量 $\boldsymbol{y}$ 方向上的投影再与向量 $\boldsymbol{y}$ 的乘积，能够反映两个向量的相似度，向量积越大，两个向量越相似。如果一个矩阵与自身的转置相乘，反映的就是一个特征值与矩阵中各个特征值的相似度，如图 4-82 所示。

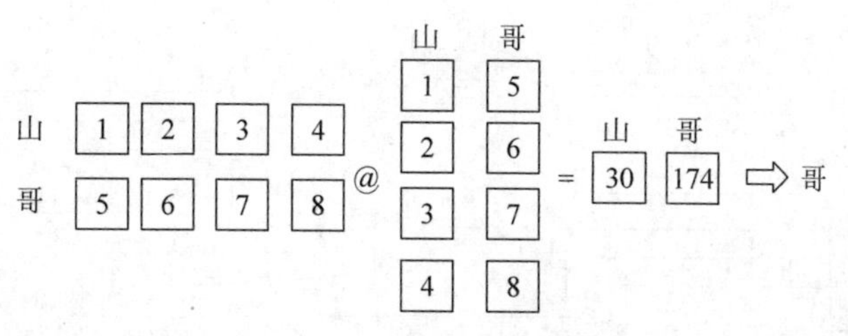

图 4-82　矩阵特征值与自身的相似度

图 4-82 中假设"山"的特征值是[1,2,3,4],"哥"的特征值是[5,6,7,8],与其自身的转置做内积,得到"山"与矩阵的相似度为 30、"哥"为 174,"哥"的值较大,说明"哥"与"山哥"之间的相似度较大,所以公式(4-17)中 $\boldsymbol{QK}^{\mathrm{T}}$ 表示 Q 与 K 的相似程度,因为这个值可能会较大,所以 $\frac{\boldsymbol{QK}^{\mathrm{T}}}{\sqrt{d_k}}$ 将值域缩小,经过 $\mathrm{Softmax}\left(\frac{\boldsymbol{QK}^{\mathrm{T}}}{\sqrt{d_k}}\right)$ 归一化到[0,1]得到注意力分数,最后与 V 相乘得到不同注意力尺度的特征。根据这一原理,自注意力机制的结构如图 4-83 所示。

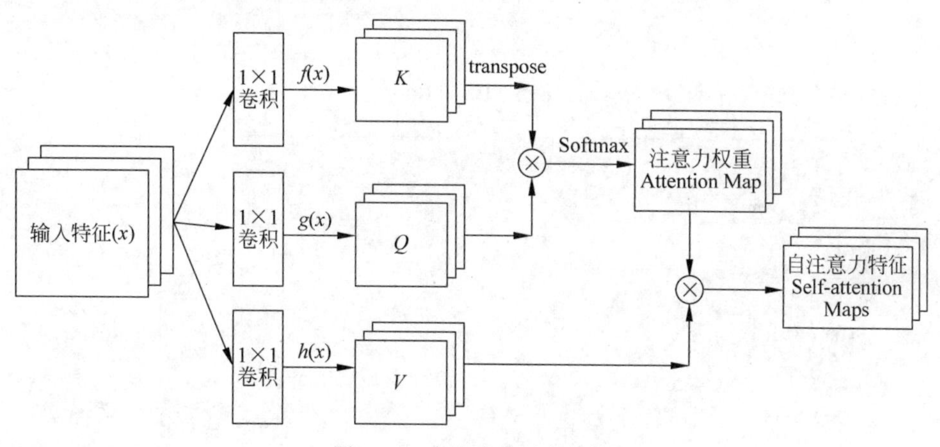

图 4-83　自注意力机制结构

输入特征(x)经过不同分支的 1×1 卷积得到 $\boldsymbol{K}$、Q、V 的特征,然后 $\boldsymbol{QK}^{\mathrm{T}}$ 做相似度计算并由 Softmax 得到注意力分数,再与 V 相乘得到不同大小的 value,最后对 value 进行加权求和得到不同注意力尺度的特征。输入特征(x)经过 1×1 卷积,卷积中的权重会通过反向传播学习调整,输出特征 $\boldsymbol{K}$、Q、V 的值会很相近,通过算法会提取出注意力特征,所以该方法称为自注意力机制。

对自注意力机制进行具体数值化计算,如图 4-84 所示。

图 4-84 中输入特征 x 为 3×3,经过 $\boldsymbol{QK}^{\mathrm{T}}$ 后 Softmax 得到注意力分数,需要保证与 V 的特征值每个都相乘,所以需要 reshape 成 3×3×1,然后 V×Softmax 值得到的是 1 个 3×3×3 的张量,然后按 axis=0 轴加权求和输出 3×3 的特征。

自注意力机制的代码如下：

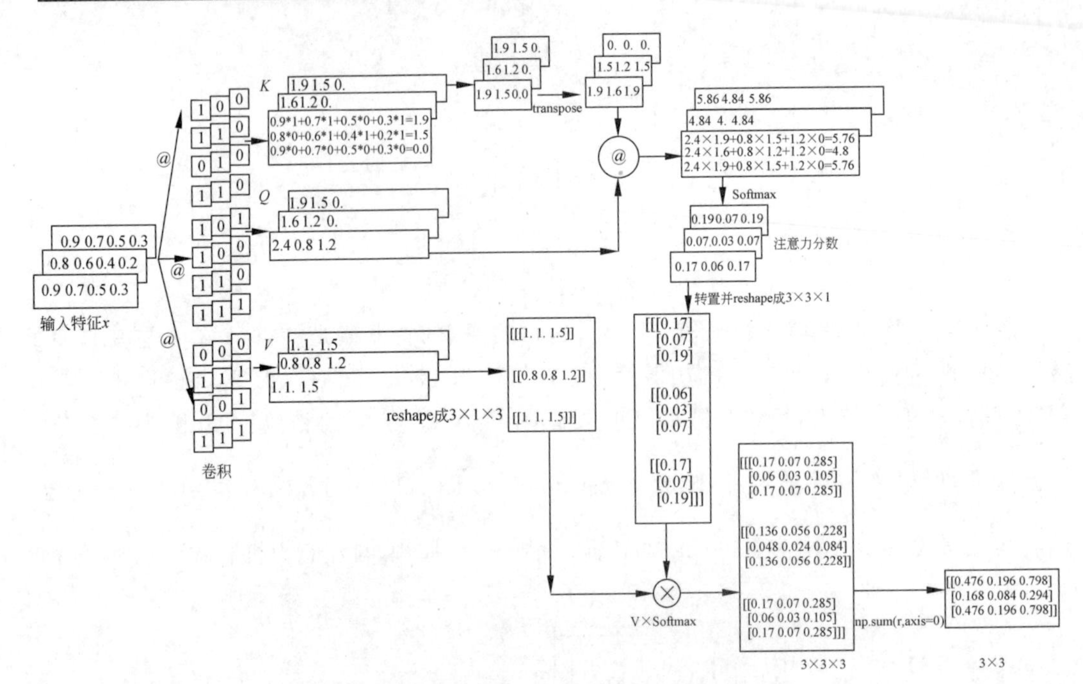

图 4-84　自注意力机制具体数值化计算

```
#第 4 章/ClassModleStudy/attention.py
class SelfAttention(layers.Layer):
    def __init__(self, filters, ratio=1 / 8,d_k=5.):
        super(SelfAttention, self).__init__()
        self.filter = filters
        #Q 和 K 的 channel 降维
        self.channel = make_divisible(filters * ratio)
        self.dk = d_k
        self.K = Conv2D(self.channel, kernel_size=1, padding='same' ,use_bias=False)
        self.Q = Conv2D(self.channel, kernel_size=1, padding='same' ,use_bias=False)
        #V * Softmax(QK.T/dk)相乘要与输入保持一致,所以 filter 要保持一致
        self.V = Conv2D(self.filter, kernel_size=1, padding='same' ,use_bias=False)
        self.softmax = layers.Softmax()

    def call(self, inputs, **kwargs):
        #获取输入时的 height 和 width
        height, width = inputs.shape[1:3]
        k = self.K(inputs)
        #统一到坐标系的维度
        k = tf.reshape(k, [-1, height * width, self.channel])
        #转置
        k_transpose = tf.transpose(k, perm=[0, 2, 1])

        q = self.Q(inputs)
        q = tf.reshape(q, [-1, height * width, self.channel])

        v = self.V(inputs)
        v = tf.reshape(v, [-1, height * width, self.filter])
```

```
#k的转置
q_k = tf.matmul(q, k_transpose)
attention = self.softmax(q_k/tf.sqrt(self.dk))

#实现 V * Softmax(QK.T/dk)
out = v[:, None] *tf.transpose(attention[:, :, None], perm=[0, 1, 3, 2])
#注意力加权相加
out = tf.reduce_sum(out, axis=1)
#与输入维度保持一致
out = tf.reshape(out, [-1, height, width, self.filter])
return out
```

如果问“卷积”,图 4-85 中有什么？那么我们知道,只要卷积采用滑动窗口局部感知的方法获取每个像素的上下文信息,就能找到例如脚、手、衣服、头等特征,卷积神经网络就可以判断这是一个小孩。这个上下文信息就是领域信息,它可以自然地表达图像的局部语义信息。

图 4-85　在公园玩树叶的小孩

如果问“卷积”,这个小孩在什么地方干什么？对于卷积来讲它记录的是像素的上下文,其实并不能获知在干什么,此时需要结合图像的其他信息,观察到手、姿态、树叶等信息才能确定小孩是在玩树叶,全图的综合信息就是全局信息,全局信息可以更好地协助局部特征进行语义描述。因为卷积是局部感知,如果用卷积获取更多全局信息,就需要更大的感受野,更多的卷积层,此时就相当于对局部的特征信息进行了整合。

但是这样并不能回答“在什么地方?”,但是如果能观察到树、草地、灯、模糊的建筑设施,同时能观察到脚、手、衣服、头、树叶、姿态等信息,就可以回答“在公园里玩树叶的小孩”,图像中一些距离较远的内容也会影响理解该图像,这种关系称为长距离依赖。长距离依赖不仅存在于图像中,也存在于视频、时间序列、语言模型等任务中。

非局部注意力机制要做的就是去捕获这种长距离依赖关系,其结构如图 4-86 所示。

图 4-86 中将输入特征(x)通过 $1\times1\times C/2$ 的卷积进行特征的压缩,然后将 $T\times H\times W\times C/2$ 压缩成 THW$\times C$,经过 QK^{T} 得到自相似度,通过 Softmax 得到注意力分数,再与

V 相乘得到自注意力特征 HWC×$C/2$，reshape 成 $T\times H\times W\times C/2$ 后经过 1×1 卷积升维成 $T\times H\times W\times C$，最后与输入特征 x 相加，残差输出 Non-local Attention。

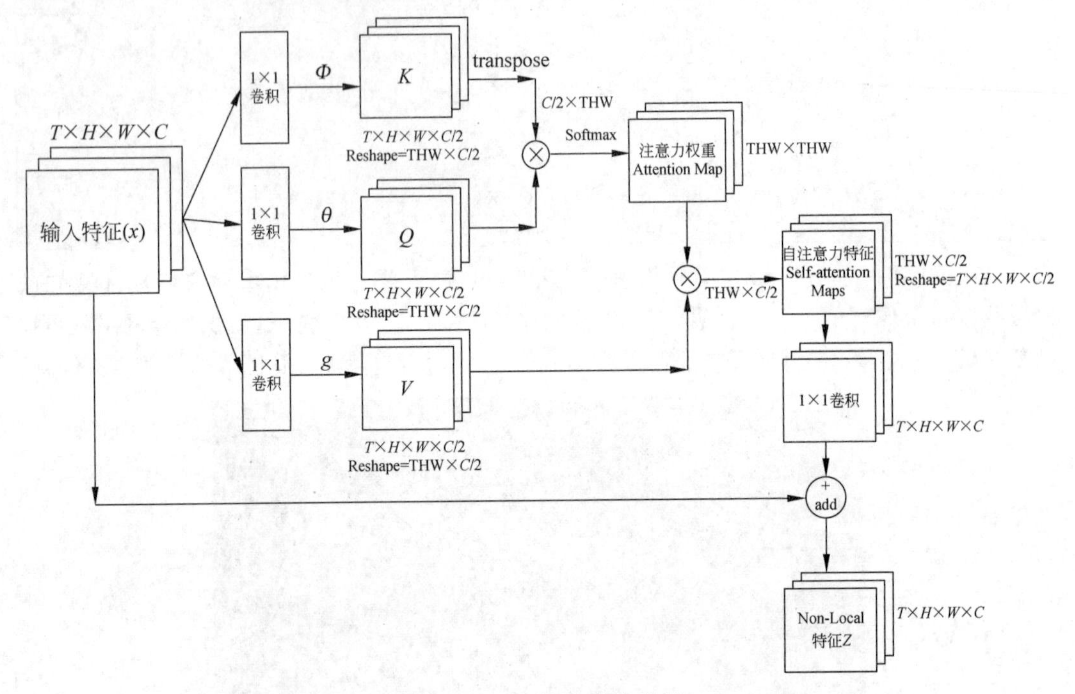

图 4-86　非局部注意力机制

Non-local Attention 的代码如下：

```
#第 4 章/ClassModleStudy/attention.py
class NonLocalAttention(layers.Layer):
    def __init__(self, filters, ratio=1/2, d_k=5.):
        super(NonLocalAttention, self).__init__()
        self.filter = filters
        #Q 和 K 的 channel 降维
        self.channel = make_divisible(filters * ratio)
        self.dk = d_k
        self.K = Conv2D(self.channel, kernel_size=1, padding='same',use_bias=False)
        self.Q = Conv2D(self.channel, kernel_size=1, padding='same',use_bias=False)
        self.V = Conv2D(self.channel, kernel_size=1, padding='same',use_bias=False)
        #1×1 实现升维,保持与输入一致,以便输出残差
        self.conv = Conv2D(self.filter, kernel_size=1, padding='same',use_bias=False)
        self.add = layers.Add()
        self.softmax = layers.Softmax()

    def call(self, inputs, **kwargs):
        #获取输入时的 height 和 width
        height, width = inputs.shape[1:3]
        k = self.K(inputs)
        #统一到坐标系的维度
```

```
        k = tf.reshape(k, [-1, height * width, self.channel])
        #转置
        k_transpose = tf.transpose(k, perm=[0, 2, 1])

        q = self.Q(inputs)
        q = tf.reshape(q, [-1, height * width, self.channel])

        v = self.V(inputs)
        v = tf.reshape(v, [-1, height * width, self.channel])
        #k 的转置
        q_k = tf.matmul(q, k_transpose)
        score = self.softmax(q_k / tf.sqrt(self.dk))
        #实现 V * Softmax(QK.T/dk)
        attention = tf.matmul(score, v)
        attention = tf.reshape(attention, [-1, height, width, self.channel])
        #实现升维
        out = self.conv(attention)
        return self.add([inputs, out])    #实现残差
```

通道域和空间域注意力机制都对网络有益，于是就出现了混合域注意力机制，例如常用的Convolutional Block Attention Module(CBAM)和Coordinate Attention(CA)注意力机制。

CBAM先进行通道域注意力，再经过空间域注意力，其结构是一个串行的过程，如图4-87所示。

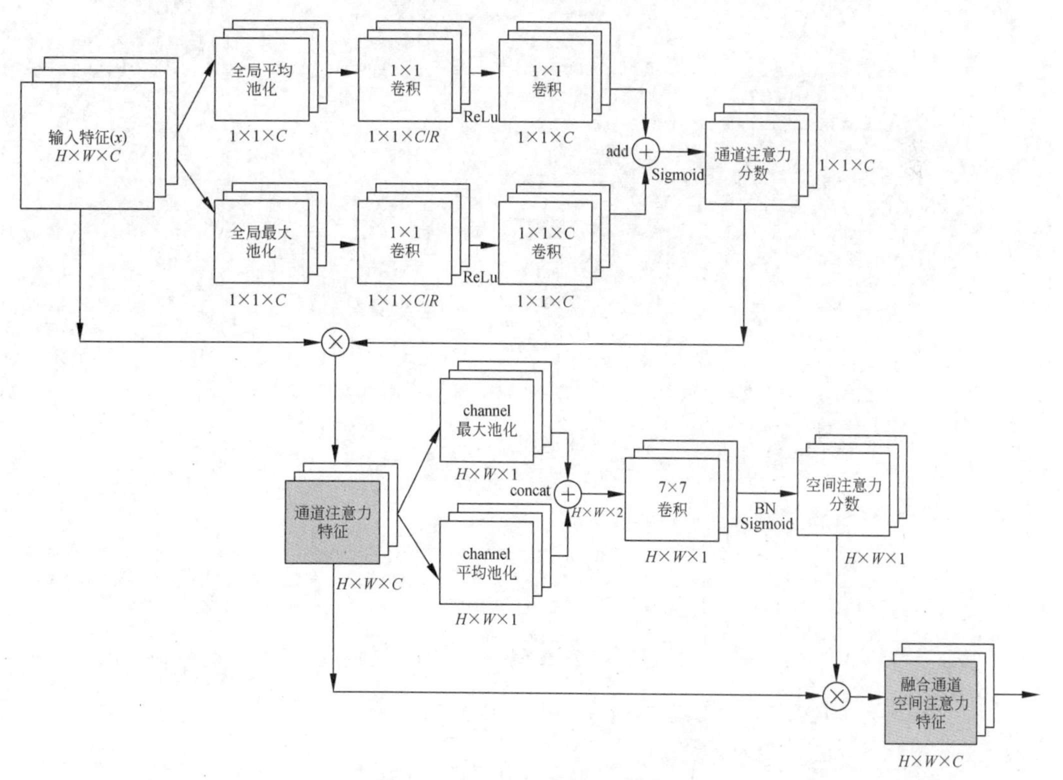

图4-87　CBAM注意力机制

图 4-87 输入特征(x)后分别经过全局平均池化和全局最大池化得到更多特征信息,并压缩成 $1\times1\times C$,经过 $1\times1\times/R$(R 原作设为 16)降维,然后通过 $1\times1\times C$ 升维,两者相加后经过 Sigmoid 变得 $1\times1\times C$ 的通道注意力分数,并与输入特征 x 后相乘得到 $H\times W\times C$ 的通道注意力特征。

将通道注意力特征,在 C 的这个维度分别进行最大池化、平均池化压缩成 $H\times W\times1$ 以保留坐标信息,然后 concat 成 $H\times W\times2$ 经过 7×7 的卷积,经 Sigmoid 后输出空间注意力分数为 $H\times W\times1$,最后与上一步通道注意力特征相乘,得到 CBAM 注意力特征。

CBAM 注意力的代码如下:

```
#第 4 章/ClassModleStudy/attention.py
class CBAMAttention(layers.Layer):
    def __init__(self, filters, ratio=1 / 16, spatial_kernel=7):
        super(CBAMAttention, self).__init__()
        self.channel = make_divisible(filters *ratio)
        self.filter = filters
        self.global_avg = GlobalAveragePooling2D()
        self.global_max = GlobalMaxPool2D()
        self.conv_list = Sequential(
            [
                Conv2D(self.channel, kernel_size=1, padding='same', use_bias=False),
                ReLU(),
                Conv2D(self.filter, kernel_size=1, padding='same', use_bias=False)
            ]
        )
        self.conv = Conv2D(self.filter, kernel_size=spatial_kernel, padding=
'same', use_bias=False)
        self.bn = layers.BatchNormalization()

    def call(self, inputs, **kwargs):
        #全局平均池化后经过 1×1×c/r->1×1×c 的卷积操作
        #全局平均池化后维度变为 1*c,所以需要 reshape 成 b *1*1*c
        avg = self.conv_list(tf.reshape(self.global_avg(inputs), [-1, 1, 1,
self.filter]))
        #全局最大池化后经过 1×1×c/r->1×1×c 的卷积操作
        max = self.conv_list(tf.reshape(self.global_max(inputs), [-1, 1, 1,
self.filter]))
        #将全局平均池化+全局最大池化,经 Sigmoid 后得到注意力分数
        channel_score = sigmoid(max + avg) #1×1×1×c
        #将注意力分数与输入相乘,得到通道注意力特征
        channel_attention = channel_score *inputs #b *h *w *c

        #按 channel 的维度进行平均和最大 pool b×h×w×1
        channel_avg = tf.reduce_mean(channel_attention, axis=-1, keepdims=True)
        channel_max = tf.reduce_max(channel_attention, axis=-1, keepdims=True)
        #将最大、平均 pool 进行 concat 后,经过 7*7 的卷积,BN 后得到空间注意力分数
        #b×h×w×c
        spatial_score = sigmoid(self.bn(self.conv(tf.concat([channel_avg,
channel_max],axis=-1))))
        #将空间注意力分数与通道注意力特征相乘,得到 CBAM 特征
        out = spatial_score *channel_attention
        return out
```

Coordinate Attention(CA)注意力机制，不仅考虑了通道注意力，并且还包含坐标位置和方向信息，使 CA 具有捕获长距离依赖信息的能力，其网络结构如图 4-88 所示。

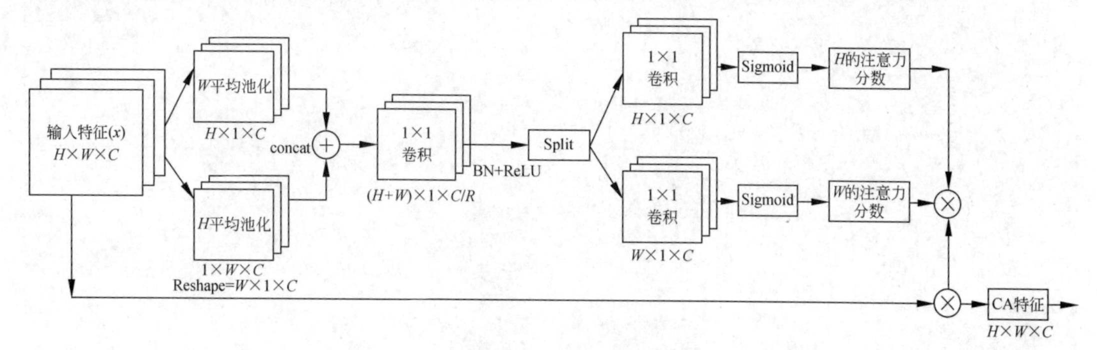

图 4-88　CA 注意力机制

图 4-88 中输入特征 $H\times W\times C$，通过对 W 维度的平均池化得到 $H\times 1\times C$，通过对 H 维度的平均池化得到 $1\times W\times C$，reshape 后得到 $W\times 1\times C$。此时，可以看成分别沿着两个方向在聚合特征，使注意力机制能够感知方向信息进而捕获长距离依赖的特征信息，然后将含有方向信息的特征 concat 变成 $(H+W)\times 1\times C$，然后使用 1×1 卷积进行降维，然后 split 成两部分经过 1×1 卷积 Sigmoid 得到 H、W 位置的注意力分数。最后，将 H 注意力分数×W 注意力分数×输入特征(x)得到 CA 特征信息。

CA 注意力机制的代码如下：

```
#第 4 章/ClassModleStudy/attention.py
class HardSigmoid(layers.Layer):
    """Hard Sigmoid 激活函数"""

    def __init__(self, **kwargs):
        super(HardSigmoid, self).__init__(**kwargs)
        self.relu6 = ReLU(max_value=6.)

    def call(self, inputs, *args, **kwargs):
        x = self.relu6(inputs + 3) * (1. / 6)
        return x

class HardSwich(layers.Layer):
    def __init__(self, **kwargs):
        super(HardSwich, self).__init__(**kwargs)
        self.hard_sigmoid = HardSigmoid()

    def call(self, inputs, **kwargs):
        x = self.hard_sigmoid(inputs) *inputs
        return x

class CoordinateAttention(layers.Layer):
    def __init__(self, filters, ratio=1 / 32):
        super(CoordinateAttention, self).__init__()
        self.channel = make_divisible(filters *ratio)
```

```
        self.filter = filters
        self.conv1 = Conv2D(self.channel, kernel_size=1, padding='same', use_
bias=False)
        self.conv2 = Conv2D(self.filter, kernel_size=1, padding='same', use_
bias=False)
        self.bn = layers.BatchNormalization()
        self.act = HardSwich()

    def call(self, inputs, **kwargs):
        height, width = inputs.shape[1:3]
        #沿 height 进行池化 b×h×1×c
        height_avg_pool = AvgPool2D(pool_size=[1, width], padding='same')(inputs)
        #沿 width 进行池化 b×1×w×c
        width_avg_pool = AvgPool2D(pool_size=[height, 1], padding='same')(inputs)
        #b×w×1×c
        width_avg_pool = tf.transpose(width_avg_pool, [0, 2, 1, 3])
        #将 height_avg_pool 和 width_avg_pool 拼接后经过 1*1 卷积->bn->act->得到 score
        #concat 后得到 b×(h+w)×1×c
        wh = self.act(self.bn(self.conv1(tf.concat([height_avg_pool, width_avg_
pool], axis=1))))
        #将 wh 分割成 w 和 h b×h(w)×1×c/r
        w, h = tf.split(wh, num_or_size_splits=2, axis=1)
        w_score = sigmoid(self.conv2(w)) #b*w*1*c
        h_score = sigmoid(self.conv2(h)) #b*h*1*c
        #将 w 和 h 轴的分数与输入数据相乘,得到 CA 注意力特征
        return inputs *w_score *h_score
```

注意：并不是增加了注意力模块模型效果就一定会变好，数据集不同及注意力模块所处的网络位置不同，效果可能会不同。增加注意力模块后，模型参数也会有一定的增加。

4.10.3 代码实战

首先，根据图 4-75 倒残差加入 SE 结构，首先封装卷积模块 CBR_HS、DWBR_HS，代码如下：

```
#第 4 章/ClassModleStudy/common.py
class CBR_HS(layers.Layer):
    #conv->bn->relu/hardswich
    def __init__(self, filters, kernel_size, strides, padding="same", is_act=
True, activation='re'):
        super(CBR_HS, self).__init__()
        self.act = layers.ReLU() if activation == "re" else HardSwich()
        self.conv = layers.Conv2D(filters, kernel_size=kernel_size, strides=
strides, padding=padding, use_bias=False)
        self.bn = layers.BatchNormalization(momentum=0.9, epsilon=1e-5)
        self.is_act = is_act

    def call(self, inputs, training=True, **kwargs):
        x = self.conv(inputs)
```

```
        x = self.bn(x, training=training)
        if self.is_act: x = self.act(x)
        return x

class DWBR_HS(layers.Layer):
    #Depthwise conv->bn->relu/hardswich
    def __init__(self, kernel_size=3, strides=1, padding="same", activation='re',
is_act=True):
        super(DWBR_HS, self).__init__()
        self.conv = layers.DepthwiseConv2D(kernel_size=kernel_size, strides=
strides, padding=padding, use_bias=False)
        self.bn = layers.BatchNormalization(momentum=0.9, epsilon=1e-5)
        self.act = layers.ReLU() if activation == "re" else HardSwich()
        self.is_act = is_act

    def call(self, inputs, training=True, **kwargs):
        x = self.conv(inputs)
        x = self.bn(x, training=training)
        return self.act(x) if self.is_act else x
```

对 CBR_HS 模块进行卷积、归一化、激活的输出，同时通过 is_act 参数来指定是否线性输出，通过 activation 指定 ReLU 或 hard-swish 激活；DWBR_HS 中间为 DW 卷积，其他同 CBR_HS 模块。

然后实现倒残差模块，代码如下：

```
#第 4 章/ClassModleStudy/common.py
class MobileInvertedBottLeNet(layers.Layer):
    #mobile 倒残差结构
    def __init__(
            self,
            input_channel,
            expend_channel,
            output_channel,
            kernel_size,
            strides=1,
            is_se=False,
            act='re',
            alpha=1.0,
            block_id=1
    ):
        super(MobileInvertedBottLeNet, self).__init__()
        #输入 channel 扩展倍数
        self.input_channel = make_divisible(input_channel *alpha)
        #中间 channel 扩展倍数
        self.expend_channel = make_divisible(expend_channel *alpha)
        #输出 channel 扩展倍数
        self.output_channel = make_divisible(output_channel *alpha)
        self.block_id = block_id                        #编号,第几个 neck
        self.is_se = is_se                              #是否为 SE 注意力机制
        self.strides = strides                          #dw 步长
```

```
        self.cbr1 = CBR_HS(self.expend_channel, 1, 1, is_act=True, activation=act)
        self.dwbr = DWBR_HS(kernel_size, strides, padding='same')
        #不进行激活函数输出
        self.cbr2 = CBR_HS(self.output_channel, 1, 1, is_act=False)
        self.se = SE(self.expend_channel, se_ratio=1 / 4)
        self.add = layers.Add()

    def call(self, inputs, training=True, **kwargs):
        #输入值
        x = inputs
        #如果不为 0,则 channel 升维。MobiLeNet v3 的第 1 个 neck 不需要升维
        if self.block_id: x = self.cbr1(inputs)
        #dw 卷积
        x = self.dwbr(x)
        #SE 注意力机制
        if self.is_se: x = self.se(x)
        x = self.cbr2(x)
        #只有步长为 1,并且输入 channel 与输出 channel 相同时才进行残差
        if self.strides == 1 and self.input_channel == self.output_channel:
            x = self.add([x, inputs])
        return x
```

因为网络结构中第 1 个 neck 不需要升维，所以当 self. block_id==0 时不需要调 self. cbr1(x)，否则进行 channel 的升维；self. dwbr(x)为残差结构中的 DW 卷积；根据 self. is_se 的结果进行注意力机制的嵌入；当前仅当步长 self. strides==1 并且输入 self. input_channel==self. output_channel 时才进行残差，否则直接线性输出，所以 self. cbr2 中的参数 is_act=False。

MobiLeNet V3 网络结构的代码如下：

```
#第 4 章/ClassModleStudy/MobiLeNet/model.py
from tensorflow.keras import Model, layers
from ClassModleStudy.common import CBR_HS, MobileInvertedBottLeNet
import tensorflow as tf

class MobiLeNetV3(Model):
    def __init__(self, num_class=7, alpha=1.0):
        super(MobiLeNetV3, self).__init__()
        self.cbr_hs = CBR_HS(16, 3, 2, activation='hs')
        #第 1 个 neck 不需要升维
        self.neck1 = MobileInvertedBottLeNet(16, 16, 16, 3, 2,
is_se=True, alpha=alpha, block_id=0)
        self.neck2 = MobileInvertedBottLeNet(16, 72, 24, 3, 2, is_se=False,
alpha=alpha, block_id=1)
        self.neck3 = MobileInvertedBottLeNet(24, 88, 24, 3, 1, is_se=False,
alpha=alpha, block_id=2)
        self.neck4 = MobileInvertedBottLeNet(24, 96, 40, 5, 2,
is_se=True, act='hs', alpha=alpha, block_id=3)
        self.neck5 = MobileInvertedBottLeNet(40, 240, 40, 5, 1, is_se=True, act=
'hs', alpha=alpha, block_id=4)
        self.neck6 = MobileInvertedBottLeNet(40, 240, 40, 5, 1, is_se=True, act=
'hs', alpha=alpha, block_id=5)
```

```
        self.neck7 = MobileInvertedBottLeNet(40, 240, 48, 5, 1, is_se=True, act=
'hs', alpha=alpha, block_id=6)
        self.neck8 = MobileInvertedBottLeNet(48, 144, 48, 5, 1, is_se=True, act=
'hs', alpha=alpha, block_id=7)
        self.neck9 = MobileInvertedBottLeNet(48, 288, 96, 5, 2, is_se=True, act=
'hs', alpha=alpha, block_id=8)
        self.neck10 = MobileInvertedBottLeNet(96, 576, 96, 5, 1, is_se=True, act=
'hs', alpha=alpha, block_id=9)
        self.neck11 = CBR_HS(576, 1, 1, activation='hs')
        self.global_avg_pool = layers.GlobalAveragePooling2D()
        #全局平均池化后再跟两个 1×1 进行预测输出
        self.cbr_hs_out = CBR_HS(576, 1, 1, activation='hs')
        self.out = CBR_HS(num_class, 1, 1, is_act=False)

    def call(self, inputs, training=False, **kwargs):
        x = self.cbr_hs(inputs)
        x = self.neck5(self.neck4(self.neck3(self.neck2(self.neck1(x)))))
        x = self.neck10(self.neck9(self.neck8(self.neck7(self.neck6(x)))))
        x = self.global_avg_pool(self.neck11(x))
        #因为全局平均池化为 1×576,所以需要 reshape 成 b×1×1×576 后才能跟 1×1 卷积
        x = layers.Reshape([1, 1, x.shape[-1]])(x)
        x = self.cbr_hs_out(x)
        return layers.Softmax()(layers.Flatten()(self.out(x)))
```

因为 MobiLeNet v3 中全局平均池化后维度变为 1×576,所以需要经过 tf.reshape()变成 1×1×576,然后跟两个 1×1 卷积进行预测分类的输出。neck1 到 neck3 使用的是 ReLU 激活函数,其他为 hard-swish。训练代码如下:

```
#第 4 章/ClassModleStudy/MobiLeNet/train.py
from ClassModleStudy.common import plt_loss, train_FERG_DB
from model import MobiLeNetV3

if __name__ == "__main__":
    net = MobiLeNetV3(num_class=7, alpha=1.0)
    #调用训练代码模板
    his = train_FERG_DB(net, img_height=224, img_width=224, epochs=5, batch_size
=16, lr=0.0001)
    #以图的形式展现训练结果
    plt_loss(his, his.params['epochs'], 'MobiLeNetV3 卡通表情识别训练结果')
```

运行结果如图 4-89 所示。

总结

Mobile 系列是轻量级网络的重要组成部分,主要使用深度可分离卷积,同时注意力机制是模型调优的重要组成部分,图像领域和工业领域常用的注意机制有 SE、CA、CBAM 等。

练习

完成 MobiLeNet v3 模型的搭建、训练、推理代码,可尝试将 SE 注意力机制更换为 CA 注意力机制,观察其结果的变化。

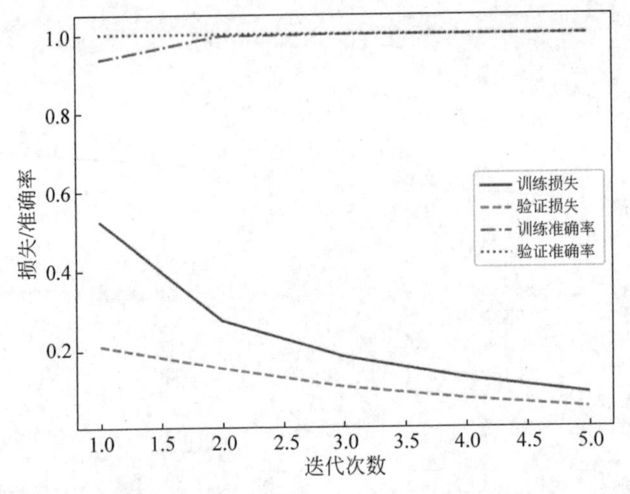

图 4-89 MobiLeNet v3 卡通表情识别结果

4.11 轻量级网络 ShuffLeNet

4.11.1 模型介绍

2017 年由 Xiangyu Zhang 等发表论文 *ShuffLeNet*：*An Extremely Efficient Convolutional Neural Network for Mobile Devices* 提出 ShuffLeNet v1。ShuffLeNet v1 主要由分组卷积构成，分组卷积虽然能够减少计算量，但是不同的组卷积之间没有信息的沟通，只是一个串联的过程，基于此提出了 Channel Shuffle 操作，如图 4-90 所示。

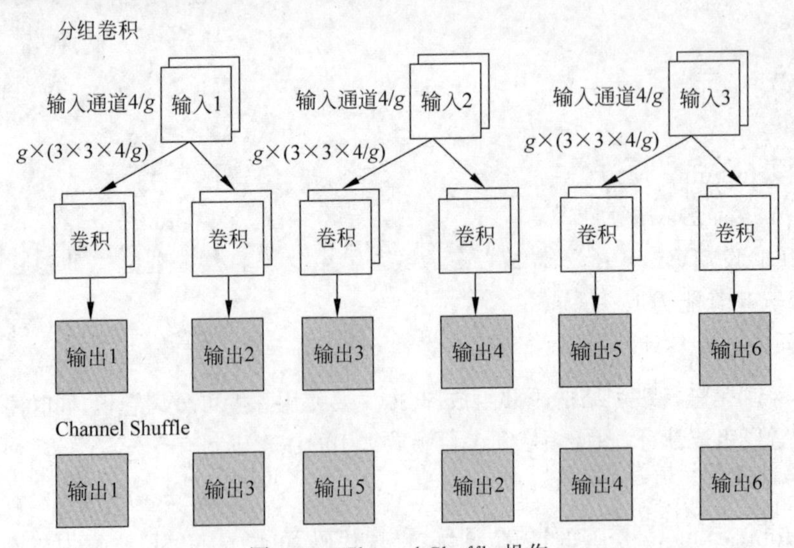

图 4-90 Channel Shuffle 操作

在分组卷积中"输出1、输出3、输出5"没有信息的沟通，通过Channel Shuffle操作让不同组的特征组合在一起，加强特征信息的沟通。

基于分组卷积、DW卷积、Channel Shuffle、残差等构建了ShuffLeNet v1模块，如图4-91所示。

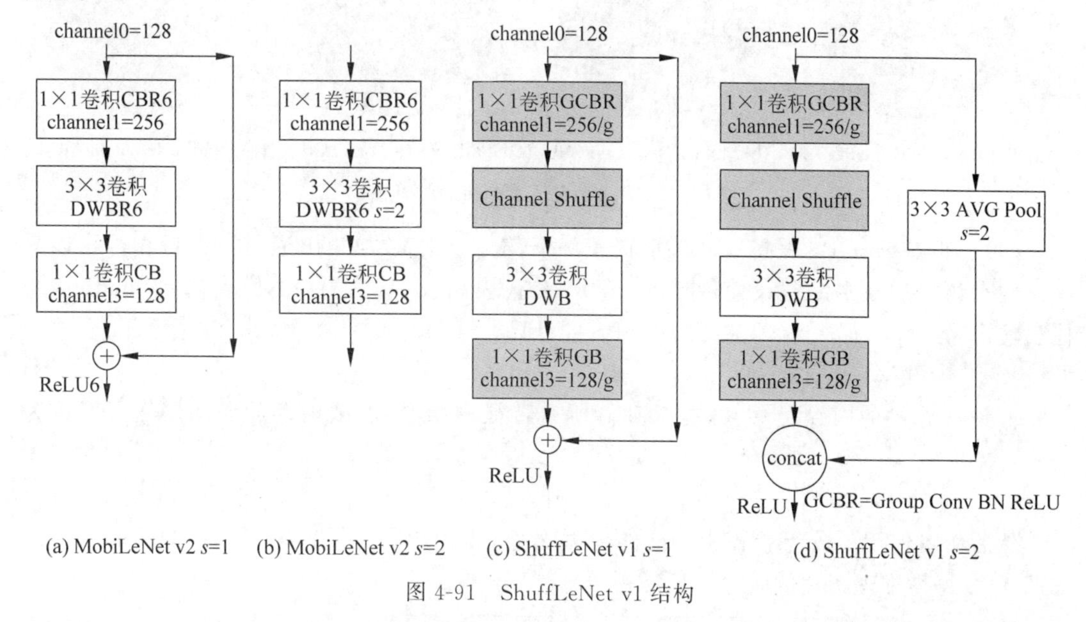

图4-91 ShuffLeNet v1结构

图4-91中ShuffLeNet v1当步长 $s=1$ 时，将1×1卷积变成分组卷积，仍然经过Channel Shuffle操作，加强分组卷积信息的沟通，再经过3×3的DW卷积，然后跟1×1的分组卷积，最后做残差操作；当 $s=2$ 时，右侧分支使用3×3的平均池化，然后做concat操作。以此模块为基础构建的ShuffLeNet v1在分组卷积 $g=8$，ImageNet数据集上top1上的错误率为32.6%，224×224的推理时间为37.8ms，同期MobiLeNet错误率为29.4%，推理时间为110.0ms。

2018年Ningning Ma等在论文 *ShuffLeNet v2: Practical Guidelines for Efficient CNN Architecture Design* 中认为评价一个高效的网络只考虑FLOPS(浮点计算量)是不恰当的，通过实验发现FLOPS与计算速度并不成正比。计算速度主要与计算机内存访问及计算机的平行计算能力有关，如果有相同的FLOPS并且并行化程度高，则计算速度就更快，并且不同的硬件其计算速度也是不相同的，如论文原图4-92中ShuffLeNet v1在GPU中Conv卷积计算运行时间占比为50%，但换到ARM平台是87%，而MobiLeNet v2 GPU是54%，ARM是89%。

图4-92中Conv指卷积、Data指数据处理部分、Elemwise为concat、add、激活函数等功能，Shuffle为信息交换功能。通过观察图中的数据可知，网络耗时最多的是卷积计算，其次是Data数据处理，再次是Elemwise等操作，所以基于此作者提出了4条高效网络的设计准则。

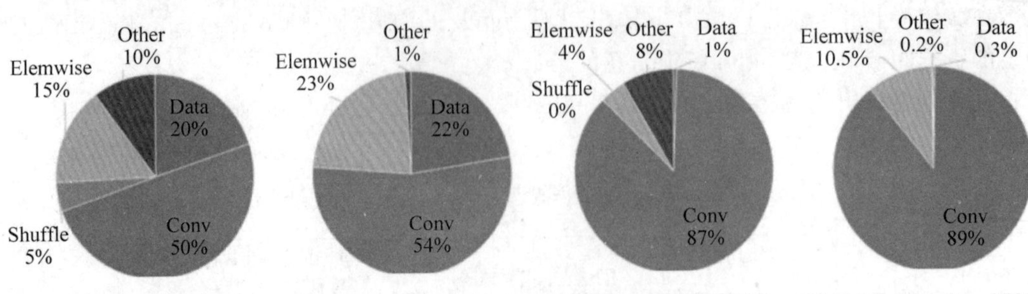

图 4-92 不同平台计算时间消耗占比

(1) G1：在 FLOPS 保持不变的前提条件下，输入通道数与输出通道数保持相等可以最小化内存访问的成本。假设使用 1×1 卷积，h 代表高，w 代表宽，c1 代表输入 channel，c2 代表输出 channel，那么 1×1 卷积 FLOPS 的计算公式为

$$B = hw\mathrm{c1c2} \tag{4-18}$$

假设计算机的内存足够大，能够存储所有特征的参数，那么内存的访问操作可以记为公式：

$$\mathrm{MAC} = hw(\mathrm{c1} + \mathrm{c2}) + \mathrm{c1c2} \tag{4-19}$$

然后作者使用了分组卷积中不同的 g 的分组，也使用了不同的输入输出 channel，得到如图 4-93 每秒推理图片的张数。

		GPU(Batches/sec.)				CPU(Images/sec.)		
g	c for ×1	×1	×2	×4	c for ×1	×1	×2	×4
1	128	2451	1289	437	64	40.0	10.2	2.3
2	180	1725	873	341	90	35.0	9.5	2.2
4	256	1026	644	338	128	32.9	8.7	2.1
8	360	634	445	230	180	27.8	7.5	1.8

图 4-93 不同分组、不同 channel 的推理图片数

从图 4-93 中可观察得知，当 $g=1$ 且输入 c 与输出 c 相同时，GPU 可推理 2451 张图片，当输出是输入的 4 倍时，GPU 只能推理 437 张图片。

将式(4-18)和式(4-19)代入：

$$\mathrm{MAC} = hw(\mathrm{c1} + \mathrm{c2}) + \frac{B}{hw}$$

因为，

$$\frac{\mathrm{c1} + \mathrm{c2}}{2} \geqslant \sqrt{\mathrm{c1c2}}$$

所以，

$$\mathrm{MAC} \geqslant 2hw\sqrt{\mathrm{c1c2}} + \frac{B}{hw} \geqslant 2\sqrt{hwB} + \frac{B}{hw} \tag{4-20}$$

所以，如果 FLOPS 更小，则 MAC 内存占用也更小，当且仅当输入通道与输出通道相等时 MAC 占用最小，则模型推理速度最快。

（2）G2：分组卷积中使用过多的组数会增加内存访问的成本。在G1的公式中，假设使用c2作为1×1的分组卷积，所以$B=hwc1\left(\frac{c2}{g}\right)$，代入得

$$\text{MAC}=hw(c1+c2)+\frac{B}{hw}=hwc1+\frac{Bg}{c1}+\frac{B}{hw} \tag{4-21}$$

当$g=1$时，内存占用最小，当g增大时内存占用会增加。图4-93中当$g=8$时只能推理634张图片，而当$g=1$时可以推理2451张图片。

（3）G3：模块结构太复杂会降低网络的并行程度（如Inception结构），降低推理速度。在作者提供的实验结果图4-94中表明，如果只有一个结构分支，则推理图片为2446张，当结构分支达到4时，则只能推理752张图片。

	GPU(Batches/sec.)			CPU(Images/sec.)		
	c=128	c=256	c=512	c=64	c=128	c=256
1-fragment	2446	1274	434	40.2	10.1	2.3
2-fragment-series	1790	909	336	38.6	10.1	2.2
4-fragment-series	752	745	349	38.4	10.1	2.3
2-fragment-parallel	1537	803	320	33.4	9.1	2.2
4-fragment-parallel	691	572	292	35.0	8.4	2.1

图4-94　模块结构不同分支数的推理图片数

图4-94中1-fragment等指以下结构，如图4-95所示。

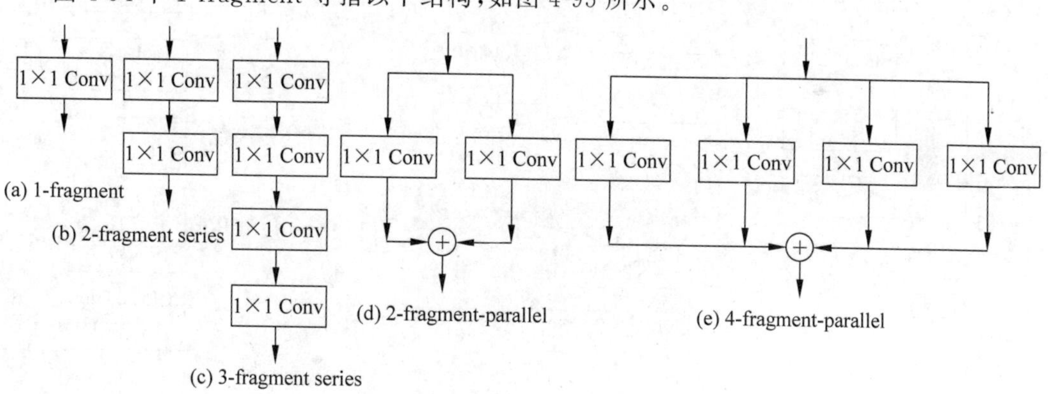

图4-95　模块结构不同分支数的推理图片数

（4）G4：Elementwise操作消耗（如ReLU、Add、Bias等）也不可忽略。作者对使用残差、ReLU函数是否使用进行实验，实验数据如图4-96所示，不使用残差和ReLU可以使推理图片达到2842张。

		GPU(Batches/sec.)			CPU(Images/sec.)		
ReLU	short-cut	c=32	c=64	c=128	c=32	c=64	c=128
yes	yes	2427	2066	1436	56.7	16.9	5.0
yes	no	2647	2256	1735	61.9	18.8	5.2
no	yes	2672	2121	1458	57.3	18.2	5.1
no	no	2842	2376	1782	66.3	20.2	5.4

图4-96　Elementwise对于推理图片的影响

基于以上 4 点作者推荐将 ShuffLeNet v2 的模块结构变更为如图 4-97 所示。

ShuffLeNet v2 中因为 G2 所以去掉了分组卷积，因为 G4 所以将 Add 残差改为 Concat 减少，因为 G3 所以模块分支设计为只有两个，因为 G1 所以将输入通道数与输出通道数设置为相等，其网络详细结构如图 4-98 所示。

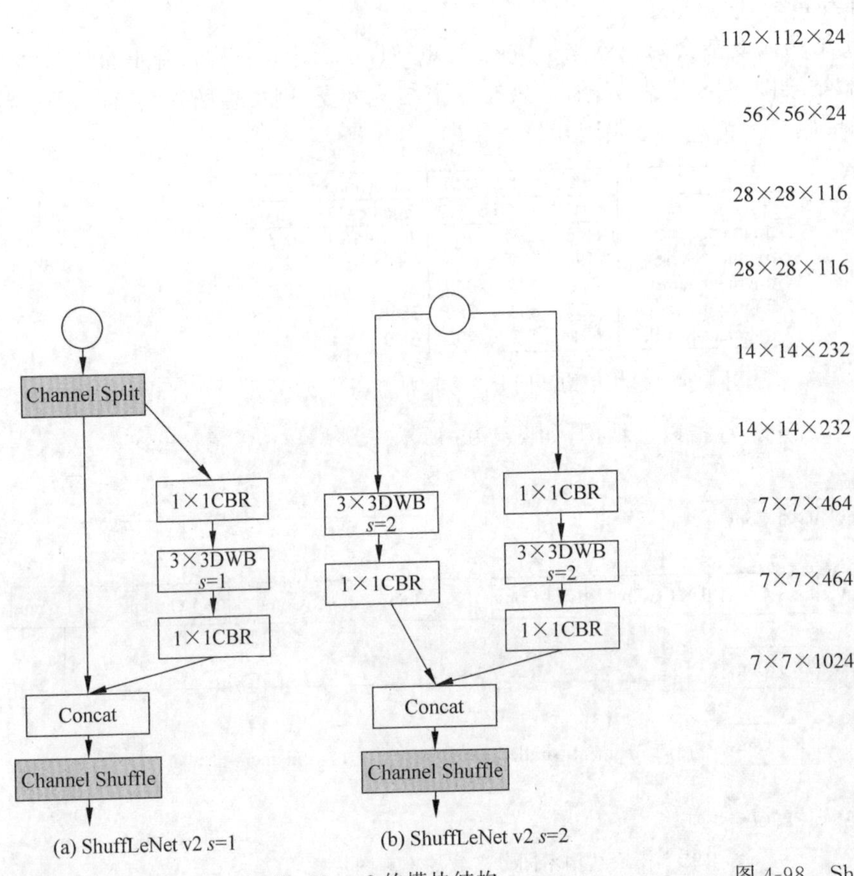

图 4-97　ShuffLeNet v2 的模块结构

图 4-98　ShuffLeNet v2 网络结构

在 ShuffLeNet v2 1.0 版本中模型结果的 Stage 如图 4-96 中的内容，当步长 $s=2$ 时进行下采样，当步长 $s=1$ 时 Shape 不变，经过全局平均池化后进行 FC 全连接以输出预测结果。

作者在论文中给出与其他网络的性能对比，如图 4-99 所示。

ShuffLeNet v2 1.0 版本，Top-1 的错误率是 30.6% 相对 ShuffLeNet v1 1.0 版本 32.6% 有所降低，推理速度 v2 是每秒 341 张图片，而 v1 是 213 张图片。从论文结果来看，ShuffLeNet v2 版本比 v1 准确率更高，推理速度更快。

Model	Complexity (MFLOPs)	Top-1 err. (%)	GPU Speed (Batches/sec.)	ARM Speed (Images/sec.)
ShuffLeNet v2 0.5×(ours)	41	**39.7**	417	**57.0**
0.25 MobiLeNet v1 [13]	41	49.4	**502**	36.4
0.4 MobiLeNet v2 [14](our impl.)*	43	43.4	333	33.2
0.15 MobiLeNet v2 [14](our impl.)*	39	55.1	351	33.6
ShuffLeNet v1 0.5×(g=3)[15]	38	43.2	347	56.8
DenseNet 0.5×[6](our impl.)	42	58.6	366	39.7
Xception 0.5×[12](our impl.)	40	44.9	384	52.9
IGCV2-0.25[27]	46	45.1	183	31.5
ShuffLeNet v2 1×(ours)	146	**30.6**	341	**24.4**
0.5 MobiLeNet v1 [13]	149	36.3	**382**	16.5
0.75 MobiLeNet v2 [14](our impl.)**	145	32.1	235	15.9
0.6 MobiLeNet v2 [14](our impl.)	141	33.3	249	14.9
ShuffLeNet v1 1×(g=3)[15]	140	32.6	213	21.8
DenseNet 1×[6](our impl.)	142	45.2	279	15.8
Xception 1×[12](our impl.)	145	34.1	278	19.5
IGCV2-0.5[27]	156	34.5	132	15.5
IGCV3-D(0.7)[28]	210	31.5	143	11.7

图 4-99　ShuffLeNet v2 性能对比

4.11.2　代码实战

首先对 Shuffle Split 和 Shuffle Channel 的实现进行封装，代码如下：

```
#第 4 章/ClassModleStudy/common.py
class ChannelSplit(layers.Layer):
    """
    当 s = 1 时，一分为两个分支，一个分支进行 1×1、dw 3×3 和 1×1 的卷积操作
    另一个分支保留原始信息，然后两者 concat
    """

    def __init__(self, num_split=2, **kwargs):
        super(ChannelSplit, self).__init__(**kwargs)
        self.num_splits = num_split

    def call(self, inputs, **kwargs):
        b1, b2 = tf.split(inputs, num_or_size_splits=self.num_splits, axis=-1)
        return b1, b2


class ChannelShuffle(layers.Layer):
    """通道混洗。加强信息之间的沟通，其实现的方法为 reshape→transpose→reshape"""

    def __init__(self, shape, groups=2, **kwargs):
        super(ChannelShuffle, self).__init__(**kwargs)
        batch_size, height, width, num_channels = shape
        #当 shuffle 时，需要指定分组的数量。分组的数量一定要是 2 的倍数
        assert num_channels % 2 == 0
        channel_per_group = num_channels //groups
        self.reshape1 = layers.Reshape((height, width, groups, channel_per_group))
        self.reshape2 = layers.Reshape((height, width, num_channels))

    def call(self, inputs, **kwargs):
```

```
        x = self.reshape1(inputs)
        x = tf.transpose(x, perm=[0, 1, 2, 4, 3])
        x = self.reshape2(x)
        return x
```

然后对步长 $s=1$ 和 $s=2$ 的模块进行封装，代码如下：

```
#第 4 章/ClassModleStudy/common.py
class ShuffleBottLeNet(layers.Layer):
    #shuffLeNet v2 中的结构
    def __init__(self, output_channel, stride=1):
        super(ShuffleBottLeNet, self).__init__()
        #输出 channel 扩展倍数
        self.output_channel = output_channel
        self.stride = stride
        self.branch_c = output_channel //2
        #第 1 个算子的分支
        self.split_1 = ChannelSplit()
        self.cbr1_1 = CBR(self.branch_c, 1, 1)
        #在 v2.0 的 block 中 dw 卷积之后，直接线性输出，没有经过 ReLU
        #DWBR_HS->dw conv->bn->relu/hardswich
        #当 is_act=Flase 时，不经过激活函数
        self.dwbr_1 = DWBR_HS(3, 1, is_act=False)
        self.cbr2_1 = CBR(self.branch_c, 1, 1)
        self.concat_1 = layers.Concatenate()
        #第 2 个分支的算子
        self.dwbr1_2 = DWBR_HS(3, 2, is_act=False)
        #CBR->conv->bn->relu
        self.cbr1_2 = CBR(self.branch_c, 1, 1)
        self.cbr21_2 = CBR(self.branch_c, 1, 1)
        self.dwbr2_2 = DWBR_HS(3, 2, is_act=False)
        self.cbr22_2 = CBR(self.branch_c, 1, 1)
        self.concat_2 = layers.Concatenate()

    def call(self, inputs, *args, **kwargs):
        if self.stride == 1:
            x1, x2 = self.split_1(inputs)
            x2 = self.cbr1_1(x2)
            x2 = self.dwbr_1(x2)
            x2 = self.cbr2_1(x2)
            x = self.concat_1([x1, x2])
            return ChannelShuffle(x.shape)(x)
        elif self.stride == 2:
            x1 = self.dwbr1_2(inputs)
            x1 = self.cbr1_2(x1)
            #第 2 个分支
            x2 = self.cbr21_2(inputs)
            x2 = self.dwbr2_2(x2)
            x2 = self.cbr22_2(x2)
            #合并
            x = self.concat_2([x1, x2])
            return ChannelShuffle(x.shape)(x)
```

ShuffleNeck1 实现步长 $s=1$ 的模块封装，ShuffleNeck2 实现步长 $s=2$ 的模块封装，根据 ShuffleBottLeNet()传入的步长，来选择 ShuffleNeck1 或者 ShuffleNeck2 的模块。

然后新建 ShuffLeNetV2()类，根据模型结构编写代码，代码如下：

```
#第 4 章/ClassModleStudy/ShuffLeNet/model.py
from tensorflow.keras import Model, layers
from ClassModleStudy.common import CBR, ShuffleBottLeNet
import tensorflow as tf

class ShuffLeNetV2(Model):
    def __init__(self, num_class=7):
        super(ShuffLeNetV2, self).__init__()
        self.num_class = num_class
        self.conv1 = CBR(24, 3, 2)
        self.maxpool = layers.MaxPool2D(3, 2, padding='same')
        #第 1 个 stage
        self.stage2_1 = ShuffleBottLeNet(output_channel=116, stride=2)
        self.stage2_2 = ShuffleBottLeNet(output_channel=116, stride=1)
        #第 2 个 stage
        self.stage3_1 = ShuffleBottLeNet(output_channel=232, stride=2)
        self.stage3_2 = ShuffleBottLeNet(output_channel=232, stride=1)
        #第 3 个 stage
        self.stage4_1 = ShuffleBottLeNet(output_channel=464, stride=2)
        self.stage4_2 = ShuffleBottLeNet(output_channel=464, stride=1)
        #最后 1 个卷积
        self.conv2 = CBR(1024, 1, 1)
        #全局平均池化
        self.global_pool = layers.GlobalAveragePooling2D()
        #预测输出头
        self.FC = layers.Dense(self.num_class)

    def call(self, inputs, training=None, mask=None):
        #卷积
        x = self.conv1(inputs)
        #池化
        x = self.maxpool(x)
        #stage2-1 232
        x = self.stage2_1(x)
        #stage2-2
        for _ in range(3): x = self.stage2_2(x)
        #stage3-1
        x = self.stage3_1(x)
        #stage3-2
        for _ in range(7): x = self.stage3_2(x)
        #stage4-1
        x = self.stage4_1(x)
        #stage4-2
        for _ in range(3): x = self.stage4_2(x)
```

```
        #卷积
        x = self.conv2(x)
        #全局平均池化
        x = self.global_pool(x)
        #预测输出
        return self.FC(x)
```

最后实现训练,代码如下:

```
#第 4 章/ClassModleStudy/MobiLeNet/train.py
from ClassModleStudy.common import plt_loss, train_FERG_DB
from model import ShuffLeNetV2

if __name__ == "__main__":
   net = ShuffLeNetV2(num_class=7)
   #调用训练代码模板
   his = train_FERG_DB(net, img_height=224, img_width=224, epochs=5, batch_size=
16, lr=0.0001)
   #以图的形式展现训练结果
   plt_loss(his, his.params['epochs'], 'ShuffLeNet v2 卡通表情识别训练结果')
```

训练结果如图 4-100 所示。

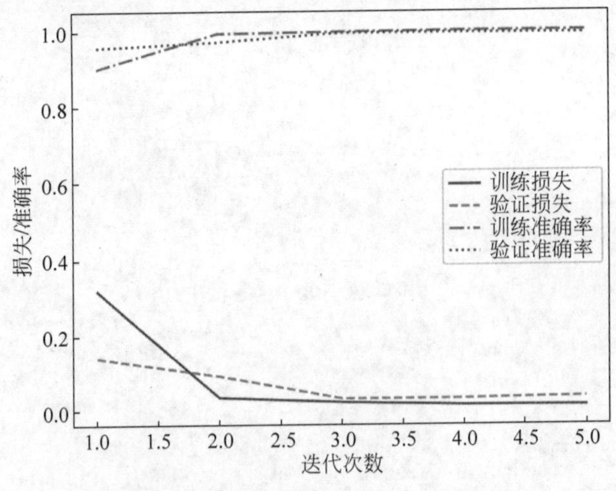

图 4-100　ShuffLeNet v2 卡通表情识别结果

总结

在 ShuffLeNet v2 中提出了 4 条高效网络的设计准则,从而使 ShuffLeNet 推理速度更快,更加适用于使用 CPU 进行推理,对于其他网络模型的设计具有重要参考意义。

练习

完成 ShuffLeNet v2 模型的搭建、训练、推理代码。

4.12 重参数网络 RepVGGNet

4.12.1 模型介绍

2021 年由 Xiaohan Ding 等发表论文 *RepVGG：Making VGG-style ConvNets Great Again*，提出了结构化重参数网络 RepVGG。

根据 4.11 节可知，直筒式卷积结构内存占用最小，推理速度最快，而类似 Inception 的多分支结构在提升网络精度方面有效，基于此作者在训练时使用多分支结构，而在预测时将卷积结构重构成直筒式结构，使训练时能够使用多分支结构提升模型性能，推理时使用直筒式结构提升推理速度，其主要结构如图 4-101 所示。

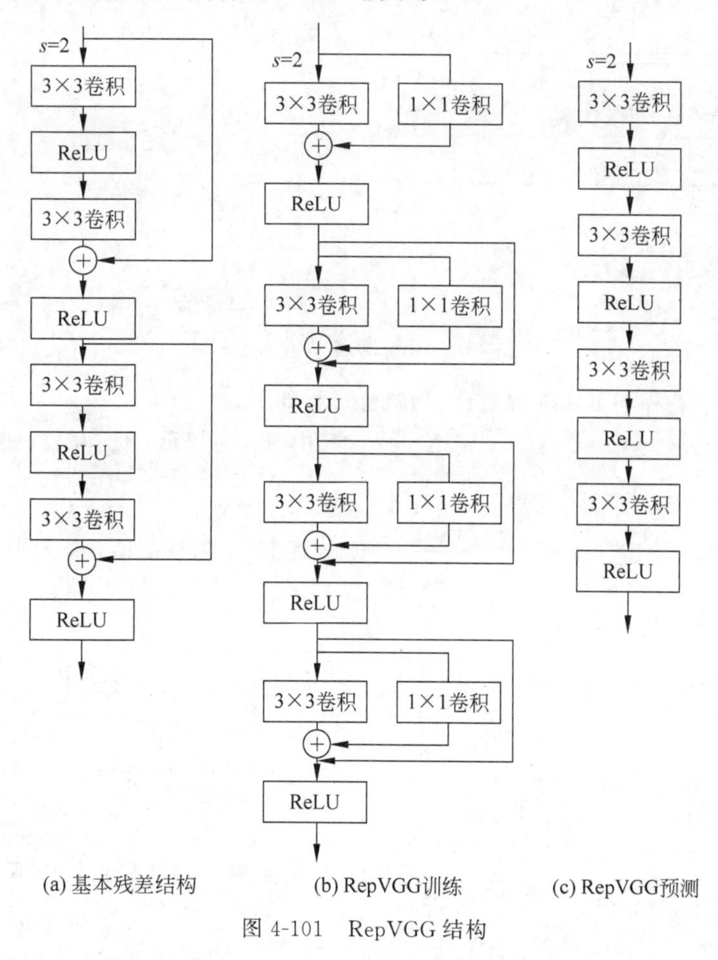

图 4-101　RepVGG 结构

最左侧是基本的残差结构，中间是 RepVGG 训练时的结构，右侧为推理时的结构。不同于残差隔两个 3×3 卷积进行相加，RepVGG 对每层都进行相加，并且残差的边路经过 1 个 1×1 卷积和远跳连接，最后将 3 条路线得到的特征值相加输入下一个模块。从图 4-101

中可知推理时 RepVGG 将 Rep 残差块重组成了 3×3 的卷积。

RepVGG 是怎样将 Rep 残差块重组成 3×3 卷积的呢？如图 4-102 所示。

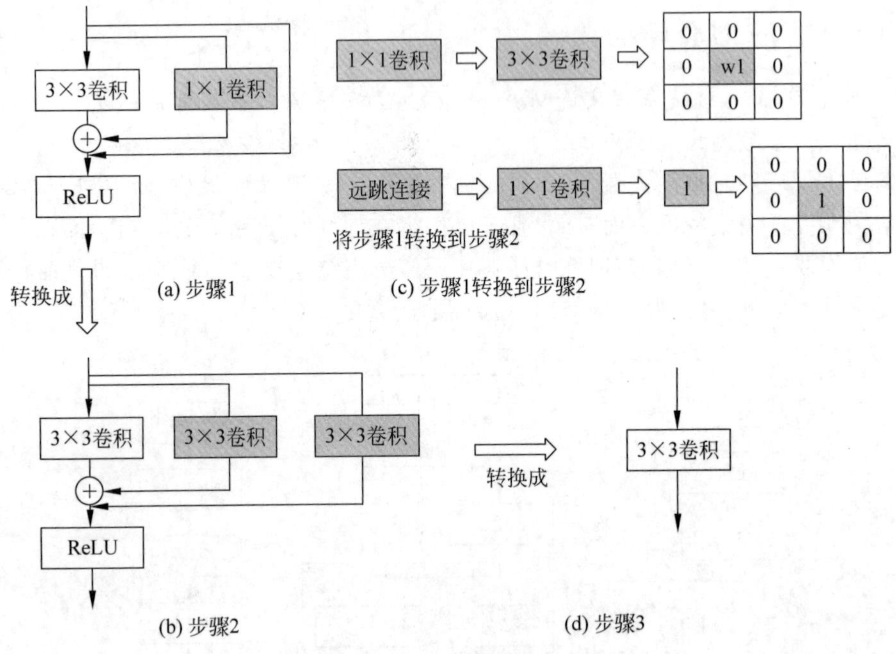

图 4-102 Rep 残差块重组过程

首先对 3×3 卷积与 BN 进行融合，其融合过程如下。

卷积计算的公式：$\text{Conv}(X_i)=W_iX_i+b_i$，其中 W_i 为权重，X_i 为上一层特征的输入，b_i 为偏置项。

BN 的计算公式：$\text{BN}_{\gamma,\beta}(X_i)=\gamma\dfrac{x_i-\mu_\beta}{\sqrt{\sigma_{\beta^2}+\epsilon}}+\beta$，其中 x_i 为特征值，μ_β 为平均值，$\sqrt{\sigma_{\beta^2}+\epsilon}$ 为方差，γ 和 β 为学习的参数。

将卷积的公式代入 BN 公式：$\text{BN}_{\gamma,\beta}(X_i)=\gamma\dfrac{W_iX_i+b_i-\mu_\beta}{\sqrt{\sigma_{\beta^2}+\epsilon}}+\beta$，化简后可得

$$N_{\gamma,\beta}(X_i)=\gamma\frac{W_iX_i}{\sqrt{\sigma_\beta^2+\epsilon}}+\left(\gamma\frac{b_i-\mu_\beta}{\sqrt{\sigma_\beta^2+\epsilon}}+\beta\right)$$

令 $W_{\text{fused}}=\gamma\dfrac{W_i}{\sqrt{\sigma_\beta^2+\epsilon}}$，$B_{\text{fused}}=\left(\dfrac{b_i-\mu_\beta}{\sqrt{\sigma_\beta^2+\epsilon}}+\beta\right)$，最终融合的结果得到以下公式：

$$N_{\gamma,\beta}(X_i)=W_{\text{fused}}\times X_i+B_{\text{fused}} \tag{4-22}$$

然后对 3×3 卷积与 1×1 卷积进行融合。具体的方法是将 1×1 卷积补 0 变成 3×3 卷积，然后在输入特征中进行插 0 操作，使两者计算后的结果相同。具体的计算过程如图 4-103 所示。

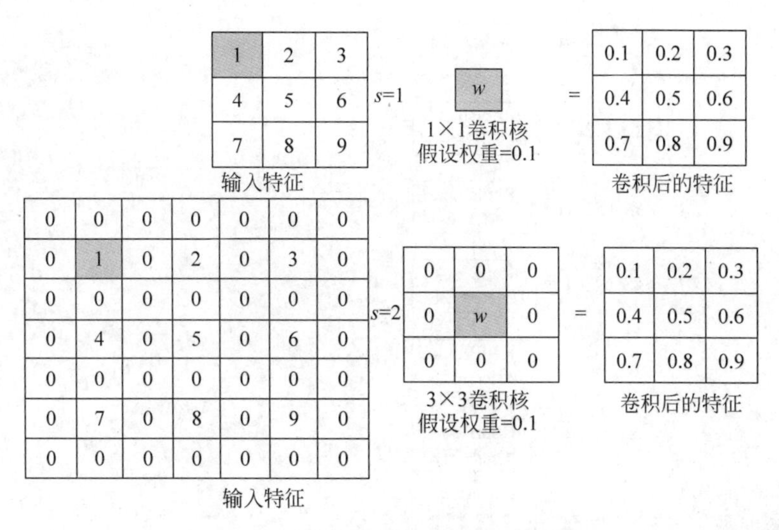

图 4-103　1×1 卷积变成 3×3 卷积

图 4-103 中首先对 1×1 卷积按步长 $s=1$ 进行计算，然后将 1×1 卷积补 0 变成 3×3 卷积的同时将输入特征插 0，并按步长 $s=2$ 进行计算，两者的结果是相同的。

因为 $X_i \times 1 = X_i$，所以可以将远跳连接等效于权重值为 1 的 1×1 卷积，然后将 1×1 卷积按图 4-102 中的方法变成 3×3 卷积，如图 4-104 所示。

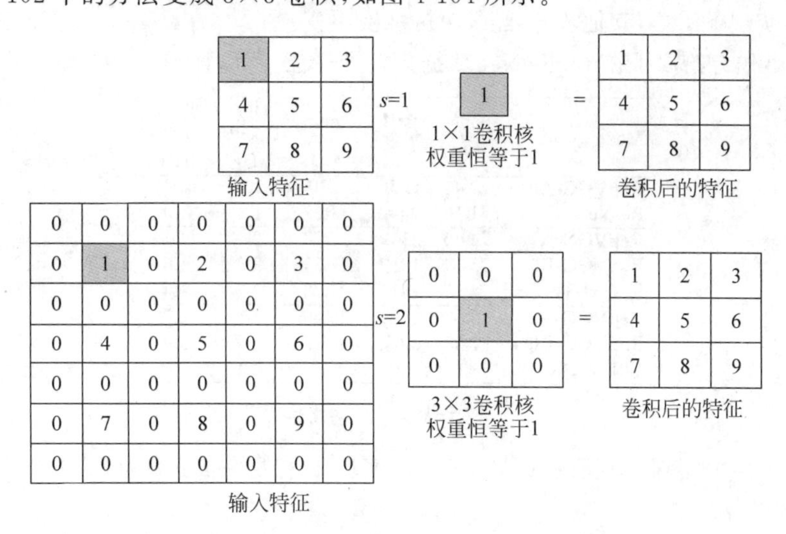

图 4-104　远跳连接等效为权重为 1 的 3×3 卷积

经过这 3 个操作实现了图 4-102 中从步骤 1 到步骤 2 的结构重组，因为此时有 3 个 3×3 的卷积，然后通加相加操作重组为步骤 3 中的 3×3 卷积。

作者以此结构构建了多个尺度的 RepVGG 模型，以 RepVGG-A 为例模型结构如图 4-105 所示。

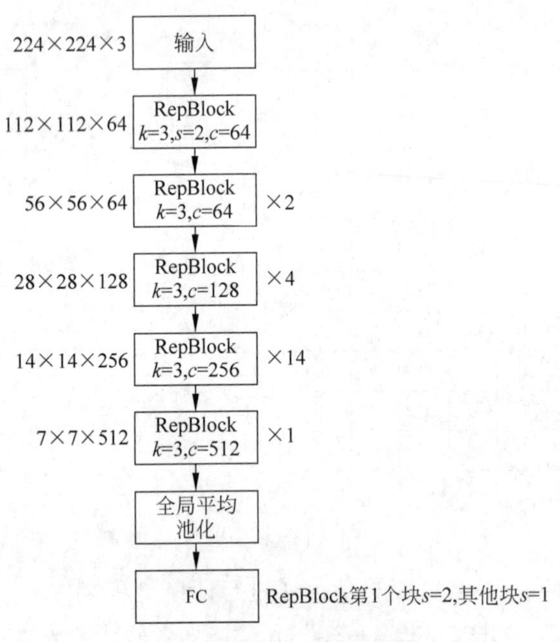

图 4-105　RepVGG 网络模块

2、4、14 和 1 指相应模块重复的次数，需要注意每个模块重复时的第 1 个 RepBlock 块的步长是 2，实现下采样，其他步长是 1，保持维度不变。

在论文中作者给出了在 ImageNet 数据集上的性能指标，如图 4-106 所示。

Model	Top-1 acc	Speed	Params (M)	Theo FLOPs (B)	Wino MULs (B)
RepVGG-A0	72.41	3256	8.30	1.4	0.7
ResNet-18	71.16	2442	11.68	1.8	1.0
RepVGG-A1	74.46	2339	12.78	2.4	1.3
RepVGG-B0	75.14	1817	14.33	3.1	1.6
ResNet-34	74.17	1419	21.78	3.7	1.8
RepVGG-A2	76.48	1322	25.49	5.1	2.7
RepVGG-Blg4	77.58	868	36.12	7.3	3.9
EfficientNet-B0	75.11	829	5.26	0.4	-

图 4-106　RepVGG-A 网络性能指标

从图 4-106 中观察可以发现 RepVGG-A0 模块 Top-1 的准备率为 72.41%，推理速度为每秒 3256 张图片。

4.12.2　代码实战

以 RepVGG 为例以下内容展示了使用 PyTorch 实现网络结构的搭建。PyTorch 与 TensorFlow 在模型搭建方面，最大的区别是在定义类时继承 nn. Module 模块，并且需要指定输入 channel 与输出 channel 的数组（TensorFlow 只需输出 channel）。

首先将 Conv 和 BN 组合成一个模块，代码如下：

```
#第 4 章/ClassModleStudy/PyTorch_RepVGGNet/model.py
import torch.nn as nn
import torch
import copy

def conv_bn(in_channels, out_channels, kernel_size, stride, padding, groups=1):
    #将 Conv 和 BN 组成一组算子
    result = nn.Sequential()
    #
    result.add_module('conv', nn.Conv2d(
        in_channels=in_channels, out_channels=out_channels,
        kernel_size=kernel_size, stride=stride,
        padding=padding, groups=groups,
        bias=False))
    result.add_module('bn', nn.BatchNorm2d(num_features=out_channels))
    return result
```

nn. Sequential()为一个容器对象,result. add_module 向该容器中添加相关网络层,以便组成一组新的算子。in_channels 为上一层的输出通道,out_channels 为当前层的输出通道,kernel_size 为卷积核,stride 为步长,padding 在卷积时如果输入特征图不够,则需要补 0。

根据图 4-101 中的结构,实现 RepVGGBlock 训练模块,代码如下:

```
#第 4 章/ClassModleStudy/PyTorch_RepVGGNet/model.py

class RepVGGBlock(nn.Module):

    def __init__(self, in_channels, out_channels, kernel_size,
                 stride=1, padding=0, dilation=1, groups=1,
                 padding_mode='zeros', deploy=False):
        #deploy = True 进行重参数
        super(RepVGGBlock, self).__init__()
        self.deploy = deploy
        self.groups = groups
        self.in_channels = in_channels
        #卷积时补 0
        padding_11 = padding - kernel_size //2
        #激活函数
        self.nonlinearity = nn.ReLU()
        #表示什么也不做的模块
        self.identity = nn.Identity()

        if deploy:
            #重参数模块为 3×3 的卷积,推理时用
            self.rbr_reparam = nn.Conv2d(
                in_channels=in_channels,
                out_channels=out_channels,
                kernel_size=kernel_size,
                stride=stride,
                padding=padding,
                dilation=dilation,
                groups=groups,
```

```
                bias=True,
                padding_mode=padding_mode
            )

        else:
            #当步长=1 且输入通道与输出通道相等时进行 BN
            self.rbr_identity = nn.BatchNorm2d(num_features=in_channels) if
out_channels == in_channels and stride == 1 else None
            #3×3 Conv->BN
            self.rbr_dense = conv_bn(
                in_channels=in_channels,
                out_channels=out_channels,
                kernel_size=kernel_size,
                stride=stride,
                padding=padding,
                groups=groups
            )
            #1×1 Conv->BN
            self.rbr_1x1 = conv_bn(
                in_channels=in_channels,
                out_channels=out_channels,
                kernel_size=1,
                stride=stride,
                padding=padding_11,
                groups=groups
            )

    def forward(self, inputs):
        #如果是重参数,则进行 3×3 的 Conv
        if hasattr(self, 'rbr_reparam'):
            return self.nonlinearity(
                    self.identity(self.rbr_reparam(inputs)))
        #如果不是重参数(即训练时),则 1 个分支进行 3×3 Conv->BN
        id_out = 0 if self.rbr_identity is None else self.rbr_identity(inputs)
        #另 1 个分支,3×3 Conv->BN +1×1 Conv->BN
        return self.nonlinearity(self.identity(self.rbr_dense(inputs) + self.
rbr_1x1(inputs) + id_out))
```

当 deploy=False 训练时，首先经过 1 个 3×3 的卷积，即 self.rbr_dense(inputs)，另 1 个分支进行 1×1 卷积，即 self.rbr_1x1(inputs)，然后如果输入 in_channels==输出 out_channels 并且 stride=1，则需要增加 BN 层，即 id_out，最后将 3 个分支进行非线性融合输出，即 self.nonlinearity()。

当 deploy=True 时，经过重参数化操作后转换成 3×3 卷积，即 self.rbr_reparam(inputs)，然后进行 self.nonlinearity()非线性输出，重参数的转换见随书源码中的 repvgg_model_convert()函数。

然后根据图 4-105 中的参数对 RepVGGBlock 进行组合，实现 RepVGG-A 的模型，代码如下：

```
#第 4 章/ClassModleStudy/PyTorch_RepVGGNet/model.py

class RepVGG(nn.Module):

   def __init__(
         self,
         num_blocks,#[2, 4, 14, 1]
         num_classes=1000, #输出类别数
         width_multiplier=None,
         deploy=False
   ):
      super(RepVGG, self).__init__()
            #传入重复的 RepBlock 模块要满足 4 个,即[0.75, 0.75, 0.75, 2.5]
      assert len(width_multiplier) == 4
            #是否启用重参数,训练时为 False
      self.deploy = deploy
      #通道数要为 64 的倍数,最终通道数为 64×[0.75]
      self.in_planes = min(64, int(64 *width_multiplier[0]))
            #初始 RepVGGBlock 对象。输入 in_channel=3,表明输入三通道图像
      self.stage0 = RepVGGBlock(
         in_channels=3,
         out_channels=self.in_planes,#输出应该是 64*0.75 = 48
         kernel_size=3,
         stride=2,
         padding=1,
         deploy=self.deploy #训练时为 False
      )
      #输出 channel,64 × 0.75 = 48,重复 2 次
      self.stage1 = self._make_stage(int(64 *width_multiplier[0]), num_blocks[0],
stride=2)
      #输出 channel,128 × 0.75 = 96,重复 4 次
      self.stage2 = self._make_stage(int(128 *width_multiplier[1]), num_blocks[1],
stride=2)
      #输出 channel,256 × 0.75 = 192,重复 14 次
      self.stage3 = self._make_stage(int(256 *width_multiplier[2]), num_blocks[2],
stride=2)
      #输出 channel,512 × 2.5 = 1280,重复 1 次
      self.stage4 = self._make_stage(int(512 *width_multiplier[3]), num_blocks[3],
stride=2)
      #全局平均池化
      self.gap = nn.AdaptiveAvgPool2d(output_size=1)
      #全连接层,输入 channel=1280,输出 num_class 个类别
      self.linear = nn.Linear(int(512 *width_multiplier[3]), num_classes)

   def _make_stage(self, planes, num_blocks, stride):
   #planes 输出通道数、num_blocks 重复次数、stride 步长
   #每个模块重复的次数,如果 num_blocks=4
   #stides = [2]+[1] ×3= [2, 1, 1, 1],即第 1 个块实现下采样,则其他模块的步长=1
      strides = [stride] + [1] * (num_blocks - 1)
      blocks = []
      for stride in strides:
         blocks.append(RepVGGBlock(
```

```
                in_channels=self.in_planes,
                out_channels=planes,
                kernel_size=3,
                stride=stride,
                padding=1,
                deploy=self.deploy)
            )
            #输入通道数为上一层的输出通道数
            self.in_planes = planes
            #组成算子模块
        return nn.Sequential(*blocks)

    def forward(self, x):
        #前向传播
        out = self.stage0(x)
        out = self.stage1(out)
        out = self.stage2(out)
        out = self.stage3(out)
        out = self.stage4(out)
        out = self.gap(out)
        out = out.view(out.size(0), -1)
        out = self.linear(out)
        return out
```

在__init__中 num_blocks=[2,4,14,1],代表每个 RepVGGBlock 重复执行的次数,第1块时其 stride=2 进行下采样,其他 RepVGGBlock 的 stride=1,所以 self.stage0 只执行1次,self.stage1 至 self.stage4 按 num_blocks 次数执行。self._make_stage()中当 num_blocks=4 时,strides=[stride]+[1]×(num_blocks-1),即[2]+[1]×3=[2,1,1,1],表示第1个块实现下采样,其他模块的步长=1,特征图的尺寸保持不变。在 forward(self,x)方法中实现了模型的前向传播。

最后实现训练,代码如下:

```
#第 4 章/ClassModleStudy/PyTorch_RepVGGNet/train_repvgg.py
from model import RepVGG
from train import pytorch_train_FERG_DB
from utils import plt_loss

if __name__ == "__main__":
    model = RepVGG(
        num_blocks=[2, 4, 14, 1],
        num_classes=7,
        width_multiplier=[0.75, 0.75, 0.75, 2.5],
        override_groups_map=None,
        deploy=False
    )
    #调用训练代码模板
    his = pytorch_train_FERG_DB(model, img_height=224, img_width=224, epochs=5,
batch_size=16, lr=0.0001)
    #以图的形式展现训练结果
plt_loss(his['result'], his['epochs'], 'RepVGG-A0 卡通表情识别训练结果')
```

PyTorch 的训练代码封装在 pytorch_train_FERG_DB()中,其实现过程与 TensorFlow 基本相同,训练结果如图 4-107 所示。

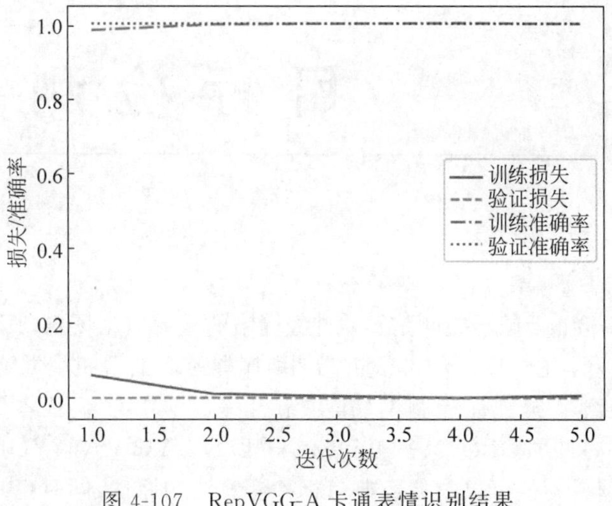

图 4-107 RepVGG-A 卡通表情识别结果

注意:RepVGGNet 使用 PyTorch 框架进行了模型的搭建、训练。

总结

RepVGG 在训练时使用多分支结构,而在预测时将卷积结构重构成直筒式结构,使在训练时能够使用多分支结构提升模型性能,在推理时使用直筒式结构提升推理速度。

练习

完成 RepVGGNet-A 模型的搭建、训练、推理代码。

第 5 章 目标检测

目标检测、物体检测及缺陷检测是卷积神经网络的重要应用场景,其主要任务是找出图像或者视频中所感兴趣的物体(目标)的位置并指明所属类别,在计算机视觉领域中占有重要地位。

本章将详细介绍经典目标检测网络,并结合深度学习框架从零重现 R-CNN、Faster R-CNN、SSD、YOLOv1、YOLOv2、YOLOv3、YOLOv4、YOLOv5、YOLOv7 模型代码。通过阅读本章内容,读者不仅可以掌握经典目标检测算法的原理,同时也具备目标检测算法代码的实现及调优的能力。

5.1 标签处理及代码

相对于图像分类任务,目标检测任务需要将真实感兴趣的目标标识出来,然后交给神经网络训练学习。标签处理包括数据标注、标签数据格式转换、标签数据读取等过程。

数据标注可使用开源工具 labelimg,安装方法为 pip install labelimg。安装成功后,在命令行中输入 labelimg 即可打开该软件。选择 Open Dir 配置训练图片的文件夹。选择 Change Save Dir 配置标注目标的保存目录,如图 5-1 所示。

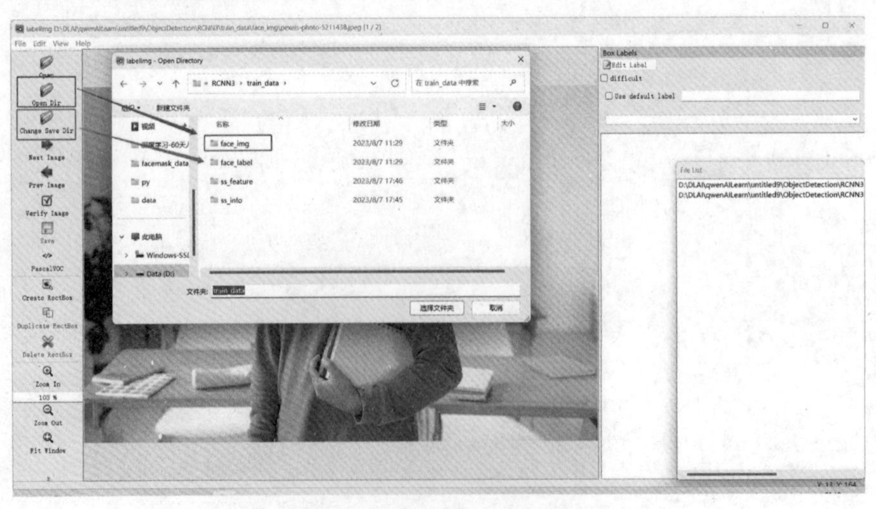

图 5-1　labelimg 选择训练文件夹和标注目标文件夹

加载图片后，选择 Create RectBox 同时默认保留标注文件格式为 PascalVoc 格式，拖动鼠标在目标区域生成一个矩形框后将矩形框的分类填写为 face_mask，如图 5-2 所示。

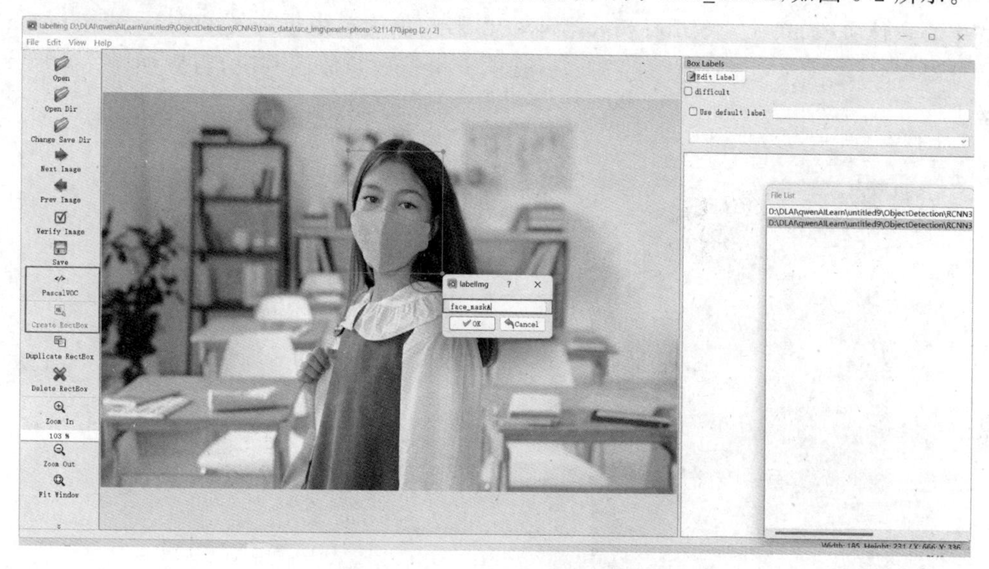

图 5-2 labelimg 数据标注方法

标注完成后会在图 5-1 设置的文件夹中生成与图片名称相同的 XML 文件，其< object >标签表示有一个目标区域，< xmin >、< ymin >为目标图像左上角坐标点的位置，< xmax >、< ymax >为目标图像右下角坐标点的位置，< name >为当前目标图像的类别，如图 5-3 所示。

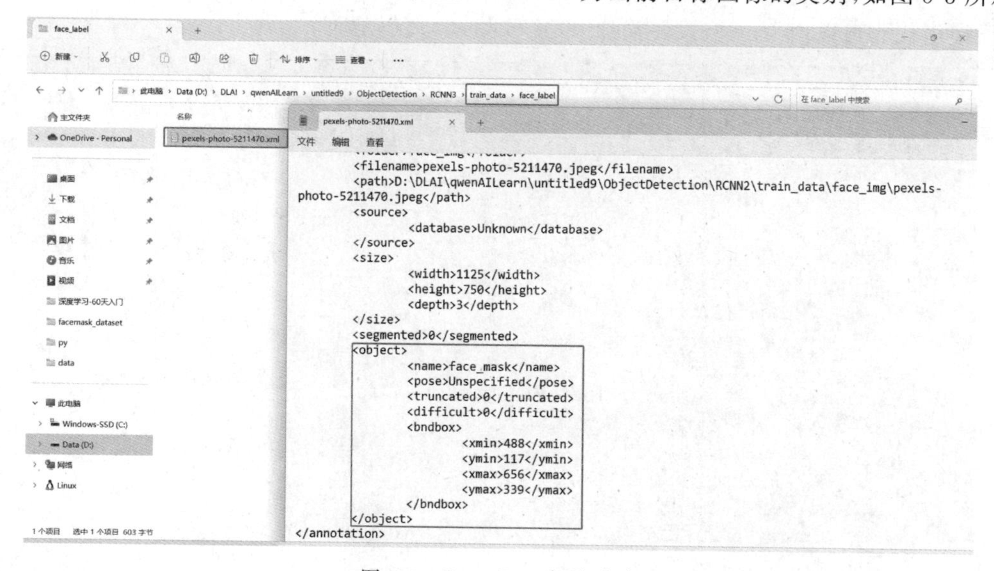

图 5-3 PascalVoc 标注内容

如果有多个感兴趣的目标和图像，就需要逐张图片对感兴趣的区域进行标注，这个过程是比较耗费时间和体力的。

如果选择 YOLO 格式，则会在标注保存文件夹中生成一个 TXT 文件，其中第 1 位为感兴趣的物体类别的下标，假设类别为[face_mask，face]则此时为 0，第 2 位和第 3 位为矩形框的中心点位置(center_x，center_y)，第 4 位和第 5 位为矩形框的宽和高(width，height)，然后针对(center_x，width)/原图的宽、(center_y，height)/原图的高，得到如图 5-4 所示的小数。

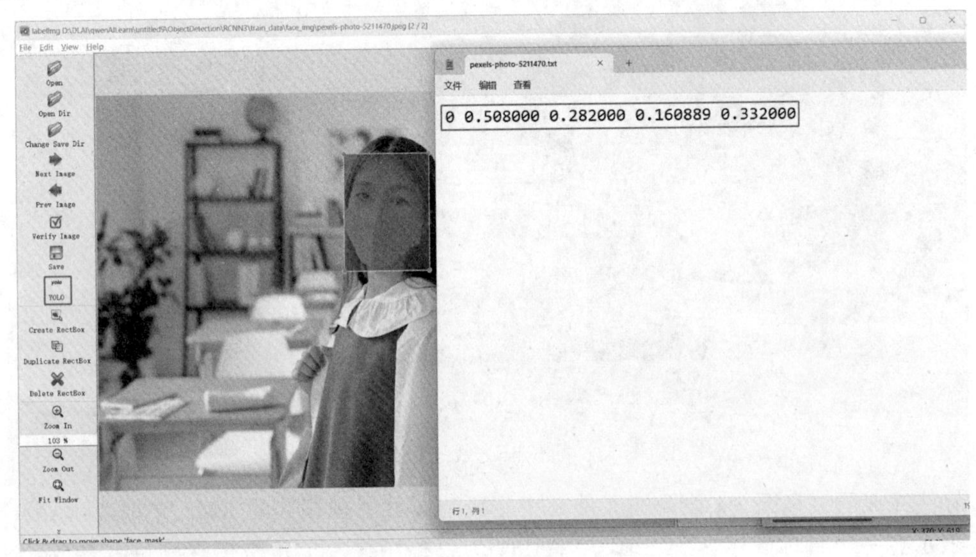

图 5-4 YOLO 标注内容

使用代码可以由 Voc 格式转换成 YOLO 格式，代码如下：

```
#第 5 章/ObjectDetection/LabelConversion/voc2yolo.py
from glob import glob
import os
import xml.etree.cElementTree as ET
import cv2
import numpy as np

def draw_box(image, boxes):
    #在原图中绘出所有的 boxes 和 label
    for i, rect in enumerate(boxes):
        #获得 box
        index, x1, y1, x2, y2 = rect[0:5].astype("int")
        #绘矩形框，需要左上、右下坐标
        cv2.rectangle(image, (x1, y1), (x2, y2), (0, 255, 0), 1, cv2.LINE_AA)
        #绘矩形框的类别
        cv2.putText(image, f"{index}", (x1 - 10, y1 - 10),
                    cv2.FONT_HERSHEY_SIMPLEX, 0.75, (0, 0, 255), 2)
    cv2.imshow('show box', image)
    cv2.waitKey(0)

def read_annotations(xml_path, all_name=[], image=None):
    #使用 etree 读取 XML 文件
```

```
    et = ET.parse(xml_path)
    element = et.getroot()
    #查找 XML 文件中所有的 object 目标区域
    element_objs = element.findall('object')
    #用来存储矩形框的区域
    results = []
    #遍历所有的 box 区域的 object 标签
    for element_obj in element_objs:
        #获得类名称
        class_name = element_obj.find('name').text
        #如果 XML 文件中类的名称与指定类的名称相同,则获取列表中类的索引值
        for i, n in enumerate(all_name):
            if n == class_name:
                index = i
                break
        #从 bndbox 中获取矩形框的左上、右下坐标并转换成 int 类型
        obj_bbox = element_obj.find('bndbox')
        #lambda 表达式,根据坐标名称获取对应内容,并转换成 float 型
        coord_xy = lambda coord_name: float(obj_bbox.find(coord_name).text)
        #组成[xmin,ymin,xmax,ymax,label index]的数组
        label = [index] + [coord_xy(name) for name in ['xmin', 'ymin', 'xmax', 'ymax']]
        results.append(label)
    #如果 image 不为 None,则可以将获取出来的 label 还原到图像中
    boxes = np.array(results)
    if image is not None:
        draw_box(image, boxes)
    return boxes

def xyxy2cxcywh(boxes, image_shape):
    #由 xmin、ymin、xmax、ymax 分别转换成 cx、cy、w、h
    #width,height = (xmax,ymax)-(xmin,ymin) 得到矩形框的 width 和 height
    boxes[..., 3:5] = boxes[..., 3:5] - boxes[..., 1:3]
    #(center_x,center_y) = (xmin,ymin + wh/2) 中心点在图像中的位置
    boxes[..., 1:3] = boxes[..., 1:3] + boxes[..., 3:5] / 2
    #分别除以 width 和 height 得到归一化后的值
    boxes[..., 1:5] = boxes[..., 1:5] / image_shape
    return boxes

if __name__ == "__main__":
    #根目录
    root = "../train_data"
    #训练图片文件夹
    jpg_path = f"{root}/face_img"
    #标注过的文件夹
    xml_path = f"{root}/face_label"
    #期望转换为 YOLO 格式的文件夹
    save_path = f"{root}/yolo_label"
    #如果指定的保存文件夹不存在,则创建 1 个
    if not os.path.exists(save_path): os.mkdir(save_path)
    #获取指定文件夹中所有以.jpeg 为后缀名的文件
    images_file = glob(os.path.join(jpg_path, '*.jpeg'))
    #遍历所有的图片
```

```
    for image_path in images_file:
        #读取图片以获取图片的 width 和 height
        #不从 XML 文件读取,这是因为有时与实际图片不一致
        image = cv2.imread(image_path)
        image_shape = image.shape[0:2][::-1]
        #组成 width,height,width,height 的数组,方便后续计算
        image_shape = np.array(image_shape + image_shape)
        #根据图片名称找到对应的 Voc XML 文件
        jpg_name = os.path.split(image_path)
        name = jpg_name[-1].split('.')[0]
        name_xml = f"{xml_path}/{name}.xml"
        #获取原始标注类别和位置[index,xmin,ymin,xmax,ymax]
        boxes = read_annotations(name_xml, all_name=['face_mask', 'face'], image=
image)
        #实现从 Voc 格式转换成 YOLO 格式
        boxes = xyxy2cxcywh(boxes, image_shape)
        #设置保存 YOLO 格式的 TXT 文件名
        save_txt = f"{save_path}/{name}.txt"
        #将 box 信息写入 save_path 中
        np.savetxt(save_txt, boxes, fmt='%.4f')
```

代码中 read_annotations()函数实现读取 XML 文件得到[index,xmin,ymin,xmax,ymax]的数组内容,并在 xyxy2cxcywh()函数中实现转换成[index,center_x,center_y,width,height]的数组。draw_box()函数主要将标注 box 还原到图像中,以便进行抽查标注信息是否正确。np.savetxt()实现整个数组保存,由于保存为 float 类型,TXT 文件中第 1 位 index 会显示为 4 位小数,这个 index 在喂入训练数据时需要转换为 int。

也可以由 YOLO 格式转换成 Voc 格式,代码如下:

```
#第 5 章/ObjectDetection/LabelConversion/YOLOv2voc.py
from glob import glob
import os
import cv2
import numpy as np

def cxywh2xyxy(boxes, image_shape):
    #实现从矩形中心点坐标转换成 xmin,ymin,xmax,ymax
    #从小数还原为图像中的位置
    box = boxes[..., 1:5].copy() *image_shape
    #(center_x,center_y) - (width,height/2)就是左上角
    xminymin = box[..., 0:2] - box[..., 2:4] / 2
    #(center_x,center_y) + (width,height/2)就是右下角
    xmaxymax = box[..., 0:2] + box[..., 2:4] / 2
    #合成[xmin,ymin,xmax,ymax]
    boxes[..., 1:5] = np.concatenate([xminymin, xmaxymax], axis=-1)
    #转换为 int 类型
    return boxes.astype('int32')

def generate_voc(boxes, all_name=[]):
    #根据 boxes 信息生成 Voc 格式的文件
    voc_str = ""
```

```
    for i, rect in enumerate(boxes.astype("int32")):
        object_str = f"""
            <object>
                <name>{all_name[0]}</name>
                <pose>Unspecified</pose>
                <truncated>0</truncated>
                <difficult>0</difficult>
                <bndbox>
                    <xmin>{rect[1]}</xmin>
                    <ymin>{rect[2]}</ymin>
                    <xmax>{rect[3]}</xmax>
                    <ymax>{rect[4]}</ymax>
                </bndbox>
            </object>
        """
        voc_str += object_str
    return f"""<annotation>{voc_str}</annotation>"""

if __name__ == "__main__":
    #根目录
    root = "../train_data"
    #训练图片文件夹
    jpg_path = f"{root}/face_img"
    #YOLO 格式的文件夹
    yolo_path = f"{root}/yolo_label"
    #期望保存为 Voc 格式的文件夹
    save_path = f"{root}/voc_label"
    #如果指定的文件夹不存在,则创建 1 个
    if not os.path.exists(save_path): os.mkdir(save_path)
    #获取指定文件夹中所有以.jpeg 为后缀名的文件
    images_file = glob(os.path.join(jpg_path, '*.jpeg'))
    #遍历所有的图片
    for image_path in images_file:
        #读取图片以获取图片的 width 和 height
        #不从 XML 文件读取,这是因为有时与实际图片不一致
        image = cv2.imread(image_path)
        image_shape = image.shape[0:2][::-1]
        #组成 width,height,width,height 的数组,方便后续计算
        image_shape = np.array(image_shape + image_shape)
        #根据图片名称找到对应的 Voc XML 文件
        jpg_name = os.path.split(image_path)
        name = jpg_name[-1].split('.')[0]
        name_txt = f"{yolo_path}/{name}.txt"
        #读取 YOLO 格式内容
        boxes = np.loadtxt(name_txt)
        #转换成[index,xmin,ymin,xmax,ymax]
        boxes_xyxy = cxywh2xyxy(boxes, image_shape)
        #创建 XML 文件
        save_xml = f"{save_path}/{name}.xml"
                #调用 generate_voc()实现 Voc 格式的写入
        with open(save_xml, 'w', encoding='utf-8') as f:
            f.write(generate_voc(boxes_xyxy, all_name=['face_mask', 'face']))
        #将 YOLO 格式还原到图像中
        draw_box(image, boxes_xyxy)
```

cxywh2xyxy()实现从 YOLO 格式转换为 xmin，ymin，xmax，ymax 的格式。同时 generate_voc()实现 XML 文件的写入，np.loadtxt()实现 TXT 文件的数组读入。

标注文件内容格式的转换为高频操作，本书的代码可以在工作中直接使用。

总结

目标检测标注工具 labeling 可标注 YOLO 格式、Voc 格式，并且实现了 YOLO 格式与 Voc 格式的互相转换。

练习

寻找一份数据集进行标注，并调试实现 YOLO 格式与 Voc 格式数据的互相转换。

5.2 开山之作 R-CNN

5.2.1 模型介绍

R-CNN 由 Ross Girshick 等在 2014 年发表的论文 *Rich Feature Hierarchies for Accurate Object Detection and Semantic Segmentation* 中首次使用深度学习进行目标检测，其网络结构和实现过程对后续模型提供了重要参考。

其网络的主要过程如图 5-5 所示，分为训练过程和预测过程。

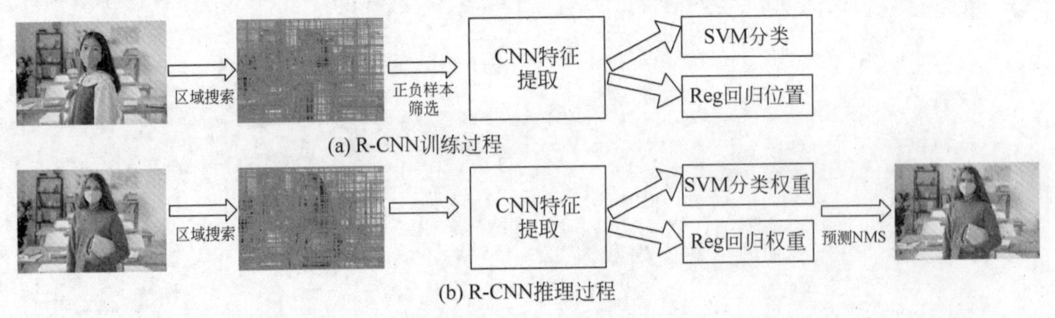

图 5-5 R-CNN 模型过程

(1) 训练时需要输入训练图片和标注 BOX，通过选择区域搜索算法生成多个目标框(取前 2000 个区域)，当区域选择框与标注 BOX 重叠面积较大时认为是有目标的框(正样本)，而重叠面积较小或无重叠时认为是无目标的框(称为负样本)。将正样本图片和负样本图片输入 CNN 中提取特征，例如使用 VGG 中的 FC 层得到 4096 维特征，然后使用 4096 维特征，通过支持向量机(SVM)进行分类训练。由于区域选择框与标注 BOX 之间有一定的偏移，所以这里使用回归算法对这个偏移进行了回归训练。

(2) 预测推理时需要输入未知图片，同样通过区域搜索算法选取 2000 个区域选择框，使用 CNN 提取特征，然后将区域选择框中的图像信息输入 SVM 权重得到预测分类，然后通过阈值设置过滤掉概率较低的区域选择框。将分类概率较高的框输入 Reg 回归权重得到区域选择框与预测框的偏移，通过区域选择框和预测偏移得到多个预测框。多个预测框之间有可能出现重叠面积较大的框，这会被认为是同一个分类。使用非极大值抑制(NMS)对其进行剔除，得到最后的预测框，将预测框和预测分类信息绘制到原图中，就得到了目标检测的结果。

整个过程步骤较多且涉及较多新概念，详细过程和相关代码将在后续几节中逐步展开。

5.2.2 代码实战选择区域搜索

选择区域搜索(Selective Search)算法，首先输入一张图片，通过图像分割的方法获得很多小的区域，然后对这些小的区域采用相似度计算的方法，将相似区域不断地进行合并，一直到无法合并为止。图像相似度计算包括颜色相似度、纹理相似度、尺度相似度、填充相似度等方法，具体可参考原论文。

选择区域搜索算法的实现可调用 OpenCV 中的 createSelectiveSearchSegmentation()对象实现，代码如下：

```
#第 5 章/ObjectDetection/R-CNN/selectivesearch.py
import cv2
import numpy as np

def start_search(image):
    #开启优化设置
    cv2.setUseOptimized(True)
    #创建选择优化搜索对象
    #pip install opencv-contrib-python
    objSrh = cv2.ximgproc.
segmentation.createSelectiveSearchSegmentation()
    #加载图片
    objSrh.setBaseImage(image)
    #快速搜索
    objSrh.switchToSelectiveSearchFast()
    #提取区域选择框
    objSrhRects = objSrh.process()
    #区域选择框由 x、y、w、h 分别转换成 xmin、ymin、xmax、ymax
    rect = objSrhRects.copy()
objSrhRects[..., 2:4] = rect[..., 0:2] + rect[..., 2:4]
#在原图绘出区域框
    #u.draw_box(image,objSrhRects)
    return objSrhRects[:2000, ...]
```

函数 start_search(image)用于实现选择区域搜索，但是 objSrhRects 提取的是左上角坐标和高宽，通过 objSrhRects[...,2:4]=rect[...,0:2]+rect[...,2:4]转换成左上、右下角坐标。

5.2.3 代码实战正负样本选择

区域选择框与标注 BOX 重叠面积占比的计算又称为交并比(Intersection over Union，IOU)，其计算公式为

$$\text{IOU}=\frac{\text{area}_{\text{intersection}}}{\text{area}_{\text{box1}}+\text{area}_{\text{box2}}-\text{area}_{\text{intersection}}} \tag{5-1}$$

当交集区域最大时，GTBOX 与 SSBOX 会重叠，此时 IOU=1；当交集区域为 0 时，IOU=0；求交集则为 $\text{area}_{\text{intersection}}=\text{abs}(\text{GTBOX}_{\text{maxx}}-\text{SSBOX}_{\text{minx}})\times\text{abs}(\text{GTBOX}_{\text{maxy}}-\text{SSBOX}_{\text{miny}})$，加绝对值 abs()是由于 GTBOX 有可能在 SSBOX 的右侧，如图 5-6 所示。

图 5-6　IOU 计算

具体的代码如下：

```
#第 5 章/ObjectDetection/R-CNN/utils.py
import numpy as np

def get_iou(boxes1, boxes2):
    #比较哪个 box 在左边，哪个 box 在右边，此时不再需要 abs
    left = np.maximum(boxes1[..., :2], boxes2[..., :2])
    right = np.minimum(boxes1[..., 2:4], boxes2[..., 2:4])
    #计算交集的 wh
    intersection = np.maximum(0.0, right - left)
    #计算交集的面积
    area_inter = intersection[..., 0] * intersection[..., 1]
    #计算每个 box 的面积
    area_box1 = (boxes1[..., 2] - boxes1[..., 0]) * (boxes1[..., 3] - boxes1[..., 1])
    area_box2 = (boxes2[..., 2] - boxes2[..., 0]) * (boxes2[..., 3] - boxes2[..., 1])
    #计算并集的面积，1e-10 保证分母为 0
    union_square = np.maximum(area_box1 + area_box2 - area_inter, 1e-10)
    #计算交并比。如果比 0 小，则为 0；如果比 1 大，则为 1
    score = np.clip(area_inter / union_square, 0.0, 1.0)
    return score
```

当 IOU>0.5 时，SSBOX 选取为正样本；当 IOU<0.3 时，SSBOX 选取为负样本，代码如下：

```
#第 5 章/ObjectDetection/R-CNN/selectivesearch.py

#根据 IOU 的得分，选择正样本和负样本
```

```
def based_iou_sample(objSrhRects, iou_array, g_index, max_threshold, is_
greater=True, image=None):
    #如果 is_greater 为 True,则选择 IOU 大于指定分数的 box 作为正样本
    #如果 is_greater 为 False,则选择 IOU 小于指定分数的 box 作为负样本
    index = np.argwhere(iou_array > max_threshold if is_greater else iou_array <
max_threshold)
    if len(index) > 0:
        #根据 index 获取 ss 的 box 信息
        ss_box = objSrhRects[index].reshape([-1, 4])
        #在原图中绘出这个区域
        u.draw_box(image, ss_box, color=[0, 0, 255])
        #大于 0.5 的框被认为是真实框,然后将这个 label 复制 N 份
        #小于 0.5 的框被认为是负样本,然后将类别置为背景,没有目标
        true_label = np.tile(np.array(g_index), [len(ss_box), 1])
        #对大于 max_threshold 的框增加 label 信息
        ss_box = np.append(ss_box,true_label, axis=-1)
        #返回区域选择框,以及样本的下标
        return ss_box, index
```

函数 based_iou_sample()根据传入的 IOU 分数通过 np.argwhere()函数得到满足条件的索引号,当大于 0.5 时为正样本,当小于 0.3 时为负样本。通过参数 g_index 来区分是正样本还是负样本,如果是负样本,则 g_index 为所有类别数+1。

调试程序,可发现 IOU>0.5 的区域选择框共有 15 个,其中最大值为 0.98190,如图 5-7 所示。

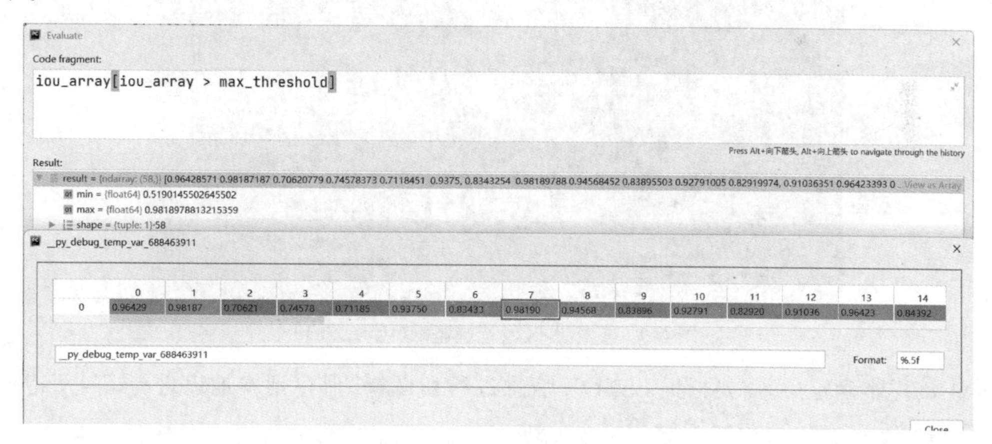

图 5-7 IOU 正样本分数

如果将 IOU>0.5 的区域选择框还原到图像中,则可发现较多框可覆盖标注 BOX。将这些区域选择框送入 CNN 提取特征,也就是正样本的特征,而那些 IOU<0.3 的区域选择框为负样本,CNN 提取的特征为负样本特征。选择负样本来进行训练的一个原因是为了让神经网络学习到某些参数,以便将非目标区域排除。

正样本区域选择框如图 5-8 所示,负样本区域选择框如图 5-9 所示,而中间区域的区域和特征信息将会被丢弃。

图 5-8　正样本区域选择框(IOU>0.5)

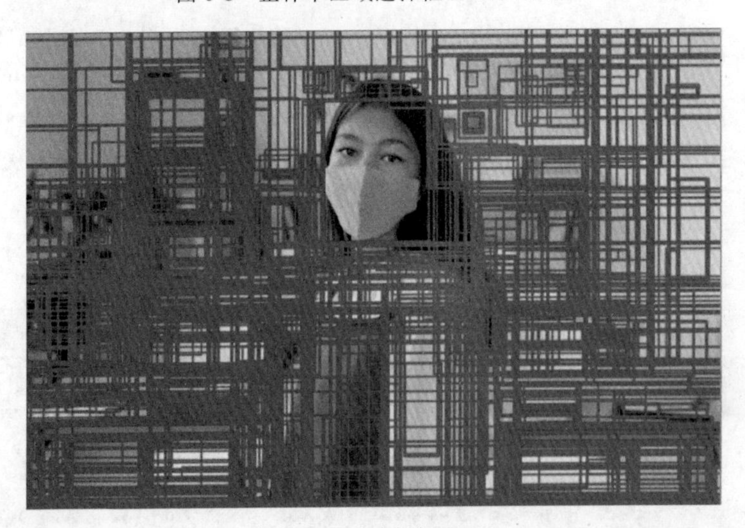

图 5-9　负样本区域选择框(IOU<0.01)

正负样本确定后,需要根据 SSBOX 信息进行切图,并保存到本地以进行 CNN 特征提取,代码如下:

```
#第 5 章/ObjectDetection/R-CNN/selectivesearch.py

#切割图像并转换到 224×224 的大小
def cut_image(image, boxes):
    #image[ymin:ymax, xmin:xmax] 切图的方法
    return [cv2.resize(image[box[0]:box[1], box[2]:box[3]], [224, 224]) for box
in boxes]

cut_positive_box = positive_box.copy()
#正样本的格式为 xmin,ymin,xmax,ymax,其下标分别为 0,1,2,3
```

```
#需要转换为 ymin,ymax,xmin,xmax,其下标分别为 1,3,0,2,并以这些值进行切图
cut_positive_box[..., [0, 1, 2, 3]] = cut_positive_box[..., [1, 3, 0, 2]]
positive_image = cut_image(img.copy(), cut_positive_box[..., 0:4])
```

因为区域选择框得到的格式为 xmin,ymin,xmax,ymax,而切图需要 ymin,ymax,xmin,xmax 格式,所以通过代码 cut_positive_box[...,[0,1,2,3]]=cut_positive_box[...,[1,3,0,2]]进行了转换。同时 cut_image()函数实现了切图,并转换到指定图像的大小。将图像大小设置为[224,224]是由于 VGG 的输入要求,如图 5-10 所示。

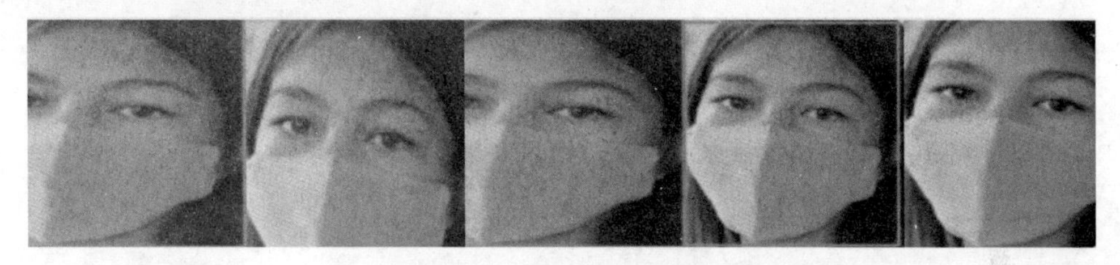

图 5-10 切图后将大小转换到 224×224

标注信息使用 5.1 节标签处理及代码中的函数 read_annotations()进行 true_box 的读取,同时随书源码中 ObjectDetection/R-CNN/selectivesearch.p 文件中的 get_positive_negative_samples()函数实现了更完整的代码。

另外在读取图片时将输入图像和标注 BOX 进行了等比例缩放。R-CNN 由于采用的是区域选择框提取,可以不进行图像的大小转换,但是为了网络的训练学习需要统一尺度及提高稳定性,本书将图像大小等比例缩放到 640×640,代码如下:

```
#第 5 章/ObjectDetection/R-CNN/utils.py
def letterbox_image(image, size, box):
    #对图片大小进行转换,使图片不失真。在空缺的地方进行 padding
    #位置不会有偏移的情况
    iw, ih = image.size
    w, h = size
    #如果刚好输入的尺寸与要求的尺寸相同,则原图返回
    if iw == w and ih == h:
        return cv2.cvtColor(np.array(image), cv2.COLOR_RGB2BGR), box
    #计算是宽还是高的比例更小
    scale = min(w / iw, h / ih)
    #按最小的比例进行缩放
    nw = int(iw * scale)
    nh = int(ih * scale)
    #缩放比例距原图的大小,因为需要上下或者左右同时缩放,所以需要整除 2
    dw = (w - nw) //2
    dh = (h - nh) //2
    #等比例缩放
    image = image.resize((nw, nh), Image.BICUBIC)
    #新建一个灰度图
    new_image = Image.new('RGB', size, (128, 128, 128))
    #在缩放的图中增加填充的像素
    new_image.paste(image, (dw, dh))
```

```
    #对 BOX 进行等比例调整
    box_resize = []
    for boxx in box:
            #对 BOX 进行等比例缩减,并加上填充的灰度图
        boxx[0] = int(boxx[0] *scale + dw)
        boxx[1] = int(boxx[1] *scale + dh)
        boxx[2] = int(boxx[2] *scale + dw)
        boxx[3] = int(boxx[3] *scale + dh)
        #略有裁剪,可不使用
        boxx[0] = np.clip(boxx[0], 0, w - 1)
        boxx[2] = np.clip(boxx[2], 0, h - 1)
        boxx[1] = np.clip(boxx[1], 0, w - 1)
        boxx[3] = np.clip(boxx[3], 0, h - 1)
        boxx = boxx.astype("int32")
        box_resize.append(boxx)
    #转回 NumPy 格式
    new_image = cv2.cvtColor(np.array(new_image), cv2.COLOR_RGB2BGR)
return new_image, np.array(box_resize)
```

函数 letterbox_image()会对输入图像进行等比例缩放,例如原图为 1125×750,按 640×640 进行缩放,则 scale=min(0.568,0.853)按 0.568 进行等比例缩放,而此时(nw,nh)=(640,426),缩放后图像的高不够 640,则需要上下各补 107 像素,即 new_image.paste(image,(0,107))。同理原图输入的标注 BOX 信息,也按 0.568 进行缩放,所以 boxx[1]×scale,但是可能像素不够,所以通过 boxx[1]×scale+107 进行填充,效果如图 5-11 所示。

图 5-11　等比例缩放图像和标注 BOX

5.2.4　代码实战特征提取

5.2.3 节已提取正样本和负样本图片、正样本 SSBOX 和标注框 GTBOX 信息，搭建 CNN 中的 VGG-16 网络并使用 ImageNet 的初始权重，代码如下：

```
#第 5 章/ObjectDetection/R-CNN/vgg_features.py
def init_vgg():
    #VGG-16 模型，从 Keras 中直接提取
    base_model = VGG-16(weights='ImageNet',include_top=True)
    base_model.summary()
    #构建模型，output 参数用于确定使用哪一个网络层作为输出
    model = Model(inputs=base_model.input, outputs=base_model.layers["fc2"].
output)
    return model

#从 ImageNet VGG-16 中得到图片的特征信息
def vgg_features(model, image):
    x = preprocess_input(image)
    #前向传播得到特征
    features = model.predict(x)
    return features
if __name__ == "__main__":
    model = init_vgg()
    get_feature_map("./train_data/ss_info/正样本图片.npy", "./train_data/ss_
feature/正样本图片特征.npy", model)
    get_feature_map("./train_data/ss_info/负样本图片.npy", "./train_data/ss_
feature/负样本图片特征.npy", model)
```

调用 keras. applications 模块下已写好的 VGG-16 模型，并同时使用 weights='ImageNet' 的初始权重，选择 fc2 层 4096 维作为输出特征，然后分别加载正样本图片. py 和负样本图片. npy，并保存特征信息。

5.2.5　代码实战 SVM 分类训练

支持向量机(SVM)是一个经典的机器学习分类方法，如图 5-12 中(a)图红色和蓝色的点显然是可以被一条直线分开的，而能够分开的线不止一条，例如(b)、(c)中的黑线 A 和 B，如果是一个平面，则称为决策面。

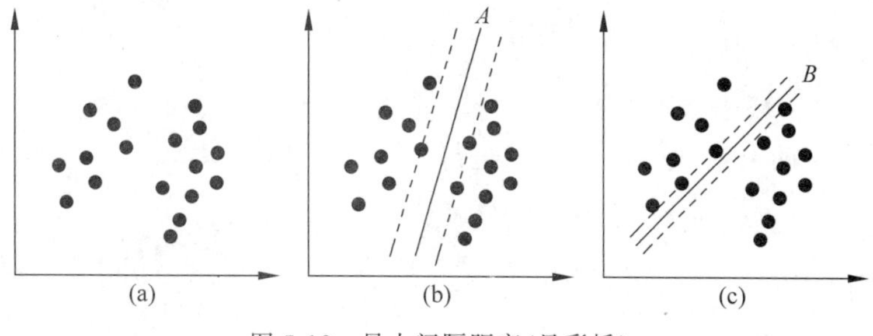

图 5-12　最大间隔距离(见彩插)

如果从决策面来观察,则可知 A 和 B 均可,但是从直觉来讲 A 优于 B,其原因是 A 的间隔距离比 B 大,所以支持向量机的核心思想就是求解能够正确划分训练数据集并且几何间隔距离最大的超平面。

支持向量机的原理推导较复杂(非重点),不过调用 sklearn 库中的 SVC()对象可轻松实现,代码如下:

```
#第 5 章/ObjectDetection/R-CNN/train.py
#SVM 训练分类器
def svm_classifier():
    #加载特征提取信息和类别
    pos_X = np.load("./train_data/ss_feature/正样本图片特征.npy")
    pos_Y = np.load("./train_data/ss_info/正样本图片分类 label.npy")
    neg_X = np.load("./train_data/ss_feature/负样本图片特征.npy")
    neg_Y = np.load("./train_data/ss_info/负样本图片分类 label.npy")
    #合并正样本和负样本
    X = np.concatenate([pos_X, neg_X], axis=0)
    #label 信息
    Y = np.concatenate([pos_Y, neg_Y], axis=0)
    #划分训练集和验证集
    x_train, x_test, y_train, y_test = train_test_split(X, Y, random_state=14)
    #SVM 线性分类器
    clf = SVC(C=1.0, kernel='linear', random_state=28, max_iter=1000,
probability=True)
    #训练
    clf.fit(x_train, y_train)
    #验证评分
    pred = clf.predict(x_test)
    print("F1-score: {0:.2f}".format(f1_score(pred, y_test, average='micro')))
    #保存 SVM 模型和参数
    joblib.dump(clf, "face_mask_svm.m")
```

对象 SVC()为一个线性可分对象,其中 C 为惩罚系数,训练完成后使用 joblib.dump()保存模型。

论文的作者在这里使用 SVM 而不是用分类网络的重要原因是由于此时只有少量的正样本,但有大量的负样本,样本数据不平衡对于 CNN 来讲会非常敏感,而 SVM 对于不平衡数据并不敏感。

5.2.6 代码实战边界框回归训练

通过 SVM 对区域选择框的背景或者目标类别进行预测,此时正样本区域选择框与标注真实框会有一定的偏差,通过以下公式进行偏移值的计算:

$$t_x = (G_x - P_x)/P_w$$
$$t_y = (G_y - P_y)/P_h$$
$$t_w = \log\left(\frac{G_w}{P_w}\right) \tag{5-2}$$

$$t_h = \log\left(\frac{G_h}{P_h}\right)$$

其中,G_x 和 G_y 表示标注 GTBOX 的中心点,P_x 和 P_y 为区域选择框 SSBOX 的中心点;G_x 和 G_y 为 GTBOX 的宽和高,P_w 和 P_h 为 SSBOX 的宽和高;t_x 和 t_y 为 GTBOX 中心点与 SSBOX 中心点的偏移,t_w 和 t_h 为 GTBOX 与 SSBOX 的宽和高的偏移。当然 t_x、t_y、t_w 和 t_h 越小越好,说明 GTBOX 与 SSBOX 的 IOU 分数较高,如图 5-13 所示。

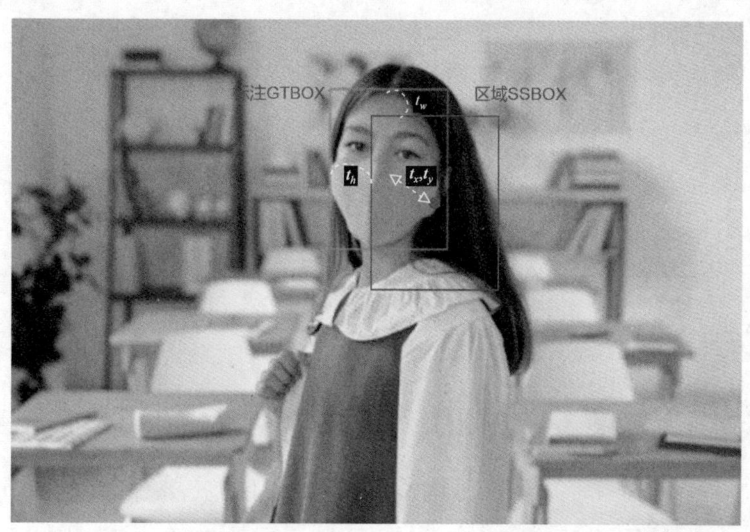

图 5-13 标注 GTBOX 和 SSBOX

计算偏移值的代码如下:

```
#第 5 章/ObjectDetection/R-CNN/train.py
def xyxy2cxcywh(boxes):
    #由 xmin,ymin,xmax,ymax 分别转换成 xmin,ymin,w,h
    #由 xmax-xmin 和 ymax-ymin 得到中心点坐标
    wh = boxes[..., 2:4] - boxes[..., :2]
    center_xy = boxes[..., :2] + wh / 2
    #合并数组
    return np.concatenate([center_xy, wh], axis=-1)

#真实框 cx,cy,w,h。区域选择框 p_cx,p_cy,w,h
def cxywh2offset(g_cxywh_boxes, p_cxywh_boxes):
    #公式(5-2)
    t_xy = (g_cxywh_boxes[..., :2] - p_cxywh_boxes[..., :2]) / p_cxywh_boxes[..., 2:4]
    t_wh = np.log(g_cxywh_boxes[..., 2:4] / p_cxywh_boxes[..., 2:4])
    #合并数组
    return np.concatenate([t_xy, t_wh], axis=-1)

def box_regression(num_class=1):
    #导入正样本的特征信息和 BOX 信息
    p_box = np.load("./train_data/ss_info/正样本 box.npy")
    p_feature = np.load("./train_data/ss_feature/正样本图片特征.npy")
```

```
    g_box = np.load("./train_data/ss_info/真样本 box.npy")
    for i in range(0, num_class):
        #类别号的下标
        index = np.where(p_box[..., -1] == i)
        #根据类别号取不同的特征信息和 box 信息并进行归一化操作
        p_class_feature = p_feature[index]
        #区域选择框,图像输入已统一到 640
        p_class_box = p_box[index] / 640
        #标注 GTBOX
        g_class_box = g_box[index] / 640
        #因为区域选择框和真实框得到的坐标是 xmin,ymin,xmax,ymax
        #为了计算偏移值需要转换成 cx,cy,w,h
        p_class_box = xyxy2cxcywh(p_class_box)
        g_class_box = xyxy2cxcywh(g_class_box)
        #计算偏移值
        offset_box = u.cxywh2offset(g_class_box, p_class_box)
        #输入 4096 维特征,输出 4 位偏移值
        x_train, x_test, y_train, y_test = train_test_split(p_class_feature,
offset_box, random_state=14)
        #sklearn 中的线性回归
        model = LinearRegression()
        #Reg 回归训练
        model.fit(x_train, y_train)
        #预测
        y_pre = model.predict(x_test)
        #计算预测值和真实值之间的平均误差
        loss = mean_squared_error(y_pre, y_test)
        print("误差", loss)
        joblib.dump(model, f"face_mask_lr{i}.m")
        return p_class_feature, offset_box
```

由于 GTBOX 和 SSBOX 格式均为 xmin,ymin,xmax,ymax,但是计算偏移时需要 cx,cy,w,h,所以通过函数 xyxy2cxcywh(boxes)进行了实现。函数 cxywh2offset(g_cxywh_boxes,p_cxywh_boxes)传入 GTBOX 和 SSBOX 的信息,套用公式(5-2)实现 GTBOX 与正样本 SSBOX 的偏移量的计算。函数 box_regression(num_class=1)读取正样本特征并调用 cxywh2offset()生成偏移量的数据,以便喂入 Reg 回归网络训练。

LinearRegression()是 sklearn 库中的逻辑回归对象,输入是正样本 4096 维特征,Y 值为 GTBOX 与 SSBOX 的偏移值数据 offset_box,当 model. predict(x_test)预测的偏移与 offset_box 最小时,说明回归权重系数已获得。回归损失使用均方差 mse()函数。

随书代码中还存在一份使用神经网络来做回归权重训练的代码,感兴趣的读者可查阅。

5.2.7 代码实战预测推理

根据训练结果得到分类权重文件 face_mask_svm. m,以及回归权重文件 face_mask_lr0. m,读入待预测图片的大小转换到训练所需的 640×640 大小,并调用区域选择搜索算法对预测图片生成 2000 个区域框,然后使用 VGG 提取特征输入分类权重,对这 2000 个区域图片的背景和低概率分类进行过滤,此时得到有预测目标的 SSBOX,代码如下:

```
#第 5 章/ObjectDetection/R-CNN/predictdetect.py
def inference_code(num_class=1,max_threshold=0.5):
    #VGG 特征提取
    model = v.init_vgg()
    #读入推理图片
    org_image = cv2.imread(
        r"../val_data/pexels-photo-5211438.jpeg")
    #由 NumPy 格式转换为 PIL 对象
    old_image = Image.fromarray(np.uint8(org_image.copy()))
    #等比例到 640×640
    img, _ = u.letterbox_image(old_image, [640, 640], [])
    #区域选择,默认得到的是 xmin,ymin,xmax,ymax
    p_box = s.start_search(img.copy())[:2000, ...]
    #根据区域选择切图
    cut_box = p_box[..., 0:4].copy()
    cut_box[..., [0, 1, 2, 3]] = cut_box[..., [1, 3, 0, 2]]
    p_img = s.cut_image(img.copy(), cut_box)
    #对切出来的图全部升一维,由原来的[w,h,c]合并成 b,w,h,c
    all_p_img = np.concatenate([np.expand_dims(p, axis=0) for p in p_img], axis=0)
    #提取特征
    features = v.vgg_features(model, all_p_img)
    #调 SVM 进行预测
    clf = joblib.load("./face_mask_svm.m")
    class_result = clf.predict_proba(features)
    #过滤置信度较低的结果
    for i in range(num_class):
        #根据分类的下标进行置信度的过滤,得到分类可能性大于 0.5 的下标
        index = np.where(class_result[..., i] >= max_threshold)
    #......接下 1 段代码......
```

在 index 处断点可知 0 为 face_mask 的分类,而 1 为背景的概率。类别概率大于 0.5 的区域选择框共有 17 个,其中最大概率为 0.96,最小概率为 0.03,index 得到满足阈值数组的下标,如图 5-14 所示。

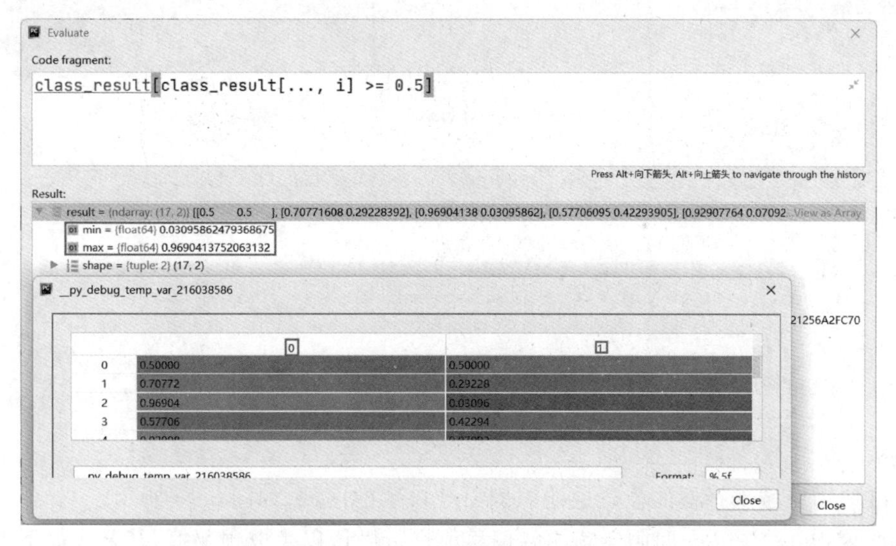

图 5-14　大于类别阈值的概率值

然后将满足分类概率的 SSBOX 和概率分数进行非极大值抑制(NMS)。NMS 先对所有分类概率进行降序排列,将最大概率的数组下标存储下来,然后将最大概率的 BOX 与其他传入的 boxes 进行 IOU 得分,如果有小于阈值的框,则认为是同类别的其他目标,否则剔除,然后对找出来的框再次进行 NMS,直到所有的框都不重叠,并返回 boxes 的下标,代码如下:

```
#第 5 章/ObjectDetection/R-CNN/predictdetect.py
def nms(boxes, nms_thresh=0.2):
    #用非极大值抑制,去掉相似框
    #当前框的分类得分
    scores = boxes[:, 4]
    #对 boxes 的概率得分从大到小进行排序,得到数组下标
    order = scores.argsort()[::-1]
    keep = []
    while order.size > 0:
        #取最大概率的下标
        i = order[0]
        #将最大概率 BOX 的下标存起来
        keep.append(i)
        #计算当前最大概率的 BOX 与其他框 boxes 的 IOU 分数
        iou_score = get_iou(boxes[i, ...], boxes[order[1:], ...])
        #只有小于指定阈值的框才会被认为是不同的框
        index = np.where(iou_score <= nms_thresh)[0]
        #将上一步的 index 存起来
        order = order[index + 1]
    return keep
```

非极大值之后使用 p_boxes=p_boxes[nms_index]获得区域选择框,调用 u. draw_box()函数实现目标区域的绘图,代码如下:

```
#第 5 章/ObjectDetection/R-CNN/predictdetect.py
def inference_code(num_class=1):
    #......接上 1 段代码......
    index = np.where(class_result[..., i] >= 0.5)
    if index[0].size:
        #得到 BOX,并归一化
        class_box = p_box[index] / 640
        #得到 BOX 当前类别的分数
        class_score = class_result[index][..., i]
        #将 BOX 和置信度合并成 n * 5
        p_boxes = np.column_stack([class_box, class_score])
        #进行非极大值抑制,得到保留 BOX 的下标
        nms_index = u.nms(p_boxes)
        #保留的区域选择 BOX
        p_boxes = p_boxes[nms_index]
        ss_img = u.draw_box(img.copy(), (p_boxes * 640).astype('int'))
```

直接使用区域选择框的位置也能识别出戴口罩的区域,如图 5-15 所示。

根据 SSBOX 与 Reg 回归器预测的偏移值进行微调以生成预测框,其公式如下:

图 5-15 区域选择框 NMS 后的目标区域

$$
\begin{aligned}
\hat{G}_x &= P_w d_x(P) + P_x \\
\hat{G}_y &= P_h d_y(P) + P_y \\
\hat{G}_w &= P_w \exp(d_w(P)) \\
\hat{G}_h &= P_h \exp(d_h(P))
\end{aligned}
\tag{5-3}
$$

其中，P_x 和 P_y 为 SSBOX 的中心点坐标，P_w 和 P_h 为宽和高；$d_x(P)$和 $d_y(P)$为 Reg 回归器预测 SSBOX 与预测框中心点的偏移值，$d_w(P)$和 $d_h(P)$为宽和高的偏移。通过式(5-3)解码得到 $\hat{G}_x$、$\hat{G}_y$、$\hat{G}_w$、$\hat{G}_h$ 预测框的中心点、宽和高值。

根据偏移值和 SSBOX 得到预测框的代码实现，代码如下：

```
#第 5 章/ObjectDetection/R-CNN/predictdetect.py

def offset2xyxy(offset_box, p_cxywh_boxes):
    p_cxy = offset_box[..., :2] *p_cxywh_boxes[..., 2:4] + p_cxywh_boxes[..., :2]
    p_wh = np.exp(offset_box[..., 2:4]) *p_cxywh_boxes[..., 2:4]
    boxes = np.concatenate([p_cxy, p_wh], axis=-1)
    #转换成 xmin,ymin,xmax,ymax
    boxes[..., :2] -= boxes[..., 2:] / 2
    boxes[..., 2:] += boxes[..., :2]
    return boxes

def inference_code(num_class=1):
    #.......接上 1 段代码.......
    #调回归器权重进行微调
    lr = joblib.load(f"./face_mask_lr{i}.m")
    pre_offset = lr.predict(features[nms_index])
    #将区域选择框转换成 cx,cy,w,h
p_boxes = u.xyxy2cxcywh(p_boxes)
#解码,从偏移值和区域选择 BOX 转换成 xmin,ymin,xmax,ymax
```

```
box = u.offset2xyxy(pre_offset, p_boxes)
rg_img = u.draw_box(img.copy(), box)
s_img = np.hstack([ss_img, rg_img])
cv2.imshow('', s_img)
cv2.waitKey(0)
```

函数 offset2xyxy(offset_box, p_cxywh_boxes)套用式(5-3)实现解码,同时将 boxes 信息转换为左上角、右下角的格式以方便绘图。在 inference_code()函数中,调用线性回归权重 face_mask_lr0.m,输入正样本的 4096 维特征信息,预测得到 pre_offset 的偏移值,然后调用 u.offset2xyxy(pre_offset, p_boxes)将偏移值转换成预测框,如图 5-16 所示。

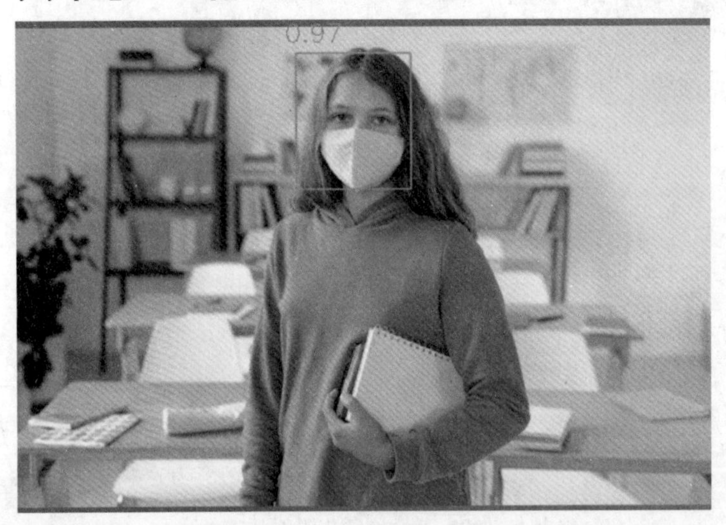

图 5-16　根据预测偏移值获得的预测框

图 5-16 中的预测框比 SSBOX 要差一些,其原因是本训练集只有 1 张图片,训练数据泛化能力不够。

总结

R-CNN 使用区域选择搜索算法,训练过程存在多次文件存储,需要消耗大量计算时间和空间资源。使用 CNN 进行特征提取,偏移值 Reg 回归预测框,IOU 选择正样本,推理的非极大值抑制对于其他网络有极大的启发作用。

练习

运行并调试本节代码,理解算法思想与代码的结合。

5.3 两阶段网络 Faster R-CNN

5.3.1 模型介绍

Ross Girshick 等在 2016 年发表的论文 *Faster R-CNN: Towards Real-Time Object*

Detection with Region Proposal Networks 中提出 Faster R-CNN，在结构上 Faster R-CNN 将特征提取、区域提取、边界框 Reg 回归、分类整合在了一个网络中，使检测速度有极大的提高，其网络结构如图 5-17 所示。

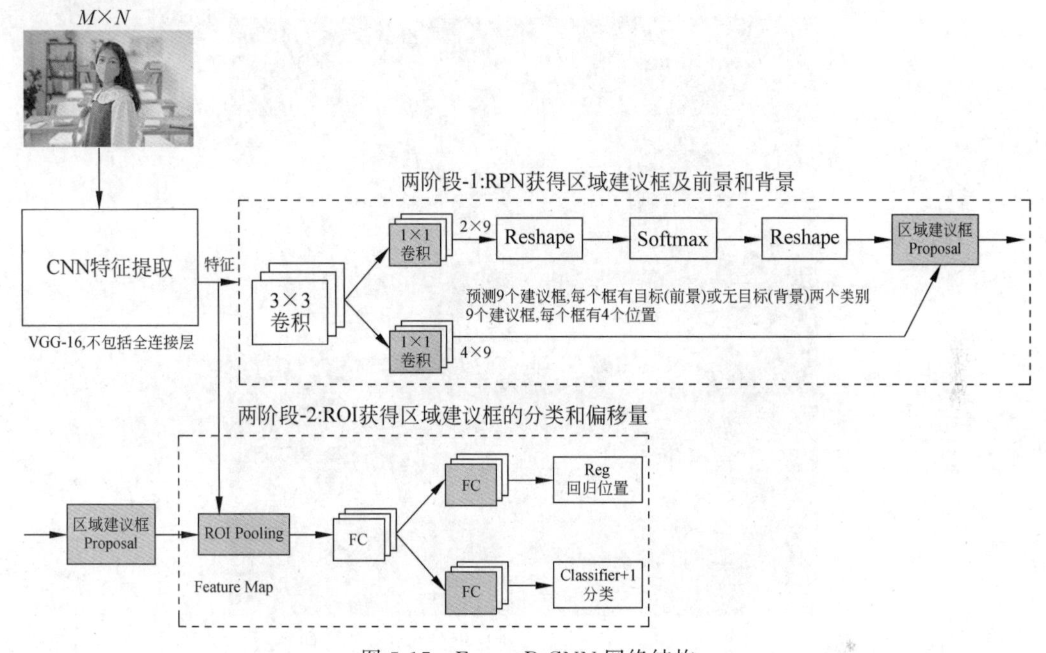

图 5-17　Faster R-CNN 网络结构

特征提取使用 VGG-16，其结构为 2-2-3-3-3-3(2 为卷积池化，最后的 3 为全连接)，但不包括最后全连接层，使用 block5_conv3 层的输出作为 RPN、ROI 网络的输入特征。

Faster-R-CNN 为 2 阶段网络，第 1 个阶段为 RPN 网络，block5_conv3 输入后经过 3×3 卷积，分成两个分支，通过 1×1 卷积输出 2×9 个特征，每个特征有 9 个建议框，每个建议框有目标或者没有目标，所以类别数为 2。另一个 1×1 卷积输出 4×9 个特征，代表每个建议框的 4 个位置。RPN 的输出，9 个建议框的有目标、无目标的分类及 9 个建议框的位置。

Faster R-CNN 中的建议框相当于 R-CNN 中的区域搜索框，不同之处在于 R-CNN 是在原图中通过选择区域搜索算法生成，而 Faster R-CNN 通过预设矩形框尺寸和比例在 block5_conv3 上生成。CNN 提取的特征点相对于原图的感受野的区域如图 5-18 所示。

根据网络特征的提取，可以发现 block5_conv3 提取的特征刚好是原图的 1/16，因此可以看成在原图中对图像每隔 16 像素进行均分，并以每个均分点为中心点，以尺寸[8,16,32]和比例[0.5,1,2]生成 9 个建议框，如图 5-19 所示。

图 5-18　特征点在原图中的感受野

图 5-19 中绿色框为真实框，红色框为生成的建议框。图中红色点坐标之间的区域相对于

block5_conv3 特征点信息，以每个红色点的坐标为中心点生成 9 个建议框，假设 block5_conv3 得到的是 40×40，那么共计生成 14 400 个建议框，则 RPN 的输出为 14 400×2 个分类，14 400×4 个框。

图 5-19　Faster R-CNN 建议框的生成(见彩插)

将 14 400 个建议框与真实框之间做 IOU，当 IOU 大于设定值时，则选择此建议框作为正样本、有目标，然后按照式(5-2)计算建议框与真实框的偏移，作为真实值与预测值之间的损失(预测值为预测框基于建议框的偏移)；因为正样本少，负样本较多，选择当 IOU 小于设定值时，将此建议框作为负样本、无目标；将两个阈值之间的建议框丢弃。正负样本数量之和可以设定为 256，如图 5-20 所示，蓝色框为真实框，绿色框为建议框，两个建议框与真实框的 IOU＝0.708，则此时有两个正样本，计算 offset 作为 y 值。

图 5-20　Faster R-CNN 正样本的选择(见彩插)

RPN 的损失函数为

$$L(\{p_i\},\{t_i\})=\frac{1}{N_{\mathrm{cls}}}\sum_i L_{\mathrm{cls}}(p_i,p_i^*)+\lambda\frac{1}{N_{\mathrm{reg}}}\sum_i p_i L_{\mathrm{reg}}(t_i,t_i^*)$$

$$L_{\mathrm{reg}}(t_i,t_i^*)=\sum_{i\in(x,y,w,h)}\mathrm{smooth}_{L_1}(t_i-t_i^*) \tag{5-4}$$

$$\mathrm{smooth}_{L_1}(t_i-t_i^*)=\begin{cases}0.5x^2 & 如果\ |x|<1\\ |x|-0.5 & 其他\end{cases}$$

其中，i 表示建议框数组的索引号，p_i 表示预测框有无目标的分类，p_i^* 代表真实框有无目标的分类，分类损失使用交叉熵损失；t_i 代表预测框相对于建议框的偏移，t_i^* 代表真实框相对于建议框的偏移，位置损失使用 smooth_{L_1} 损失。

回归损失可使用 L_1、L_2 损失，L_1 的导数为 $\frac{\partial_{L_1}(x)}{\partial x}=\begin{cases}1 & 如果\ x\geqslant 0\\ -1 & 其他\end{cases}$，$L_2$ 的导数为 $\frac{\partial_{L_2}(x)}{\partial x}=2x$，而 smooth_{L_1} 的导数为 $\frac{\partial_{\mathrm{smooth}_{L1}}(x)}{\partial x}=\begin{cases}x & 如果|x|<1\\ \pm 1 & 其他\end{cases}$，当 x 增大时 L_2 损失对 x 的导数也增大，这就导致训练的初期，当预测值与真实值的差异过大时，损失函数对预测值的梯度较大，训练不稳定。当 x 变小时，L_1 对 x 的导数为常数，在训练后期当预测值与真实值的差异较小时，L_1 损失对预测值的导数的绝对值仍然为 1，如果此时学习率不变，则损失函数将在稳定值附近波动，难以继续收敛以达到更高精度，而 smooth_{L_1} 损失则避免了 L_2、L_1 的缺点，如图 5-21 所示。

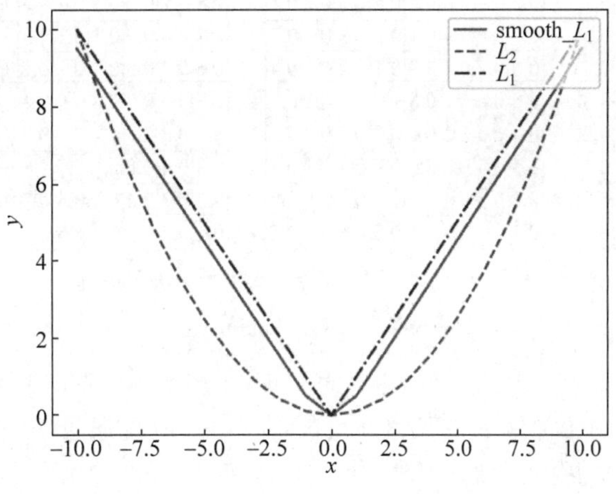

图 5-21 L_1、L_2、smooth_{L_1} 损失

从图 5-21 中可知 smooth_{L_1} 在远离坐标点时下降得非常快，类似 L_1，而离坐标点较近时又类似 L_2 有一个转折的平滑效果，下降会慢一些。

第 2 阶段为 ROI 网络，ROI 网络的输入为 block5_conv3 的特征信息，以及 RPN 网络输出的预测建议框和有无目标的分类信息。图 5-17 中 Proposal 是 RPN 预测框信息，但是并没有全部输入 ROI 网络，而是将 RPN 输出信息解码成预测框，然后在预测框与建议框之间根据有无目标的概率做非极大值抑制，选出 2000 个重叠率不高的框作为 Proposal，然后对 Proposal 的区域进行 ROI Pooling 以得到 7×7 的特征信息，输入全连接后再次微调 BOX 的位置信息，并预测目标区域是什么的分类信息，如图 5-22 所示。

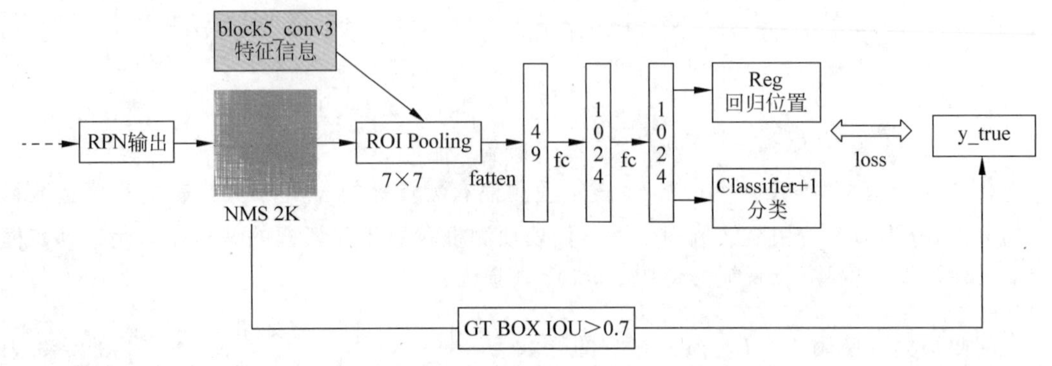

图 5-22　ROI 网络过程

图 5-22 中 y_true 为真实值，训练时需要将真实 BOX 的信息与选出来重叠率较低的 RPN 框做 IOU，当大于阈值时为正样本，当小于阈值时为负样本，正负样本之和可自定义设置，例如 128。

ROI Pooling 对输入的特征进行提取后形成一个固定的区域，Proposal 投影之后左上角的位置为[0,3]，右下角为[7,8]，然后划分成 2×2 的区域，对区域中的最大值进行提取，输出为 2×2 的特征，如图 5-23 所示。

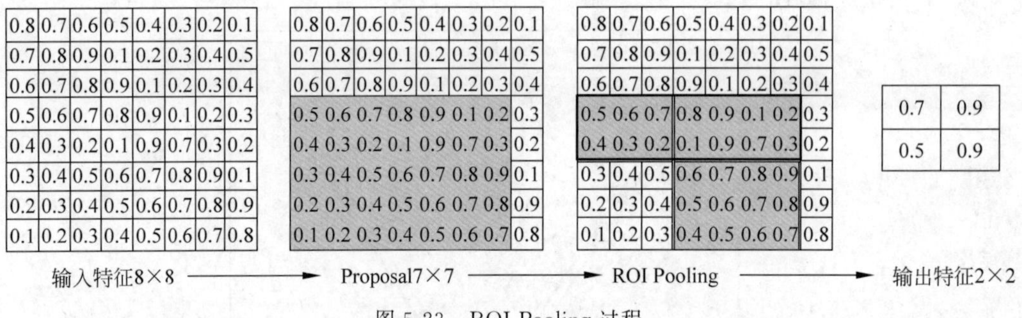

图 5-23　ROI Pooling 过程

因为 Faster R-CNN 要求的输入图像的尺度为 $M \times N$，原图映射到特征图的大小为 $[M/16, N/16]$，不同的图像输入的尺度 M、N 值不同，但是通过 ROI Pooling 可以得到相同大小特征图的输出。因为其划分比例为 3∶4，同时也可以看到是不同尺度最大池化信息的融合，这对于分类或回归是有益的。

ROI Pooling 后进入全连接，并通过 Reg 回归再次精修位置，通过 Softmax 进行位置的回归，其损失函数同式(5-4)。

5.3.2　代码实战 RPN、ROI 模型搭建

根据图 5-17 中的结构描述，在 CNN 特征提取时使用 VGG-16 模型，同时使用 ImageNet

的权重信息作为初始权重。将 VGG-16 中 block5_conv3 的输出作为 RPN 网络的输入，得到建议框的位置 out_rpn_offset 和有无目标 out_rpn_clf，然后将 block5_conv3 的输出及 roi_input 的输入作为 ROI 网络的输入，输出分类预测和回归预测信息。

前向传播的代码如下：

```
#第 5 章/ObjectDetection/FasterR-CNN/model.py
import tensorflow as tf
from keras.applications.VGG-16 import VGG-16
from keras.layers import Input, Conv2D, Reshape, \
    Layer, Flatten, Dense, TimeDistributed
from keras import Model


def Faster R-CNN(input_shape=(640, 640, 3),
             roi_input_shape=(None, 4),
             num_anchors=9,
             num_class=2 + 1
             ):
    #图片输入 shape
    inputs = Input(shape=input_shape, name='image_input')
    #ROI Pooling 的输入维度
    roi_input = Input(shape=roi_input_shape, name='roi_input')

    #调用 Keras 自带的 VGG-16 作为 backbone，并且不要全连接层，并使用 ImageNet 的权重
    #假设输入为 600 × 600 × 3
    base_model = VGG-16(weights=None, include_top=False)
    base_model.load_weights(
        "../R-CNN/VGG-16_weights_tf_dim_ordering_tf_kernels.h5",
        by_name=True,
        skip_mismatch=True
    )
    #重新构建 backbone 网络，使用 VGG-16 的 block5_conv3 作为特征信息的输出
    base_model = Model(inputs=base_model.input, outputs=base_model.get_layer
('block5_conv3').output)
    #得到 block5_conv3 的特征
    backbone_feature = base_model(inputs)
    #得到 backbone 的输出(None,None,512)，作为 RPN 的输入
    #RPN 的作用：根据先验框，得到建议框
    out_rpn_clf, out_rpn_offset = rpn_proposal(backbone_feature, num_anchors)
    #RPN 模块的网络
    rpn_net = Model(inputs=inputs, outputs=[out_rpn_clf, out_rpn_offset])
    #第 2 个阶段，根据输入的 Feature Map 特征信息，对区域 BOX 进行 offset 微调，同时对
    #BOX 属于哪个类别进行预测
    out_class, out_regbox = roi_pooling_box_cls(backbone_feature, roi_input,
num_class)
    #预测时为整个网络：输入图片和 ROI 框的信息，以及输出类别和 Reg BOX 偏移值
    all_net = Model(inputs=[inputs, roi_input], outputs=[out_class, out_regbox])
    return rpn_net, all_net
```

RPN 网络的实现封装在函数 rpn_proposal(backbone_feature, num_anchors)中，代码如下：

```
#第5章/ObjectDetection/FasterR-CNN/model.py
def rpn_proposal(backbone_feature, num_anchors=9):
    #候选区域网络。在R-CNN区域选择算法的基础上进行改进,每个像素生成9个建议框,有目
    #标或无目标类别
    #RPN以任意大小的图像作为输入,输出一组矩形的目标Proposals
    #每个Proposals都有一个目标得分,即有目标还是没有目标 None *None *512
    x = Conv2D(512, (3, 3), padding='same', activation='relu',
            name='rpn_conv_3x3')(backbone_feature)
    #在Feature Map的基础上,每个特征点输出9个Anchor,每个Anchor有4个位置
    #None × None × 36
    p_box_x = Conv2D(4 *num_anchors, (1, 1), padding='same',
                activation='linear', name='rpn_box_conv_1x1')(x)
    #每个BOX分为有目标或者没有目标两个类别
    #None × None × 18
    p_conf_x = Conv2D(2 *num_anchors, (1, 1), padding='same',
                activation='softmax', name='rpn_cnf_conv_1x1')(x)
    #(h×w×2×num_anchors,1),即每个像素只有一个概率,则有无目标
    out_rpn_clf = Reshape((-1, 1), name="rpn_p_conf")(p_conf_x)
    #(h×w×num_anchors,4),即每个像素有9个Anchor,每个Anchor有4个位置
    out_rpn_offset = Reshape((-1, 4), name="rpn_p_box")(p_box_x)
    return out_rpn_clf, out_rpn_offset
```

VGG-16网络层block5_conv3的特征信息,经过3×3的卷积后,分别经过1×1的p_box_x位置预测,1×1的p_conf_x有无目标Softmax的预测。假设输入为640×640,则out_rpn_clf成[40×40×2×9,1]=[28800,1],out_rpn_offset为[40×40×9,4]=[14400,4]。

ROI网络封装在roi_pooling_box_cls(backbone_feature,roi_input,num_class)函数中,输入为block5_conv3的特征信息及RPN网络预测的Proposal,将其值赋给roi_input变量,经过全连接后输出分类out_class和out_regbox位置信息,代码如下:

```
#第5章/ObjectDetection/FasterR-CNN/model.py
def roi_pooling_box_cls(feature_map_input, roi_input, num_class=2 + 1):
    #实现roi_pool的过程
    roi_pool = RoiPooling()([feature_map_input, roi_input])
    #根据roi_pool使用两个全连接层,分别输出BOX的偏移和类别的概率
    #不加TimeDistributed得到的是None,None。加了之后得到的是None,None,25088
    #TimeDistributed实现了在每个num_rois上面进行一个全连接操作,实现多对多的功能
    x = TimeDistributed(Flatten(name='flatten'))(roi_pool)
    #batch_size,h,w,1024
    x = Dense(1024, activation='relu', name='fc1')(x)
    x = Dense(1024, activation='relu', name='fc2')(x)
    #分类的概率
    out_class = Dense(num_class, activation='softmax')(x)
    #每个类别的4个位置
    out_regbox = Dense(4 *(num_class - 1), activation='linear')(x)
    return [out_class, out_regbox]
```

out_class中包含"分类+有无目标"的概率,假设类别为2,Proposal为2000个,则out_class输出是2000×3,out_regbox为每个类别的BOX信息,输出为2000×8。

ROI Pooling在RoiPooling()([feature_map_input, roi_input])中实现,主要调用

tf.image.crop_and_resize()函数实现 7×7 区域的 Pooling，代码如下：

```
#第 5 章/ObjectDetection/FasterR-CNN/model.py
class RoiPooling(Layer):
    #Region of Interest，将不同大小的特征图 ROI Pooling 到同一个尺寸
    def __init__(self):
        super(RoiPooling, self).__init__()
        self.pool_size = (7, 7)

    def build(self, input_shape):
        self.out_channel = input_shape[0][3]

    def compute_output_shape(self, input_shape):
        input_shape2 = input_shape[1]
        return None, input_shape2[1], self.pool_size[0], self.pool_size[1],
self.out_channel

    def call(self, inputs, *args, **kwargs):
        #输入 block5_conv3 特征图及 RPN
        #40×40×512, 2000×4
        feature_map, roi_x = inputs
        #区域选择框的数量 batch,2k
        batch_size, num_roi = tf.shape(feature_map)[0], tf.shape(roi_x)[1]
        #假设 batch_size=2，则[0,1]-->[[0],[1]]
        index = tf.expand_dims(tf.range(0, batch_size), 1)
        #假设 num_roi=2，则[[0],[1]]-->[[0,0],[1,1]]
        index = tf.tile(index, (1, num_roi))
        #由[[0,0],[1,1]]变回[0,0,1,1]
        index = tf.reshape(index, [-1])
        #2000×7×7×512
        roi_pooling_feature = tf.image.crop_and_resize(
            feature_map,                     #特征图
            tf.reshape(roi_x, [-1, 4]),      #BOX 的个数，每个有 4 个位置信息
            index,                           #roi_x 与特征图下标的对应关系
            self.pool_size                   #裁剪到特征图的大小
        )
        #batch_size × 2000 × 7 × 7 × 512
        output = tf.reshape(
            roi_pooling_feature,
            (batch_size, num_roi, self.pool_size[0], self.pool_size[1], self.out_
channel)
        )
        return output
```

如代码输入的是 40×40×512 的特征，2000×4 的 Proposal，经过 ROI Polling 后输出是 2000×7×7×512。整个网络模块搭建的思路是按功能、按算子分别进行封装，然后组建成 Faster R-CNN 模型，并根据需要可调用 RPN 或者 ROI 网络。

5.3.3 代码实战 RPN 损失函数及训练

代码实现损失函数通常需要如图 5-24 所示的步骤。

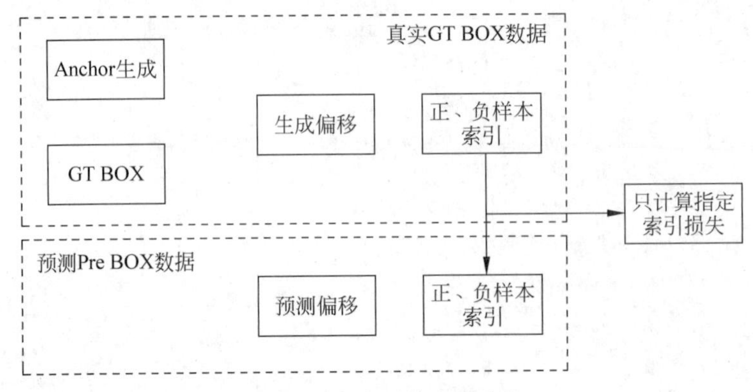

图 5-24 损失函数实现过程

因为要做损失,所以需要先计算 y 值,y 值需要在特征图上计算建议框(Anchor)与真实框 IOU,并将满足 IOU 分数的 Anchor 计算偏移,并记录有目标的 Anchor 的索引 index,及选取为负样本的 index,然后在前向传播预测后,根据 index 信息从预测值中取对应的预测偏移值、预测正负样本的分类信息,按损失函数公式进行计算,然后网络会根据损失函数进行反向传播并学习权重参数。整个过程实现较复杂,但比较关键的是如何计算 y 值。

首先看 Anchor 的生成,代码如下:

```
#第 5 章/ObjectDetection/FasterR-CNN/anchors.py
import numpy as np
import utils as u

#生成 Anchor 默认的各种尺寸
def generate_anchors(base_size=16, ratios=[0.5, 1, 2], anchor_scales=[8, 16, 32]):
    #M×N/16,映射到特征图上的大小,如果目标较小或者 Feature Map 较小,则此值需要调节
    #需要跟 Feature Map 相等
    py, px = base_size / 2., base_size / 2.
    #Anchor 的 3 种比例
    num_ratios = len(ratios)
    #3 种 Anchor 的尺寸
    num_scales = len(anchor_scales)
    #初始全为 0 的矩阵 3× 3 = 9 个 Anchor
    anchor_base = np.zeros([num_ratios * num_scales, 4], dtype=np.float32)
    #每个以不同比例生成不同 h 和 w 的 Anchor
    for i in range(num_ratios):
        for j in range(num_scales):
            #Anchor 的 height 和 width
            #h=16×8×sqrt(0.5)=90。对于小目标来讲,此值较大。需要调节比例或者尺寸
            h = base_size * anchor_scales[j] * np.sqrt(ratios[i])
            #w=16×8×1/sqrt(0.5)=181
            w = base_size * anchor_scales[j] * np.sqrt(1. / ratios[i])
            index = i * num_scales + j
            #计算每个框的 xmin, ymin, xmax, ymax
            anchor_base[index, 0] = py - h / 2.
            anchor_base[index, 1] = px - w / 2.
            anchor_base[index, 2] = py + h / 2.
```

```
            anchor_base[index, 3] = px + w / 2.
        return anchor_base


def mapping_anchor_2_original_image(
        anchor_base,          #基本 Anchor 的尺寸
        feature_hw_size,      #特征图的尺寸
        feature_stride=16,    #M/16
        anchor_num=9,
        image=None,
        gt_boxes=None
):
    #feature_hw_size 特征图的尺寸
    h, w = feature_hw_size
    #每隔 16 生成一个坐标信息,h×feature_stride 即原图的大小。也就是原图被分成了 16 份
    #每隔 16,将原图分为 40 份,宽和高各为 640。也就是此时 1 像素,映射到原图为 16×16 的区域
    shift_y = np.arange(0, h * feature_stride, feature_stride)
    shift_x = np.arange(0, w * feature_stride, feature_stride)
    #组成(x,y)的坐标信息
    shift_x, shift_y = np.meshgrid(shift_x, shift_y)

    shift = np.stack((shift_y.ravel(), shift_x.ravel(),
                      shift_y.ravel(), shift_x.ravel()), axis=1)
    #在每个坐标中都成 9 个 Anchor,并且 9 个 Anchor 根据坐标的位置进行相应调整
    #1×9×4 + 1×64×4
    anchor = anchor_base.reshape([1, anchor_num, 4]) + \
          shift.reshape([1, shift.shape[0], 4]).transpose([1, 0, 2])
    #以第 693 个坐标点为中心,绘 Anchor
    #u.draw_anchor(image, shift_x, shift_y, anchor[693, ...],gt_boxes)
    #reshape 成每个特征点都有 4 个位置,即(num_box,4)
    anchor = anchor.reshape([shift.shape[0] * anchor_num, 4]).astype(np.float32)
    return anchor
```

在函数 generate_anchors(base_size=16,ratios=[0.5,1,2],anchor_scales=[8,16,32])中根据经验值将 Anchor 的比例先验设置为[0.5,1,2],然后每个比例预测尺寸为[8,16,32],通过 base_size×anchor_scales[j]×np.sqrt(ratios[i])和 base_size×anchor_scales[j]×np.sqrt(1./ratios[i])来计算 Anchor 的大小。每个坐标点都会生成 9 个 Anchor。

在 mapping_anchor_2_original_image()函数中 feature_hw_size 为特征图的尺寸,将原图划分为 np.arange(0,h×feature_stride,feature_stride)份,假设输入为 640×640,则为 np.arange(0,40×16,16),然后沿 x 轴和 y 轴进行复制,并且在每个坐标中对 Anchor 共 9 个框进行复制,如果可视化,则可得图 5-18。

然后根据图 5-24 中的结构,计算 offset、正样本、负样本,代码如下:

```
#第 5 章/ObjectDetection/FasterR-CNN/data_processing.py
    def assign(
            self, true_boxes,
            pos_threshold=0.7,
            neg_threshold=0.3,
```

```
        num_sample=256,        #正样本+负样本，一共只能有 256 个
        image=None,
        old_boxes=None
):
    anchor_num, feature_stride = 9, 16
    feature_hw_size = np.array(self.input_shape) //feature_stride
    #各种 Anchor 的比例，得到的是 xmin,ymin,xmax,ymax
    anchor_base = a.generate_anchors(base_size=feature_stride)
    #以 640×640 输入时，在每幅图中生成 14400×4 个 BOX 的信息
    anchors = a.mapping_anchor_2_original_image(
        anchor_base, feature_hw_size,
        feature_stride, anchor_num, image,
        gt_boxes=old_boxes
    )
    #Anchor 归一化
    anchors = anchors / self.input_shape[0]
    #初始化 1 个 14400×4 为 0 的 BOX 信息
    true_boxes_assign = np.zeros([anchors.shape[0], 4])
    #[14400,2]，初始概率为[0,1]。0 代表背景的概率，1 代表前景的概率
    true_boxes_assign_clf = np.zeros([anchors.shape[0], 2])
    #遍历每个 GT BOX，计算 IOU 后的偏移
    for i in range(len(true_boxes)):
        #计算 Anchor 与 GT BOX 的 IOU
        iou_score = u.get_iou(true_boxes[i], anchors)
        #区域搜索出来的 BOX 与真实框>0.7 的 BOX 信息，当为正样本数据时正样本有目标，
        #所以类别为 1
        #得到的是满足条件的索引 index
        pos_index = np.argwhere(iou_score >= pos_threshold)
        if not pos_index.shape[0]:
            #如果没有一个 IOU 的值大于 0.5，则获取最大的那个 BOX，并置为有目标
            pos_index = np.argmax(iou_score)
            num_pos = 1
        else:
            num_pos = pos_index.shape[0]
        #根据正样本的 index 取出正样本的 Anchor
        positive_box = anchors[pos_index]
        #正样本可视化
        #anchor_img = u.draw_box(image,positive_box.reshape([-1,4])*640)
        #GT BOX 可视化
        #u.draw_box(anchor_img,true_boxes[i][0:4].reshape([-1,4])*640,
        #color=[255,0,0])
        #offset 计算时，需要转换成 cx,cy,w,h
        a_class_box = u.xyxy2cxcywh(positive_box)        #正样本
        g_class_box = u.xyxy2cxcywh(true_boxes[i])       #GT BOX
        #计算偏移值
        offset_box = u.cxywh2offset(g_class_box, a_class_box)
        #正样本的偏移值，更新到 true_boxes_assign 中
        true_boxes_assign[pos_index] = offset_box
        #RPN 不用读取分类信息。正样本的 np.array([1, 0])代表前景概率为 1，背景概率为 0
        true_boxes_assign_clf[pos_index] = np.array([1, 0])
        #负样本<0.3 并且 256-正样本的数量为负样本
        neg_index = np.argwhere(iou_score < neg_threshold)
```

```
        neg_num = abs(num_sample - num_pos)
        neg_index = neg_index[:neg_num]
        true_boxes_assign_clf[neg_index] = np.array([0, 1])
    #reshape 成[14400×2,1]
    true_boxes_assign_clf = np.reshape(true_boxes_assign_clf, [-1, 1])
    #输出[14400×4],[14400×2,1]
    return true_boxes_assign, true_boxes_assign_clf, anchors
```

代码中 Anchors 的生成调用 mapping_anchor_2_original_image()函数,然后遍历每个 true_boxes 通过 get_iou(true_boxes[i],anchors)计算 IOU 和得分。根据 pos_index=np.argwhere(iou_score>=pos_threshold)得到满足阈值数组的 index,然后由 positive_box=anchors[pos_index]取出正样本的信息,由 offset_box=u.cxywh2offset(g_class_box,a_class_box)得到正样本的偏移值。最后将编码的值赋给 true_boxes_assign,true_boxes_assign_clf 得到真实的偏移值和有无目标的分类概率。

RPN 损失函数的构建,代码如下:

```
#第 5 章/ObjectDetection/FasterR-CNN/loss.py
import tensorflow as tf

def l1_smooth_loss(y_true, y_pred):
    """回归损失"""
    abs_loss = tf.abs(y_true - y_pred)
    sq_loss = 0.5 * (y_true - y_pred) **2
    l1_loss = tf.where(tf.less(abs_loss, 1.0), sq_loss, abs_loss - 0.5)
    return tf.reduce_sum(l1_loss, -1)


def cross_entropy_loss(y_true, y_pred):
    """交叉熵损失"""
    y_pred = tf.maximum(y_pred, 1e-8)
    softmax_loss = -tf.reduce_sum(y_true *tf.math.log(y_pred), axis=-1)
    return softmax_loss

def rpn_loss(y_pre, true_box, true_clf):
    batch_size = true_box.shape[0]
    #由 NumPy 转换成 Tensor 类型
    true_box = tf.cast(true_box, dtype=tf.float32)
    true_clf = tf.cast(true_clf, dtype=tf.float32)
    #预测值 (b,14400,4) (b,28800,1)
    pre_box, pre_clf = y_pre[1], y_pre[0]
    #当 y_true 传过来时,true_clf 已包括正样本和负样本,并且均有设置
    #因为 true_clf 默认为[0,0],当有目标时是[1,0],当没有目标时是[0,1]
    #reshape 后变成[[0],[0]],[[1],[0]],[[0],[1]],
    #而交叉熵-y_ture *log(y_pre),如果 y_true 为 0,则最后的值为 0,所以不再取正样本
    clf_loss = tf.reduce_mean(
        cross_entropy_loss(true_clf, pre_clf)
    )
    #因为如果是负样本,则 true_box 对应的值为 0
    pos_mask = true_box[..., :4] != tf.convert_to_tensor([0., 0., 0., 0.])
```

```
    #只选正样本的偏移值
    box_loss = tf.reduce_mean(
        l1_smooth_loss(true_box[pos_mask], pre_box[pos_mask])
    )
    total = clf_loss + box_loss
    return total / batch_size
```

函数 l1_smooth_loss()封装了 Smooth 损失，cross_entropy_loss()则封装了交叉熵损失，rpn_loss(y_pre，true_box，true_clf)对 RPN 有无目标进行分类损失和正样本 offset 的回归损失。损失函数的构建需要用深度学习框架构建，否则在反向传播时求出的梯度值有可能为 None。rpn_loss()构建的复杂、简易度受 assign()函数的影响。

训练时只需读取 DataProcessingAndEnhancement()类中的 generate()方法，根据参数设置传输数据，代码如下：

```
#第 5 章/ObjectDetection/FasterR-CNN/data_processing.py
  def generate(self, isTraining=True, isRoi=False):
      #训练时将训练集打乱，不训练时使用验证集
      shuffle(self.train_lines)
      lines = self.train_lines if isTraining else self.val_lines
      #batch 存储相关字段
      inputs = []
      true_box_list = []
      true_clf_list = []
      roi_offset_list = []
      roi_class_list = []
      #循环处理数据集中的数据
      for row in lines:
          #将图像和 BOX 缩放到指定的尺寸
          img, y = self.get_image_processing_results(row)
          if len(y) != 0:
              boxes = np.array(y[:, :4], dtype=np.float32)
              #对 label 进行归一化
              old_boxes = boxes.copy()
              boxes = boxes / np.array(self.input_shape[0:2] + self.input_shape[0:2])
              #对 label 构建 one_hot 编码
              one_hot_label = np.eye(self.num_classes)[np.array(y[:, 4], np.int32)]
              #将 max xy - min xy <0 的 label 过滤掉
              if ((boxes[:, 3] - boxes[:, 1]) <= 0).any() and ((boxes[:, 2] - boxes
[:, 0]) <= 0).any():
                  continue
              #组成 xmin, ymin, xmax, ymax, [0,1]
              y = np.concatenate([boxes, one_hot_label], axis=-1)
          true_boxes_assign, true_boxes_assign_clf, anchors = self.assign(y,
image=img, old_boxes=old_boxes)
          if isRoi:
              roi_offset_y, roi_class = self.roi_generate(img, anchors, y)
              roi_offset_list.append(roi_offset_y)
              roi_class_list.append(roi_class)
          inputs.append(img)
          true_box_list.append(true_boxes_assign)
```

```
        true_clf_list.append(true_boxes_assign_clf)
        #按 batch_size 传输 targets
        if len(inputs) == self.batch_size:
            tmp_inp = np.array(inputs, dtype=np.float32)

            if isRoi:
                tmp_roi = np.array(roi_offset_list)
                tmp_roi_class = np.array(roi_class_list)
                roi_offset_list = []
                roi_class_list = []
                inputs = []
                yield tmp_inp, tmp_roi, tmp_roi_class
            else:
                tmp_box = np.array(true_box_list)
                tmp_clf = np.array(true_clf_list)
                true_box_list = []
                true_clf_list = []
                inputs = []
                yield tmp_inp, tmp_box, tmp_clf
```

方法 generate()主要根据 isRoi 的设定返回 RPN 的训练数据或者 ROI 的训练数据，如果是 RNP 数据，则通过 self. assign()方法得到 Anchor 与 GT BOX 的 offset 和 class 分类信息。self. get_image_processing_results()仅实现了输入尺寸 $M\times N$ 的调节及图像的归一化功能。当 len(inputs)==self. batch_size 相等时，通过 yield 生成器返回 tmp_inp 图片、tmp_boxanchor 与 GT BOX 的 offset、tmp_clf 有无目标的分类信息。

训练代码相对容易，只需读取 generate()的数据，然后调用 train_step()实现反向传播，代码如下：

```
#第 5 章/ObjectDetection/FasterR-CNN/rpn_train.py
    def train_step(model, features, true_box, true_clf):
        with tf.GradientTape() as tape:
            #置信度是 batch×28 800×1
            #位置是 batch×14 400×4
            predictions = model(features, training=True)
            #传入 RPN 预测的分类和位置信息，并且传入 true_box 信息
            loss = rpn_loss(predictions, true_box, true_clf)
        #求梯度
        gradients = tape.gradient(loss, model.trainable_variables)
        #反向传播
        optimizer.apply_gradients(zip(gradients, model.trainable_variables))
        #更新 loss 信息
        train_loss.update_state(loss)
        train_metric.update_state(true_clf, predictions[0])
        global_steps.assign_add(1)
```

代码 loss=rpn_loss(predictions,true_box,true_clf)中的 predictions 包含 pre_box 预测框基于 Anchor 的偏移，以及 pre_clf 预测框有无目标的概率，此时维度要跟 GT BOX 的信息保持一致，这样就更方便 rpn_loss()函数进行计算。更多更详细的代码可参考随书代码。

5.3.4 代码实战 ROI 损失函数及训练

经过 RPN 的训练此时 14 400×4 个预测框基于 Anchor 的偏移，根据图 5-22 的结构需要选出 2000 个重叠率较低的框，然后将 2000 个框与 GT BOX 再进行 1 次偏移作为 roi_loss()中的 y_true，此过程封装在代码 roi_generate()中，代码如下：

```
#第5章/ObjectDetection/FasterR-CNN/data_processing.py

    def roi_generate(self, img, anchors, true_box, input_shape=[640, 640, 3]):
        rpn_model, all_model = Faster R-CNN(
            input_shape=input_shape,
            roi_input_shape=[None, 4],
            num_anchors=9,
            num_class=2 + 1
        )
        #加载上一步训练的权重
        rpn_model.load_weights('./weights/last.h5')
        #输入图片的特征，输出
        pre_rpn_cls, pre_rpn_box = rpn_model(np.expand_dims(img, axis=0))
        #如果启动 ROI，则根据 y 值和上一次的权重得到区域选择框
        #得到 2000 个 roi(1,2000,4)
        pre_roi = pre_roi_2_box(pre_rpn_box, pre_rpn_cls, anchors, np.array
(self.input_shape[0:2]))
        #u.draw_box(img *255, pre_roi.reshape([2000, 4]) *640)
        #构建 ROI 网络的 y 值 2000×8
        roi_y = gt_box_2_roi(true_box, pre_roi)
        return roi_y
```

代码 rpn_model.load_weights()调用 RPN 网络的权重，然后前向传播得到 pre_rpn_cls，pre_rpn_box 的输出，然后通过 pre_roi_2_box()函数得到 2000 个建议框，gt_box_2_roi()函数得到 ROI 的 y_true。

函数 pre_roi_2_box()的实现，代码如下：

```
#第5章/ObjectDetection/FasterR-CNN/proposal.py
import numpy as np
import utils as u
import tensorflow as tf

def pre_roi_2_box(rpn_box_loc, rpn_box_score,
                  anchor, img_size,
                  num_pre=12000, nms_thresh=0.7, ss_num=2000):
    #根据 RPN 网络输入 BOX Loc 和 BOX Score，以及 Anchor
    #选择输出 2000 个训练 rois。注意此时没有 GT BOX 的事情
    #因为 Anchor 是 xmin,ymin,xmax,ymax 转换成 x,y,w,h 的形式
    anchor = u.xyxy2cxcywh(anchor)
    batch = rpn_box_loc.shape[0]
    #因为 RPN 输出的是 offset 值，所以需要将其解码成 xmin,ymin,xmax,ymax
    roi_box = u.offset2xyxy(rpn_box_loc.NumPy(), anchor)
    #对 roi_box 中的 BOX 进行裁剪
```

```
        roi_box[:, slice(0, 4, 2)] = np.clip(roi_box[:, slice(0, 4, 2)], 0, img_size[0])
        roi_box[:, slice(1, 4, 2)] = np.clip(roi_box[:, slice(1, 4, 2)], 0, img_size[1])
        #########################
        #得到满足目标大小的置信度分数[1, 0]
        rpn_box_score = np.reshape(rpn_box_score, [batch, -1, 2])
        pos_score = rpn_box_score[..., 0] #得到前景的分数
        #对置信度的分数进行降序排列
        order = tf.argsort(pos_score, direction='DESCENDING').NumPy()
        #取前 12000 个预测框
        order = order.ravel()[:num_pre]
        #对 ROI 进行过滤,只要 120000 个置信度较大的框
        roi = roi_box[..., order, :]
        score = pos_score[..., order]
        #使用 NMS 去掉交并比>0.7 的框,留下重叠率不高的框,最多只有 2000 个
        #non_max_suppression()得到的是 ROI 中的 index
        keep = tf.image.non_max_suppression(
            np.reshape(roi, roi.shape[1:]),
            np.reshape(score, score.shape[1:]),
            max_output_size=ss_num,
            iou_threshold=nms_thresh
        )
        #利用 Anchor 和 RPN 网络预测出来的 offset,选取可能有目标最大的前 2000 个框,作为建议框
        roi = roi[..., keep, :]
        return roi
```

因为 RPN 预测输出的是 offset,所以调用 offset2xyxy()进行解码以得到左上、右下坐标,此时 BOX 信息较多,然后根据 pos_score=rpn_box_score[...,0]有无目标概率的得分进行非极大值抑制,从而保留重叠率不高的框,共计 2000 个,所以此函数的输出为 2000×4,代码如下:

```
#第 5 章/ObjectDetection/FasterR-CNN/proposal.py
def gt_box_2_roi(gt_box, roi, sample_num=128, pos_iou_thresh=0.7, neg_iou_
thresh=0.1, num_class=2 + 1):
    #RPN 产生了 2000 个区域选择框,但是并没有将 2000 个框全用作训练,而是将 2000 个框与
GT BOX 做 IOU,一共选择 128 个正样本
    #假设 IOU>0.7 的正样本选择 32 个。IOU<0.3 的负样本选择 128-32=96 个
    roi = np.reshape(roi, roi.shape[1:])
    #初始为 0 的数组,用来保存计算后的值
    true_offset = np.zeros([roi.shape[0], 4 * (num_class - 1)])
    true_class = np.zeros([roi.shape[0], num_class])

    for box in gt_box:
        #如果此时的 BOX 类别假设为[1,0,0]
        #如果此时的 BOX 类别假设为[0,1,0],offset 则更新到 1 的位置
        #再次计算每个 GT BOX 与 ROI 的 IOU 值
        iou_score = u.get_iou(box[:4], roi[..., :4])
        #如果 IOU>0.7,则全记为正样本
        pos_iou = iou_score > pos_iou_thresh
        if not sum(pos_iou):
            pos_index = np.argmax(iou_score)
        else:
```

```
            pos_index = np.argwhere(pos_iou == True)
        num_pos = pos_index.size
        #如果 IOU<0.3,则为负样本
        neg_iou = iou_score < neg_iou_thresh
        neg_index = np.argwhere(neg_iou == True)
        if pos_index.size != 1:
            pos_index = pos_index[:num_pos]
        #负样本的数量
        neg_index = neg_index[:sample_num - num_pos]
        #取出正样本的 BOX
        pos_box = roi[pos_index]
        pos_box = np.reshape(pos_box, [-1, 4])
        #计算 cx,cy,w,h
        pos_box = u.xyxy2cxcywh(pos_box)
        gt_box = u.xyxy2cxcywh(gt_box)
        #然后对正样本的 BOX 进行编码
        roi_offset = u.cxywh2offset(box, pos_box)
        for i in range(num_class - 1):
            if box[4 + i] == 1:
                start = i * 4
                #正样本的 offset
                true_offset[pos_index.flat, start:4 + start] = roi_offset
        #负样本只填充负样本的分类信息
        neg_class = np.zeros(num_class)
        neg_class[-1] = 1 #[0,0,1]代表负样本
        #将计算后的值赋给 true_class
        true_class[neg_index, :] = neg_class
        true_class[pos_index, :] = box[4:]
    return true_offset, true_class
```

函数 gt_box_2_roi()实现将输入的 2000 个建议框与 GT BOX 之间做 IOU,如果 IOU>0.7,则为正样本并计算 offset,如果 IOU<0.3,则为负样本,则正负样本之和为 128(可调)。因为 true_offset 的输出为 2000×4×(num_class−1),在此函数为 2000×8,即每个框可能为两个分类的偏移,所以当 if box[4+i]==1 时,当前分类才赋值 true_offset[pos_index.flat,start:4+start]=roi_offset。

接下来是损失函数的构建,代码如下:

```
#第 5 章/ObjectDetection/FasterR-CNN/loss.py
def roi_loss(y_pre, roi, roi_class, num_class=2 + 1):
    #预测信息 2000×3 和 2000×8
    out_class, out_regbox = y_pre
    #真实值 2000×8 和 2000×3
    y_true = tf.cast(roi, tf.float32)
    true_cls = tf.cast(roi_class, tf.float32)
    batch_size = y_true.shape[0]
    #只要位置信息
    true_box = y_true[..., :8]
    #因为 true_cls 默认为[0,0,0],负样本是[0,0,1],正样本是[1,0,0]
    #所以-y_true * log(y_pred),不用提取出负样本的分类
    clf_loss = tf.reduce_mean(
```

```
        cross_entropy_loss(true_cls, out_class)
    )
    #按每个类别去求正样本 BOX offset 的损失
    box_loss = 0
    for i in range(num_class - 1):
        #因为如果是负样本,true_box 对应的值为 0,所以只选正样本的偏移值
        mask = tf.reshape(true_cls[..., i], [batch_size, -1, 1])
        ind = tf.where(mask == 1)
        #根据 index 取值
        pos_true_box = tf.gather_nd(true_box, ind)
        pos_out_box = tf.gather_nd(out_regbox, ind)
        box_loss += tf.reduce_mean(
            l1_smooth_loss(pos_true_box, pos_out_box)
        )
    total = clf_loss + box_loss
    return total / batch_size
```

函数 roi_loss()与 rpn_loss()与此类似,不同之处在于求解了预测 BOX 的分类信息,同时实现时按每个类别的 offset 损失进行求和。

训练代码 train_step()与 RPN 类似,不同之处在于 predictions=model([features,roi_box[...,:4]],training=True)时传入图像信息及 2000 个 Proposal 信息。更多更详细的代码可参考随书代码。

5.3.5 代码实战预测推理

Faster R-CNN 的推理首先要经过 RPN 网络得到 14 400×2 个分类、14 400×4 个框,再经过 NMS 选出 2000×4 个框,然后送入 ROI 网络,假设分类数量为 2,则类别概率为 2000×3(2+1 前景/背景),每个类别的 BOX 信息为 2000×8(有目标的 BOX 和没有目标的 BOX),其详细的推理代码如下:

```
#第 5 章/ObjectDetection/FasterR-CNN/detected.py
class Detected():
    def __init__(self, rpn_path, roi_path, input_size):
    #读取模型和权重
        self.rpn_model = load_model(
            rpn_path,
            custom_objects={'rpn_loss': rpn_loss}
        )
        self.roi_model = load_model(
            roi_path,
            custom_objects={'roi_loss': roi_loss}
        )
        self.confidence_threshold = 0.5             #有无目标置信度
        self.class_prob = [0.5, 0.5]                #两个类别的分类阈值
        self.nms_threshold = 0.5                    #NMS 的阈值
        self.input_size = input_size

    def generate_anchor(self):
        #默认 Anchor 的生成
```

```
        anchor_num, feature_stride = 9, 16
        feature_hw_size = np.array(self.input_size) //feature_stride
        #各种 Anchor 的比例, 得到的是 xmin,ymin,xmax,ymax
        anchor_base = a.generate_anchors(base_size=feature_stride)
        #以 640×640 输入时,在每幅图中生成 14400×4 个 BOX 的信息
        anchors = a.mapping_anchor_2_original_image(
            anchor_base, feature_hw_size,
            feature_stride, anchor_num
        )
        #Anchor 归一化
        anchors = anchors / self.input_size[0]
        self.anchors = anchors

    def readImg(self, img_path=None):
        #读取要预测的图片
        img = cv2.imread(img_path)
        #将图片转换为 640×640
        self.img, _ = u.letterbox_image(img, self.input_size, [])

    def rpn_forward(self):
        #读取图片以进行前向传播
        img_tensor = tf.expand_dims(self.img / 255.0, axis=0)
        #前向传播
        self.output = self.rpn_model.predict(img_tensor)
        self.old_img = self.img.copy()

    def rpn_nms2k(self):
        #预测置信度的结果为 14400×2
        pre_rpn_cls = tf.cast(self.output[0], dtype=tf.float32)
        #预测框的偏移
        pre_rpn_box = tf.cast(self.output[1], dtype=tf.float32)
        #pre_roi_2_box 已集成根据预测框进行解码操作,并根据置信度的得分进行 NMS 去重,
        #得到 2000×4 个框
        #14400× 4, 28800×1
        pre_roi = pre_roi_2_box(pre_rpn_box, pre_rpn_cls, self.anchors,
np.array(self.input_size[0:2]), isReturnAnchor=True)
        #喂入 ROI 网络此时的 BOX 信息
        self.pre_roi = pre_roi[0]
        #喂入 ROI 网络此时 BOX 对应的 Anchor 值
        self.pre_anchor = pre_roi[1]

    def roi_forward(self):
        img_tensor = tf.expand_dims(self.img / 255.0, axis=0)
        #ROI 前向传播,得到 2000×3 和 2000×8
        self.output = self.roi_model.predict([img_tensor, self.pre_roi])
        #分类+置信度的值
        self.pre_class = self.output[0]
        #BOX 的值,但是输出为 2000×8,其中有 4 位是没有目标的框
        #此时仍为偏移值
        self.pre_box = self.output[1]

    def _roi_box_decode(self, pre_box):
```

```
        #转换成 cx,cy,w,h。只取喂入 ROI 网络的 2000 个 Anchor
        anchor = utils.xyxy2cxcywh(self.pre_anchor)
        #解码操作封装在 offset2xyxy 函数中
        decode_box = utils.offset2xyxy(pre_box, anchor)
                #裁剪值域在[0,1]
        decode_box = np.clip(decode_box, 0, 1)
        return decode_box

    def classification_filtering(self):
        #首先根据置信度过滤
        #因为在训练时 [0,0,1]代表负样本,所以只有最后 1 位小于阈值时才代表有目标
        #即背景的概率越小越好
        obj_mask = self.pre_class[..., -1] < self.confidence_threshold
        #将 ROI 输出的预测值 BOX 与 RPN 输出的最佳 2000 个框进行解码操作
        #并且只传前 4 个位置的,因为后面 4 个位置的是背景所处的位置
        boxes = self._roi_box_decode(self.pre_box[..., :4])
        #根据类别的概率过滤
        for i, p in enumerate(self.class_prob):
            if len(obj_mask):
                #根据 obj_mask 过滤分类
                cls_obj = self.pre_class[obj_mask]
                #类别的得分要超过阈值
                cls_obj_mask = cls_obj[..., i] > p
                classification = cls_obj[cls_obj_mask]
                #根据置信度和类别的 mask 过滤出对应 boxes 中的信息
                class_boxes = boxes[obj_mask][cls_obj_mask]
                class_score = classification[..., i]
                #将 BOX 与分类得分合并
                box = np.concatenate([class_boxes, np.reshape(class_score,
[-1, 1])], axis=-1)
                #NMS 得到的索引
                index = u.nms(box, nms_thresh=self.nms_threshold)
                #根据索引取出 boxes 信息
                box = class_boxes[index]
                #绘图
                img = u.draw_box(self.old_img, box)
        u.show(img)
if __name__ == "__main__":
    rpn_path = "./weights/last"
    roi_path = "./weightsRoi/RoiLast"
    img_path = "../val_data/pexels-photo-5211438.jpeg"
    #实例化对象
    det = Detected(rpn_path, roi_path, [640, 640])
    #读取预测图片
    det.readImg(img_path)
    #生成 Anchor
    det.generate_anchor()
    ####################
    #RPN 网络前向传播
    det.rpn_forward()
    #RPN 网络解析得到 2000 个框
    det.rpn_nms2k()
```

```
#####################
#ROI 前向传播
det.roi_forward()
#对最后的结果解码
det.classification_filtering()
```

因为 Faster R-CNN 是两阶段网络，所以需要分别加载 RPN、ROI 的权重进行预测。rpn_forward(self)实现读取图片的前向传播，然后调用 rpn_nms2k(self)得到 2000 个候选框，并同时得到此时相对应的 2000 个 Anchor 的值。将 ROI 的 2000 个框在 roi_forward(self)进行预测 self. roi_model. predict([img_tensor, self. pre_roi])，获得分类和置信度的输出 self. pre_class 及最终的 BOX 信息 self. pre_box。在 classification_filtering(self)方法中根据置信度为前景的概率和分类得分进行过滤，并对 self. pre_box 进行由偏移值转向位置值的解码操作，最后经过 NMS 非极大值抑制到得最终的结果。

总结

Faster R-CNN 延续了 R-CNN 中的思想，分为两个阶段。第 1 个阶段，RPN 网络用来提取 2000 个前景、背景框；第 2 个阶段，网络输入特征信息和 2000 个框，进行分类＋背景及 BOX 的微调。

练习

运行并调试本节代码，理解算法的设计与代码的结合。

5.4 单阶段多尺度检测网络 SSD

5.4.1 模型介绍

SSD 是于 2015 年由 Wei Liu 等在发表的论文 *SSD：Single Shot MultiBox Detector* 中提出的单阶段、多检测头的目标检测网络，其网络结构如图 5-25 所示。

在特征提取阶段使用 VGG-16(结构为 2—2—3—3—3—3，2 卷积，1 池化，最后 3 为全连接)，并选取 Conv4_3(第 4 个层的最后 1 个卷积)输出 38×38×512 作为 classifier1 检测头的输入。当时没有使用 BN 归一化，而 Conv4_3 提取的特征值较大，所以使用 Normalization 归一化以防止梯度爆炸。

将 VGG-16 中的全连接 FC6 层更换为卷积，并使用空洞卷积，空洞率为 6×6，增大感受野。SSD 的源码在 Conv5_3 后的 Pooling 由 2×2－s2 更换为 3×3－s1，原有 FC6 在 Conv5_3 上的感受野为 14×14，而 FC6 由原来的 7×7 变成 3×3 后，使用空洞卷积 $3+(3+2(n-1)-1)\times1=14$，算出来 $n=5.5$，向上取整为 6，使 FC6 的感受野与 VGG-16 时保持一致，然后将 FC7 更换为 1×1×1024 卷积，得到输出 19×19×1024 作为 classifier2 检测头的输入。

在 VGG-16 的基础上添加 Extra Layer，Conv8_2 由 1×1×256、3×3×512－s2 输出 10×10×512 作为 classifier3 检测头的输入；同理 Conv9_2 由 1×1×128、3×3×256－s2 输出为 5×5×256 作为 classifier4 检测头的输入；Conv10_2 由 1×1×128、3×3×256－s1

并且卷积 valid 丢弃，输出为 3×3×256 作为 classifier5 检测头的输入；Conv11_2 与此类似，输出为 1×1×256 作为 classifier6 检测头的输入。

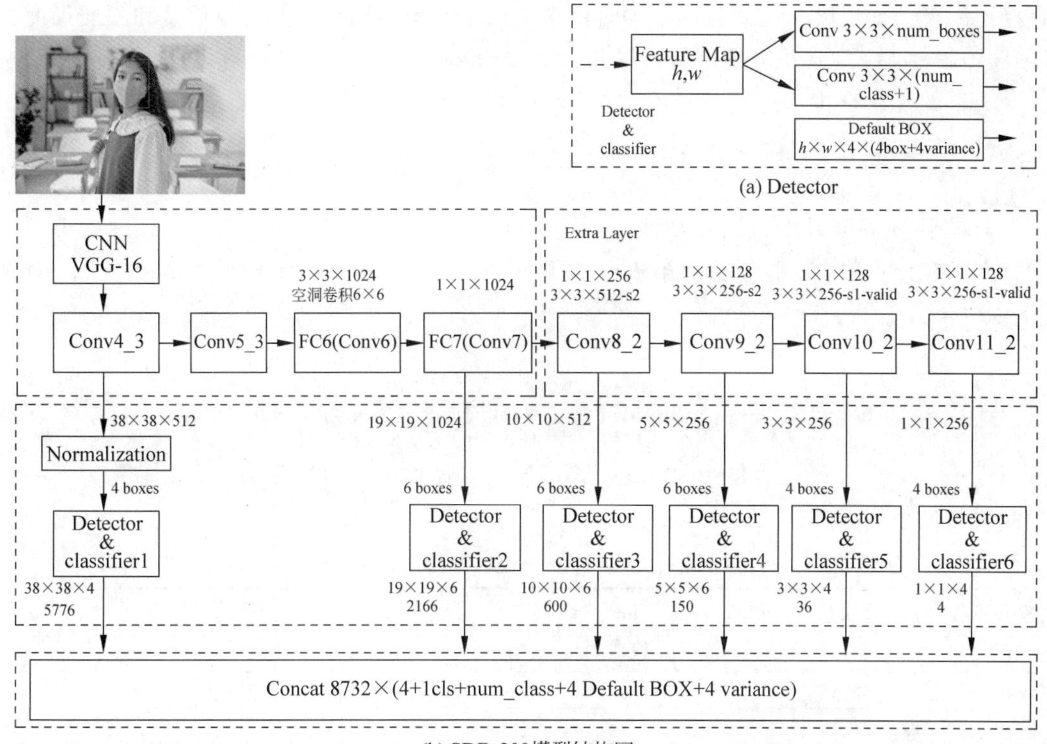

图 5-25 SSD 网络结构

共计 6 个特征作为 6 个检测头的输入，并且由 classifier1 检测小目标、classifier6 检测大目标（classifier6 的感受野最大，所以检测大目标），中间的检测头次之。

得到特征信息后，分别输入两个 3×3 卷积，如 classifier1 输出 38×38×4，即 BOX 的 4 个偏移位置信息，38×38×(1+num_class)的分类信息，其中 1 代表 BOX 有无目标。另外一个分支输出 Default BOX 指在 38×38 的特征层上生成 4 个建议框，即每个特征图对应到原图约 300/38=7.9 的区域，那么一共就有 38×38×4=5776 个建议框，并随着 classifier1 进行输出，SSD 检测头有 4 个，分别是位置偏移值、置信度（有无目标）及分类概率、Default BOX。Default BOX 本来是 5776×4，但是因为 offset 的值较小，作者对(tx,ty)/variance 0.1、(tw,th)/variance 0.2 进行了放大，防止梯度消失，所以其输出变成了 5776×8。

建议框为 classifier1 分配 4 个，为 classifier2、classifier3、classifier4 分配 6 个，为 classifier5、classifier6 分配 4 个，共计 38×38×4+19×19×6+10×10×6+5×5×6+3×3×4+1×1×4=8732 个建议框，所以 SSD 的输出为 8732×4 个框，8732×(1+num)分类概率，8732×8 个 Default BOX，合并后输出为 8732×(4+1cls+num+4Default BOX+4variance)。

综上，SSD 拥有 6 个检测头，其特征信息使用 Conv4_3、Conv7、Conv8_2、Conv9_2、

Conv10_2、Conv11_2 的输出，每个特征信息分配 4、6、6、6、4、4 个框，直接输出 8732×(4+1cls+num+4Default BOX+4variance)，不再经过 Faster R-CNN 中的 RPN 选出建议框，直接预测极大地提高了检测速度。同时由于使用了 6 个不同尺度的特征信息和不同尺度的 Default BOX，3×3 卷积分别进行位置和分类预测，对于模型的精度也有极大的提高。

SSD 的建议框生成过程首先经过以下公式计算最小、最大高宽的尺寸。

$$S_k = S_{\min} + \frac{S_{\max} - S_{\min}}{m-1}(k-1), \quad k \in [1, m] \tag{5-5}$$

公式中 m 指有几个检测头，所以 $m=6$，同时原文初始设定 $S_{\min}=0.2$，$S_{\max}=0.9$。

实际在计算时，由于考虑到第 1 个检测头 $S_k=0.2\times300=60$ 作为 max_size_1，而将 $\text{min\_size}_1=60\times1/2=30$，所以求其他检测头的尺寸时设置 $m=5$，故 $\frac{S_{\max}-S_{\min}}{m-1}=\frac{0.9-0.2}{5-1}\approx0.17$，然后 $\text{max\_size}_2=(0.2+0.17\times1)\times300=111$，$\text{max\_size}_3=162$，$\text{max\_size}_4=213$，$\text{max\_size}_5=264$，$\text{max\_size}_6=315$，再将上一个检测头的最大值作为下一个检测头的最小值，见表 5-1。

表 5-1　SSD 建议框默认尺寸

特征层名称	特征层尺寸	min_size(k)	max_size(k)	比　例	step
Conv4_3	38×38	30	60	$1:1, 1:\sqrt{2}, \sqrt{2}:1, 1:1$	8
Conv7	19×19	60	111	$1:1, 1:\sqrt{2}, \sqrt{2}:1, 1:\sqrt{3}, \sqrt{3}:1, 1:1$	16
Conv8_2	10×10	111	162	$1:1, 1:\sqrt{2}, \sqrt{2}:1, 1:\sqrt{3}, \sqrt{3}:1, 1:1$	32
Conv9_2	5×5	162	213	$1:1, 1:\sqrt{2}, \sqrt{2}:1, 1:\sqrt{3}, \sqrt{3}:1, 1:1$	64
Conv10_2	3×3	213	264	$1:1, 1:\sqrt{2}, \sqrt{2}:1, 1:1$	100
Conv11_2	1×1	264	315	$1:1, 1:\sqrt{2}, \sqrt{2}:1, 1:1$	300

第 1 个 1∶1 的比例，宽和高 wh=[30,30]；1∶$\sqrt{2}$ 时为 wh=$[30\times\sqrt{2}, 30\times1/\sqrt{2}]$=[42,21]；$\sqrt{2}$∶1 时 wh=$[30\times1/\sqrt{2}, 30\times\sqrt{2}]$=[21,42]，最后 1 个 1∶1 为 wh=$\sqrt{30\times60}:\sqrt{30\times60}$=42∶42，其他层与此类似，最后得到建议框的宽和高见表 5-2。

表 5-2　SSD 建议框宽和高默认值

特征层名称	特征层尺寸	建议框宽和高
Conv4_3	38×38	[30,30]、[42,21]、[21,42]、[42,42]
Conv7	19×19	[60,60]、[84,42]、[42,84]、[103,34]、[34,103]、[81,81]
Conv8_2	10×10	[111,111]、[156,78]、[78,156]、[192,64]、[64,192]、[134,134]
Conv9_2	5×5	[162,162]、[229,114]、[114,229]、[280,93]、[93,280]、[185,185]
Conv10_2	3×3	[213,213]、[301,150]、[150,301]、[237,237]
Conv11_2	1×1	[264,264]、[373,186]、[186,373]、[288,288]

与 Faster R-CNN 建议框生成略有不同的是均分不是从图像中的(0,0)坐标开始的，而是采用[0.5×step，300－0.5×step]均分到特征图的尺寸份，例如 Conv7 则为[0.5×16，

300－0.5×16]均分成 19 份。

假设 19×19 特征图中的每个像素映射回原图，x 轴和 y 轴均有 19 个均分点，每两个点之间相差 16×16，并且建议框需要在每个 x 和 y 坐标点为中心生成 6 个红色建议框，然后计算红色框与绿色框之间的 IOU，当 IOU≥0.5 时 Anchor 为正样本，其他为负样本。绿色为真实标注框，可视化效果如图 5-26 所示。

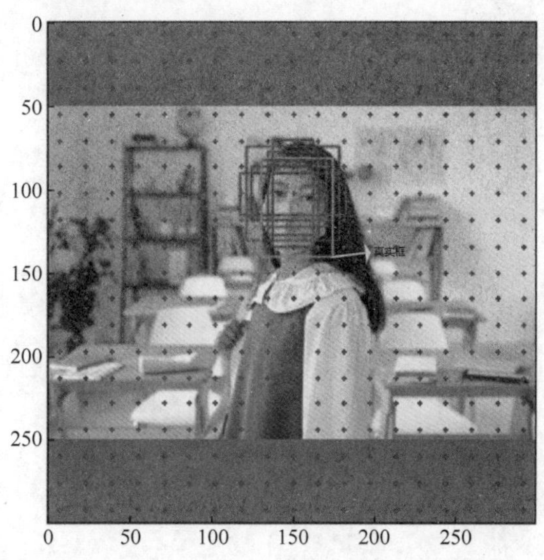

图 5-26　Conv7 特征图中生成的建议框(见彩插)

在计算建议框与真实框之间的偏移时，其计算公式仍采用式(5-2)，得到偏移值后 t_x/variance1、t_y/variance1、t_w/variance2、t_h/variance2，其中 variance1＝0.1，variance2＝0.2，对偏移值进行了放大。在推理时使用式(5-3)，然后对于预测值乘以 variance1、variance2 进行相同比例的缩小还原。

SSD 的损失函数分为位置损失、分类误差损失：

$$
\begin{aligned}
&L(x,c,l,g)=\frac{1}{N}(L_{\text{conf}}(x,c)+\alpha L_{\text{loc}}(x,l,g)) \\
&L_{\text{conf}}(x,c)=-\sum_{i\in \text{Pos}}^{N} x_{ij}^{p}\log(\hat{c}_i^p)-\sum_{i\in \text{Neg}}^{N}\log(\hat{c}_i^0) \\
&\hat{c}_i^p=\frac{\exp(c_i^p)}{\sum_p \exp(c_i^p)}
\end{aligned}
\tag{5-6}
$$

其中，N 为真实边界框配对的建议框数量，对于 x，如果 1 个建议框与真实边界框配对为 1，否则为 0；c 为真实物体的预测值；l 用于预测边界框中心位置的长、宽；g 为真实边界框中心位置的长、宽。$L_{\text{loc}}(x,l,g)$ 使用的是 Smooth 损失函数；$L_{\text{conf}}(x,c)$ 包含正样本的损失和负样本的损失，并使用 Softmax 求解分类 $\hat{c}_i^p$。

因为正样本较少，负样本较多，所以在求解损失时将负样本按背景(无目标)概率从大到小进行排序,控制负样本与正样本的比例为 3∶1。整个网络的实现,可参见 5.4.2 节。

5.4.2 代码实战模型搭建

整个模型的实现可以分为 3 部分,VGG-16 作为特征提取部分,extra_layer 作为 VGG 补充特征提取,detect_head 作为检测头进行多尺度检测并作为输出。

重构 VGG-16 的代码如下:

```
#第 5 章/ObjectDetection/TensorFlow_SSD_Detected/backbone/vggssd.py
def vgg(input_tensor):
    net = {}
    #默认不使用 BN
    is_bn = False
    #将输入内容放入 net 字典中
    net['input'] = input_tensor
    #原 VGG 的配置结构,数字代表输出 channel,M 代表池化
    vgg_config = [
        64, 64, 'M',
        128, 128, 'M',
        256, 256, 256, 'M',
        512, 512, 512, 'M',
        512, 512, 512, 'M'
    ]
    #输入内容
    x = net['input']
    #对配置文件进行解析
    for i, channel in enumerate(vgg_config):
        if channel != 'M':
            #ConvBnRelu 已封装
            x = ConvBnRelu(filters_channels=channel,
                           kernel_size=[3, 3],
                           strides=[1, 1],
                           padding='same',
                           is_bn=is_bn,
                           is_relu=True
                           )(x)
            #每次将卷积内容放入字典中
            net['conv_{}'.format(i + 1)] = x
        else:
            #如果不是最后 1 个池化,则默认为 2×2-s2
            if i != len(vgg_config) - 1:
                pool_size = [2, 2]
                strides = [2, 2]
            else:
                pool_size = [3, 3] #pool5,最后一个池化为 3×3-s=1
                strides = [1, 1]
            x = layers.MaxPooling2D(pool_size=pool_size, strides=strides,
padding='same')(x)
            net['max_pool_{}'.format(i + 1)] = x
    return net
```

重构并通过配置生成模型是为了调优时增加卷积深度和宽度。在构建模型时使用字典来管理各个网络层。ConvBnRelu()是卷积、BN、归一化的封装,is_bn 默认不开启。

在 VGG 的基础上添加 extra_layer 的代码如下：

```
#第 5 章/ObjectDetection/TensorFlow_SSD_Detected/backbone/vggssd.py
def extra_layer(input_tensor):
    """VGG 额外增加的一些 Net """
    net = {}
    is_bn = False
    x = input_tensor
    #格式为[channel, kernel_size,strides,padding]
    extra_config = [
        [1024, [3, 3], [6, 6], 'same'],                #fc6
        [1024, [1, 1], [1, 1], 'same'],                #fc7

        [256, [1, 1], [1, 1], 'same'],                 #conv8_1
        [512, [3, 3], [2, 2], 'same'],                 #conv8_2

        [128, [1, 1], [1, 1], 'same'],                 #conv9_1
        [256, [3, 3], [2, 2], 'same'],                 #conv9_2

        [128, [1, 1], [1, 1], 'same'],                 #conv10_1
        [256, [3, 3], [1, 1], 'valid'],                #conv10_2

        [128, [1, 1], [1, 1], 'same'],                 #conv11_1
        [256, [3, 3], [1, 1], 'valid'],                #conv11_2
    ]
    #读配置文件,生成 extra_layer
    for i, cnf in enumerate(extra_config):
        if i == 0:
            #第 1 个是空洞卷积,[6, 6],dilation_rate 指定空洞率
            x = ConvBnRelu(cnf[0], cnf[1], strides=[1, 1],
                        dilation_rate=cnf[-2], padding=cnf[-1],
                        is_bn=is_bn, is_relu=True)(x)
        else:
            x = ConvBnRelu(cnf[0], cnf[1], cnf[2], cnf[-1],
                        is_bn=is_bn, is_relu=True)(x)
        net['conv_extra_{}'.format(i)] = x
    return net

def vgg_extra(input_tensor):
    """对 VGG 和 extra_layer 进行整合"""
    net1 = vgg(input_tensor)
    net2 = extra_layer(net1['max_pool_18'])
    net1.update(net2)
    return net1
```

函数 extra_layer()根据配置文件增加网络层，并且第 1 个网络层的卷积为空洞卷积，vgg_extra(input_tensor)将 VGG 和 extra_layer 进行了整合，构建成整个特征提取网络。net1['max_pool_18']也就是 VGG 的第 5 个卷积后的池化层，如图 5-27 所示。

对 net1 和 net2 进行合并后一共有 29 个网络层，其中网络图与代码的对应关系为 conv4_3＝conv_13、conv7＝conv_extra_1、conv8_2＝conv_extra_3、conv9_2＝conv_extra_5、conv10_2＝conv_extra_7、conv11_2＝conv_extra_9，如图 5-28 所示。

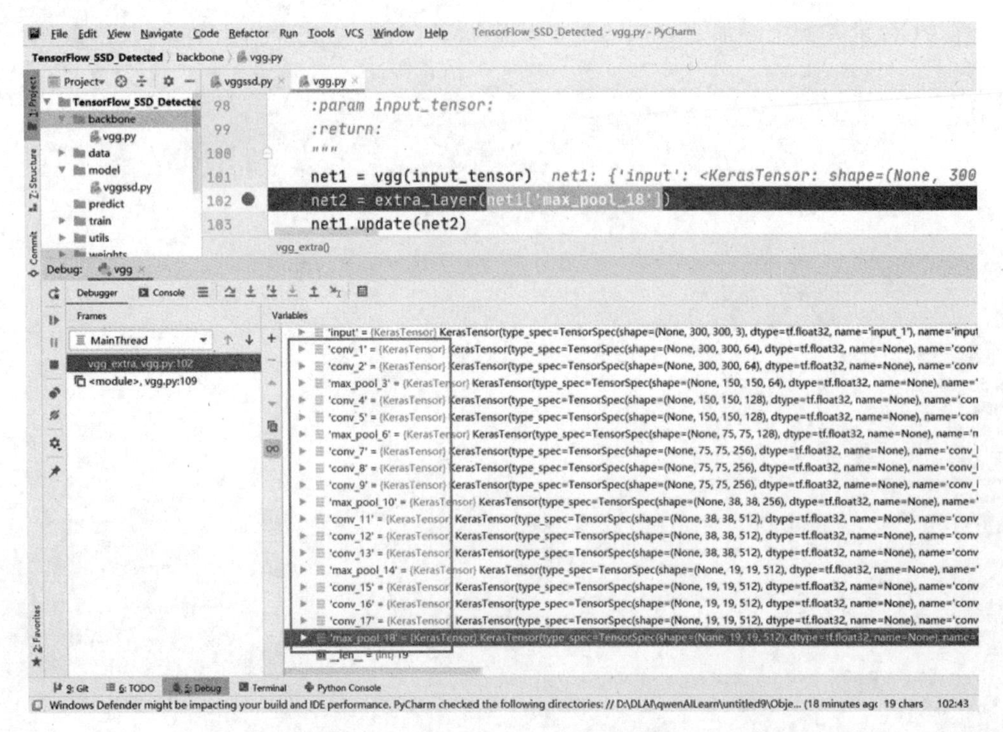

图 5-27 max_pool_18 所处网络层

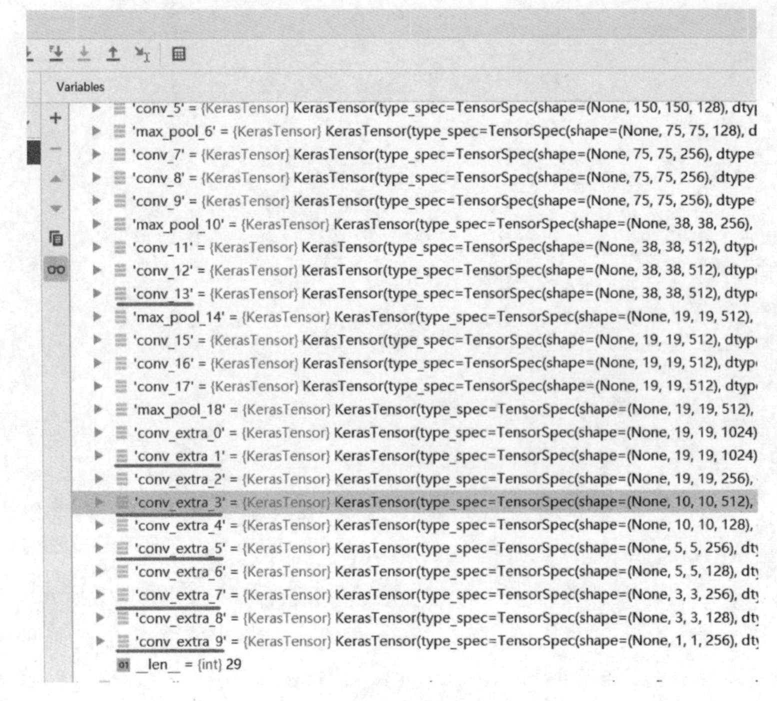

图 5-28 SSD 选择的特征层

检测头的构建分别用了两个 3×3 卷积，并且传入当前特征建议框的宽高比，将在 DefaultBox()中自动生成对应的建议框，代码如下：

```
#第 5 章/ObjectDetection/TensorFlow_SSD_Detected/model/detect_head.py
def vgg_detect_head(input_tensor, num_priors, num_classes,
                    box_layer_name, min_max_size, ratios,
                    img_size=[300, 300]):
    """
    VGG 的检测头，一个用来预测偏移值，一个用来预测分类，另外一个是 Default BOX
    :param input_tensor: 预测的卷积层
    :param num_priors: Default BOX 的数量
    :param num_classes: 类别的数量
    :param box_layer_name: 预测的卷积层的名称
    :param min_max_size: 锚框的最小和最大尺寸
    :param ratios: 建议框的比例
    :param img_size: 输入图像的尺寸
    :return:
    """
    net = {}
    x = input_tensor
    #用来检测 location 的偏移值
    net[box_layer_name + "_loc"] = ConvBnRelu(num_priors *4,
                                              kernel_size=[3, 3],
                                              strides=[1, 1],
                                              padding='same',
                                              is_bn=False,
                                              is_relu=False)(x)
    #location 位置摊平
    net[box_layer_name + "_loc" + "_flat"] = layers.Flatten()(net[box_layer_name +
"_loc"])
    #用来检测每个 default_box 的 classes 得分
    net[box_layer_name + "_conf"] = ConvBnRelu(num_priors *num_classes,
                                               kernel_size=[3, 3],
                                               strides=[1, 1],
                                               padding='same',
                                               is_bn=False,
                                               is_relu=False)(x)
    net[box_layer_name + "_conf" + "_flat"] = layers.Flatten()(net[box_layer_
name + "_conf"])
    #Default BOX 的生成
    net[box_layer_name + "_default_box"] = DefaultBox(min_max_size, ratios, img_
size)(x)
    return net
```

代码中 num_priors * 4 为每个检测头预测 4 个偏移位置，在 num_priors * num_classes 中，num_priors 为当前检测头分配的建议框的数量，num_classes 为预测的 1+分类数，1 代表有无目标的分类。DefaultBox(min_max_size，ratios，img_size)传入的当前特征图，生成的建议框的尺寸和比例将在预测时输出建议框的详细信息，具体实现见 5.4.3 节。layers. Flatten()操作是为了对多检测头的输出进行合并。box_layer_name 为检测头输出的字典名称。

将上面的操作在 vgg_ssd_300(input_shape，num_classes=21)中进行整合，就能实现 SSD 的前向传播，代码如下：

```
#第 5 章/ObjectDetection/TensorFlow_SSD_Detected/model/vggssd.py
def vgg_ssd_300(input_shape, num_classes=21):
    """构建 SSD 模型"""
    #输入
    input_tensor = layers.Input(shape=input_shape)
    #输入宽、高
    img_w, img_h = input_shape[0], input_shape[1]
    net = vgg_extra(input_tensor) #vgg+extra layer
    #对 conv4_3 的输入进行归一化
    net['conv_13_norm'] = Normalize()(net['conv_13']) #38 × 38 × 512,即 conv_4_3
    #依赖的 layer,Default BOX 的数量,输出层的名称,[最小尺寸,最大尺寸],[比例]
    detect_layer = {
        'conv_13_norm': [4, 'conv_13_norm', [30, 60], [2], [38, 38]],
        'conv_extra_1': [6, 'fc7_mbox', [60, 111], [2,3], [19, 19]],
        'conv_extra_3': [6, 'conv8_2_mbox', [111, 162], [2, 3], [10, 10]],
        'conv_extra_5': [6, 'conv9_2_mbox', [162, 213], [2,3], [5, 5]],
        'conv_extra_7': [4, 'conv10_2_mbox', [213, 264], [2], [3, 3]],
        'conv_extra_9': [4, 'conv11_2_mbox', [264, 315], [2], [1, 1]],
    }
    #detect_layer 为配置的检测头的 key 名称
    for k, v in detect_layer.items():
        #传入检测头的信息
        pre_head = vgg_detect_head(net[k], v[0], num_classes, v[1], v[2], v[3],
[img_w, img_h])
        net.update(pre_head)
    #将位置合并在一起
    net['mbox_loc'] = layers.Concatenate(axis=1)(
        [net[v[1] + "_loc_flat"] for v in detect_layer.values()]
    )
    #将置信度合并在一起
    net['mbox_conf'] = layers.Concatenate(axis=1)(
        [net[v[1] + "_conf_flat"] for v in detect_layer.values()]
    )
    #将分类合并在一起
    net['mbox_default_box'] = layers.Concatenate(axis=1)(
        [net[v[1] + "_default_box"] for v in detect_layer.values()]
    )

    #location 8732 *4
    net['mbox_loc'] = layers.Reshape([-1, 4])(net['mbox_loc'])
    #conf 8732 *num_classes
    net['mbox_conf'] = layers.Reshape([-1, num_classes])(net['mbox_conf'])
    #Softmax 会互斥
    net['mbox_conf'] = layers.Activation('softmax')(net['mbox_conf'])
    #将预测值合并在一起,8732 *33 = 8732 *[4 + 21 + 4 Default BOX+ 4variances]
    net['predictions'] = layers.Concatenate(axis=2)([
        net['mbox_loc'],
        net['mbox_conf'],
        net['mbox_default_box']
    ])
    return Model(inputs=net['input'], outputs=net['predictions'])
```

代码中 detect_layer 格式分别为传入的特征层 net 字典的名称、Default BOX 的数量、输出层的名称、建议框[最小尺寸,最大尺寸]、建议框[比例]、特征层的大小。通过 for k,v in detect_layer. items()遍历特征层并进行检测头的构建，然后将检测头的输出在 net['mbox_loc']中进行合并，并进行维度的 Reshape，如图 5-29 所示。

```
net['mbox_loc'] = layers.Concatenate(axis=1)(
    [net[v[1] + "_loc_flat"] for v in detect_layer.values()]
)
#将置信度合并在一起
net['mbox_conf'] = layers.Concatenate(axis=1)(
    [net[v[1] + "_conf_flat"] for v in detect_layer.values()]
)
#将分类合并在一起
net['mbox_default_box'] = layers.Concatenate(axis=1)(
    [net[v[1] + "_default_box"] for v in detect_layer.values()]
```

```
vgg_ssd_300()
Variables
'conv10_2_mbox_conf' = {KerasTensor} KerasTensor(type_spec=TensorSpec(shape=(None, 3, 3, 84), dtype=tf.float32, name=No
'conv10_2_mbox_conf_flat' = {KerasTensor} KerasTensor(type_spec=TensorSpec(shape=(None, 756), dtype=tf.float32, name=No
'conv10_2_mbox_default_box' = {KerasTensor} KerasTensor(type_spec=TensorSpec(shape=(None, 36, 8), dtype=tf.float32, name
'conv11_2_mbox_loc' = {KerasTensor} KerasTensor(type_spec=TensorSpec(shape=(None, 1, 1, 16), dtype=tf.float32, name=Non
'conv11_2_mbox_loc_flat' = {KerasTensor} KerasTensor(type_spec=TensorSpec(shape=(None, 16), dtype=tf.float32, name=None
'conv11_2_mbox_conf' = {KerasTensor} KerasTensor(type_spec=TensorSpec(shape=(None, 1, 1, 84), dtype=tf.float32, name=No
'conv11_2_mbox_conf_flat' = {KerasTensor} KerasTensor(type_spec=TensorSpec(shape=(None, 84), dtype=tf.float32, name=Nor
'conv11_2_mbox_default_box' = {KerasTensor} KerasTensor(type_spec=TensorSpec(shape=(None, 4, 8), dtype=tf.float32, name=
'mbox_loc' = {KerasTensor} KerasTensor(type_spec=TensorSpec(shape=(None, 34928), dtype=tf.float32, name=None), name='c
'mbox_conf' = {KerasTensor} KerasTensor(type_spec=TensorSpec(shape=(None, 183372), dtype=tf.float32, name=None), name
'mbox_default_box' = {KerasTensor} KerasTensor(type_spec=TensorSpec(shape=(None, 8732, 8), dtype=tf.float32, name=None)
```

图 5-29 各个检测头输出 Flat

最后对位置、置信度的预测及 Default BOX 进行合并，输出为 8732×33，如图 5-30 所示。

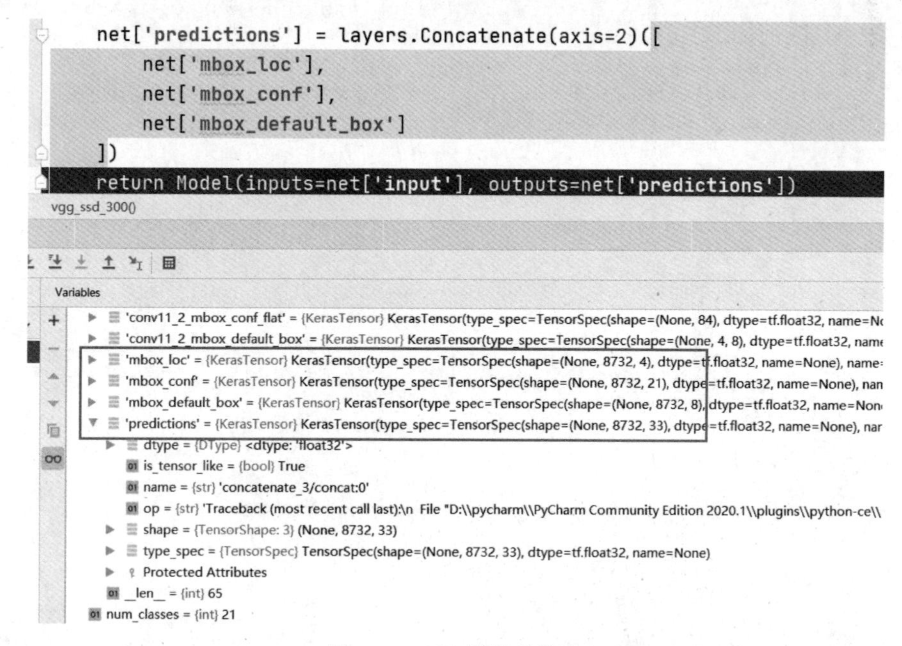

图 5-30 SSD 网络的输出

5.4.3 代码实战建议框的生成

代码生成 Default BOX 需要根据表 5-1 中的比例进行计算，假设输入为 38×38×512 的特征层，则比例为 1∶1、1∶$\sqrt{2}$、$\sqrt{2}$∶1、1∶1，然后根据输入的 min_size 和 max_size 生成 Default BOX 的宽和高，然后将原图分为 300/38 份，使用均分指令将原图生成 38×38 的坐标点（将这个坐标点看成特征较上的每个点），在每个坐标点根据计算出来的尺寸，计算 Default BOX 的左上、右下角位置的坐标，详细的代码如下：

```
#第 5 章/ObjectDetection/TensorFlow_SSD_Detected/utils/default_box.py
class BaseDefaultBox(object):
    def __init__(
            self,
            min_max_size: list,                          #Anchor 的最大和最小尺寸
            ratios: list,                                #比例
            img_size: list = [300, 300],                 #原图大小
            variances: list = [0.1, 0.1, 0.2, 0.2],      #缩放尺寸
            **kwargs):
        #属性初始化
        self.variances = np.array(variances)
        self.ratios = aspect_ratios(ratios)
        self.img_size = img_size
        self.min_max_size = min_max_size
        super(BaseDefaultBox, self).__init__(**kwargs)

    def call(self, feature_map, *args, **kwargs):
        #Feature Map 中的 w 和 h
        feature_map_width, feature_map_height = feature_map[0], feature_map[1]
        #原图大小
        img_width, img_height = self.img_size[0], self.img_size[1]
        #存放 Default Box 的 w 和 h
        box_width, box_height = [], []
        #根据 self.ratios 属性选择生成 Anchor 的 w 和 h
        for ar in self.ratios:
            #假设输入为 38×38 的特征，则第 1 个比例为 30：30
            if ar == 1.0 and len(box_width) == 0:
                box_width.append(self.min_max_size[0])
                box_height.append(self.min_max_size[0])
            elif ar == 1.0 and len(box_width) > 0:
                #第 2 个 1：1 比例为 sqrt(30 × 60)
                box_width.append(np.sqrt(self.min_max_size[0] *self.min_max_
size[1]))
                box_height.append(np.sqrt(self.min_max_size[0] *self.min_max_
size[1]))
            elif ar != 1.0:
                #第 3 个为 30 *sqrt(2): 30 *1/sqrt(2),即 1：2
                #第 4 个为 30 *1/sqrt(2) : 30*sqrt(2) 即 2：1
                box_width.append(self.min_max_size[0] *np.sqrt(ar))
                box_height.append(self.min_max_size[0] / np.sqrt(ar))
        #求 BOX 中心点
        box_widths = 0.5 *np.array(box_width)
```

```
box_heights = 0.5 * np.array(box_height)
#映射到 Feature Map 中的比例,即 300/38=7.8
step_x = self.img_size[0] / feature_map_width
step_y = self.img_size[1] / feature_map_height

#在原图中均分为 38 份,生成每个坐标点
lin_x = np.linspace(0.5 * step_x, img_width - 0.5 * step_x, feature_map_width)
lin_y = np.linspace(0.5 * step_y, img_width - 0.5 * step_y, feature_map_height)
#得到 38 × 38 和 38 × 38 坐标点
centers_x, centers_y = np.meshgrid(lin_x, lin_y)
#得到 1444×1 和 1444×1
centers_x, centers_y = centers_x.reshape(-1, 1), centers_y.reshape(-1, 1)
                                                                  #变成一维
#每个先验框需要两个(centers_x, centers_y),前一个用来计算左上角,后一个用来
#计算右下角
default_box = np.concatenate([centers_x, centers_y], axis=1)  #1444 × 2
#再复制一份
num_default_box = len(self.ratios)
#先沿 x 轴复制 1 倍,再沿 y 轴复制 2 × 4 和 1444 × 16,两个位置预测 xmin,ymin,
#xmax,ymax,共 4 个锚框
default_box = np.tile(default_box, (1, 2 * num_default_box))   #1444×16

#将锚框各个比例的值更新到 Default BOX 中
default_box[:, 0::4] = default_box[:, 0::4] - box_widths        #xmin
default_box[:, 1::4] = default_box[:, 1::4] - box_heights       #ymin
default_box[:, 2::4] = default_box[:, 2::4] + box_widths        #xmax
default_box[:, 3::4] = default_box[:, 3::4] + box_heights       #ymax

#转换成浮点数
default_box[:, ::2] = default_box[:, ::2] / self.img_size[0]
default_box[:, 1::2] = default_box[:, 1::2] / self.img_size[1]
#38 × 38 × 4 原比例为(1444,16) = 1444 × 4 个框 × 4 个位置,reshape 之后就是
#5776 × 4 个位置
default_box = default_box.reshape([-1, 4])
#将那些位置信息为负数的值转换成 0
default_box = np.minimum(np.maximum(default_box, 0.0), 1.0)
#将 variances 信息复制 Default BOX 这么多份
variances = np.tile(self.variances, (len(default_box), 1))
#将 default_box 和 variances 合并
default_box = np.concatenate([default_box, variances], axis=1)
return default_box
```

代码中 step_x,step_y 在这里取 7.8,意味着 38×38 的特征图在原图的区域是 7.8,np.linspace(0.5 * step_x,img_width－0.5 * step_x,feature_map_width)表明 Default BOX 的起点从(0,0)偏移了 0.5 * step_x;生成的 BOX 信息归一化后,再通过 default_box＝np.concatenate([default_box,variances],axis＝1)对缩放因子[0.1,0.1,0.2,0.2]进行了合并。

然后根据 BaseDefaultBox 类的实现,封装 DefaultBox(layers.Layer)算子,使在搭建模型时使用 net[box_layer_name＋"_default_box"]＝DefaultBox(min_max_size,ratios,img_size)(x)进行调用,并合并到网络层中。DefaultBox(layers.Layer)的代码如下:

```
#第 5 章/ObjectDetection/TensorFlow_SSD_Detected/utils/default_box.py
class DefaultBox(layers.Layer):
    def __init__(
            self,
            min_max_size: list,                    #Anchor 设置的最小尺寸和最大尺寸
            ratios: list,                          #比例
            img_size: list = [300, 300],           #输入图像尺寸
            variances: list = [0.1, 0.1, 0.2, 0.2], #缩放因子
            **kwargs):
        super(DefaultBox, self).__init__()
        #默认框生成类实例化
        self.base = BaseDefaultBox(
            min_max_size,
            ratios,
            img_size,
            variances,
            **kwargs
        )

    def call(self, inputs, *args, **kwargs):
        if hasattr(inputs, '_keras_shape'):
            input_shape = inputs._keras_shape
        elif hasattr(K, 'int_shape'):
            input_shape = K.int_shape(inputs)
        #根据特征图的尺寸调用 self.base 对象,并生成默认框,假设为 38×38,则输出为[5776,8]
        default_box = self.base.call([input_shape[2], input_shape[1]])
        #增维并转换成 Tensor 格式[1,5776,8]
        default_box_tensor = K.expand_dims(tf.cast(default_box, dtype=
tf.float32), 0)
        #在每个 batch_size 中都复制 1 份
        pattern = [tf.shape(inputs)[0], 1, 1]
        #[b,5776,8]
        prior_boxes_tensor = tf.tile(default_box_tensor, pattern)
        return prior_boxes_tensor
```

DefaultBox(layers. Layer)用于获得建议框,接下来就需要在建议框与真实标注框之间做 IOU 的计算,并将 IOU>0.5 的样本设置为正样本,以此计算真实框与建议框的偏移,作为 y 值,代码如下:

```
#第 5 章/ObjectDetection/TensorFlow_SSD_Detected/utils/default_box.py
def assign_boxes(targets, row):
    """
    在图像上生成 Default BOX 及 y 值,构建成 8732 ×33
    :param targets: GT 格式 xmin,ymin,xmax,ymax,one_hot
    :return:
    """
    #生成一个初始为 0 的 8732×8 的矩阵
    assignment = np.zeros((self.num_priors, 4 + self.num_classes + 4 + 4))
    #是否为背景,默认都为背景
    assignment[:, 4] = 0.
    #如果没有目标,则返回全是 0 的 Tensor
```

```
    if len(targets) == 0: return assignment

    #对真实的 BOX 进行编码
    #在 true_encode_box()函数内实现编码
    encoded_boxes = np.apply_along_axis(true_encode_box, 1,
                                        targets[:, :4],
                                        self.priors_box,
                                        self.iou_overlap_threshold,
                                        True,
                                        row)
    #reshape 成[batch_size,8732,5]
    encoded_boxes = encoded_boxes.reshape(-1, self.num_priors, 5)

    #取重合程度最大的先验框,并且获取这个先验框的 index
    #-1 的位置是 IOU 的得分,即 IOU 最大得分的 BOX 的 index
    best_iou = encoded_boxes[:, :, -1].max(axis=0)
    best_iou_mask = best_iou > 0
    best_iou_idx = encoded_boxes[:, :, -1].argmax(axis=0)
    best_iou_idx = best_iou_idx[best_iou_mask]
    #有物体的先验框的个数
    assign_num = len(best_iou_idx)
    #保留重合程度最大的先验框的应该有的预测结果
    encoded_boxes = encoded_boxes[:, best_iou_mask, :]

    #把有物体的 BOX 更新到 assignment 中[8732, 4 + self.num_classes + 4 + 4]
    assignment[:, :4][best_iou_mask] = encoded_boxes[best_iou_idx, np.arange
(assign_num), :4]
    assignment[:, 4][best_iou_mask] = 1. #4 为背景的概率,当然为 0;损失要求有样本是 1
    assignment[:, 5:-8][best_iou_mask] = targets[best_iou_idx, 4:]   #one_hot
    return assignment
```

在函数 assign_boxes()中实现了真实框与建议框进行 IOU 的计算并获得正样本的代码,其中 np.apply_along_axis 为关键代码,即将 targets[:, :4]真实框的位置信息和 self.priors_box 建议框信息,根据 self.iou_overlap_threshold 的阈值 0.5 在 true_encode_box()函数中进行编码操作,然后将 true_encode_box()得到的正样本信息赋值到 assignment 中,包括位置 assignment[:, :4][best_iou_mask]=encoded_boxes[best_iou_idx,np.arange(assign_num), :4]、有无目标 assignment[:, 4][best_iou_mask]=1、分类信息 assignment[:, 5:-8][best_iou_mask]=targets[best_iou_idx, 4:]。

计算真实框与建议框的偏移在 true_encode_box()函数中实现,其核心仍然是调用式(5-3)完成,代码如下:

```
#第 5 章/ObjectDetection/TensorFlow_SSD_Detected/utils/default_box.py
def true_encode_box(
      true_box,                                    #真实 BOX
      priors_box,                                  #Default BOX
      overlap_threshold=0.5,                       #阈值
      is_there_any_object=True,
            row=""
```

```
):
    #对 GT BOX 进行>0.5 选择为正样本
    iou_score = iou(priors_box, true_box)
    #将 Tensor 初始为 8732 × 5
    encoded_box = np.zeros([len(priors_box), 4 + is_there_any_object]) #[8732,5]
    #将先验框得到的 IOU 与阈值进行对比
    assign_mask = iou_score >= overlap_threshold
    if not assign_mask.any():
        assign_mask[iou_score.argmax()] = True
    if is_there_any_object:
        #将 iou_score 大于阈值的 BOX 的得分设置到 encoded_box 中,即有物体的概率
        encoded_box[:, -1][assign_mask] = iou_score[assign_mask]
    #得到有物体的 BOX 位置信息
    assigned_priors = priors_box[assign_mask] #[any_object_total_num_priors, 8]
    #因为输入的格式为 xmin,ymin,xmax,ymax,所以要转换为中心点来计算
    box_center = 0.5 * (true_box[:2] + true_box[2:])    #xmin,ymin + xmax,ymax
    box_wh = true_box[2:] - true_box[:2]                 #xmax,ymax - xmin,ymin

    #计算先验框的中心点
    assigned_priors_center = 0.5 * (assigned_priors[:, :2] + assigned_priors[:, 2:4])
    assigned_priors_wh = assigned_priors[:, 2:4] - assigned_priors[:, :2]

    #encoded_box 为[8732, 5],先验框与真实框 IOU 大于 0.5 的,真实框与先验框中心点的偏移值
    encoded_box[:, :2][assign_mask] = box_center - assigned_priors_center
    #dx 和 dy,代入公式进行计算
    encoded_box[:, :2][assign_mask] /= assigned_priors_wh
    #variances 是[0.1,0.1,0.2,0.2],[-4,-2]就是[0.1,0.1]
    encoded_box[:, :2][assign_mask] /= assigned_priors[:, -4:-2]
    #dw 和 dh 的偏移值,代入 log 函数进行计算
    encoded_box[:, 2:4][assign_mask] = np.log(box_wh / assigned_priors_wh)
    #[-2:]就是[0.2,0.2]
    encoded_box[:, 2:4][assign_mask] /= assigned_priors[:, -2:]
    #encoded_box 为[8732,5]
    return encoded_box.ravel()
```

代码中 assigned_priors[:, −4:−2]为 variances 参数中的[0.1,0.1],assigned_priors[:, −2:]为 variances 参数中的[0.2,0.2],在编码时这里对值进行了放大。

然后新建 DataProcessingAndEnhancement(object)类,并创建 generate(self,isTraining = True)方法,使训练时得到 x 的图片数据 img,y 的数据(真实框与建议框的偏移及分类信息),代码如下:

```
#第 5 章/ObjectDetection/TensorFlow_SSD_Detected/data/data_processing.py
class DataProcessingAndEnhancement(object):
    def generate(self, isTraining=True):
        while True:
            if isTraining:
                shuffle(self.train_lines)
                lines = self.train_lines
            else:
                lines = self.val_lines
```

```
        inputs, targets = [], []
        if len(lines):
            rnd_row = random.choice(lines)
        else:
            rnd_row = None
        label_result = {x: 0 for x in range(self.num_classes - 1)}
        for row in lines:
            #读图片和 BOX 的数据
            img, y = self.get_image_processing_results(row, rnd_row, isTraining)

            if len(y) != 0:
                boxes = np.array(y[:, :4], dtype=np.float32)
                #BOX 归一化
                boxes[:, 0] = boxes[:, 0] / self.input_shape[1]   #xmin
                boxes[:, 1] = boxes[:, 1] / self.input_shape[0]   #ymin
                boxes[:, 2] = boxes[:, 2] / self.input_shape[1]   #xmax
                boxes[:, 3] = boxes[:, 3] / self.input_shape[0]   #ymax
                #构建 one_hot 编码
                one_hot_label = np.eye(self.num_classes - 1)[np.array(y[:, 4],
np.int32)]                                                       #[[1,0],[0,1]]
                if ((boxes[:, 3] - boxes[:, 1]) <= 0).any() and ((boxes[:, 2] -
boxes[:, 0]) <= 0).any():
                    continue
                #GT 格式 xmin,ymin,xmax,ymax,[0,1]
                y = np.concatenate([boxes, one_hot_label], axis=-1)
            y = self.assign_boxes(y, row)          #将 y 值统一到 8732×8 这个维度,
                                                   #并且是编码后的内容
            inputs.append(img)
            targets.append(y)

            #按 batch_size 传输 targets
            if len(targets) == self.batch_size:
                tmp_inp = np.array(inputs, dtype=np.float32)
                tmp_targets = np.array(targets)
                inputs = []
                targets = []
                #注释以下语句,就可以进行调试
                yield preprocess_input(tmp_inp), tmp_targets
```

代码中 self.assign_boxes()实现将 y 值统一到 8732×33 这个维度,np.concatenate([boxes,one_hot_label],axis=−1)增加了真实的分类 one_hot 编码值,if len(targets)== self.batch_size 成立时返回指定 batch_size 的数据。

5.4.4 代码实战损失函数的构建及训练

根据式(5-6)实现 SSD 的损失函数的构建,代码如下:

```
#第 5 章/ObjectDetection/TensorFlow_SSD_Detected/utils/loss.py
class MultiAngleLoss(object):
    """
    SSD 损失函数的构建
```

```
    """
    def __init__(self,
                 num_classes=21,
                 negative_sample_ratio=3.0,
                 difficult_sample_top=300,
                 total_regression_loss_ratio=1.,
                 total_classification_loss_ratio=1.0
                 ):
        """
        :param num_classes: 类别数
        :param negative_sample_ratio: 负样本：正样本的比例,SSD 默认为 3：1
        :param difficult_sample_top: 当一个正样本都没有时设置负样本挖掘数量
        :param total_regression_loss_ratio: 最后回归损失的比例
        :param total_classification_loss_ratio: 最后分类损失的比例
        """
        super(MultiAngleLoss, self).__init__()
        self.num_classes = num_classes - 1
        self.negative_sample_ratio = negative_sample_ratio
        self.difficult_sample_top = difficult_sample_top
        self.total_regression_loss_ratio = total_regression_loss_ratio
        self.total_classification_loss_ratio = total_classification_loss_ratio
    def compute_loss(self, y_true, y_pred):
        batch_size = tf.shape(y_true)[0]
        num_boxes = tf.cast(tf.shape(y_true)[1], tf.float32)   #8732

        #conf_loss[0][conf_loss[0]!=0],每个框的损失
        conf_loss = self._softmax_loss(y_true[:, :, 4:-8], y_pred[:, :, 4:-8])
        #每个框的回归损失
        loc_loss = self._l1_smooth_loss(y_true[:, :, :4], y_pred[:, :, :4])
        #统计不为背景的 BOX 的数量
        true_box_num = tf.reduce_sum(y_true[:, :, 4], axis=-1)

        #每个框的回归损失,然后求和
        regression_loss_per_box = tf.reduce_sum(loc_loss *y_true[:, :, 4], axis=1)
        #每个框的分类损失,然后求和
        classification_loss_of_each_box = tf.reduce_sum(conf_loss *y_true[:, :, 4],
axis=1)
        #进行正负样本比例的调节,负样本是正样本的 3 倍
        select_negative_samples_num = tf.minimum(self.negative_sample_ratio *
true_box_num, num_boxes - true_box_num)
        #select_negative_samples_num,有可能一个也没有,因为 true_box_num 可能为 0,
        #即一个正样本也没有
        #与 0 进行比较,得到[True,True,...]或者[True,True,...,False]的情况
        negative_samples_num_mask = tf.greater(select_negative_samples_num, 0)
        #tf.reduce_any 在张量的维度上计算元素的 "逻辑或",如果所有的结果为 False,
        #则 has_min 为 0;只要有一个为 True,则为 1
        #如果图片中没有一个正样本,则 select_negative_samples_num 为 0,也就没有负样本
        #但是这时应该有很多负样本,所以我们要处理一下
        is_a_negative_sample = tf.cast(tf.reduce_any(negative_samples_num_mask),
tf.float32)

        select_negative_samples_num = tf.concat(axis=0, values=[
            select_negative_samples_num,   #如果一个正样本都没有,则此项为 0
```

```
            [(1 - is_a_negative_sample) *self.difficult_sample_top]
                                        #那么此时补 difficult_sample_top;
        ])
        #求本次 batch 中负样本数的平均数
        batch_avg_select_negative_samples_num = tf.reduce_mean(
            tf.boolean_mask(
                select_negative_samples_num,
                tf.greater(select_negative_samples_num, 0)
            )
        )
        #将负样本数转换为 int32 数据类型
        batch_avg_select_negative_samples_num = tf.cast(batch_avg_select_
negative_samples_num, tf.int32

        #取出来预测的最大的分类损失
        pre_max_confs = tf.reduce_max(y_pred[:, :, 5:-8], axis=2)
        #得到 true label 中负样本的置信度损失的下标
        _, top_negative_sample_loss_index = tf.nn.top_k(
            pre_max_confs *(1 - y_true[:, :, 4]),
            k=batch_avg_select_negative_samples_num
        )
        #获取难例样本的损失
        #tf.gather 根据难例样本的下标取 conf_loss 中的损失
        negative_sample_loss = tf.gather(tf.reshape(conf_loss, [-1]), top_
negative_sample_loss_index)
        #对难例样本的损失求总数
        total_negative_sample_loss = tf.reduce_sum(negative_sample_loss, axis=1)

        #如果为真,则为 x,否则为 y
        true_box_num = tf.where(tf.not_equal(true_box_num, 0), true_box_num, tf.
ones_like(true_box_num))
        total_true_box_num = tf.reduce_sum(true_box_num)
        total_conf_loss = tf.reduce_sum(classification_loss_of_each_box) + tf.
reduce_sum(total_negative_sample_loss)
        #总分类损失 = (正样本的损失 + 难例样本的损失) / (正样本数 + 难例样本数)
        #total_conf_loss = total_conf_loss / (total_batch_select_negative_
        #samples_num + total_true_box_num)
        #N 在论文中取正样本的个数
        total_conf_loss = total_conf_loss / (0 + total_true_box_num)
        #总回归损失 = 回归损失 / 正样本数
        total_loc_loss = tf.reduce_sum(regression_loss_per_box) / total_true_box_num
        #总损失 = 总分类损失 + alpha *总回归损失,论文中 alpha 是 1
        total_loss =\
            self.total_classification_loss_ratio *total_conf_loss + self.total_
regression_loss_ratio *total_loc_loss
        return total_loss
```

损失函数的构建,主要在 compute_loss(self,y_true,y_pred)中实现,传入 y_true 为真实框,即上文 assign_boxes(targets,row)函数中处理过的编码值 8732×33,y_pred 预测值也为 8732×33。self._softmax_loss(y_true[:,:,4:-8],y_pred[:,:,4:-8]),4:-8 为置信度+分类的位置,此处使用交叉熵损失。self._l1_smooth_loss(y_true[:,:,:4],y_pred[:,:,:4])为回归 Smooth 损失。在 true_box_num=tf.reduce_sum(y_true[:,:,4],axis=-1)中 4 这个位

置为置信度，因为只有为 1 时才代表有物体，所以 true_box_num 为有目标的正样本的数量。tf.reduce_sum(loc_loss * y_true[:,:,4],axis=1)只对有目标的损失进行求和运算。

损失函数的第 2 部分是实例难例样本的挖掘，即根据正样本的数量，求 3 倍负样本的损失值。batch_avg_select_negative_samples_num 为考虑特殊情况后的负样本的数量。pre_max_confs=tf.reduce_max(y_pred[:,:,5:-8],axis=2)取出预测的分类概率，pre_max_confs * (1-y_true[:,:,4])中因为(1-y_true[:,:,4])得到的是没有目标的分类，所以整个语句得到的就是没有目标但是分类预测最大的概率，并通过 tf.nn.top_k()获得其难例样本的概率排前 k 的索引下标 top_negative_sample_loss_index，然后根据 negative_sample_loss=tf.gather(tf.reshape(conf_loss,[-1]),top_negative_sample_loss_index)获得置信度和分类的损失，求和后将正样本、难例样本的损失相加得到 total_conf_loss，最后实现位置损失+正样本分类损失+负样本分类损失。

训练代码调用 model.fit()函数，将相关参数传入即可，主要代码如下：

```
#第 5 章/ObjectDetection/TensorFlow_SSD_Detected/train/main.py
    #构建模型
    model = vgg_ssd_300(input_shape, num_classes)
    model.build([1, 300, 300, 3])
    model.compile(
        optimizer=Adam(learning_rate=init_lr),
        loss=MultiAngleLoss(num_classes).compute_loss,
        run_eagerly=True,     #是否启用调试模型
    )
    model.fit(
        data_object.generate(True),
        steps_per_epoch=num_train                  //BATCH_SIZE,
        validation_data=data_object.generate(False),
        validation_steps=num_val                   //BATCH_SIZE,
        epochs=end_EPOCH,
        initial_epoch=init_EPOCH,
        callbacks=[logging, checkpoint]
    )
```

更多更详细的代码可参考随书代码。

5.4.5 代码实战预测推理

SSD 的推理流程包括加载模型、加载推理图片进行置信度阈值过滤、偏移值解码，以及 NMS 非极大值抑制等操作，详细的代码如下：

```
#第 5 章/ObjectDetection/TensorFlow_SSD_Detected/detected/detected.py
class Detected():
   def __init__(self, model_path, input_size):
       self.model = load_model(
          model_path,
          custom_objects={'compute_loss': MultiAngleLoss(3).compute_loss}
       ) #读取模型和权重
```

```
        self.confidence_threshold = 0.5                        #有无目标置信度
        self.class_prob = [0.5, 0.5]                           #两个类别的分类阈值
        self.nms_threshold = 0.5                               #NMS 的阈值
              self.input_size = input_size

    def readImg(self, img_path=None):
        img = cv2.imread(img_path)                             #读取要预测的图片
        #将图片转到 300×300
        self.img, _ = u.letterbox_image(img, self.input_size, [])
        self.old_img = self.img.copy()

    def forward(self):
        #升成 4 维
        img_tensor = tf.expand_dims(self.img, axis=0)
        #前向传播
        self.output = self.model.predict(img_tensor)

    def confidence_filtering(self):
        #根据置信度的阈值进行过滤,如果大于>0.5,则为有目标
        self.targeted = self.output[self.output[..., 4] > self.confidence_threshold]

    def classification_filtering(self):
        #根据类别的概率过滤
        for i, p in enumerate(self.class_prob):
            #第 i 个类别
            classification = self.targeted[self.targeted[..., 5 + i] > p]
            if len(classification):
                #偏移值
                offset_box = classification[..., :4]
                #置信度
                confidence = classification[..., [4]]
                #default box
                default_box = classification[..., -8:-4]
                #缩放因子
                variances = classification[..., -4:]
                #对该类别的框进行解码
                #因为 Default BOX 为 xmin,ymin,xmax,ymax,所以需要转换成 cx,cy,w,h
                default_box = u.xyxy2cxcywh(default_box)
                boxes = u.offset2xyxy(offset_box, default_box, variances)
                boxes = np.concatenate([boxes, confidence], axis=-1)
                #NMS 得到的索引
                index = u.nms(boxes, nms_thresh=self.nms_threshold)
                #根据索引取出 boxes 信息
                boxes = boxes[index]
                #绘图
                self.old_img = u.draw_box(self.old_img, boxes)
                u.show(self.old_img)
if __name__ == "__main__":
    d = Detected('../weights')                                 #读权重
    d.readImg('../../val_data/pexels-photo-5211438.jpeg')      #预测图片
    d.forward()                                                #前向传播
    d.confidence_filtering()                                   #置信度过滤
d.classification_filtering()                                   #分类过滤并进行解码,NMS
```

在 Detected 类中按功能进行了区分，readImg()用来读取图片，并调用 letterbox_image 转换成 SSD 的 300×300 大小，然后在 forward()中进行前向传播以得到预测值，在 confidence_filtering()中进行置信度阈值的过滤，将大于 0.5 的 self.targeted 传递给 classification_filtering()进行类别的概率过滤，将满足类别概率的 boxes 进行解码 u.offset2xyxy 和 u.nms 非极大值抑制操作，从而得到最终预测目标。

总结

SSD 有 6 个检测头，每个检测头分配不同的先验框，共计分配 8732 个先验框进行预测，具有多尺度、密集锚框检测的特点。在损失函数构建中，采用了难例样本挖掘的技术。

练习

运行并调试本节代码，理解算法的设计与代码的结合，重点梳理本算法的实现方法。

5.5 单阶段速度快的检测网络 YOLOv1

5.5.1 模型介绍

YOLOv1 由 Joseph Redmon 等在 2015 年发表的论文 *You Only Look Once: Unified, Real-Time Object Detection* 中提出，是一个单阶段目标检测网络，其主要特点为速度快。YOLOv1 的主要结构如图 5-31 所示。

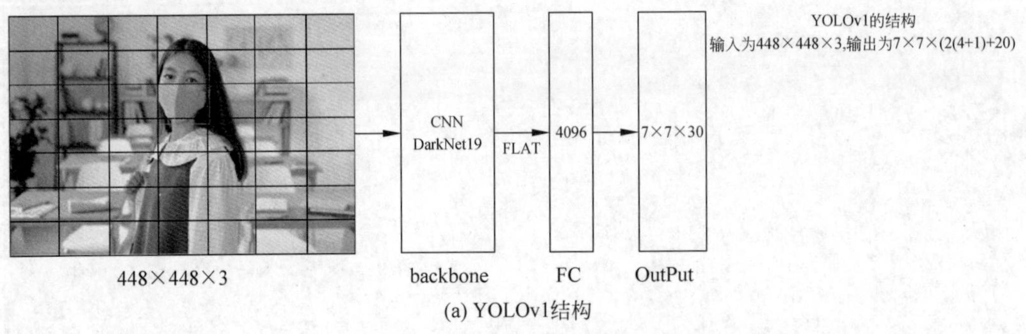

(a) YOLOv1结构

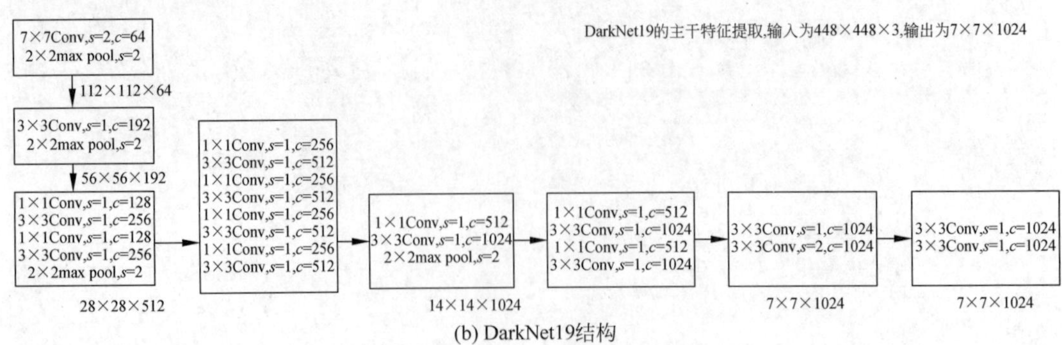

(b) DarkNet19结构

图 5-31　YOLOv1 网络结构

YOLOv1 其主干特征提取网络由 DarkNet19 构成，DarkNet19 的主要特点是由 3×3 和 1×1 卷积构成，共计 8 个块，提取到最后的特征层为 7×7×1024，然后将 7×7×1024 摊平送入 4096 维的全连接层，预测输出为 7×7×(2×(4 个位置+1 有无目标的概率)+20 个类别的概率)=7×7×30。将 7×7×30 中的特征映射到原图就相对于将原图划分成 7×7 个格子，每个格子对应到输出就是每个特征点，这个特征点有 30 个通道，代表预测 30 个特征的输出。

YOLOv1 没有建议框(先验框)，由真实标注物体所在格子进行预测，也就是标注物体所在格子为正样本，每个标注物体所在格子最多预测两个物体，所以 YOLOv1 一共只能预测 7×7×2=98 个目标，而没有标注物体的格子全都是负样本，如图 5-32 所示。

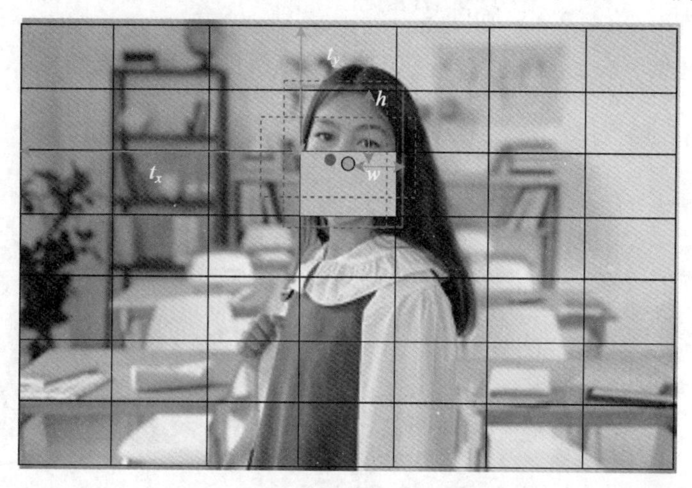

图 5-32　YOLOv1 正样本的选择

图 5-32 中口罩所处中心点为(center_x,center_y)，其所在格子的左上角行 $i=3$、列 $j=2$，所以 t_x=center_x$-i$，t_y=center_y$-j$，则 t_x 和 t_y 为相对于原点(0,0)的偏移，w、h 为标签目标的宽和高，所以 y_true 为 t_x,t_y,w,h；y_true 只在高亮背景中有目标，其他格式都没有目标，所以当前格子的置信度为 1，其他为 0，当前格子也负责预测分类的概率。

预测值 y_pred 也为 t_x,t_y,w,h，每个格子预测两个偏移值，也就是两个预测框，每个预测框预测前背景、分类的概率。在求损失时，只有高亮背景的格子为正样本，其他格子全是负样本，当预测值 y_true 与真实框 y_pred 很接近时，说明网络预测接近目标。需要注意的是 YOLOv1，宽和高 wh 是直接预测，t_x、t_y 是相对于(0,0)点的偏移。没有设置建议框的这种方法被称为 Anchor Free。

在损失函数方面，位置损失、置信度损失(有无目标)、分类损失均使用均方差，其公式如下：

$$\mathrm{Loss}=\lambda_{\mathrm{coord}}\sum_{i=0}^{s^2}\sum_{j=0}^{B}1_{\mathrm{ij}}^{\mathrm{obj}}[(x_i-\hat{x}_i)^2+(y_i-\hat{y}_i)^2]+$$
$$\lambda_{\mathrm{coord}}\sum_{i=0}^{s^2}\sum_{j=0}^{B}1_{\mathrm{ij}}^{\mathrm{obj}}[(\sqrt{w_i}-\sqrt{\hat{w}_i})^2+(\sqrt{h_i}-\sqrt{\hat{h}_i})^2]+$$

$$\sum_{i=0}^{s^2}\sum_{j=0}^{B}1_{ij}^{obj}[(C_i-\hat{C}_i)^2]+\lambda_{noobi}\sum_{i=0}^{s^2}\sum_{j=0}^{B}1_{ij}^{noobj}[(C_i-\hat{C}_i)^2]+$$
$$\sum_{i=0}^{s^2}1_{ij}^{obj}\sum_{c\in 1}[(p_i(c)-\hat{p}_i(c))^2] \tag{5-7}$$

其中，λ_{coord} 为平衡系数，原作者将其设置为 5。1_{ij}^{obj} 代表有物体的格子。x_i 代表真实 tx，$\hat{x}_i$ 代表预测 tx。$\sqrt{w_i}$ 为真实宽，$\sqrt{\hat{w}_i}$ 为预测宽，由于 w、h 的值较大，所以开方是为了防止梯度爆炸。1_{ij}^{noobj} 代表没有物体，所以置信度损失由有物体的损失＋没有物体的损失构成，而没有物体的损失占比应该更小，所以超参数 $\lambda_{noobi}=0.5$，最后 $p_i(c)$ 为真实物体的分类，$\hat{p}_i(c)$ 为预测物体的分类。ij 代表每个格子。整个网络的实现，可参见 5.5.2 节。

5.5.2 代码实战模型搭建

YOLOv1 模型前向传播的代码如下：

```
#第 5 章/ObjectDetection/TesnsorFlow_YOLO_V1_Detected/backbone/yolo_v1.py
def yolo_v1(input_shape, CLASS_NUM=20, BOX_NUM=2, GRID_NUM=7):
    #输入
    input_tensor = layers.Input(shape=input_shape)
    #默认不启用 BN
    is_bn = False
    #DarkNet19 的结构,Conv 代表卷积,MaxP 即最大池化
        #[算子类型,核,步长,channel]
    darknet_config = [
        ['Conv', 7, 2, 64],
        ['MaxP', 2, 2],                      #112 × 112 × 64

        ['Conv', 3, 1, 192],
        ['MaxP', 2, 2],                      #56 × 56 × 192

        ['Conv', 1, 1, 128],
        ['Conv', 3, 1, 256],
        ['Conv', 1, 1, 256],
        ['Conv', 3, 1, 512],
        ['MaxP', 2, 2],                      #28 × 28 × 512

        ['Conv', 1, 1, 256],
        ['Conv', 3, 1, 512],
        ['Conv', 1, 1, 256],
        ['Conv', 3, 1, 512],
        ['Conv', 1, 1, 256],
        ['Conv', 3, 1, 512],
        ['Conv', 1, 1, 256],
        ['Conv', 3, 1, 512],

        ['Conv', 1, 1, 512],
        ['Conv', 3, 1, 1024],
        ['MaxP', 2, 2],                      #14 × 14 × 1024
```

```
        ['Conv', 1, 1, 512],
        ['Conv', 3, 1, 1024],
        ['Conv', 1, 1, 512],
        ['Conv', 3, 1, 1024],

        ['Conv', 3, 1, 1024],
        ['Conv', 3, 2, 1024],                    #7 × 7 × 1024

        ['Conv', 3, 1, 1024],
        ['Conv', 3, 1, 1024],                    #7 × 7 × 1024
    ]
    x = input_tensor
    #解析配置文件,生成模型结构
    for i, c in enumerate(darknet_config):
        if c[0] == 'Conv':
            x = ConvBnLeakRelu(c[3], c[1], c[2], is_bn=is_bn, padding='same', is_
relu=True)(x)
        elif c[0] == 'MaxP':
            x = layers.MaxPooling2D(c[1], c[2], padding='same')(x)
    x = layers.Flatten()(x)                      #7 × 7 × 1024
    #全连接后激活函数使用 LeakyReLU
    x = layers.LeakyReLU(0.1)(layers.Dense(4096)(x))
    x = layers.DropOut(0.5)(x)                   #防止过拟合
    #输出为 1470 维特征向量
    x = layers.Dense(GRID_NUM * GRID_NUM * (5 * BOX_NUM + CLASS_NUM))(x)
    x = layers.DropOut(0.5)(x)                   #防止过拟合
    x = tf.sigmoid(x)                            #Sigmoid 是为了将值域控制在 0~1
    #将 1470 维特征向量变成 7×7×30,7×7 是把原图像划为 7×7 个格子和(4+1)×2+20 个类别
    x = tf.reshape(x, [-1, GRID_NUM, GRID_NUM, 5 * BOX_NUM + CLASS_NUM])
    return Model(inputs=input_tensor, outputs=x)
```

特征提取使用配置文件 darknet_config 来构建 DarkNet19,具体实现在 for 循环中。ConvBnLeakRelu()即 Conv→BN→LeakRelu 的封装。通过 DarkNet19 后 layers.Flatten()得到 50 176 个神经元,然后输出 layers.Dense(GRID_NUM * GRID_NUM * (5 * BOX_NUM+CLASS_NUM))(x)变成 1470 维特征,对这 1470 维特征归一化后 reshape 成 7×7×30,即每个特征点代表每个格子,每个格子预测(4+1)×2 个框和有无目标,并预测 20 个分类的概率。

5.5.3 无建议框时标注框编码

虽然 YOLOv1 为 Anchor Free,但仍然需要将标注框编码成 tx,ty,w,h 作为 y_true,以便与预测 y_pred 做损失,代码如下:

```
#第 5 章/ObjectDetection/TesnsorFlow_YOLO_V1_Detected/utils/tools.py
def yolo_v1_true_encode_box(true_boxes, CLASS_NUM=20, BOX_NUM=2, GRID_NUM=7):
    """
    将图片编码成 YOLOv1 的输出格式
    :param GRID_NUM:
    :param BOX_NUM: 默认为两个
```

```
    :param CLASS_NUM: 预测的类别数
    :param true_boxes: NumPy类型[center_x,center_y,w,h,object] YOLO格式
    :return: 7 × 7 × (5 × 2 + 20)
    """
    #初始 7×7×30 为 0 的 Tensor
    target_box = np.zeros([GRID_NUM, GRID_NUM, (5 *BOX_NUM + CLASS_NUM)])
    #cell_size = 1.0 /7 每个格子的宽、高,归一化
    cell_size = 1.0 / GRID_NUM
    if len(true_boxes) == 0:
        return target_box
    boxes_wh = true_boxes[:, 2:4]                        #gt wh
    boxes_cxy = true_boxes[:, :2]                        #gt cx,cy
    box_label = true_boxes[:, -1]                        #gt label
    for ibox in range(true_boxes.shape[0]):
        center_xy, wh, label = boxes_cxy[ibox], boxes_wh[ibox], int(box_label[ibox])
        #得到 1/ S=7 格子中的相对位置。-1.0 是为了排除当前格子
        ij = np.ceil(center_xy / cell_size) - 1.0
        i, j = int(ij[0]), int(ij[1])                    #第几个格子中
        #由第 i、第 j 个格子的中心点进行预测
        grid_xy = ij *cell_size                          #获得格子的左上角 xy
        #(bbox 中心坐标 - 网络左上角的坐标) / 网格大小 = tx,ty
        grid_p_center_xy = (center_xy - grid_xy) / cell_size
        for k in range(BOX_NUM):
            s = 5 *k
            target_box[j, i, s:s + 2] = grid_p_center_xy
            target_box[j, i, s + 2:s + 4] = wh
            target_box[j, i, s + 4] = 1.0                #置信度,有没有物体
        #i,j 正样本格子的 label
        target_box[j, i, 5 *BOX_NUM + label] = label
    return target_box #输出类型为 7 × 7 × 30,与预测的结果保持一致
```

代码中 cell_size＝1.0/GRID_NUM,1/7＝0.14 为每个格子的大小。true_boxes 传进来时就是 center_x,center_y,w,h。ij＝np.ceil(center_xy/cell_size)－1.0,$i=3$、$j=2$ 即标注框的中心点落在此格子中(见图 5-32),那么在计算损失时正样本为 $i=3$、$j=2$,其他样本均为负样本。

在得到 $i=3$、$j=2$ 后,计算当前格子的左上角 xy 的坐标 grid_xy＝ij * cell_size,然后通过(center_xy－grid_xy)/cell_size 得到标注中心点相对于(0,0)点的偏移。target_box[j,i,s:s+2]＝grid_p_center_xy,即 target_box[2,3,0:2]＝grid_p_center_xy 为 tx,ty 的偏移值,target_box[2,3,2:4]＝wh 为 wh 的宽和高。因为每个格子预测两个物体,所以当 $k=1$ 时,target_box[2,3,5:7]＝grid_p_center_xy,也就是当前格子赋两个相同值作为 y_true。

然后再次编写 generate(self,isTraining＝True)函数实现,数据的传输,代码如下:

```
#第 5 章/ObjectDetection/TesnsorFlow_YOLO_V1_Detected/data/data_processing.py
class DataProcessingAndEnhancement(object):
    def generate(self, isTraining=True):
            while True:
        if isTraining:
            shuffle(self.train_lines)
```

```
                lines = self.train_lines
            else:
                lines = self.val_lines
            inputs, targets = [], []
            if len(lines):
                rnd_row = random.choice(lines)
            else:
                rnd_row = None

            for row in lines:
                #读取 Image 和 BOX 信息
                img, y = self.get_image_processing_results(row, rnd_row, isTraining)
                #将 Voc 格式转换为 YOLO 格式
                y = self.voc_label_convert_to_yolo(y)
                #将 GT BOX 编码成与预测值相同的格式
                y = yolo_v1_true_encode_box(y, self.CLASS_NUM, self.BOX_NUM,
self.GRID_NUM)
                #存储图片和编码后的 BOX 信息
                inputs.append(img)
                targets.append(y)
                #按 batch_size 传输 targets
                if len(targets) == self.batch_size:
                    tmp_inp = np.array(inputs, dtype=np.float32)
                    tmp_targets = np.array(targets)
                    inputs = []
                    targets = []
                    #注释以下语句,就可以进行调试了
                    yield preprocess_input(tmp_inp), tmp_targets
```

因为本数据集的默认格式为 Voc,所以需要调用 self.voc_label_convert_to_yolo(y)实现由 Voc 格式转 YOLO 格式,代码类似 5.1 节标签处理及代码,更多更详细的内容可参考随书代码。

5.5.4 代码实现损失函数的构建及训练

第 1 步,根据式(5-7)实现正样本格子 i、j 位置的损失、置信度的损失及分类概率的损失,同时对 1 负样本给予较小权重的损失,详细的代码如下:

```
#第 5 章/ObjectDetection/TesnsorFlow_YOLO_V1_Detected/utils/loss.py
class MultiAngleLoss(object):
    def __init__(self, CLASS_NUM=20, BOX_NUM=2, GRID_NUM=7, coord=5.0, no_obj=0.5):
        """
        YOLOv1 的 loss 计算
        """
        self.CLASS_NUM = CLASS_NUM                              #类别数
        self.BOX_NUM = BOX_NUM                                  #每个格子的预测数
        self.GRID_NUM = GRID_NUM                                #一共有多少个格子
        self.coord = coord                                      #坐标损失的系数
        self.no_obj = no_obj                                    #不包含物体的损失系数
        self.output_dim = 5 * self.BOX_NUM + self.CLASS_NUM     #输出维度,即 30
```

```
        super(MultiAngleLoss, self).__init__()

    def compute_loss(self, y_true, y_pred):
        """
        计算损失的函数
        :param y_true:
        :param y_pred:
        :return:
        """
        #(1)第 1 部分,获取有物体和没有物体的 mask,有物体为 true,没有物体为 false
        batch_size = tf.shape(y_true)[0]
        #true 值中有物体的框
        get_object_mask = y_true[:, :, :, 4] > 0
        #true 值中没有物体的框
        get_no_object_mask = y_true[:, :, :, 4] == 0
        #(2)第 2 部分,根据第 1 部分获得的 mask,从预测值获取有物体 get_pre_object_mask、
        #没有物体的 get_pre_no_object_mask
        #及有物体的 bbox_pre、class_pre

        #扩维成 b,7,7,30,好获得整组的值
        get_object_mask = tf.tile(np.expand_dims(get_object_mask, -1), [1, 1, 1,
self.output_dim])
        get_no_object_mask = tf.tile(np.expand_dims(get_no_object_mask, -1), [1,
1, 1, self.output_dim])

        #从预测框中获得置信度框的内容,y_pred[get_object_mask] get_object_mask
        #有物体的 mask
        get_pre_object_mask = tf.reshape(y_pred[get_object_mask], [-1, self.
output_dim])
        #获取两个预测框的值,因为预测输出为 2×(4+1)+20
        bbox_pre = tf.reshape(get_pre_object_mask[..., :5 * self.BOX_NUM], [-1, 5])
        #类别信息的值
        class_pre = get_pre_object_mask[..., 5 * self.BOX_NUM:]

        #获取没有物体的格子 y_pred[get_no_object_mask] get_no_object_mask 没有物体
        #的 mask
        get_pre_no_object_mask = tf.reshape(y_pred[get_no_object_mask], [-1,
self.output_dim])
        #(3)第 1 部分只是获得置信度的 mask,接下来需要根据 mask 获取标注物体为正样本的
        #get_true_object_mask,bbox_true
        #及 class_true,和没有物体的 get_true_no_object_mask

        #y_true[get_object_mask],有物体的 mask。get_true_object_mask 为标注框,
        #有物体的部分
        get_true_object_mask = tf.reshape(y_true[get_object_mask], [-1, self.
output_dim])
        #标注框,有物体的 BOX 信息
        bbox_true = tf.reshape(get_true_object_mask[..., :5 * self.BOX_NUM], [-1, 5])
        #标注框,有物体的分类信息
        class_true = get_pre_object_mask[..., 5 * self.BOX_NUM:]
        #y_true[get_no_object_mask],标注没有物体的 mask
```

```
        get_true_no_object_mask = tf.reshape(y_true[get_no_object_mask], [-1,
self.output_dim])

        #(4)根据预测没有物体的 get_pre_no_object_mask,分别去获取没有物体的预测
        #no_obj_pre_conf
        #及没有物体的标注 no_obj_true_conf

        #初始为 0,没有物体的 mask,此时背景为 1
        get_pre_conf_no_object_mask = np.zeros(get_pre_no_object_mask.shape)
        for b in range(self.BOX_NUM):
            #起到的作用是默认先选择所有的没有物体的格子,为了方便从 get_true_no_
            #object_mask 和 get_pre_no_object_mask 中取值
            get_pre_conf_no_object_mask[:, 4 + b *5] = 1
        #取出来没有物体预测的格子
        no_obj_pre_conf = tf.gather(get_pre_no_object_mask, get_pre_conf_no_
object_mask.astype(int))
        #取出来没有物体真实的格子
        no_obj_true_conf = tf.gather(get_true_no_object_mask, get_pre_conf_no_
object_mask.astype(int))

        #所有没有物体的格子的损失
        loss_no_obj = tf.reduce_sum(tf.losses.MSE(no_obj_pre_conf, no_obj_true_conf))
        #(5)从预测的 BOX 中获取与真实框最大的 IOU,取最大的为有物体的 mask,然后让预测值
        #与真实值之间做损失
        coord_response_mask = np.zeros(bbox_true.shape)
        coord_not_response_mask = np.ones(bbox_true.shape)
        bbox_target_iou = np.zeros(bbox_true.shape)

        #从预测的 BOX 中获取与真实框最大的 IOU,遍历 batch 下所有的有物体的格子。因为每
        #个格子预测两个框,所以 step 是 2
        for i in range(0, bbox_true.shape[0], self.BOX_NUM):
            pre_box = bbox_pre[i:i + self.BOX_NUM]    #[0,2],预测的每个格子的 2 个框都
                                                      #进行计算
            pre_xy = np.zeros(pre_box.shape)
            #因为预测出来的是 cx,cy,w,h,并且缩放了,所以要还原到原图中,以便进行 IOU 的比较
            #算出 x1,y1,x2,y2
            pre_xy[:, :2] = pre_box[:, :2] / float(self.GRID_NUM) - 0.5 *pre_box[:, 2:4]
            pre_xy[:, 2:4] = pre_xy[:, :2] / float(self.GRID_NUM) + 0.5 *pre_xy[:, 2:4]
            #当为真实值时,因为编码时每个格子的 2 个框赋的值是一样的,所以取 1 个就可以了
            target_true = bbox_true[i]
            target_true = tf.reshape(target_true, [-1, 5])
            true_xy = np.zeros_like(pre_xy)
            #因为传的是 cx,cy,w,h,但是计算 IOU 要使用 x1,y1,x2,y2,所以要转换一下
            true_xy[:, :2] = target_true[:, :2] / float(self.GRID_NUM) - 0.5 *
target_true[:, 2:4]
            true_xy[:, 2:4] = target_true[:, :2] / float(self.GRID_NUM) + 0.5 *
target_true[:, 2:4]
            #获取预测框与真实框之间的 IOU
            get_iou = iou(pre_xy, true_xy)
            #得到所有框中最大的 max
            max_iou, max_index = np.max(get_iou), np.argmax(get_iou)
            coord_response_mask[i + max_index] = 1 #将有物体的 IOU 位置设置为 1,默认为 0
```

```
            coord_not_response_mask[i + max_index] = 0    #将没有物体的设置为 0,默认
                                                          #为 1,为 1 是为了方便取值
            bbox_target_iou[i + max_index, 4] = max_iou   #将 IOU 的值赋为置信度
            #(6)计算损失

        #根据有目标的 mask 取出 pre 的 BOX 值。因为 batch 中设置的都是第 1 个格子有目标,
        #所以这里都是第 1 个格子
        bbox_pred_response = tf.reshape(tf.gather(bbox_pre, coord_response_
mask.astype(int)), [-1, 5])
        bbox_target_response = tf.reshape(tf.gather(bbox_true, coord_response_
mask.astype(int)), [-1, 5])
        target_iou = tf.reshape(bbox_target_iou[coord_response_mask.astype(int)],
[-1, 5])

        #计算 x 和 y 的损失
        loc_loss_xy = tf.reduce_sum(
            tf.losses.MSE(bbox_pred_response[:, :2],
                          bbox_target_response[:, :2])
        )
        #计算 w 和 h 的损失
        loc_loss_wh = tf.reduce_sum(
            tf.losses.MSE(tf.sqrt(bbox_pred_response[:, 2:4]),
                          tf.sqrt(bbox_target_response[:, 2:4]))
        )
        #位置损失
        loc_loss = loc_loss_xy + loc_loss_wh

        #计算置信度损失。预测的置信度与 true 值和 pre 的 IOU 越接近越好
        loss_obj = tf.reduce_sum(tf.losses.MSE(bbox_pred_response[:, 4], target_
iou[:, 4]))
        #分类损失
        class_loss = tf.reduce_sum(
            tf.losses.MSE(class_pre,
                          class_true)
        )
        #位置损失+ 有物体的置信度损失+ 没有物体所有格子的损失+ 有物体的分类损失
        loss = self.coord * loc_loss + \
            tf.cast(loss_obj, dtype=tf.float32) + \
            self.no_obj * loss_no_obj + class_loss
        #总损失/batch_size
        loss = loss / tf.cast(batch_size, dtype=tf.float32)
        return loss
```

此损失计算的代码较长,共由 6 个步骤构成,首先 y_true[:,:,:,:,4]>0 获取有目标的 get_object_mask,y_true[:,:,:,:,4]==0 没有目标的 get_no_object_mask,如图 5-33 所示。

第 2 步,根据 get_object_mask、get_no_object_mask 到 y_pred 预测值中获取有目标的置信度、预测 BOX、预测分类信息,并得到预测值为有物体的 get_pre_object_mask、没有物体的 get_pre_no_object_mask,如图 5-34 所示。

第 3 步,根据第 1 步中的 get_object_mask 计算真实值的 BOX、置信度、分类信息的值,并获取没有目标 get_true_no_object_mask 的值,如图 5-35 所示。

```
get_object_mask = y_true[:, :, :, 4] > 0  get_object_mask: tf.Tensor(\n[[[
# true值中没有物体的框
get_no_object_mask = y_true[:, :, :, 4] == 0  get_no_object_mask: tf.Tensoi
```

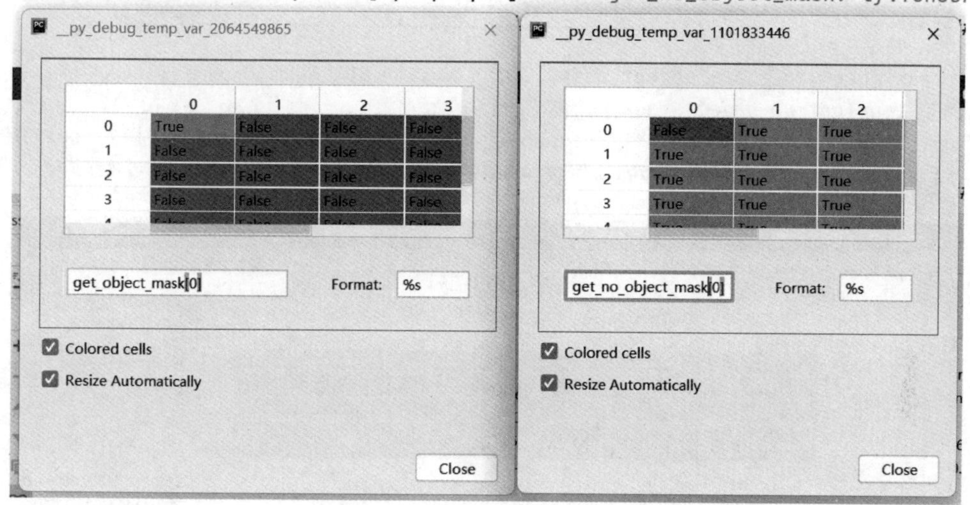

图 5-33　y_true 有无目标 mask

```
# 从预测框中得到置信度框的内容,y_pred[get_object_mask] get_object_mask有物体的mask
get_pre_object_mask = tf.reshape(y_pred[get_object_mask], [-1, self.output_d
# 获取2个预测框的值.因为预测输出为 2*(4+1)+20
bbox_pre = tf.reshape(get_pre_object_mask[..., :5 * self.BOX_NUM], [-1, 5])
# 类别信息的值
class_pre = get_pre_object_mask[..., 5 * self.BOX_NUM:]  class_pre: tf.Tenso

# 获取没有物体的格子 y_pred[get_no_object_mask] get_no_object_mask没有物体的mask
get_pre_no_object_mask = tf.reshape(y_pred[get_no_object_mask], [-1, self.ou
```

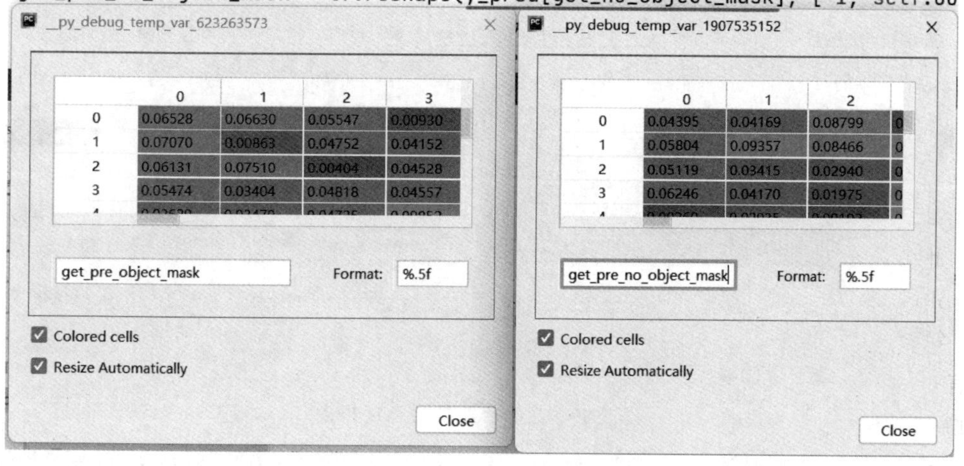

图 5-34　y_pred 有无目标 BOX 信息

```
# y_true[get_object_mask]，有物体的mask。get_true_object_mask为标注框，有物体的部分
get_true_object_mask = tf.reshape(y_true[get_object_mask], [-1, self.output_
#标注框，有物体的BOX信息
bbox_true = tf.reshape(get_true_object_mask[..., :5 * self.BOX_NUM], [-1, 5]
#标注框，有物体的分类信息
class_true = get_pre_object_mask[..., 5 * self.BOX_NUM:]  class_true: tf.Ten
# y_true[get_no_object_mask]，标注没有物体的mask
get_true_no_object_mask = tf.reshape(y_true[get_no_object_mask], [-1, self.o
################################################################################
# 初始为0，没有物体的mask，此时背景为1
```

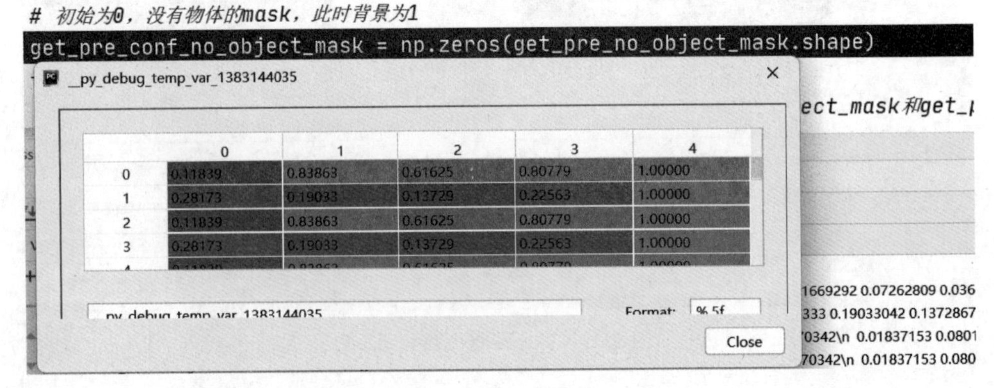

图 5-35　y_true 有无目标 BOX 信息

第 4 步，根据预测没有物体的 get_pre_no_object_mask，分别获取没有物体的预测 no_obj_pre_conf 及没有物体的标注 no_obj_true_conf，然后计算所有没有目标的置信度损失 loss_no_obj＝tf. reduce_sum(tf. losses. MSE(no_obj_pre_conf，no_obj_true_conf))，如图 3-36 所示。

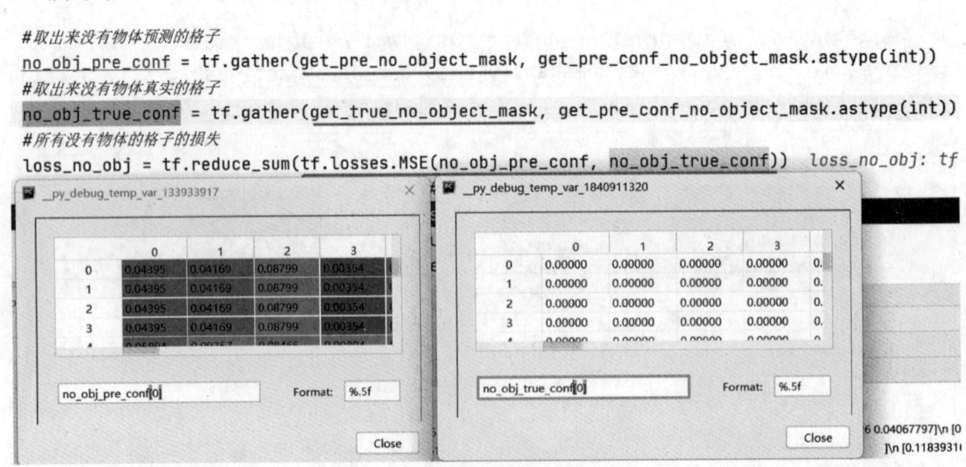

图 5-36　没有目标的置信度损失

第 5 步，从预测的 BOX 中获取与真实 BOX 最大的 IOU，取最大 IOU 作为有物体的 mask，然后让预测值与真实值之间做损失。具体是将 pre_box 与 bbox_true 计算 IOU，取

最大的那个 IOU 所在的 BOX 作为有目标，然后根据 coord_response_mask 取出预测值的 bbox_pred_response，以及真实值 bbox_target_response，如图 5-37 所示。

```
max_iou, max_index = np.max(get_iou), np.argmax(get_iou)  max_iou: 0.000655899
coord_response_mask[i + max_index] = 1  #有物体的IOU位置，设置为1。默认为0
coord_not_response_mask[i + max_index]= 0 #没有物体的，设置为0，默认为1，为1表示有目标
bbox_target_iou[i + max_index, 4] = max_iou  #将IOU的值赋为置信度

#根据有目标的mask取出pre的BOX值。因为batch中设置的都是第1个格子有目标，所以这里都是第1个格子
bbox_pred_response = tf.reshape(tf.gather(bbox_pre, coord_response_mask.astype(int
bbox_target_response = tf.reshape(tf.gather(bbox_true, coord_response_mask.astype(
```

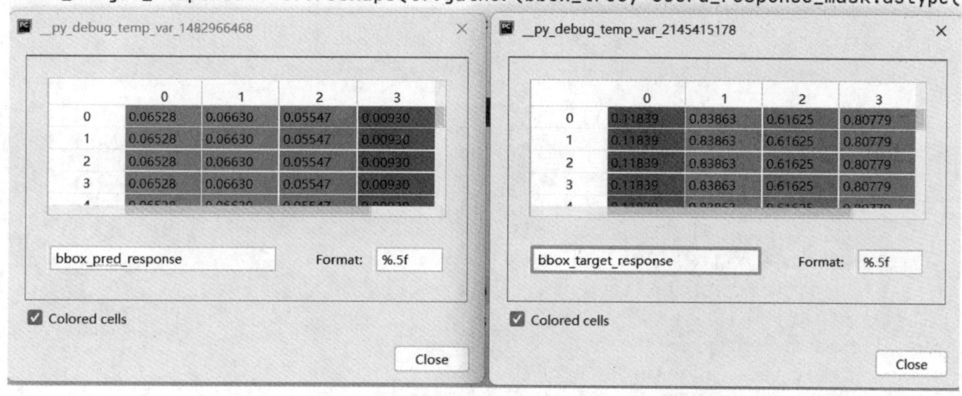

图 5-37 BOX 损失取预测 BOX 与真实 BOX 的最大 IOU

第 6 步，根据前面的步骤按式(5-7)计算损失，完成"5×位置损失＋有物体的置信度损失＋0.5×没有物体所有格子的损失＋有物体的分类损失"。

训练代码可使用 model.fit()函数完成，更多更详细的代码可参考随书代码。

5.5.5 代码实战预测推理

YOLOv1 的推理流程包括加载模型、加载推理图片进行置信度阈值过滤、偏移值解码，以及 NMS 非极大值抑制等操作，详细的代码如下：

```
#第 5 章/ObjectDetection/TesnsorFlow_YOLO_V1_Detected/detected/detected.py
class Detected():
  def __init__(self, model_path, input_size):
      self.model = load_model(
          model_path,
          custom_objects={'compute_loss': MultiAngleLoss(3).compute_loss}
      ) #读取模型和权重
      self.confidence_threshold = 0.5                     #有无目标置信度
      self.class_prob = [0.5, 0.5]                        #两个类别的分类阈值
      self.nms_threshold = 0.5                            #NMS 的阈值
      self.input_size = input_size

  def readImg(self, img_path=None):
      img = cv2.imread(img_path)                          #读取要预测的图片
```

```
        #将图片转换到 448×448
        self.img, _ = u.letterbox_image(img, self.input_size, [])
        self.old_img = self.img.copy()

    def forward(self):
        #升成 4 维
        img_tensor = tf.expand_dims(self.img / 255.0, axis=0)
        #前向传播
        self.output = self.model.predict(img_tensor)

    def confidence_filtering(self):
        #根据置信度的阈值进行过滤，如果大于>0.5，则为有目标
        #YOLOv1 的输出是 7×7×[(4+1) ×2+2]，所以这里应该输出 7×7×2
        self.targeted = np.concatenate([self.output[..., [4]], self.output
[..., [4 + 5]]], axis=-1)

    def classification_filtering(self):
        self.S = 7                                          #划分的格子数
        self.B = 2                                          #每个格子预测两个框
        cell_size = 1.0 / float(self.S)                     #每个格子的 size 为 1/7
        #用来存储筛选后的内容
        boxes, labels, confidences, class_scores = [], [], [], []
        #遍历每个格子
        for i in range(self.S):
            for j in range(self.S):
                #遍历每个格子中的两个框
                for b in range(self.B):
                    #分类得分
                    class_score = self.output[..., j, i, 5 * self.B:]
                    #取最大的分类得分的下标
                    class_label = np.argmax(class_score)
                    #最大的分类得分值
                    score = class_score[..., class_label]
                    #当前格子的置信度
                    conf = self.targeted[..., j, i, b]
                    #每个格子最后的得分为置信度*分类得分
                    prob = score * conf
                    #如果小于阈值，则跳过
                    if float(prob) < self.confidence_threshold:
                        continue
                    #当前预测 BOX 信息
                    box = self.output[..., j, i, 5 * b:5 * b + 4]
                    #每个格子点的归一化坐标
                    x0y0_normalized = np.array([i, j]) * cell_size
                    #解码操作，x+i，y+j 即预测的 cxcy
                    xy_normalized = box[..., :2] * cell_size + x0y0_normalized
                    #YOLOv1 直接预测 wh
                    wh_normalized = box[..., 2:]
                    #合并 cxcyxy
                    cxcywh = np.concatenate([xy_normalized, wh_normalized], axis=-1)
                    #由 cx，cy，w，h 转换成 xmin，ymin，xmax，ymax
                    xyxy_box = u.cxcy2xyxy(cxcywh)
```

```
                #对结果进行存储
                boxes.append(xyxy_box)
                labels.append([class_label])
                confidences.append(conf)
                class_scores.append(class_score)

        #对于得到的 BOX 信息,按类别进行非极大值抑制
        if len(boxes) > 0:
            #由 list 合并成 NumPy
            boxes_normalized_all = np.stack(boxes, 1)
            class_labels_all = np.stack(labels, 1)
            confidences_all = np.stack(confidences, 1)
            class_scores_all = np.stack(class_scores, 1)
            #遍历每个类别
            for label in range(len(self.class_prob)):
                #如果 class_labels_all==label,则取当前 label 中的信息
                mask = class_labels_all == label
                #如果都不是当前 label,则跳过
                if np.sum(mask) == 0: continue
                #当前 label 的 boxes
                boxes_mask = boxes_normalized_all[mask]
                #当前 label 的 class_labels。reshape 是由于得到的是(50,)变成(50,1),
                #方便后面计算
                class_labels_mask = class_labels_all[mask].reshape([-1, 1])
                #当前 label 的 confidences
                confidences_mask = confidences_all[mask].reshape([-1, 1])
                #当前 label 的 class_scores
                class_scores_mask = class_scores_all[mask][..., label].reshape([-1, 1])
                #合并
                cat_boxes = np.concatenate([boxes_mask, confidences_mask], axis=-1)
                #NMS
                index = u.nms(cat_boxes, self.nms_threshold)
                #绘框
                self.old_img = u.draw_box(self.old_img, cat_boxes[index])
        #最后结果
        u.show(self.old_img)


if __name__ == "__main__":
    d = Detected(r'../weights', input_size=[448, 448])
    d.readImg('../../val_data/pexels-photo-5211438.jpeg')

    d.forward()
    d.confidence_filtering()
    d.classification_filtering()
```

相对于其他模型,推理的改变在 confidence_filtering(self)方法中,这是由于 YOLOv1 的输出是 7×7×[(4box+1 置信度)×2 个框+2 个分类],所以 self.output 输出 7×7×12,获取置信度结果在第 4、第 9 位,如图 5-38 所示。

在 classification_filtering(self)中实现了根据每个格子每个框 prob=score * conf,置信度×分类概率的得分得到 box=self.output[...,j,i,5 * b:5 * b+4]信息,然后根据当前所

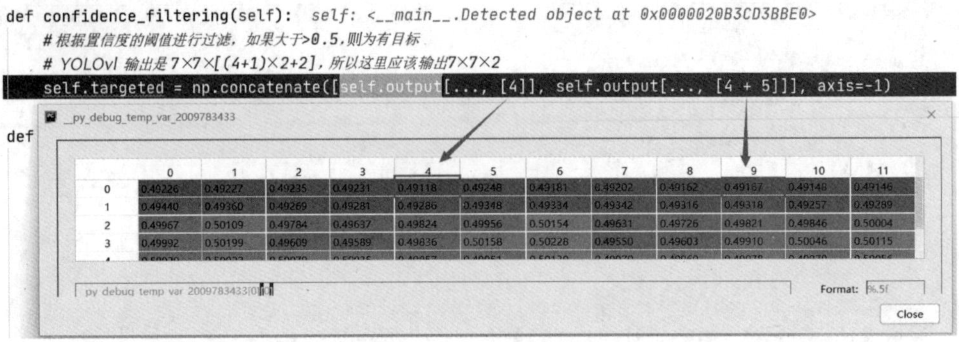

```
def confidence_filtering(self):  self: <__main__.Detected object at 0x0000020B3CD3BBE0>
    #根据置信度的阈值进行过滤，如果大于>0.5，则为有目标
    # YOLOvl 输出是7×7×[(4+1)×2+2]，所以这里应该输出7×7×2
    self.targeted = np.concatenate([self.output[..., [4]], self.output[..., [4 + 5]]], axis=-1)
def
```

_py_debug_temp_var_2009783433

	0	1	2	3	4	5	6	7	8	9	10	11
0	0.49226	0.49227	0.49235	0.49231	0.49118	0.49248	0.49181	0.49202	0.49162	0.49167	0.49148	0.49146
1	0.49440	0.49360	0.49269	0.49281	0.49286	0.49348	0.49334	0.49342	0.49316	0.49318	0.49257	0.49289
2	0.49967	0.50109	0.49784	0.49637	0.49824	0.49956	0.50154	0.49631	0.49726	0.49821	0.49846	0.50004
3	0.49992	0.50199	0.49609	0.49589	0.49836	0.50158	0.50228	0.49550	0.49603	0.49910	0.50046	0.50115

py debug temp var 2009783433[0][0]　Format: %.5f　Close

图 5-38　YOLOv1 置信度取值

在的格子进行 xy_normalized＝box[...，:2] * cell_size＋x0y0_normalized 解码操作，得到预测的 cx，cy。由于 YOLOv1 采用的是无锚框机制，所以它直接预测的是 wh_normalized＝box[...，2:]，然后将解码的内容存储到 boxes 中，如图 5-39 所示。

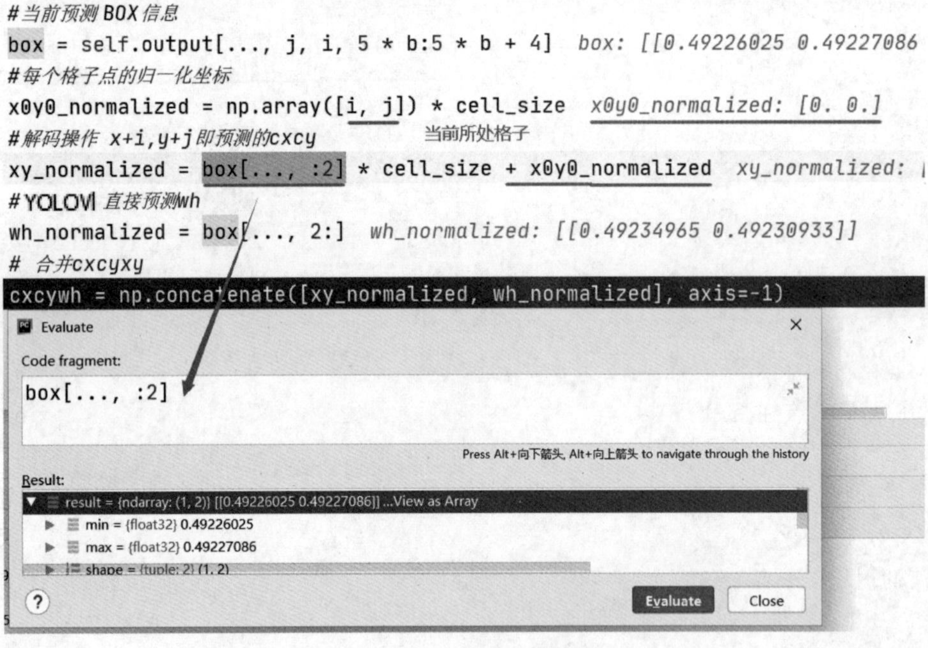

```
#当前预测BOX信息
box = self.output[..., j, i, 5 * b:5 * b + 4]   box: [[0.49226025 0.49227086
#每个格子点的归一化坐标
x0y0_normalized = np.array([i, j]) * cell_size   x0y0_normalized: [0. 0.]
                                   当前所处格子
#解码操作 x+i,y+j即预测的cxcy
xy_normalized = box[..., :2] * cell_size + x0y0_normalized   xy_normalized:
#YOLOVl 直接预测wh
wh_normalized = box[..., 2:]   wh_normalized: [[0.49234965 0.49230933]]
#  合并cxcyxy
cxcywh = np.concatenate([xy_normalized, wh_normalized], axis=-1)
```

Evaluate

Code fragment:

```
box[..., :2]
```

Press Alt+向下箭头, Alt+向上箭头 to navigate through the history

Result:

```
▼ result = {ndarray: (1, 2)} [[0.49226025 0.49227086]] ...View as Array
    ▶ min = {float32} 0.49226025
    ▶ max = {float32} 0.49227086
    ▶ shape = {tuple: 2} (1, 2)
```

Evaluate　Close

图 5-39　YOLOv1 解码操作

因为 class_labels_all 中存储了所有解码后的 label，如果 mask＝class_labels_all＝＝label，则表明只取当前 label 的信息，经过 u. nms(cat_boxes，self. nms_threshold)非极大值抑制，最后得到当前类别的 cat_boxes[index]，以及最后预测的 BOX 信息，如图 5-40 所示。

总结

YOLOv1 为 Anchor Free 机制，通过划分 7×7 个格子，由 GT 落入某个格子的中心点来预测目标，其主要特点为速度快。

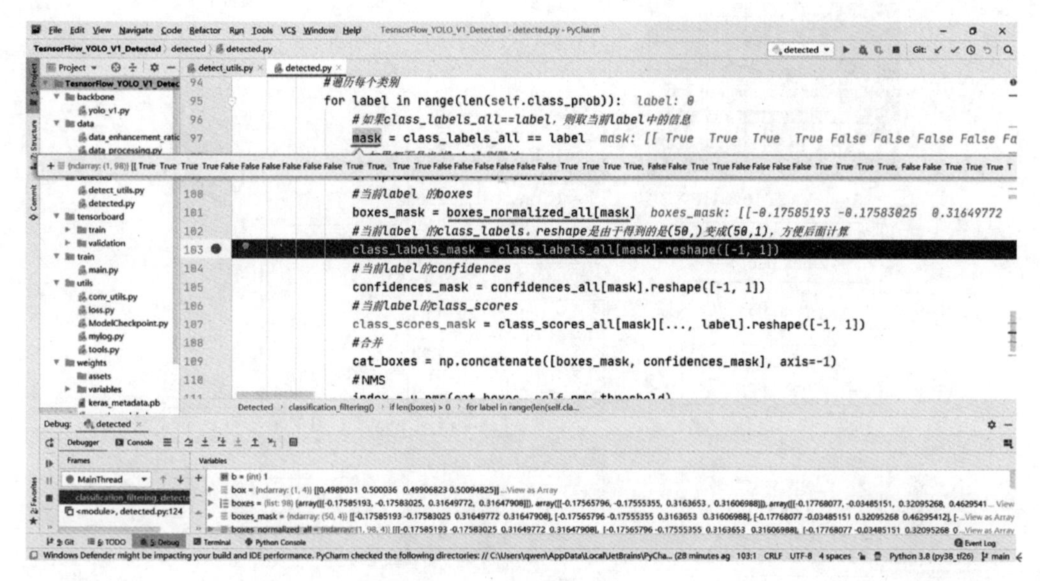

图 5-40　mask 过滤当前 label

练习

运行并调试本节代码，理解算法的设计与代码的结合，重点梳理本算法的实现方法。

5.6　单阶段速度快的检测网络 YOLOv2

5.6.1　模型介绍

YOLOv2 由 Joseph Redmon 等在 2016 年论文 *YOLO9000：Better，Faster，Stronger* 中提出，其主要特点在 YOLOv1 的基础上引入了 BN 归一化、建议框和 PassThrough Layer 层，其网络结构如图 5-41 所示。

在主干特征提取方面将 YOLOv1 的 7×7 卷积替换成 3×3 的卷积，并且在每个卷积后面跟了 BN 归一化。深层的语义信息较丰富，而浅层的几何信息较丰富，为了提高多尺度特征信息的融合，作者在这里使用了 PassThrough 层，经过 PassThrough 层的特征信息与 13×13×1024 进行 Concate，从而得到 13×13×1280 维特征。

PassThrough 具体的实现对输入的 26×26×64 每两个尺度进行重组，促进特征和通道信息的合并交流，如图 5-42 所示。

图 5-42 中输入为 4×3×3，对每两个通道且每个通道的每两组进行拼接，通道与通道之间的信息进行了重组，加强了通道之间的交流，并且没有权重参数。

在 Concate 之后经过 3×3 卷积，然后用 1×1 卷积代替全连接，输出为(4+1+20)×5，每个特征点输出 4 个预测 BOX 相对于 Anchor 的偏移，并且预测 BOX 有无目标，20 个分类的概率，每个特征分配 5 个建议框，所以输出为 13×13×125 维向量。

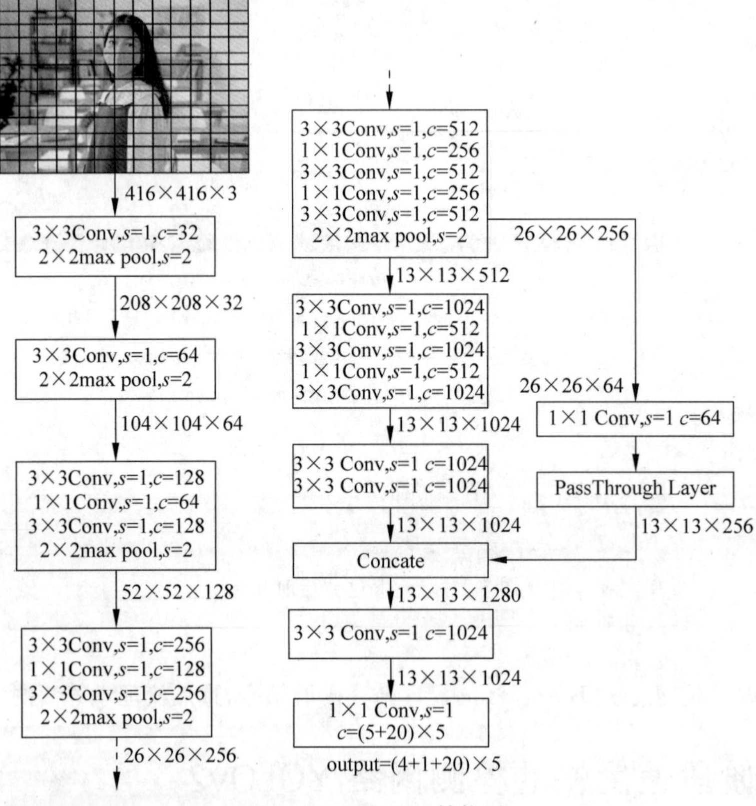

图 5-41 YOLOv2 结构图

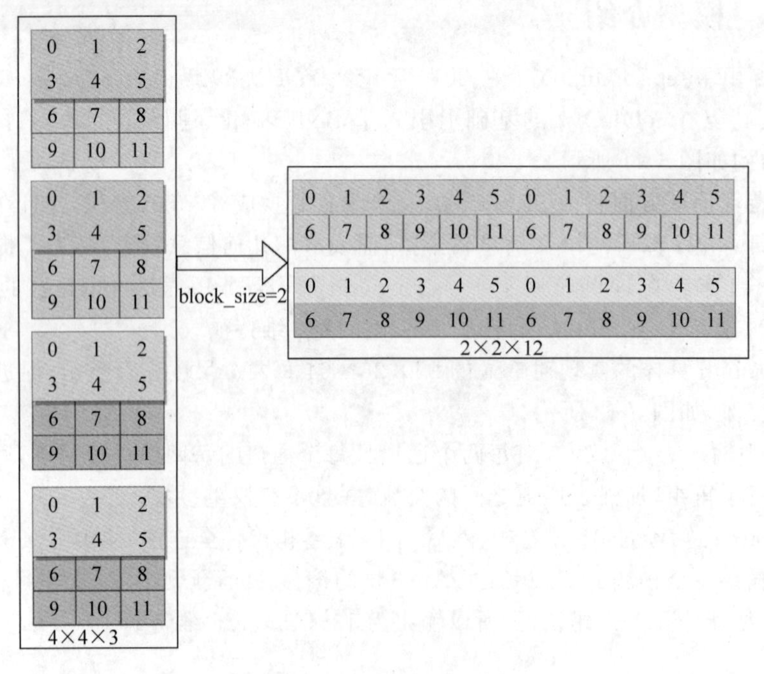

图 5-42 PassThrough 结构

YOLOv2 在计算 t_x、t_y 时仍然是基于(0,0)坐标的偏移，但是在计算 t_w、t_h 时是基于每个像素生成的建议框与 GT BOX 标注框 wh 的偏移，每个坐标点生成 5 个尺寸的建议框，如图 5-43 所示。

图 5-43　YOLOv2 的正样本

YOLOv2 将原图等比例缩放至 416×416，其输出为 13×13×125，则相对于将原图划分为 13×13 个格子，每个格子对于特征图的输出。每个特征点生成 5 个建议框，这 5 个建议框使用以下公式计算 t_x、t_y、t_w、t_h。

$$
\begin{aligned}
t_x &= G_x - j \\
t_y &= G_y - i \\
t_w &= \log\left(\frac{G_w}{P_w}\right) \\
t_h &= \log\left(\frac{G_h}{P_h}\right)
\end{aligned}
\tag{5-8}
$$

其中，G_x、G_y 代表标注 BOX 所在图像的中心点；j、i 为标注 BOX 所在的格子；P_w、P_h 为预设建议框的宽和高。从公式 5-8 可知 YOLOv2 正样本的选择与标注 BOX 所在的格子及建议框有关，在挑选建议框时将标注 BOX 与建议框计算 IOU，并挑选最大 IOU 作为正样本的 Anchor。

在损失函数方面，仍由正样本 BOX 损失＋正样本置信度损失＋负样本置信度损失＋分类损失构成，其公式可表述如下：

$$
\begin{aligned}
\text{Loss} = \sum_{i=0}^{W}\sum_{j=0}^{H}\sum_{k=0}^{A} & \lambda_{\text{noobj}} 1_{\text{ijk}}^{\text{noobj}}[(C_i - \hat{C}_i)^2] + \lambda_{\text{obj}} 1_{\text{ijk}}^{\text{obj}}[(C_i - \hat{C}_i)^2] + \\
& \lambda_{\text{coord}} 1_{\text{ijk}}^{\text{obj}}[(\text{Box}_i - \widehat{\text{Box}_i})^2] + \lambda_{\text{class}} 1_{\text{ijk}}^{\text{obj}}[(p_i(c) - \hat{p}_i(c))^2]
\end{aligned}
\tag{5-9}
$$

其中，C_i 为标注 BOX 与 Anchor 之间 IOU 得分值，如果大于设定的阈值，则为正样本，如果小于设定的预测值，则为负样本。$\hat{C}_i$ 为预测的置信度概率值。Box_i 为标注 BOX 与 Anchor 的偏移，$\widehat{\text{Box}}_i$ 为预测的偏移值。$p_i(c)$为标注 BOX 的分类概率，$\hat{p}_i(c)$为预测物体的分类概率。$\sum_{i=0}^{W}\sum_{j=0}^{H}\sum_{k=0}^{A}$ 表明需要遍历每个格子及建议框。

5.6.2 代码实战模型搭建

模型代码实现参考图 5-41 完成，PassThrough 层的实现可调用 tf.nn.space_to_depth()，详细的代码如下：

```
#第 5 章/ObjectDetection/TesnsorFlow_YOLO_V2_Detected/backbone/yolo_v2.py
def yolo_v2(input_shape, class_num=20, anchor_num=5):
    """YOLOv2 由两部分构成，一部分是 DarkNet19，另一部分是 YOLO 增加的
    :param anchor_num:Anchor 的数量
    :param class_num:分类的数量
    :param input_shape:输入 shape
    :return:
    """
    input_tensor = layers.Input(shape=input_shape)
    is_bn = True                                    #使用 BN
    net = {}
    #结构配置文件
    #[类型, kernel_size, strides, out_channel, 'same']
    darknet_config = [
        ['Conv', 3, 1, 32, 'same'],          #416 × 416
        ['MaxP', 2, 2, 32, 'same'],

        ['Conv', 3, 1, 64, 'same'],          #208 × 208
        ['MaxP', 2, 2, 64, 'same'],

        ['Conv', 3, 1, 128, 'same'],         #104 × 104
        ['Conv', 1, 1, 64, 'same'],
        ['Conv', 3, 1, 128, 'same'],
        ['MaxP', 2, 2, 128, 'same'],

        ['Conv', 3, 1, 256, 'same'],         #52 × 52
        ['Conv', 1, 1, 128, 'same'],
        ['Conv', 3, 1, 256, 'same'],
        ['MaxP', 2, 2, 256, 'same'],

        ['Conv', 3, 1, 512, 'same'],         #26 × 26
        ['Conv', 1, 1, 256, 'same'],
        ['Conv', 3, 1, 512, 'same'],
        ['Conv', 1, 1, 256, 'same'],
        ['Conv', 3, 1, 512, 'same'],
        ['MaxP', 2, 2, 512, 'same'],

        ['Conv', 3, 1, 1024, 'same'],        #13 × 13
```

```
        ['Conv', 1, 1, 512, 'same'],
        ['Conv', 3, 1, 1024, 'same'],
        ['Conv', 1, 1, 512, 'same'],
        ['Conv', 3, 1, 1024, 'same'],            #Conv-18, 13 × 13
    ]
    #concate 前的两个 3×3 卷积
    yolo_add = [
        ['Conv', 3, 1, 1024, 'same'],
        ['Conv', 3, 1, 1024, 'same'],
    ]
    x = input_tensor
    net['input'] = x
    #解析配置文件
    for i, c in enumerate(darknet_config):
        if c[0] == 'Conv':
            #已封装好的 conv→bn→leak_relu
            x = ConvBnLeakRelu(c[3], c[1], c[2], is_bn=is_bn, padding=c[4],
 is_relu=True)(x)
        elif c[0] == 'MaxP':
            #池化
            x = layers.MaxPooling2D(c[1], c[2], padding=c[4])(x)
        net[i] = x
    #构建两个 3×3 卷积
    for j, c in enumerate(yolo_add):
        if c[0] == 'Conv':
            x = ConvBnLeakRelu(c[3], c[1], c[2], is_bn=is_bn, padding=c[4],
 is_relu=True)(x)
        net[j + i] = x
    #将通过 pass_through 得到的 13×13×256 与 13×13×1024 进行合并
    x = layers.concatenate([x, pass_through(net[16], is_bn)], axis=-1)
    net['pass_concatenate'] = x
    x = ConvBnLeakRelu(1024, 3, 1, is_bn=is_bn, padding='same', is_relu=True)(x)
    net['pass_next'] = x
    #输出 13 × 13 × (4+1+num_class) ×anchor,即 13 × 13 × 125
    #直接位置预测,(4+1+20) ×5
    net['output'] = ConvBnLeakRelu((4 + 1 + class_num) *anchor_num, 1, 1, is_bn=
False, is_relu=False)(x)
    #构建模型
    return Model(inputs=net['input'], outputs=net['output'])

def pass_through(x, is_bn=True, channel=64):
    #pass_through 的封装,由 space_to_depth()函数实现该功能
    cx = ConvBnLeakRelu(channel, 1, 1, is_bn=is_bn, padding='same', is_relu=
True)(x)                                          #26 × 26 × 512
    cx = tf.nn.space_to_depth(cx, 2)              #13×13×256 pass through 起到的作用
                                                  #是在各个通道中每隔两个进行合并交流
return cx
```

代码中 darknet_config 为主干网络的配置，ConvBnLeakRelu()为封装好的卷积、BN、LeakRelu 激活函数，pass_through(x，is_bn＝True，channel＝64)实现 PassThrough 层，网络最后的输出为 13×13×(4＋1＋num_class)×anchor。

5.6.3 代码实战聚类得到建议框宽和高

YOLOv2 建议框的宽和高通过聚类得到，其判断距离的公式为

$$d(\text{box},\text{centroid}) = 1 - \text{IOU}(\text{box},\text{centroid}) \tag{5-10}$$

其中，d 代表距离，box 代表标注 BOX，centroid 为聚簇中心，1－IOU(box，centroid)，因为 IOU 越大说明两个 BOX 越近为 1，所以最小距离接近 0，核心代码如下：

```
#第 5 章/ObjectDetection/TesnsorFlow_YOLO_V2_Detected/utils/get_anchors.py
class AnchorKmeans(object):
    """聚类实现，获取 Anchor 的宽和高"""

    def __init__(self, k, max_iter=300, random_seed=None):
        self.k = k                              #设置几个中心点
        self.max_iter = max_iter                #最多迭代多少次
        self.random_seed = random_seed          #随机种子
        self.n_iter = 0
        self.anchors_ = None
        self.labels_ = None
        self.ious_ = None

    def fit(self, boxes):
        """得到 anchors"""
        assert self.k < len(boxes), "K 必须少于 BOX 的数量"
        #迭代次数，保证每次从 0 开始
        if self.n_iter > 0: self.n_iter = 0
        #随机种子
        np.random.seed(self.random_seed)
        #boxes 的数量
        n = boxes.shape[0]
        #从现有 Anchor 中随机选择 K 个 Anchor 作为初始点
        self.anchors_ = boxes[np.random.choice(n, self.k, replace=True)]
        #label 标签
        self.labels_ = np.zeros((n,))
        #开始聚类
        while True:
            #每迭代 1 次 self.n_iter+1
            self.n_iter += 1
            #迭代的次数要小于设置的总次数
            if self.n_iter > self.max_iter: break
            #将其他 BOX 与随机选择的中心点 Anchor 做 IOU
            self.ious_ = self.iou(boxes, self.anchors_)
            #距离 1-IOU→0，如果离得很近，则说明 BOX 与 Anchor 趋近于 1
            distances = 1 - self.ious_
            #取最小距离的下标
            cur_labels = np.argmin(distances, axis=1)
            #如果最小距离的下标与分配的下标一致，则停止
            if (cur_labels == self.labels_).all():
                break
            #更新 Anchor 的位置
            for i in range(self.k):
```

```
            self.anchors_[i] = np.mean(boxes[cur_labels == i], axis=0)
        self.labels_ = cur_labels
if __name__ == "__main__":
    xml_dir = "../face_mask/facemask_dataset_annotations"
    jpg_dir = '../face_mask/facemask_dataset'
    #获取所有标注的 boxes
    boxes = parse_xml(xml_dir, jpg_dir)
    #设置 k=5
    model = AnchorKmeans(5, random_seed=1000)
    model.fit(boxes)
    #获得聚类结果
    print(model.anchors_)
```

代码 distances＝1－self.ious_,表示所有 BOX 与聚簇中心越近，且当 if(cur_labels==self.labels_).all():break 时，如果最小距离的下标与分配的下标一致，则停止。如果将 K 的数量从 2 设置到 20，则可以获取 K 越大其 Anchor 与 GT BOX 的 IOU 得分越高，如图 5-44 中 $K=15$ 时平均 IOU＝0.75。增大 K 的设置可以提高精度，但同时会降低网络的推理速度，所以可以根据实际情况进行选择。

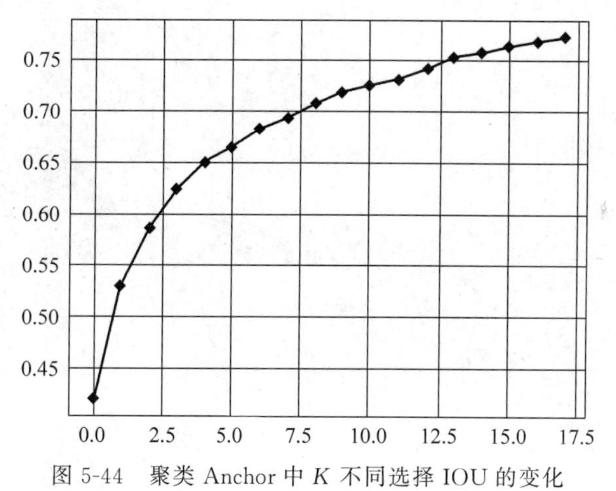

图 5-44 聚类 Anchor 中 K 不同选择 IOU 的变化

5.6.4 代码实战建议框的生成

首先，代码需要先计算 GT BOX 在哪个格子中，然后计算 GT BOX 与 Anchor 的 IOU 值，取最大 IOU 值所在的 Anchor 作为正样本，然后在 GT BOX 与挑选出来的 Anchor 之间计算偏移，参考代码如下：

```
#第 5 章/ObjectDetection/TesnsorFlow_YOLO_V2_Detected/utils/tools.py
def yolo_v2_true_encode_box(true_boxes,
                            anchors=None,
                            input_size=(416, 416)):
    """
    YOLOv2 的真值编码。需要编码成 GRID_NUM * GRID_NUM * (4 + 1 + class_num) * anchor_num
```

```
    :param true_boxes:
    两个维度:
    第 1 个维度:一张图片中有几个实际框
    第 2 个维度:[cx, cy, w, h, class],x 和 y 是框中心点坐标,w 和 h 是框的宽度和高度
    x,y,w,h 均是除以图片分辨率(原始图片尺寸 416*416)得到的[0,1]范围的比值
    :param anchors: 实际 anchor boxes 的值,论文中使用了 5 个。[w,h]都是相对于 grid
cell 的比值
    :param input_size: true box 中的输入图片的尺寸要与预测时的保持一致
    :return:
    true_boxes2 13 × 13 × 5 × 4,返回它是为了方便背景损失的计算
    detectors_mask 13 × 13 × 5 × 1,置信度
    matching_true_boxes 13 × 13 × 5 × 5,GT BOX 基于 Anchor 的偏移
    """
    if anchors is None:
        #默认设置的 Anchor 大小,需要乘以 13
        anchors = [[1.08, 1.19], [3.42, 4.41], [6.63, 11.38], [9.42, 5.11], [16.62,
10.52]]
        #anchors×13 为真实 Anchor
        anchors = np.array(anchors) *(input_size[0] //32)
    #输入图像的尺寸
    height, width = input_size
    assert height % 32 == 0 #必须是 32 的倍数
    assert width % 32 == 0
    #特征图的尺寸,也就是 13×13
    conv_height, conv_width = height //32, width //32
    #true box 的数量
    num_box_params = true_boxes.shape[1]
    #5 个 Anchor
    NumPy_boxes = len(anchors)
    #初始一个 Tensor,用来存储 BOX 正样本值 13×13×5×4
    true_boxes2 = np.zeros(
        [conv_height, conv_width, NumPy_boxes, 4], dtype=np.float32
    )
    #初始一个 Tensor,用来存储置信度值 13×13×5×1
    detectors_mask = np.zeros(
        [conv_height, conv_width, NumPy_boxes, 1], dtype=np.float32
    )
    #13 × 13 × 5 × len(true_boxes), BOX+置信度,跟 true boxes 数保持一致
    matching_true_boxes = np.zeros(
        [conv_height, conv_width, NumPy_boxes, num_box_params], dtype=np.float32
    )
    #对所有的 true_boxes 进行编码
    for box in true_boxes:
        #置信度
        box_class = box[4:5]        #这样得到的是一个数组。如果是 box[4],则得到的是一个
                                    #具体值
        #box = (13 × x,13 × y, 13 × w, 13 × h) 换算成相对 grid cell 的值
        #[0.5078125 0.36830357 0.14955357 0.19642857] × 13
        #[6.6015625 4.78794637 1.94419637 2.55357137]
        box = box[0:4] *np.array([
            conv_width, conv_height, conv_width, conv_height
        ])
        box_true = box.copy()
```

```
        #向下取整,计算中心点落在哪个格子中
        i = np.floor(box[1]).astype('int')                    #i=4
        j = np.floor(box[0]).astype('int')                    #j=6

        best_iou = 0
        best_anchor = 0
        #将 true_box 与每个 Anchor 进行 IOU,并取最佳的 Anchor 作为正样本
        for k, anchor in enumerate(anchors):
            #true box 的 wh 1/2
            box_maxes = box[2:4] *0.5
            box_mines = -box_maxes
            #Anchor BOX 的 wh 1/2
            anchor_maxes = anchor *0.5
            anchor_mines = -anchor_maxes
            #将真实 wh 与 Anchor 之间进行 IOU 的计算,并获取最佳 IOU 是哪个 Anchor
            intersect_mines = np.maximum(box_mines, anchor_mines)
            intersect_maxes = np.minimum(box_maxes, anchor_maxes)
            intersect_wh = np.maximum(intersect_maxes - intersect_mines, 0.)
            intersect_area = intersect_wh[0] *intersect_wh[1]
            box_area = box[2] *box[3]
            anchor_area = anchor[0] *anchor[1]
            iou_score = intersect_area / (box_area + anchor_area - intersect_area)
            #比较当前 Anchor 的 IOU 是否比上一次的 IOU 值大,如果大,则是最佳 IOU,并记录是
            #第几个 k
            if iou_score > best_iou:
                best_iou = iou_score
                best_anchor = k
        if best_iou > 0: #当前示例 best_iou=0.85,k=2
            #detectors_mask 为 13×13×5×1,将最佳 Anchor 的置信度设置为 1.0
            #detectors_mask[4,6,2]=1.0
            detectors_mask[i, j, best_anchor] = 1.0
            #true_boxes2 为 13×13×5×4,即 BOX 的值
            #true_boxes2[4,6,2]=[6.6015625 4.78794637 1.94419637 2.55357137]
            true_boxes2[i, j, best_anchor] = box_true
            #套公式,算偏移。cx - j, cy - i, np.log(w/w *),np.log(h/h *)
            adjusted_box = np.array([
                box[0] - j, #gt_cx-j,gt_cy-i
                box[1] - i,
                np.log(box[2] / anchors[best_anchor][0]),    #gt_w/anchors[2][0],
                                                             #即 gt_w/a_w
                np.log(box[3] / anchors[best_anchor][1]),    #gt_h/anchors[2][1],
                                                             #即 gt_w/a_h
                box_class                                    #GT BOX 的 label 下标
            ], dtype=np.float32)
            #matching_true_boxes 13 × 13 × 5 × 5
            #matching_true_boxes[4,6,2] = offset 值
            matching_true_boxes[i, j, best_anchor] = adjusted_box
            #返回 GT BOX,Anchor 正样本的置信度,GT BOX 基于 Anchor 的偏移
    return true_boxes2, detectors_mask, matching_true_boxes

def test_yolo_v2_true_encode():
    data = np.array(
        [[0.5078125, 0.36830357, 0.14955357, 0.19642857, 1.]]
```

```
        , dtype=np.float32
    )
    return yolo_v2_true_encode_box(data)
```

代码中 box=box[0:4] * 13,box[0:4]是归一化后的值,乘以 13 得到[6.6015625 4.78794637 1.94419637 2.55357137],即从归一化值后放大到 13×13 的特征图的大小,向下取整则所在格子为 $i=4$、$j=6$。box_maxes=box[2:4] * 0.5 得到 GT BOX 的中心点,因为计算 IOU 还需要左上角的值,所以 box_mines=-box_maxes,得到 best_iou=0.8,best_anchor=2,那么将 detectors_mask[4,6,2]=1.0,表示第 $i=4$、$j=6$ 格子中的第 2 个 Anchor 置信度为 1,然后选择第 2 个 Anchor 并通过公式求与 GT BOX 的偏移值并赋给 adjusted_box,最后 matching_true_boxes[4,6,2]=adjusted_box,返回正样本的偏移值。

得到正样本的偏移值后,仍然像 YOLOv1 中的 DataProcessingAndEnhancement(object)类的 generate(self,isTraining=True)喂给训练数据,具体代码实际基本一致,详细可参考随书代码。

5.6.5 代码实现损失函数的构建及训练

根据式(5-9)实现正样本格子 i、j、k 位置的损失、置信度的损失及分类概率的损失,核心代码如下:

```
#第 5 章/ObjectDetection/TesnsorFlow_YOLO_V2_Detected/utils/loss.py
class MultiAngleLoss(object):
    def yolo_v2_loss(self, y_true, y_pre):
        """
        求 YOLOv2 损失
        :param y_pre: model 预测值,需要解码
        :param y_true: 真实值
        :return:
        """
        #预测值
        #(1)对预测出来的 offset 解码成 xmin,ymin,xmax,ymax,conf,class_score
        #对解码出来的值跟 GT BOX 之间做 IOU,如果 IOU>阈值,则为正样本置信度,否则为负样
        #本置信度
        yolo_output = y_pre
        true_boxes = y_true[..., :4]              #真实坐标 b × 13 × 13 × 5 × 4
        detectors_mask = y_true[..., 4:5]         #置信度 b × 13 × 13 × 5 × 1
        matching_true_boxes = y_true[..., 5:]     #anchor b × 13 × 13 × 5 × 5
        #对预测出来的值进行 offset 解码,解码成 xmin,ymin,xmax,ymax
        boxes, pre_box_confidence, pre_box_class_props = yolo_v2_head_decode(
            yolo_output,
            self.anchors,
            self.num_classes,
            self.input_size
        )
        #预测置信度 b×13×13×5×1
        pre_box_confidence = tf.reshape(
            pre_box_confidence, [-1, self.conv_height, self.conv_width, self.num_
anchors, 1]
```

```
        )
        #预测分类 b×13×13×5×20
        pre_box_class_props = tf.reshape(
            pre_box_class_props, [-1, self.conv_height, self.conv_width, self.
num_anchors, self.num_classes]
        )
        #预测 xmin,ymin,xmax,ymax
        pre_boxes_xy = boxes[..., :2]
        pre_boxes_wh = boxes[..., 2:4]
        ##########################################
        #(2)再从 pre_y 中取出偏移值,是为了计算预测偏移与真实 BOX 偏移之间的损失
        #将 YOLO 输出的[b, 13, 13, 125]转换为[b, 13, 13, 5, 25],5 是 Anchor 数量
        yolo_out_shape = yolo_output.shape[1: 3] #[b, 13, 13, 125]
        features = tf.reshape(
            yolo_output,
            [-1, yolo_out_shape[0], yolo_out_shape[1], self.num_anchors, self.
num_classes + 5]
        )
        #预测出来的是偏移值 xy 限制了值域,只能在 0~1
        pre_d_boxes = tf.concat([tf.nn.sigmoid(features[..., 0:2]), features[...,
2:4]], axis=-1)
        ##########################################
        #(3)由 xmin,ymin,xmax,ymax 转换成 cx,cy,w,h,这是为了计算 IOU
        #维度调整
        pre_boxes_xy = tf.reshape(pre_boxes_xy, [-1, self.conv_height, self.conv_
width, self.num_anchors, 2])
        pre_boxes_wh = tf.reshape(pre_boxes_wh, [-1, self.conv_height, self.conv_
width, self.num_anchors, 2])
        #预测 cx,cy,w,h,为了计算 IOU
        pre_box = tf.concat([
            (pre_boxes_xy - pre_boxes_wh) *0.5,
            (pre_boxes_xy + pre_boxes_wh) *0.5,
        ], axis=-1)
        ##########################################
        #(4)当 GT BOX 与 pre box 之间 IOU>阈值时,才认为 object_detections 有目标
        #算一下 true 的 xmin,ymin,xmax,ymax,方便做 IOU 的计算
        true_box = tf.concat([
            (true_boxes[..., 0:2] - true_boxes[..., 2:4]) *0.5,
            (true_boxes[..., 0:2] + true_boxes[..., 2:4]) *0.5,
        ], axis=-1)
        #得到预测框与真实框之间的 IOU 得分
        iou_result = iou(pre_box, true_box)   #1 × 13 × 13 × 5
        #获得最大得分,即 13 *13 *5 *1
        iou_score = tf.expand_dims(tf.reduce_max(iou_result, axis=-4), axis=-1)
        #过滤 IOU 要大于指定的置信度
        object_detections = iou_score > self.overlap_threshold
        object_detections = tf.cast(object_detections, dtype=iou_score.dtype)
        ##########################################
        #(5)根据公式计算损失
        #当预测框的 IOU 与真实框中的 IOU 小于 0.6 时都为背景
        #没有目标物体的损失 1 - object_detections 预测没有目标,1 - detectors_mask
        #真实没有目标
        #tf.square(0-pre_box_confidence)没有目标的置信度
```

```
        no_object_loss = self.no_obj_scale *(
                1 - object_detections
        ) *(1 - detectors_mask) *tf.square(0-pre_box_confidence)

        #有目标物体的损失
        object_loss = self.obj_scale * detectors_mask * tf.square(1 - pre_box_
confidence)
        #置信度损失=没有目标物体的损失+有目标物体的损失
        confidence_loss = tf.reduce_sum(object_loss + no_object_loss)
        #分类损失
        matching_classes = tf.cast(matching_true_boxes[..., 4], 'int32')
        matching_classes = tf.one_hot(matching_classes, self.num_classes)
        classification_loss = tf.reduce_sum(
            self.class_scale *detectors_mask *tf.square(matching_classes - pre_
box_class_props)
        )
        #boxes loss,计算的是偏移值之间的误差
        box_loss = tf.reduce_sum(
            self.coordinates_scale * detectors_mask * tf.square(matching_true_
boxes[..., 0:4] - pre_d_boxes)
        )
        #所有损失
        total_loss = (confidence_loss + classification_loss + box_loss) *0.5
        return total_loss
```

此损失计算的代码较长，共由 5 个步骤构成。第 1 步，对预测出来的 y_pre 进行解码，y_pre 输出的是预测值基于 Anchor 的偏移，根据以下公式可计算出预测框的左上、右下坐标。

$$
\begin{aligned}
P_x &= \mathrm{Sigmoid}(t_x) + c_x \\
P_y &= \mathrm{Sigmoid}(t_y) + c_y \\
P_w &= A_w \mathrm{e}^{t_w} \\
P_h &= A_h \mathrm{e}^{t_h}
\end{aligned}
\tag{5-11}
$$

其中，t_x、t_y 为预测的偏移值，c_x、c_y 为每个 13×13 的坐标值，t_w、t_h 为预测框与 Anchor 的偏移值。A_w、A_h 为 Anchor 的 w、h。P_x、P_y、P_w、P_h 为解码出来的 c_x、c_y、w、h，其具体的代码如下：

```
#第 5 章/ObjectDetection/TesnsorFlow_YOLO_V2_Detected/utils/tools.py
def yolo_v2_head_decode(features, anchors=None, num_classes=20, input_size=
(416, 416)):
    """
    YOLO 预测边界框的中心点相对于网格左上角的偏移值,而每个网格有 5 个 Anchor,然后套用
公式便可得到实际位置
    features:       预测出来的值 conv。预测出来的是偏移值 [None, 13, 13, (4 + 1 + num_
classes) *5]
    anchors:        Anchor 的 widths 和 heights
    num_classes:    分类数
    :return:
    boxes : 返回 xmin,ymin,xmax,ymax,这是为了方便求背景损失的计算,实际上求位置的损
失没用这个返回值
```

```
    box_confidence: 置信度
    box_class_props: 类别,类别是进行了 Softmax 的
    """
    height, width = input_size
    assert height % 32 == 0                    #必须是 32 的倍数
    assert width % 32 == 0
    conv_height, conv_width = height //32, width //32
    if anchors is None:
        anchors = [[1.08, 1.19], [3.42, 4.41], [6.63, 11.38], [9.42, 5.11], [16.62,
10.52]]
    #即 anchors×13
    anchors = np.array(anchors) *(input_size[0] //32)
    anchor_size = tf.constant(len(anchors))
    #将输入的 b×13×13×125 reshape 成 b × 169×5×25
    features = tf.reshape(features, [features.shape[0], conv_height * conv_
width, anchor_size, num_classes + (4 + 1)])
    #因为预测出来的是相对于该左上角的偏移值,Sigmoid 函数归一化到(0,1)之间
    xy_offset = tf.nn.sigmoid(features[..., 0:2])
    #置信度
    box_confidence = tf.sigmoid(features[..., 4:5])
    #wh 偏移
    wh_offset = tf.exp(features[..., 2:4])
    #类别进行 Softmax 输出
    box_class_props = tf.nn.softmax(features[..., 5:])

    #在 feature 上面生成 anchors
    height_index = tf.range(conv_height, dtype=tf.float32)
    width_index = tf.range(conv_width, dtype=tf.float32)
    #得到网格 13×13
    x_cell, y_cell = tf.meshgrid(height_index, width_index)
    #和上面[h,w,num_anchors,num_class+5]对应
    x_cell = tf.reshape(x_cell, [1, -1, 1]) #
    y_cell = tf.reshape(y_cell, [1, -1, 1])

    #根据网格求坐标位置,套公式
    bbox_x = (x_cell + xy_offset[..., 0])                   / conv_height
    bbox_y = (y_cell + xy_offset[..., 1])                   / conv_width
    bbox_w = (anchors[:, 0] *wh_offset[..., 0])             / conv_height
    bbox_h = (anchors[:, 1] *wh_offset[..., 1])             / conv_width

    #由 cx,cy,w,h 转换成 xmin, ymin, xmax, ymax
    boxes = tf.stack(
        [
            bbox_x - bbox_w / 2,
            bbox_y - bbox_h / 2,
            bbox_x + bbox_w / 2,
            bbox_y + bbox_h / 2
        ],
        axis=3
    )
    return boxes, box_confidence, box_class_props
```

在解码代码中 x_cell 和 y_cell 为每个格子的坐标，x_cell+xy_offset[...,0]为每个偏移值解码后的 x 值，除以 conv_height=13 是为了归一化操作，如图 5-45 所示。

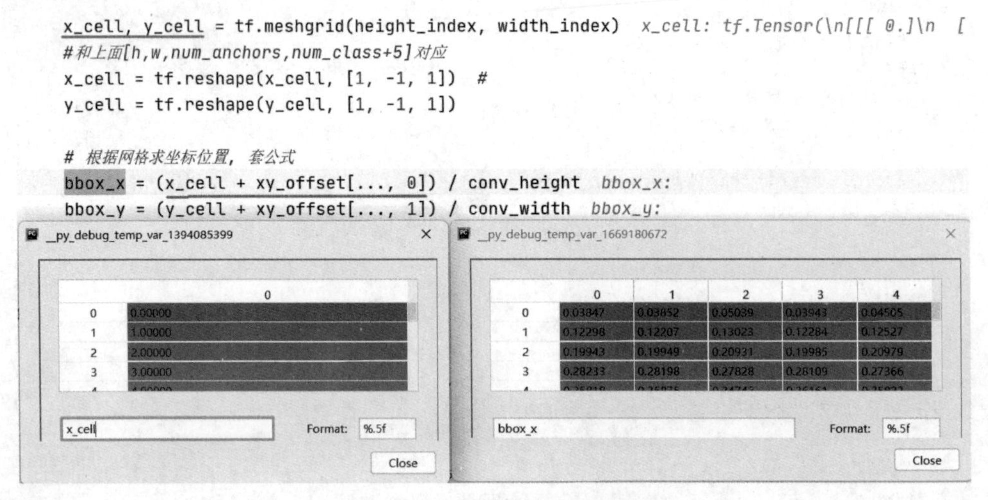

图 5-45 预测偏移值解码

损失函数的第 2、第 3 步是从预测值中得到预测的偏移值 pre_d_boxes=tx,ty,tw,th，同时从预测解码 boxes 中得到 pre_box=cx,cy,w,h，以便计算真实 BOX 与预测 BOX 的 IOU，如图 5-46 所示。

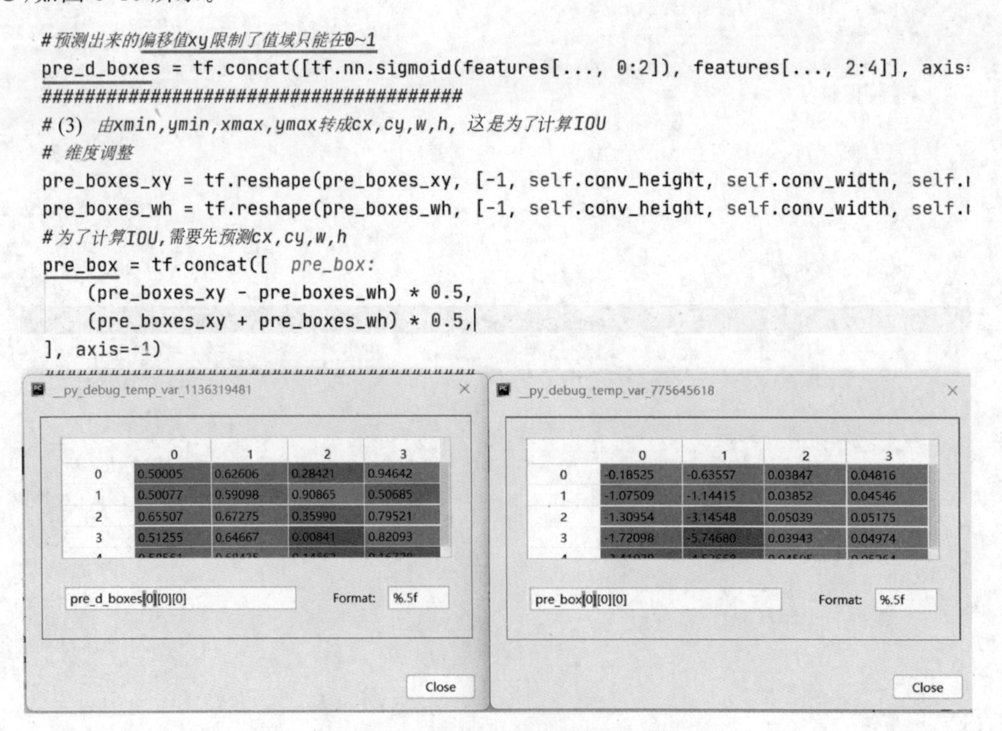

图 5-46 从预测值中获取预测偏移值

第 4 步，将解码出来的 pre_boxes 与 true_boxes 做 IOU，当 IOU＞0.5 时置为正样本 mask 的 object_detections，如图 5-47 所示。

```
true_box = tf.concat([   true_box:
    (true_boxes[..., 0:2] - true_boxes[..., 2:4]) * 0.5,
    (true_boxes[..., 0:2] + true_boxes[..., 2:4]) * 0.5,
], axis=-1)
#得到预测框与真实框之间的IOU得分
iou_result = iou(pre_box, true_box)  # 1 * 13 * 13 * 5  iou_result: tf.
0.0384654  0.04815853]\n  [-1.0750892  -1.144149  0.03852105  0.04545987]\n  [-1.3095378  -3.1454785  0.05038989  0.0517497
iou_score = tf.expand_dims(tf.reduce_max(iou_result, axis=-4), axis=-1)
#过滤IOU要大于指定的置信度
                                        0.5
object_detections = iou_score > self.overlap_threshold  object_detectio
object_detections = tf.cast(object_detections, dtype=iou_score.dtype)
```

图 5-47 IOU＞0.5 时的 mask

第 5 步，根据式(5-11)计算没有目标的置信度损失、有目标的置信度损失、分类损失和偏移值之间的损失，如图 5-48 所示。

```
no_object_loss = self.no_obj_scale * (  no_object_loss:
        1 - object_detections
) * (1 - detectors_mask) * tf.square(0-pre_box_confidence)

#有目标物体的损失
object_loss = self.obj_scale * detectors_mask * tf.square(1 - pre_box_confidence)  object_loss:
#置信度损失=没有目标物体的损失+有目标物体的损失
confidence_loss = tf.reduce_sum(object_loss + no_object_loss)  confidence_loss: tf.Tensor(325.76306,
#分类损失
matching_classes = tf.cast(matching_true_boxes[..., 4], 'int32')  matching_classes:
matching_classes = tf.one_hot(matching_classes, self.num_classes)
classification_loss = tf.reduce_sum(  classification_loss: tf.Tensor(0.93039364, shape=(), dtype=floc
    self.class_scale * detectors_mask * tf.square(matching_classes - pre_box_class_props)
)
# boxes loss, 计算的是偏移量之间的误差
box_loss = tf.reduce_sum(  box_loss: tf.Tensor(0.4022428, shape=(), dtype=float32)
    self.coordinates_scale * detectors_mask * tf.square(matching_true_boxes[..., 0:4] - pre_d_boxes)
)
# 所有损失
total_loss = (confidence_loss + classification_loss + box_loss) * 0.5  total_loss:
```

图 5-48 损失计算

在训练方面，先使用 224×224 训练 DarkNet19 的权重并迁移到 YOLOv2 的主干中，然后使用不同尺度的大小进行训练，例如 320、352、608，然后在 416 上进行微调，这样训练将使模型具备不同分辨率图像的泛化能力，稳健性更强，关键代码如下：

```
#第 5 章/ObjectDetection/TesnsorFlow_YOLO_V2_Detected/utils/train_YOLOv2.py
if __name__ == "__main__":
    input_shape = [
        (320, 320, 3),
        (352, 352, 3),
```

```
        (608, 608, 3),
        (416, 416, 3)
    ]
    #每个尺度的训练次数
    train_EPOCH = [10, 20, 30, 50]
    #每个尺度训练的学习率
    learn = [1e-3, 1e-4, 1e-5, 1e-5]
    old_epoch = 0                        #上一次的 epoch num
    old_name = 0                         #上一次训练的权重名
    #构建不同尺度的图形模型训练
    for im_shape, epoch, lr, save_dir in zip(input_shape, train_EPOCH, learn,
save_weights):
        model = yolo_v2(im_shape, num_classes, is_class=False)
        #如果已经有训练好的其他尺度的权重文件,则作为下一个尺度的初始权重
        if not old_name:
            exist_weights = f"../weights/last_{old_name}.h5"
            if os.path.exists(exist_weights):
                model.load_weights(exist_weights, by_name=True, skip_mismatch=True)
        else:
            #第 1 个尺寸调入分类的训练权重
            model.load_weights('../weights/darknet_10_224.h5', by_name=True,
skip_mismatch=True)
        #每隔 3 个 epoch 设置检测点并保存最优模型
        checkpoint = ModelCheckpoint(
            save_dir,
            monitor='val_loss',
            save_weights_only=False,
            save_best_only=True,
            period=check_step_epoch
        )
        #设置优化器
        model.compile(
            optimizer=Adam(learning_rate=lr),
            loss=MultiAngleLoss(num_classes=num_classes, input_size=im_shape
[:2]).yolo_v2_loss,
            run_eagerly=False,      #是否启用调试模型
        )
        #加载数据类
        data_object = DataProcessingAndEnhancement(
            train_lines,
            val_lines,
            num_classes,
            batch_size=BATCH_SIZE,
            input_shape=im_shape[0:2]
        )
        #训练
        model.fit(
            data_object.generate(True),
            steps_per_epoch=num_train //BATCH_SIZE,
            validation_data=data_object.generate(False),
            validation_steps=num_val //BATCH_SIZE,
            epochs=epoch,
```

```
            initial_epoch=old_epoch,
            callbacks=[logging, checkpoint]
        )
        old_epoch = epoch
        old_name = im_shape[0]
        #保存模型
        model.save(f"../weights/last_{old_name}.h5")
```

更多更详细的代码可参考随书代码。

5.6.6 代码实战预测推理

YOLOv2 引入了 5 个 Anchor 并根据式(5-11)对预测值进行解码操作,代码如下:

```
#第 5 章/ObjectDetection/TensorFlow_Yolo_V2_Detected/detected/detected.py
class Detected():
    def __init__(self, model_path, input_size):
        self.model = load_model(
            model_path,
            custom_objects={'compute_loss': MultiAngleLoss(2).compute_loss}
        ) #读取模型和权重
        self.confidence_threshold = 0.01                    #有无目标置信度
        self.class_prob = [0.5, 0.5]                        #两个类别的分类阈值
        self.nms_threshold = 0.5                            #NMS 的阈值
        self.input_size = input_size
        self.anchors = np.array([
            (1.3221, 1.73145), (3.19275, 4.00944),
            (5.05587, 8.09892), (9.47112, 4.84053), (11.2364, 10.0071)
        ])

    def readImg(self, img_path=None):
        img = cv2.imread(img_path)                          #读取要预测的图片
        #将图片转换到 300×300
        self.img, _ = u.letterbox_image(img, self.input_size, [])
        self.old_img = self.img.copy()

    def forward(self):
        #升成 4 维
        img_tensor = tf.expand_dims(self.img / 255.0, axis=0)
        #前向传播
        self.output = self.model.predict(img_tensor)
        self.h, self.w = self.output.shape[1:3] #1*13*13*(4+1+num_class)*5

    def generate_anchor(self):
        #使用均分的方法生成格子。一共有 h×w 个格子,即 169 个格子
        self.lin_x = np.ascontiguousarray(np.linspace(0, self.w - 1, self.w).
repeat(self.h)).reshape(self.h *self.w)
        self.lin_y = np.ascontiguousarray(np.linspace(0, self.h - 1, self.h).
repeat(self.w)).T.reshape(self.h *self.w)
        #得到锚框的 w 和 h,因为一共有 5 个 Anchor,所以 reshape 成[1,5,1]
        self.anchor_w = np.ascontiguousarray(self.anchors[..., 0]).reshape([-1,
5, 1])
```

```
        self.anchor_h = np.ascontiguousarray(self.anchors[..., 1]).reshape([-1,
5, 1])

    def _decode(self):
        #对预测出来的结果解码
        #YOLOv2 的输出(4+1+num_class)×5 个 Anchor
        self.b = self.output.shape[0]
        #变成[b,5,(4+1+2),169)
        logits = np.reshape(self.output, [self.b, 5, -1, self.h *self.w])
        self.result = np.zeros(logits.shape)
        #套用公式进行解码,Sigmoid()函数进行值域限制
        #根据 Anchor 得到 cx 和 cy
        self.result[:, :, 0, :] = (tf.sigmoid(logits[:, :, 0, :]).NumPy() + self.
lin_x) / self.w
        self.result[:, :, 1, :] = (tf.sigmoid(logits[:, :, 1, :]).NumPy() + self.
lin_y) / self.h
        #根据 Anchor 得到 w 和 h
        self.result[:, :, 2, :] = (tf.exp(logits[:, :, 2, :]).NumPy() *self.anchor_
w) / self.w
        self.result[:, :, 3, :] = (tf.exp(logits[:, :, 3, :]).NumPy() *self.anchor_
h) / self.h
        #得到置信度
        self.result[:, :, 4, :] = tf.sigmoid(logits[:, :, 4, :]).NumPy()
        #得到分类
        self.result[:, :, 5:, :] = tf.nn.softmax(logits[:, :, 5:, :]).NumPy()

    def classification_filtering(self):
        self._decode()
        #取最大的分类得分的下标,此时 result 为[b,5,(4+1+num_class),169]
        class_score = self.result[:, :, 5:, :]           #[1,5,num_class,169]
        class_label = np.argmax(class_score, axis=2)    #[1,5,169]
        #最大的分类得分值
        score = np.max(class_score, axis=2)             #[1,5,169]
        #当前格子的置信度
        conf = self.result[:, :, 4, :]                  #[1,5,169]
        #每个格子最后的得分为置信度*分类得分
        prob = score *conf                              #[1,5,169]
        #根据阈值,得到评分
        score_mask = prob > self.confidence_threshold   #[1,5,169]
        if np.sum(score_mask.reshape([-1])) > 0:
            #只有转置成[b×5×(4+1+num_class) ×169],才能根据 score_mask 进行取值,
            #最后需要得到[b,5,169,7]中的 7
            self.result = np.reshape(self.result, [self.b, 5, self.h *self.w, -1])
            boxes = self.result[score_mask][..., :4] #根据 score_mask 过滤得到 boxes
            cls_scores = prob[score_mask] #根据 score_mask 过滤得到 cls_scores
            idx = class_label[score_mask] #根据 score_mask 过滤得到 labels index
            #根据类别进行非极大值抑制
            for label in range(len(self.class_prob)):
                #如果 class_labels_all==label,则取当前 label 中的信息
                mask = idx == label
                #如果都不是当前 label,则跳过
                if np.sum(mask) == 0: continue
```

```
                #由 cx,cy,w,h 转换成 xmin,ymin,xmax,ymax
                xyxy_box = u.cxcy2xyxy(boxes[mask][..., :4])
                cat_boxes = np.concatenate([xyxy_box, cls_scores[mask].reshape
([-1, 1])], axis=-1)
                #NMS
                index = u.nms(cat_boxes, self.nms_threshold)
                #绘框
                self.old_img = u.draw_box(self.old_img, cat_boxes[index])

            #最后结果
            u.show(self.old_img)


if __name__ == "__main__":
    d = Detected(r'../weights/chk416', input_size=[416, 416])
    d.readImg('../../val_data/pexels-photo-5211438.jpeg')
    d.forward()
    d.generate_anchor()
    d.classification_filtering()
```

在__init__()中增加 self. anchors 以描述先验框的 wh 值，forward()前向传播后得到的 shape 为[b,13,13,(4+1+num_class) * 5]，如图 5-49 所示。

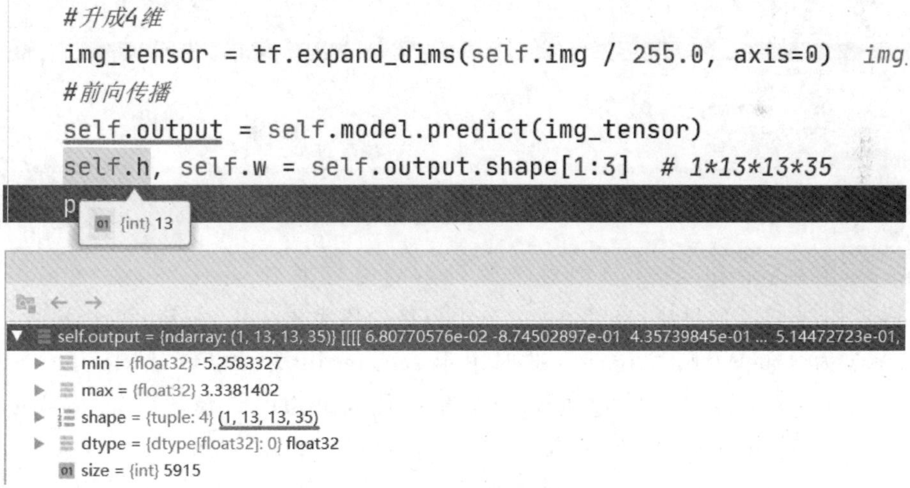

```
#升成4维
img_tensor = tf.expand_dims(self.img / 255.0, axis=0)  img
#前向传播
self.output = self.model.predict(img_tensor)
self.h, self.w = self.output.shape[1:3]  # 1*13*13*35
p
   {int} 13

self.output = {ndarray: (1, 13, 13, 35)} [[[[ 6.80770576e-02 -8.74502897e-01  4.35739845e-01 ... 5.14472723e-01,
    min = {float32} -5.2583327
    max = {float32} 3.3381402
    shape = {tuple: 4} (1, 13, 13, 35)
    dtype = {dtype[float32]: 0} float32
    size = {int} 5915
```

图 5-49　YOLOv2 前向传播输出

在 generate_anchor()中根据 np. linspace(0,self. w−1,self. w)指令将输入图像均分为 13 份，为了组成完整坐标，所以 repeat(self. h)重复了 13 次，摊平后 reshape(self. h * self. w)得到 169 个坐标并存储在 self. lin_x 变量中，以同样的方法得到 self. lin_y。在 self. anchor_w、self. anchor_h 分别得到锚框的宽和高，如图 5-50 所示。

_decode()首先将 self. output 由[b,13,13,(4+1+num_class) * 5]变成[b,5,(4+1+num_class),13 * 13]，然后根据 self. result[:, :, 0, :]的位置对于预测的 offset 值根据式(5-11)进行解码，并采用 tf. sigmoid()函数进行值域限定，如图 5-51 所示。

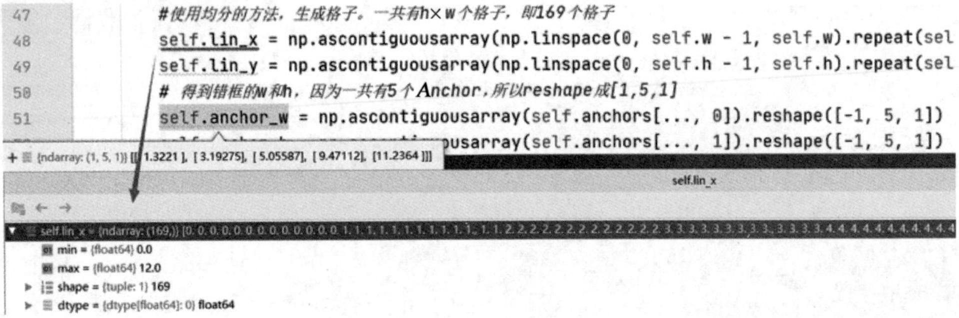

```
47        #使用均分的方法，生成格子。一共有h×w个格子，即169个格子
48        self.lin_x = np.ascontiguousarray(np.linspace(0, self.w - 1, self.w).repeat(sel
49        self.lin_y = np.ascontiguousarray(np.linspace(0, self.h - 1, self.h).repeat(sel
50        # 得到错框的w和h，因为一共有5个Anchor，所以reshape成[1,5,1]
51        self.anchor_w = np.ascontiguousarray(self.anchors[..., 0]).reshape([-1, 5, 1])
                                            usarray(self.anchors[..., 1]).reshape([-1, 5, 1])
+ {ndarray: (1, 5, 1)} [[[ 1.3221 ], [ 3.19275], [ 5.05587], [ 9.47112], [11.2364 ]]]
self.lin_x
self.lin_x = {ndarray: (169,)} [0. 0. 0. 0. 0. 0. 0. 0. 0. 0. 0. 0. 0. 1. 1. 1. 1. 1. 1. 1. 1. 1. 1. 1. 1. 1. 2. 2. 2. ...
    min = {float64} 0.0
    max = {float64} 12.0
    shape = {tuple: 1} 169
    dtype = {dtype[float64]: 0} float64
```

图 5-50　YOLOv2 画格子

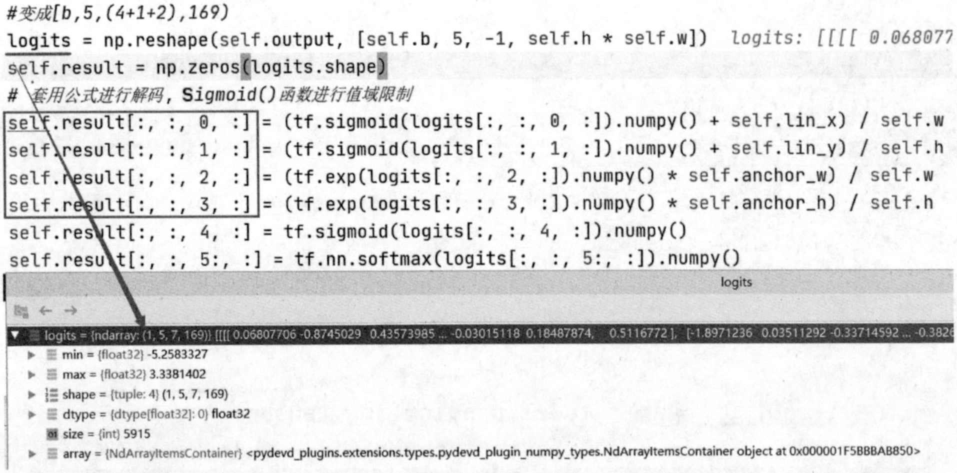

```
#变成[b,5,(4+1+2),169)
logits = np.reshape(self.output, [self.b, 5, -1, self.h * self.w])  logits: [[[[ 0.068077
self.result = np.zeros(logits.shape)
# 套用公式进行解码，Sigmoid()函数进行值域限制
self.result[:, :, 0, :] = (tf.sigmoid(logits[:, :, 0, :]).numpy() + self.lin_x) / self.w
self.result[:, :, 1, :] = (tf.sigmoid(logits[:, :, 1, :]).numpy() + self.lin_y) / self.h
self.result[:, :, 2, :] = (tf.exp(logits[:, :, 2, :]).numpy() * self.anchor_w) / self.w
self.result[:, :, 3, :] = (tf.exp(logits[:, :, 3, :]).numpy() * self.anchor_h) / self.h
self.result[:, :, 4, :] = tf.sigmoid(logits[:, :, 4, :]).numpy()
self.result[:, :, 5:, :] = tf.nn.softmax(logits[:, :, 5:, :]).numpy()
logits
logits = {ndarray: (1, 5, 7, 169)} [[[[ 0.06807706 -0.8745029  0.43573985 ... -0.03015118 0.18487874,  0.5116772 ], [-1.8971236 0.03511292 -0.33714592 ... -0.3826
    min = {float32} -5.2583327
    max = {float32} 3.3381402
    shape = {tuple: 4} (1, 5, 7, 169)
    dtype = {dtype[float32]: 0} float32
    size = {int} 5915
    array = {NdArrayItemsContainer} <pydevd_plugins.extensions.types.pydevd_plugin_numpy_types.NdArrayItemsContainer object at 0x000001F5BBBAB850>
```

图 5-51　YOLOv2 解码操作

在 classification_filtering()中对于解码出的内容进行 score_mask = prob > self.confidence_threshold 的过滤，当置信度×分类概率的值大于 self.confidence_threshold 时，score_mask 为 True，否则为 False。此时由于 self.result 为[b,5,(4+1+num_class),169]，而 score_mask 为[1,5,169]，所以需要将 self.result 的 shape 转换 np.reshape(self.result,[self.b,5,self.h * self.w, −1])变成[b,5,169,(4+1+num_class)]，从而 boxes=self.result[score_mask][...,:4]得到过滤后的 boxes 内容，继而得到分类概率 cls_scores、置信度概率 cls_scores，如图 5-52 所示。

遍历 for label in range(len(self.class_prob))每个类别的下标，当 mask=idx==label 时从 boxes 中取出当前类别的 boxes，然后通过 xyxy_box=u.cxcy2xyxy(boxes[mask][...,:4])，由 cx,cy,w,h 转换成 xmin,ymin,xmax,ymax，最后进行 NMS，从而得出预测框的内容，如图 5-53 所示。

总结

YOLOv2 采用 Anchor 机制，并划分为 13×13 个格子，每个格子有 5 个 Anchor，由 GT

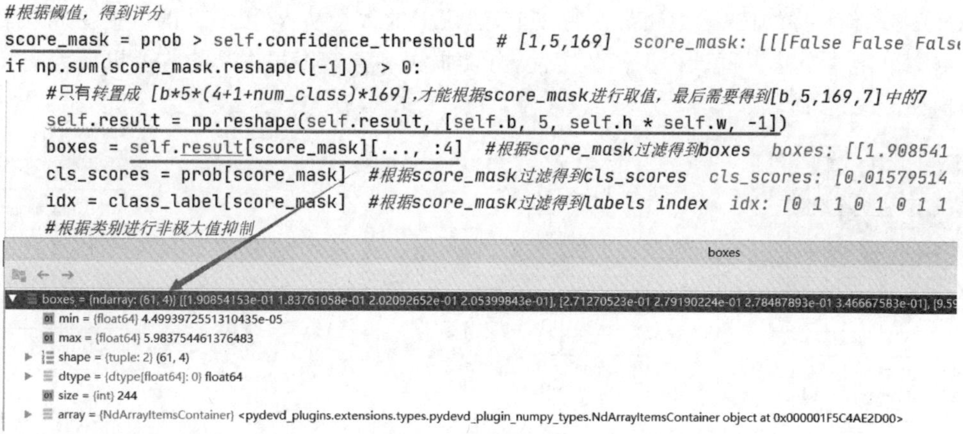

```
#根据阈值，得到评分
score_mask = prob > self.confidence_threshold  # [1,5,169]  score_mask: [[[False False Fals
if np.sum(score_mask.reshape([-1])) > 0:
    #只有转置成 [b*5*(4+1+num_class)*169],才能根据score_mask进行取值，最后需要得到[b,5,169,7]中的7
    self.result = np.reshape(self.result, [self.b, 5, self.h * self.w, -1])
    boxes = self.result[score_mask][..., :4]  #根据score_mask过滤得到boxes  boxes: [[1.908541
    cls_scores = prob[score_mask]  #根据score_mask过滤得到cls_scores  cls_scores: [0.01579514
    idx = class_label[score_mask]  #根据score_mask过滤得到labels index  idx: [0 1 1 0 1 0 1 1
    #根据类别进行非极大值抑制
boxes
boxes = {ndarray: (61, 4)} [[1.90854153e-01 1.83761058e-01 2.02092652e-01 2.05399843e-01], [2.71270523e-01 2.79190224e-01 2.78487893e-01 3.46667583e-01], [9.5
    min = {float64} 4.4993972551310435e-05
    max = {float64} 5.983754461376483
    shape = {tuple: 2} (61, 4)
    dtype = {dtype[float64]: 0} float64
    size = {int} 244
    array = {NdArrayItemsContainer} <pydevd_plugins.extensions.types.pydevd_plugin_numpy_types.NdArrayItemsContainer object at 0x000001F5C4AE2D00>
```

图 5-52 YOLOv2 过滤大于条件概率值

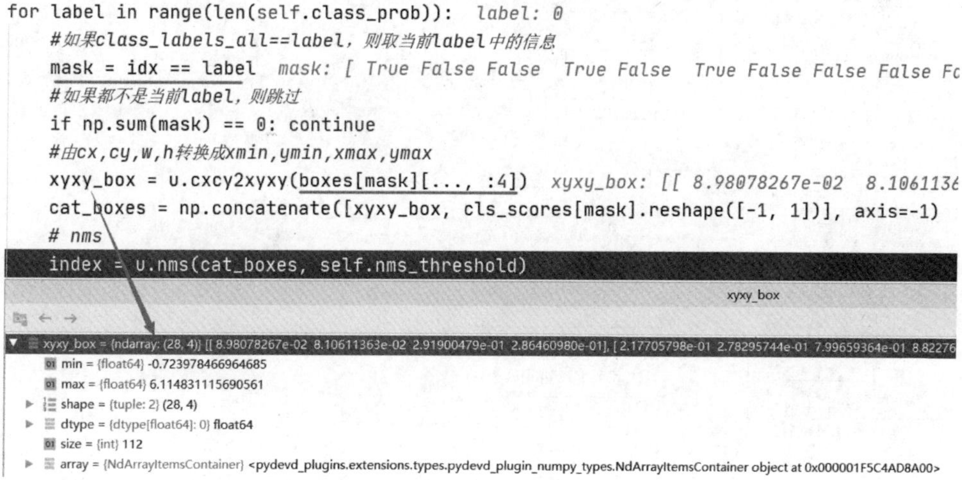

```
for label in range(len(self.class_prob)):  label: 0
    #如果class_labels_all==label，则取当前label中的信息
    mask = idx == label  mask: [ True False False  True False  True False False False Fa
    #如果都不是当前label，则跳过
    if np.sum(mask) == 0: continue
    #由cx,cy,w,h转换成xmin,ymin,xmax,ymax
    xyxy_box = u.cxcy2xyxy(boxes[mask][..., :4])  xyxy_box: [[ 8.98078267e-02  8.1061136
    cat_boxes = np.concatenate([xyxy_box, cls_scores[mask].reshape([-1, 1])], axis=-1)
    # nms
    index = u.nms(cat_boxes, self.nms_threshold)
xyxy_box
xyxy_box = {ndarray: (28, 4)} [[ 8.98078267e-02 8.10611363e-02 2.91900479e-01 2.86460980e-01], [ 2.17705798e-01 2.78295744e-01 7.99659364e-01 8.82276
    min = {float64} -0.723978466964685
    max = {float64} 6.114831115690561
    shape = {tuple: 2} (28, 4)
    dtype = {dtype[float64]: 0} float64
    size = {int} 112
    array = {NdArrayItemsContainer} <pydevd_plugins.extensions.types.pydevd_plugin_numpy_types.NdArrayItemsContainer object at 0x000001F5C4AD8A00>
```

图 5-53 YOLOv2 根据类别索引过滤

BOX 落在某个格子所在的最佳 Anchor 进行预测。Anchor 的生成采用聚类算法获得，该方法在 YOLO 系列得到延续。

练习

运行并调试本节代码，理解算法的设计与代码的结合，重点梳理本算法的实现方法。

5.7 单阶段速度快多检测头网络 YOLOv3

5.7.1 模型介绍

YOLOv3 由 Joseph Redmon 等在 2018 年发表的论文 *YOLOv3: An Incremental Improvement* 中提出，其主要特点是在 YOLOv2 的基础上将主干特征提取网络换成

DarkNet-53，使用了 FPN。检测头变为 3 个，建议框的数量从 YOLOv2 的 5 个变成 9 个，每个检测头分配 3 个建议框。损失函数方面，分类损失和置信度损失从均方差更换为交叉熵，其结构如图 5-54 所示。

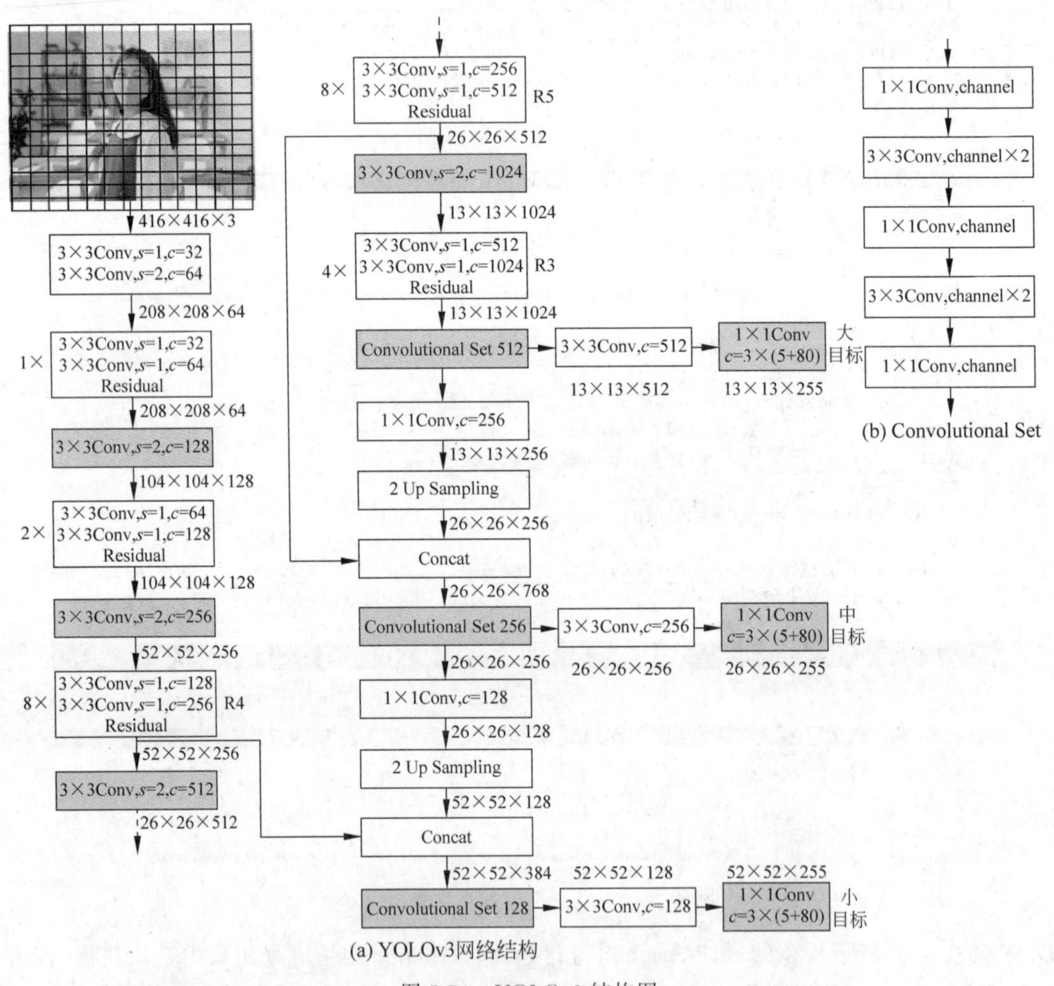

图 5-54 YOLOv3 结构图

YOLOv3 在主干特征提取层引入了 Residual 差结构，可以使网络更稀疏、网络层数更深，以便更好地提取特征信息，如图 5-54 所示，分别重复 1、2、8、8、4 次。输出特征 13×13×1024 后经过 Convolutional Set 512 层，接 3×3、1×1 卷积输出 13×13×3 建议框(4 偏移位置+1 置信度+80 个分类)作为第 1 个检测头，检测大目标(13×13×255 的感受野最大，所以检测大目标)。

Convolutional Set 512 后经过 1×1 卷积、2 倍 Up Sampling 上采集得 26×26×256，与 R5 的输出 Concat 得 26×26×768，然后 Convolutional Set 256，接 3×3、1×1 卷积输出 26×26×3 建议框(4 偏移位置+1 置信度+80 个分类)作为第 2 个检测头，检测中目标。

Convolutional Set 256 后经过 1×1 卷积、2 倍 Up Sampling 上采集得 52×52×128，与 R5 的输出 Concat 得 26×26×384，然后 Convolutional Set 128，接 3×3、1×1 卷积输出 52×52×3 建议框(4 偏移位置+1 置信度+80 个分类)作为第 3 个检测头，检测小目标。

在 CNN 结构中深层网络语义特征信息丰富，浅层特征几何信息丰富，在目标检测任务中特征提取是一个很重要的问题，深层网络虽然能够得到丰富的语义特征信息，但是由于特征图的尺寸较小，所以包含的几何信息较少，不利于物体的位置检测，潜层网络虽然包含了丰富的几何信息，但是图像的语义信息较少，又不利于图像的分类预测，这个问题尤其在小目标检测任务中表现尤为明显。

回顾 Faster R-CNN、YOLOv1 等只使用最深层的特征图信息，单尺度特征图限制了模型的检测能力，尤其是那些较小的样本或者数量较少的建议框尺寸。SSD 利用卷积的层次结构，从 VGG 中的 Conv4_3、Conv7、Conv8_2、Conv9_2、Conv10_2、Conv11_2 得到多尺度特征信息，该方法虽然能提高精度并且检测速度略有下降，但由于没有使用更加深层的特征信息，所以对于检测小目标仍然不够稳健，如图 5-55 所示。

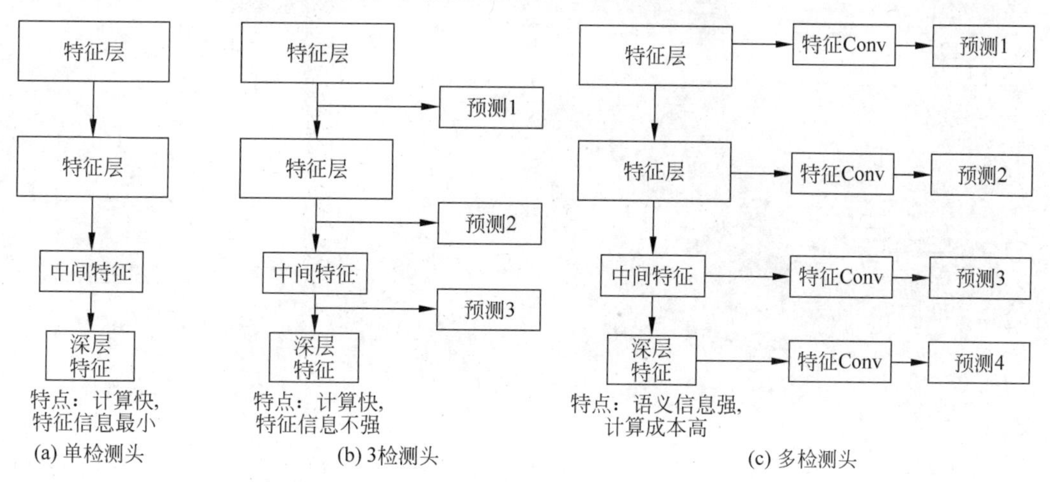

图 5-55 不同的检测头结构

FPN 在 SSD 的结构上不仅使用了深层特征图的信息，并且浅层网络的特征信息也被使用，并通过自底向上、自顶向下及横向连接的方式对这些特征图的信息进行整合，在提升精度的同时检测速度也没有较大降低，其结构如图 5-56 所示。

FPN 自上而下、由底而上进行了特征信息的融合，将语义信息与几何信息进行融合，有助于小目标的特征信息提取，能够显著地提高小目标的检测能力。

在 FPN 进行上采样时，可选择最近邻插值、双线性插值的方法放大图像。最近邻插值在放大图时补充的像素是最近邻的像素值，由于方法简单，所以处理速度很快，但是当放大图像时会有锯齿，画质较低，如图 5-57 所示。

已知 $A=(x_0,y_0)$、$B=(x_1,y_1)$，将 (x_0,y_0) 和 (x_1,y_1) 连成一条直线，求区间 (x_0,x_1) 上某一点 x 在该直线上的 y 值，这个求解过程就是单线性插值的过程，如图 5-58 所示。

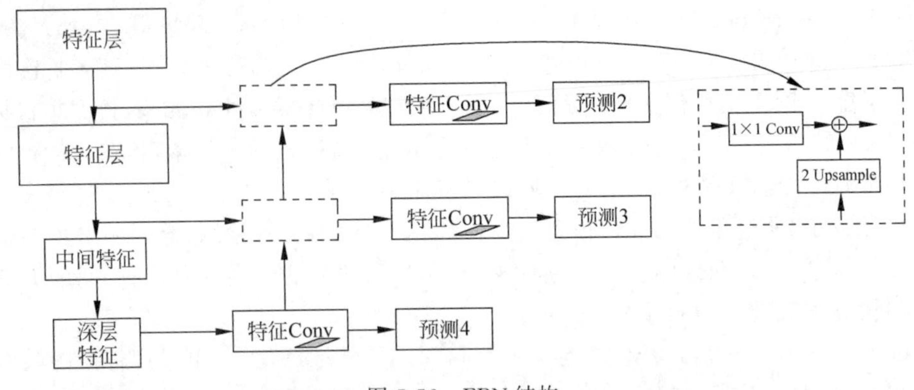

图 5-56　FPN 结构

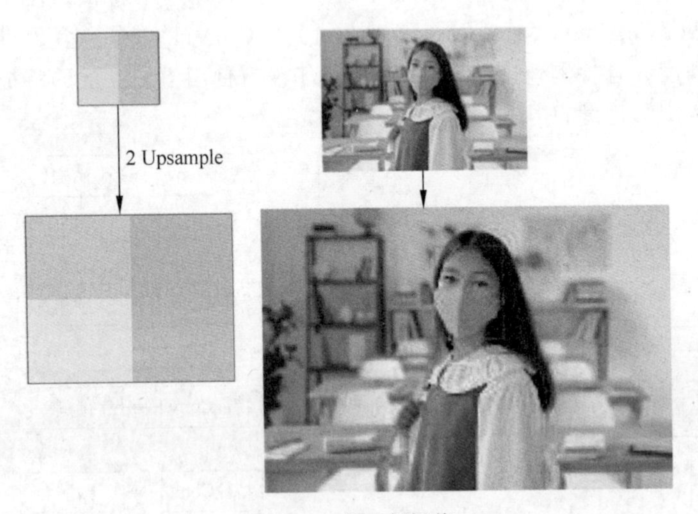

图 5-57　最近邻插值

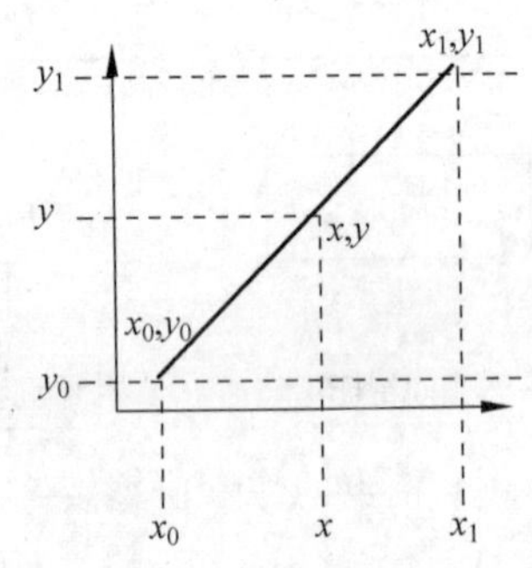

图 5-58　单线性插值

因为$\frac{y_1-y}{x_1-y}=\frac{y-y_0}{x-x_0}$，所以 $y=y_0+\frac{x_1-y}{x_1-x}(x-x_0)=\frac{x-x_0}{x_1-x_0}y_1+\frac{x_1-x}{x_1-x_0}y_0$，同样已知 y 也可求出 x，代码如下：

```
#第 5 章/ObjectDetection/TesnsorFlow_YOLO_V3_Detected/utils/interp.py
def interp():
    #x 值
    x = np.arange(0, 10, 0.5)
    y = x **2
    #插入点为 xvals
    x_vals = np.linspace(0, 10, 5)
    y_interp = np.interp(x_vals, x, y)
    for ix,iy in zip(x_vals,y_interp):
        plt.text(ix,iy+1,f"{ix},{iy}")
    plt.plot(x, y, 'o',label="xy 轴")
    plt.plot(x_vals, y_interp, '-x',label='线性插值')
    plt.legend()
    plt.show()
```

调用代码，运行后单线性插值如图 5-59 所示。

在卷积图像中图像的维度是三维的，在进行上采样时就需要用到双线性插值，如图 5-60 所示。

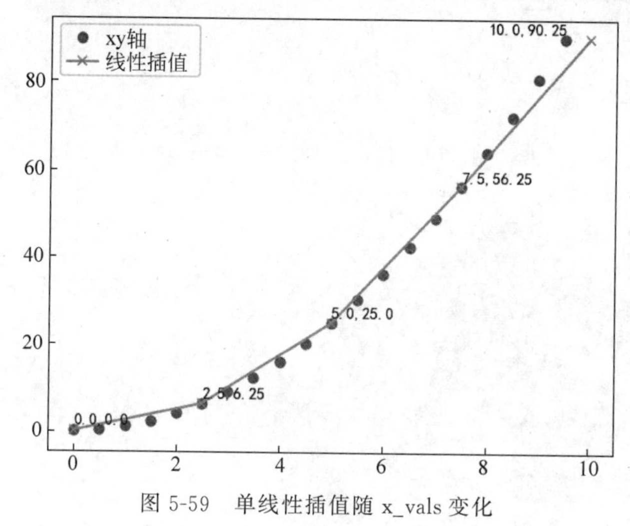

图 5-59　单线性插值随 x_vals 变化

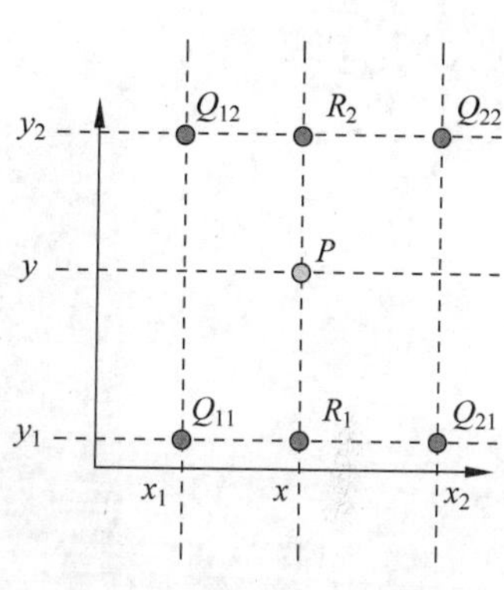

图 5-60　双线性插值算法

所谓双线性插值，原理与线性插值相同，如图 5-60 中已知点 $Q_{11}=(x_1,y_1)$、$Q_{12}=(x_1,y_2)$、$Q_{21}=(x_2,y_1)$和 $Q_{22}=(x_2,y_2)$共四个点的值，求函数 $f(x,y)$在 $P=(x,y)$的值。

首先做两次线性插值，分别求出点 $R_1=(x,y_1)$和 $R_2=(x,y_2)$的像素值，然后用这两个点再做 1 次线性插值以求出 $P=(x,y)$的像素值。对于 Q_{11} 和 Q_{12} 来讲，它们连成线的纵坐标是相同的，所以可以忽略这个纵坐标的影响(但这个影响是存在的，所以线性插值是近似的)，而用当前点的像素值直接代替纵坐标，所以 $Q_{11}=(x,f(Q_{11}))$。$f(R_1)\approx\frac{x-x_1}{x_2-x_1}f(Q_{21})+\frac{x_2-x}{x_2-x_1}f(Q_{11})$，$f(R_2)\approx\frac{x-x_1}{x_2-x_1}f(Q_{22})+\frac{x_2-x}{x_2-x_1}f(Q_{12})$，$f(P)\approx\frac{y-y_1}{y_2-y_1}f(R_2)+\frac{y_2-y}{y_2-y_1}f(R_1)=\left[\frac{y-y_1}{y_2-y_1},\frac{y_2-y}{y_2-y_1}\right]\begin{bmatrix}f(Q_{22}), & f(Q_{12})\\ f(Q_{22}), & f(Q_{12})\end{bmatrix}\left[\frac{x-x_1}{x_2-x_1},\frac{x_2-x}{x_2-x_1}\right]^{\mathrm{T}}$，在双线性插值上采样时 $x_2-x_1=1$、$y_2-y_1=1$。

双线性插值是一种比较好的图像缩放算法，它充分利用一源图像中虚拟的点四周的 4 个真实存在的像素来共同决定目标图像中的一像素，这样所生成的新图效果更好，过渡更自然，边缘更光滑，如图 5-61 所示的效果比图 5-57 要好得多。

YOLOv3 虽然有 3 个检测头，但是它正样本的选取仍然跟 YOLOv2 一样，即从检测头 13×13、26×26、52×52 生成的 Anchor 与标注 BOX 做 IOU，取最大的 IOU 得分作为正样本，其他样本作为负样本。稍有不同之处，3 个检测头分配的建议框尺寸有变化，13×13 分配大一些的建议框，如[116,90]、[156,198]、[373,326]；26×26 分配中等大小的建议框，如[30,61]、[62,45]、[59,119]；52×52 分配小目标的建议框，如[10,13]、[16,30]、[33,23]。这些建议框仍然通过聚类得到，一共有 9 个 Anchor。

图 5-61　双线性插值效果图

在损失函数方面，将 YOLOv2 中的分类损失更改为二元交叉熵、位置损失，仍然使用均方差损失，置信度损失使用交叉熵，其公式如下：

$$\text{Loss}=\lambda_{\text{coord}}\sum_{i=0}^{s^2}\sum_{j=0}^{B}l_{i,j}^{\text{obj}}[(b_x-\hat{b}_x)^2+(b_y-\hat{b}_y)^2+(b_w-\hat{b}_w)^2+(b_h-\hat{b}_h)^2]+\sum_{i=0}^{s^2}\sum_{j=0}^{B}l_{i,j}^{\text{obj}}[-\log(p_c)]+\lambda_{\text{noobj}}\sum_{i=0}^{s^2}\sum_{j=0}^{B}l_{i,j}^{\text{noobj}}[-\log(1-p_c)]+\sum_{i=1}^{n}\text{BCE}(\hat{c}_i,c_i)\tag{5-12}$$

其中，$l_{i,j}^{\text{obj}}$ 表示每个格子中有目标；b_x、b_y、b_w、b_h 代表真实 BOX 的位置，$\hat{b}_x$、$\hat{b}_y$、$\hat{b}_w$、$\hat{b}_h$ 代表预测的 BOX 位置；$-\log(p_c)$表示有目标的置信度、$-\log(1-p_c)$代表没有目标的置信度；$\text{BCE}(\hat{c}_i,c_i)$为二次交叉熵，即逻辑回归的损失函数，每个类为 0 或者 1；原作者论文并没有说明具体的损失函数，不同的代码作者所使用的损失函数可能有所不同。

5.7.2　代码实战模型搭建

因为 YOLOv3 使用了残差结构，所以需要对 Residual 进行实现，代码如下：

```
#第 5 章/ObjectDetection/TensorFlow_Yolo_V3_Detected/utils/conv_utils.py

class ResidualBlock(layers.Layer):
```

```
    def __init__(self, num_filters, num_blocks, **kwargs):
        """
        基本残差模块,即 a 3 × 3 channels,接着 b 1 × 1 channels/2,再 c 3 × 3 channels
        然后 a + c
        :param num_filters: channels 数
        :param num_blocks: 重复次数
        """
        super(ResidualBlock, self).__init__(**kwargs)
        #3×3 卷积,s=2 因为 padding=same,所以尺寸不变
        self.conv1 = ConvBnLeakRelu(num_filters, kernel_size=3, strides=2)
        #1×1 卷积,s=1,outchannel 是输入的 1/2
        self.block1 = ConvBnLeakRelu(num_filters //2, kernel_size=1, strides=1,
padding='valid')
        #1×1 卷积,s=1
        self.block2 = ConvBnLeakRelu(num_filters, kernel_size=3, strides=1)
        self.add = layers.Add()
        self.num_blocks = num_blocks

    def call(self, inputs, *args, **kwargs):
        #输入的卷积
        x1 = self.conv1(inputs)
        #残差可能会执行多次
        for i in range(self.num_blocks):
            #1×1
            y = self.block1(x1)
            #1×1
            y = self.block2(y)
            #进行 add 残差
            x1 = self.add([x1, y])
        return x1
```

ConvBnLeakRelu 中已对 Conv、BN、LeakRelu 进行了封装。self.num_blocks 用来控制残差重复的次数。

然后根据结构图 5-54 实现 DarkNet-53,代码如下:

```
#第 5 章/ObjectDetection/TensorFlow_Yolo_V3_Detected/backbone/YOLOv3.py
def darknet_body(input_x):
    """根据结构图实现 DarkNet-53 部分"""
    #input 416 × 416 × 3
    x = ConvBnLeakRelu(32, 3, 1)(input_x)
    #重复的次数是 1,2,8,8,4
    #-> 208 × 208 × 64
    x = ResidualBlock(64, 1)(x)
    #-> 104 × 104 × 128
    x = ResidualBlock(128, 2)(x)
    #-> 52 × 52 × 256
    result4 = ResidualBlock(256, 8)(x)
    #-> 26 × 26 × 512
    result5 = ResidualBlock(512, 8)(result4)
    #-> 13 × 13 × 1024
    result6 = ResidualBlock(1024, 4)(result5)
    return result4, result5, result6
```

因为 YOLOv3 使用 13×13×1024 作为第 1 个检测头，使用 26×26×512、52×52×256 作为 FPN 上采样 Concat 层，所以输出为 result4、result5 和 result6。

在图 5-54 中还存在 Convolutional Set 结构，该结构主要是由 1×1、3×3、1×1、3×3、1×1 卷积构成的，代码如下：

```
#第 5 章/ObjectDetection/TensorFlow_Yolo_V3_Detected/utils/conv_utils.py
class ConvolutionSet(layers.Layer):
    """
    即 1 × 1 -> 3 × 3 -> 1 × 1 -> 3 × 3 -> 1 × 1
    """
    def __init__(self, number_filters, **kwargs):
        super(ConvolutionSet, self).__init__(**kwargs)
        self.conv1x1_1 = ConvBnLeakRelu(number_filters, kernel_size=1, strides=1)
        self.conv3x3_2 = ConvBnLeakRelu(number_filters *2, kernel_size=3,
strides=1)
        self.conv1x1_3 = ConvBnLeakRelu(number_filters, kernel_size=1, strides=1)
        self.conv3x3_4 = ConvBnLeakRelu(number_filters *2, kernel_size=3,
strides=1)
        self.conv1x1_5 = ConvBnLeakRelu(number_filters, kernel_size=1, strides=1)

    def call(self, inputs, *args, **kwargs):
        x = self.conv1x1_1(inputs)
        x = self.conv3x3_2(x)
        x = self.conv1x1_3(x)
        x = self.conv3x3_4(x)
        x = self.conv1x1_5(x)
        return x
```

在检测头方面由 3×3、1×1 卷积组成，并且 1×1 卷积的输出采用线性模型，不经过 ReLU 也不经过 BN，代码如下：

```
#第 5 章/ObjectDetection/TensorFlow_Yolo_V3_Detected/utils/conv_utils.py
class Yolo_V3_Pre_Head(layers.Layer):
    """预测的头部，即先经过一个 3×3 卷积，再经过一个 1×1 卷积"""
    def __init__(self, number_filters, classes_filters, **kwargs):
        super(Yolo_V3_Pre_Head, self).__init__(**kwargs)
        self.conv3 = ConvBnLeakRelu(number_filters, kernel_size=3, strides=1)
        self.out = ConvBnLeakRelu(classes_filters, kernel_size=1, is_relu=
False, is_bn=False, strides=1)

    def call(self, inputs, *args, **kwargs):
        x = self.conv3(inputs)
        x = self.out(x)
        return x
```

基于以上结构的封装，然后组合在 yolo-v3()函数中实现前向传播，代码如下：

```
#第 5 章/ObjectDetection/TensorFlow_Yolo_V3_Detected/backbone/yolo_v3.py
def yolo_v3(input_shape, anchors=None, class_num=80, is_class=False):
    if anchors is None:
```

```
        anchors_mask = np.array([
            [[10, 13], [16, 30], [33, 23]],                          #小目标
            [[30, 61], [62, 45], [59, 119]],                         #中目标
            [[116, 90], [156, 198], [373, 326]]                      #大目标
        ])
    else:
        anchors_mask = anchors
    #输入层
    input_x = layers.Input(shape=input_shape, dtype='float32')
    #调用 DarkNet-53
    r4, r5, r6 = darknet_body(input_x)
    #经过基本的 DarkNet 后,先经过一个 ConvolutionalSet,进入第 1 个预测
    x = ConvolutionSet(512)(r6)
    #第 1 个预测结果,小目标
    p5 = Yolo_V3_Pre_Head(512, len(anchors_mask[0]) * (class_num + 4 + 1))(x)
                                          #1 个置信度 + 4 个 bbox,每个 anchors 有 3 个尺度
    #第 1 个上采样
    p5_1x1 = ConvBnLeakRelu(256, kernel_size=1, strides=1)(p5)
    p5_up = layers.UpSampling2D(size=2)(p5_1x1)                 #2 倍上采样
    x = layers.Concatenate()([p5_up, r5])                       #26×26
    x = ConvolutionSet(256)(x)
    #第 1 个预测,中目标
    p6 = Yolo_V3_Pre_Head(256, len(anchors_mask[1]) * (class_num + 4 + 1))(x)
    #第 1 个上采样
    p6_1x1 = ConvBnLeakRelu(128, kernel_size=1, strides=1)(p6)
    p6_up = layers.UpSampling2D(size=2)(p6_1x1)
    x = layers.Concatenate()([p6_up, r4])                       #52×52
    x = ConvolutionSet(128)(x)
    #第 2 个预测,大目标
    p7 = Yolo_V3_Pre_Head(128, len(anchors_mask[2]) * (class_num + 4 + 1))(x)
    return Model(inputs=input_x, outputs=[p5, p6, p7])
```

建议框 anchors_mask 默认初始设置为 9 个,其中每个检测头为 3 个。r4、r5 和 r6 分别对应 52×52×256、26×26×512 和 13×13×1024,所以 r6 经过 ConvolutionSet 后成为 p5 的输入,p5 的输出为 len(anchors_mask[0]) * (class_num+4+1) * 512; p5 经过 2 倍上采样(默认为最近邻),然后由 layers. Concatenate()([p5_up,r5])得到 26×26×768 并经过 ConvolutionSet 后成为 p6 的输入,p6 的输出为 len(anchors_mask[0]) * (class_num+4+1) * 256; p6 经过 2 倍上采样,然后由 layers. Concatenate()([p6_up,r4])得到 52×52×384,并经过 ConvolutionSet 后成为 p7 的输入,p7 的输出为 len(anchors_mask[0]) * (class_num+4+1) * 128。

5.7.3 代码实战建议框的生成

将真实框转换为 GT BOX 与 Anchor 的偏移,其实现思路是遍历每个检测头和每个 Anchor 以获取所在 j、i 格子最佳 IOU 得分的 Anchor 作为正样本,计算偏移后赋给 true_box,代码如下:

```
#第 5 章/ObjectDetection/TensorFlow_Yolo_V3_Detected/utils/tools.py

def YOLOv3_v3_true_encode_box(
    true_boxes,                                   #传 GT BOX
    anchors=None,                                 #新的 Anchor
    class_num=80,                                 #类别数
    input_size=(416, 416),                        #默认尺寸
    ratio=[32, 16, 8]                             #416//13, 416//26, 416//52
):
    #将 3 个检测头 Anchor 与 GT BOX 最大 IOU 设为正样本
    if anchors is None:
        anchors = np.array([
            [[10, 13], [16, 30], [33, 23]],       #小目标
            [[30, 61], [62, 45], [59, 119]],      #中目标
            [[116, 90], [156, 198], [373, 326]]   #大目标
        ])
    #Anchor 的数量
    detected_num = anchors.shape[0]
    #用来存储 3 个检测头正样本结果
    grid_shape = []
    #用来存储 3 个检测头 GT BOX 结果
    true_boxes2 = []
    #升维
    true_boxes = np.expand_dims(true_boxes, axis=0)
    #batch size
    batch_size = true_boxes.shape[0]
    #每个检测头分配为 0 的 grid 矩阵
    for i in range(detected_num):
        #input_size[0] //ratio[i] = 13,获得格子数
        grid = np.zeros(
            [batch_size, input_size[0] //ratio[i], input_size[1] //ratio[i],
             len(anchors[i]), 4 + 1 + class_num], dtype=np.float32)
        grid_shape.append(grid)
        #
        true_boxes2.append(
            np.zeros([batch_size, input_size[0] //ratio[i], input_size[1] //ratio[i],
                      len(anchors[0]), 4 + 1 + class_num], dtype=np.float32))
    #3 个检测头的 grid
    grid = grid_shape
    #遍历每个 GT BOX
    for box_index in range(true_boxes.shape[1]):
        #根据下标取是第几个 GT BOX
        box = true_boxes[:, box_index, :]
        box = np.squeeze(box, axis=0) #降维
        box_class = box[4].astype('int32')
        best_choice = {}
        #将每个 GT BOX 与每个检测头中的 Anchor 进行 IOU 的计算
        for index in range(0, 3, 1):
            #即 input_size[0] //ratio[i] = 13
            ratio_input = grid[index].shape[-3]
            #因为 BOX 是归一化的值 × 13,放大到 13×13 的特征图的尺寸
            box2 = box[0:4] *np.array([
```

```
                ratio_input, ratio_input, ratio_input, ratio_input
            ])
            #内存复制
            box_true = box2.copy()
            box2 = box2.copy()
            #算是第 ij 个格子
            i = np.floor(box2[1]).astype('int')
            j = np.floor(box2[0]).astype('int')
            #如果 ij>13 就继续下一个 BOX,说明这个 BOX 可能标注错误
            if i > ratio_input or j > ratio_input: continue
            #最佳 IOU
            best_iou = 0
            #最佳 Anchor; 接下来,将 true_box 与 Anchor 进行 IOU,并取最佳 Anchor 用来与
            #预测的内容进行 loss
            best_anchor = 0
            #遍历每 1 个 Anchor 以获取最佳 IOU 作为正样本
            for k, anchor in enumerate(anchors[2 - index]):
                #wh center
                box_maxes = box2[2:4] *0.5
                box_mines = -box_maxes
                #
                anchor_maxes = anchor *0.5
                anchor_mines = -anchor_maxes
                #将真实 wh 与 Anchor 之间进行 IOU 的计算,并获取最佳 IOU 是哪个 Anchor
                intersect_mines = np.maximum(box_mines, anchor_mines)
                intersect_maxes = np.minimum(box_maxes, anchor_maxes)
                intersect_wh = np.maximum(intersect_maxes - intersect_mines, 0.)
                intersect_area = intersect_wh[0] *intersect_wh[1]
                box_area = box2[2] *box2[3]
                anchor_area = anchor[0] *anchor[1]
                #IOU 得分
                iou_score = intersect_area / (box_area + anchor_area - intersect_area)
                #如果 IOU 得分>上 1 次的得分,就更改 best_iou,并记录 k
                if iou_score > best_iou:
                    best_iou = iou_score
                    best_anchor = k
            #记录下来与最佳 best_iou 相关的参数
            #[历史 GT BOX, 第几个检测头, 最佳 IOU 得分,GT BOX,格子 j,格子 i,最佳第几个
            #Anchor,历史 GT BOX,历史 GT BOX]
            best_choice[best_iou] = [box_true, index, best_iou, box_index, j, i,
best_anchor, box2, box]
        #按最佳 IOU 的得分进行排序,获得最大 IOU
        best_iou_choice = best_choice[sorted(best_choice.keys(), reverse=True)[0]]
        #从 best_iou_choice 中获得最佳 box,j,i 的信息
        box = best_iou_choice[-1]
        j = best_iou_choice[-5]
        i = best_iou_choice[-4]
        box2 = best_iou_choice[-2]
        index = best_iou_choice[1]
        best_anchor = best_iou_choice[-3]
        box_true = best_iou_choice[0]
        #对最佳 box2 进行偏移值的求解
```

```
            adjusted_box = np.array([
                box2[0] - j,
                box2[1] - i,
                np.log(box2[2] / anchors[2 - index][best_anchor][0]),
                np.log(box2[3] / anchors[2 - index][best_anchor][1]),
            ], dtype=np.float32)
            #检测头[第几个][...,j 格子,i 格子,最佳 anchor,:4] = 偏移值
            grid[index][..., j, i, best_anchor, 0:4] = adjusted_box
            #置信度
            grid[index][..., j, i, best_anchor, 4] = 1
            #分类信息 box_class 分类的下标,假设为 1,则为 2
            grid[index][..., j, i, best_anchor, 5 + box_class] = 1

            #获取真值,方便后面计算损失时做 IOU
            true_boxes2[index][..., j, i, best_anchor, 0:4] = box_true
        return true_boxes2, grid

    def test_yolo_v3_true_encode_box():
        data = np.array(
            [[0.50, 0.47, 0.05, 0.12, 0], [0.50, 0.47, 0.05, 0.12, 1]]
            , dtype=np.float32
        )
        return YOLOv3_v3_true_encode_box(data)
```

代码中 for i in range(detected_num)初始化设置 3 个 grid,即设置 3 个检测头为 0 的矩阵,用来存放正样本相关值。for box_index in range(true_boxes.shape[1])循环每个 true box(可能有多个),for index in range(0, 3, 1)循环每个检测头,for k, anchor in enumerate(anchors[2−index])循环每个检测头分配的第 k 个 Anchor,best_iou=iou_score 得到最佳的得分,best_anchor=k 得到最佳 Anchor,best_choice[best_iou]=[box_true,index,best_iou,box_index,j,i,best_anchor,box2,box]记录下来与最佳 best_iou 相关的参数,因为 best_choice 是一个字典,如果 best_iou 有相同得分,则将会被替换掉,如果没有,则会保留下来。

3 个检测头和 Anchor 循环结束后,best_iou_choice=best_choice[sorted(best_choice.keys(),reverse=True)[0]]根据 best_choice.keys()即 IOU 的得分进行降序排列,从而得到当前最佳的 best_iou_choice 信息,如图 5-62 所示。

然后根据 best_iou_choice 中的信息,计算 adjusted_box 偏移值,并同时赋给 grid[index]相关值。此时就实现某个 GT BOX 从 3 个检测头中得到最佳 IOU 分配到某个检测头中,如图 5-63 所示。

代码中的 GT BOX 将分配到第 2 个检测头(26×26),第 $j=26$、$i=24$ 的第 1 个 Anchor 为正样本,其他都为负样本。

得到正样本的偏移值后,仍然像 YOLOv1 中的 DataProcessingAndEnhancement(object)类的 generate(self,isTraining=True)喂给训练数据,具体代码实现基本一致,详细可参考随书代码。

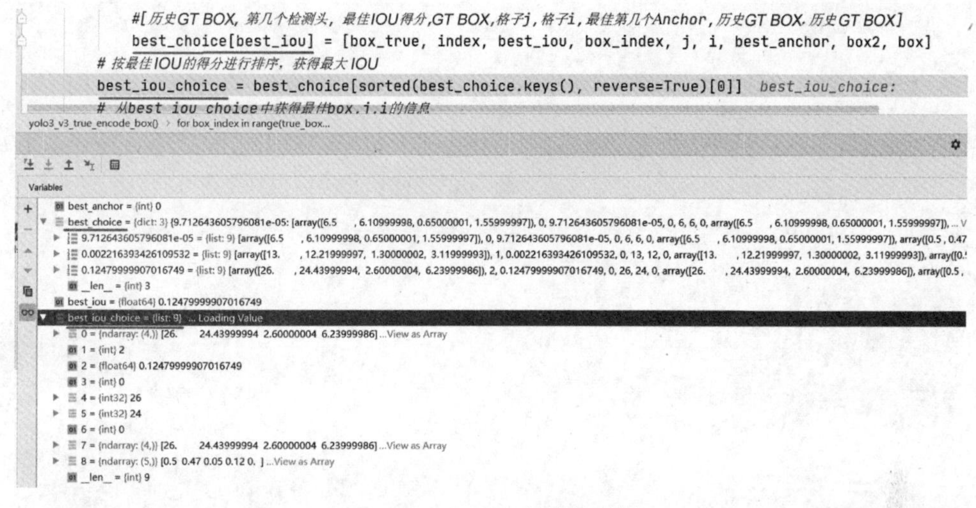

图 5-62　best_iou 的计算示例

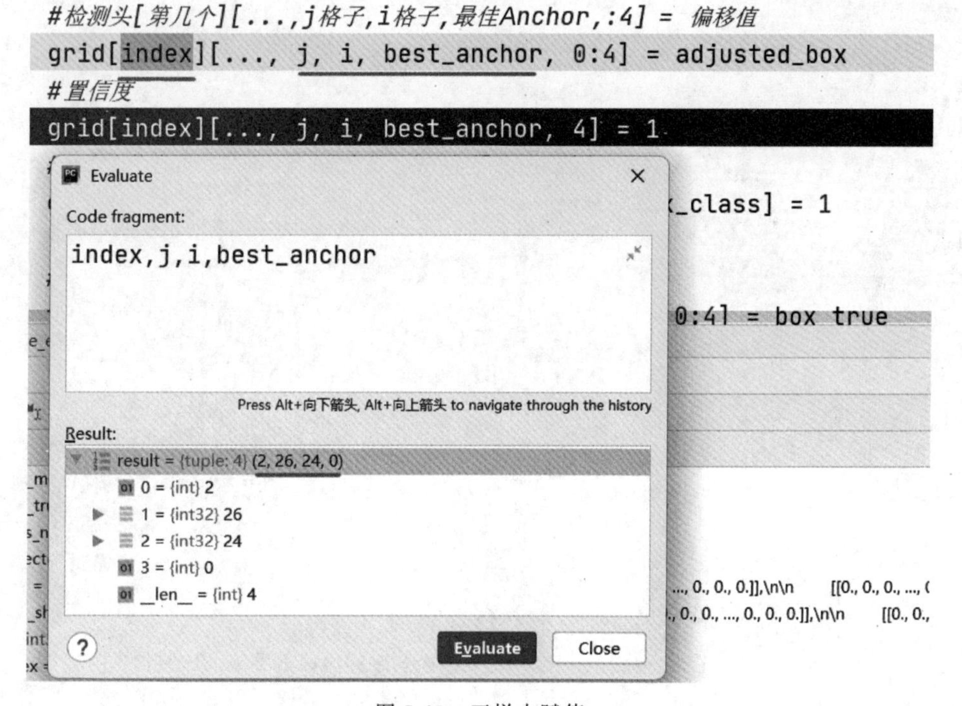

图 5-63　正样本赋值

5.7.4　代码实现损失函数的构建及训练

YOLOv3 的损失函数与 YOLOv2 的损失函数基本类似，不同之处在于需要将 3 个检测头的损失相加，另外在计算分类损失时使用二元交叉熵损失，关键代码参考如下：

```
#第 5 章/ObjectDetection/TensorFlow_Yolo_V3_Detected/utils/loss.py
class MultiAngleLoss(object):
    def __init__(self, wh_scale=0.5, overlap_threshold=0.5, num_layers=3, num_
class=80, anchors=None):
        #置信度
        self.overlap_threshold = overlap_threshold
        #wh 损失的比例
        self.wh_scale = wh_scale
        self.num_layers = num_layers
        self.num_class = num_class
        if anchors is None:
            self.anchor = np.array([
                [[10, 13], [16, 30], [33, 23]],                      #小目标
                [[30, 61], [62, 45], [59, 119]],                     #中目标
                [[116, 90], [156, 198], [373, 326]]                  #大目标
            ])
        else:
            self.anchor = anchors

    def yolo_v3_loss(self, y_true, y_pred, true_box2):
        #true_box2 传入的是原标签值
        #有几个检测头
        num_layers = self.num_layers
        #将数据类型转换成 Tensor
        mf = tf.cast(y_pred[0].shape[0], dtype=y_pred[0].dtype)
        loss = 0 #累加每个检测头中的损失
        for index in range(num_layers):
            y_pre1 = y_pred[index]                                   #第 1 个检测头
            #维度从 b×13×13×255 转换为 b×13×13×3×85,3 是 Anchor,85 是 4+1+80
            y_pre = tf.reshape(
                y_pre1,
                [
                    -1, y_pre1.shape[1], y_pre1.shape[2],
                    len(self.anchor[0]), 4 + 1 + self.num_class
                ])
            #所有的同层 true_box concat 在一起
            true_boxes = true_box2[0][index]
            #传入的是编码过的 3 个检测头的真值 tx,ty,tw,th
            true_grid = y_true[0][index]
            #获取置信度信息,在真值中只有一个检测头是用来预测的
            object_mask = true_grid[..., 4:5]
            #true_class_probs = true_grid[..., 5:] #获取分类信息
            #预测返回的是 xmin,ymin, xmax, ymax 位置信息,置信度信息,分类信息
            pre_boxes, pre_box_confidence, pre_box_class_props = yolo_v3_head_
decode(y_pre, 2 - index,num_class=self.num_class)
            #因为 true_box 中的位置信息是 cx,cy,w,h,所以要转换成 xmin,ymin,xmax,
            #ymax 做 IOU 计算
            #计算 true 的 xmin,ymin,xmax,ymax 以方便做 IOU 的计算
            true_box = tf.concat([
                (true_boxes[..., 0:2] - true_boxes[..., 2:4]) *0.5,
                (true_boxes[..., 0:2] + true_boxes[..., 2:4]) *0.5,
```

```
            ], axis=-1)
        #2 - (w × h) 如果 wh 较少,则惩罚学习小框。因为值放大了
        #如果小目标较多,则可以在 box_loss_scale 的基础上再乘以 1.5
        box_loss_scale = 2 - true_boxes[..., 2] *true_boxes[..., 3]
        #将预测值与真值之间做 IOU,通过 IOU 进行过滤
        iou_result = iou(pre_boxes, true_box)
        iou_score = tf.reduce_max(iou_result, axis=-4)
        iou_score = tf.expand_dims(iou_score, axis=-1)
        #预测值与真值之间的 IOU,如果大于指定阈值,但是又不是最大 IOU 的内容则会被
        #忽略处理
        #真实的框只有一个,而小于阈值的都作为负样本
        object_detections = iou_score > self.overlap_threshold
        object_detections = tf.cast(object_detections, dtype=iou_score.dtype)
        #没有物体的损失,即背景的损失
        #置信度损失,预测框的 IOU 与真实框中的 IOU 小于 0.6 的都为背景
        no_object_loss = (1 - object_detections) *(
            1 - object_mask) *tf.expand_dims(tf.keras.losses.binary_
crossentropy(
            object_mask, pre_box_confidence, from_logits=True), axis=-1)
        object_loss = object_mask *tf.expand_dims(tf.keras.losses.binary_
crossentropy(
            object_mask, pre_box_confidence, from_logits=True), axis=-1)
        confidence_loss = object_loss + no_object_loss

        #有文章说使用 binary_crossentropy 有助于抑制 exp 指数溢出,所以这里更改了
        #位置损失
        #https://github.com/qqwweee/keras-yolov3/blob/master/yolov3/model.py
        xy_loss = object_mask *tf.expand_dims(box_loss_scale, axis=-1) *tf.
expand_dims(
            tf.keras.losses.binary_crossentropy(
                true_grid[..., :2], y_pre[..., :2], from_logits=True
            ), axis=-1)

        wh_loss = object_mask *tf.expand_dims(box_loss_scale, axis=-1) *tf.
square(
            true_grid[..., 2:4] - y_pre[..., 2:4]) *self.wh_scale
        #分类损失,使用 binary_crossentropy 交叉熵
        class_loss = object_mask *tf.expand_dims(tf.keras.losses.binary_
crossentropy(
            true_grid[..., 5:], y_pre[..., 5:], from_logits=True), axis=-1)
        #统计损失
        xy_loss = tf.reduce_sum(xy_loss) / mf
        wh_loss = tf.reduce_sum(wh_loss) / mf
        confidence_loss = tf.reduce_sum(confidence_loss) / mf
        class_loss = tf.reduce_sum(class_loss) / mf
        #将 3 个检测头的损失合并在一起
        loss += xy_loss + wh_loss + confidence_loss + class_loss
    return loss
```

代码中 for index in range(num_layers)循环了 3 个检测头以进行损失的计算。yolo_v3_head_decode()解码函数的实现逻辑与 YOLOv2 是一致的,求 GT BOX 与 Anchor BOX 的偏移,然后将 iou_result=iou(pre_boxes, true_box)预测 BOX 与真实 BOX 做 IOU,根据

object_detections＝iou_score＞self. overlap_threshold 得到预测有目标，然后由 1－object_detections 得到预测，没有目标，1－object_mask 是真值，有目标，所以将 binary_crossentropy (object_mask，pre_box_confidence)做交叉熵后得到没有目标的置信度损失。将有目标和没有目标相加 confidence_loss＝object_loss＋no_object_loss，得到总的置信度损失。类别损失这里使用交叉熵损失，class_loss＝object_mask * binary_crossentropy(true_grid[...，5：]，y_pre[...，5：])，最后将 xy_loss＋wh_loss＋confidence_loss＋class_loss 相加得到总损失。求损失的整体过程跟 YOLOv2 类似，故不再赘述。训练脚本也无重要更新，更多更详细的代码可参考随书代码。

5.7.5 代码实战预测推理

YOLOv3 的推理过程与 YOLOv2 基本相同，不同之处在于 YOLOv3 需要遍历 3 个检测层的结果，并根据不同的检测层生成不同的锚框，详细的代码如下：

```
#第 5 章/ObjectDetection/TensorFlow_Yolo_V3_Detected/detected/detected.py
class Detected():
    def __init__(self, model_path, input_size):
        self.model = load_model(
            model_path,
            custom_objects={'compute_loss': MultiAngleLoss(2).compute_loss}
        ) #读取模型和权重
        self.confidence_threshold = 0.5                          #有无目标置信度
        self.class_prob = [0.5, 0.5]                             #两个类别的分类阈值
        self.nms_threshold = 0.5                                 #NMS 的阈值
        self.input_size = input_size
              #锚框初始值
        self.anchors = np.array([
            [[10, 13], [16, 30], [33, 23]],
            [[30, 61], [62, 45], [59, 119]],
            [[116, 90], [156, 198], [373, 326]]
        ]) / 416

    def readImg(self, img_path=None):
        img = cv2.imread(img_path)                               #读取要预测的图片
        #将图片转换到 416×416
        self.img, _ = u.letterbox_image(img, self.input_size, [])
        self.old_img = self.img.copy()

    def forward(self):
        #升成 4 维
        img_tensor = tf.expand_dims(self.img / 255.0, axis=0)
        #前向传播
        self.output = self.model.predict(img_tensor)

    def _generate_anchor(self):
        #根据当前预测 layer 得到特征图的 height、width
        self.h, self.w = self.output[self.currentLayer].shape[1:3] #1×13×13×21
        #使用均分的方法生成格子。一共有 h×w 个格子，即 169 个格子
```

```
        self.lin_x = np.ascontiguousarray(np.linspace(0, self.w - 1, self.w).
repeat(self.h)).reshape(self.h * self.w)
        self.lin_y = np.ascontiguousarray(np.linspace(0, self.h - 1, self.h).
repeat(self.w)).T.reshape(self.h * self.w)
        #得到锚框的 w 和 h,因为一共有 3 个 Anchor,所以 reshape 成[1,3,1]
        self.anchor_w = np.ascontiguousarray(self.anchors[2 - self.currentLayer]
[..., 0] * (416 //self.w)).reshape([-1, 3, 1])
        self.anchor_h = np.ascontiguousarray(self.anchors[2 - self.currentLayer]
[..., 1] * (416 //self.h)).reshape([-1, 3, 1])

    def _decode(self):
        #根据当前预测 layer 对预测出来的结果解码
        #YOLOv3 的输出(4+1+num_class) ×3 个 Anchor
        self.b = self.output[self.currentLayer].shape[0]
        #变成[b,3,(4+1+2),169)
        logits = np.reshape(self.output[self.currentLayer], [self.b, 3, -1,
self.h * self.w])
        self.result = np.zeros(logits.shape)
        #套用公式进行解码,Sigmoid()函数进行值域限制
        self.result[:, :, 0, :] = (tf.sigmoid(logits[:, :, 0, :]).NumPy() + self.
lin_x) / self.w
        self.result[:, :, 1, :] = (tf.sigmoid(logits[:, :, 1, :]).NumPy() + self.
lin_y) / self.h
        self.result[:, :, 2, :] = (tf.exp(logits[:, :, 2, :]).NumPy() * self.anchor_
w) / self.w
        self.result[:, :, 3, :] = (tf.exp(logits[:, :, 3, :]).NumPy() * self.anchor_
h) / self.h
        self.result[:, :, 4, :] = tf.sigmoid(logits[:, :, 4, :]).NumPy()
        self.result[:, :, 5:, :] = tf.nn.softmax(logits[:, :, 5:, :]).NumPy()

    def classification_filtering(self):
        #根据 3 个预测 layer 对预测结果进行解析
        all_boxes = []
        all_cls_scores = []
        all_idx = []
        for i in range(3):
            #当前预测 layer
            self.currentLayer = i
            #根据当前预测 layer 进行 Anchor 的生成
            self._generate_anchor()
            #根据当前预测 layer 进行解码操作,解码后此时的内容存储在 self.result 属性中
            self._decode()
            #取最大的分类得分的下标,此时 result 为[b,3,(4+1+num_class),169]
            class_score = self.result[:, :, 5:, :]          #[1,3,num_class,169]
            class_label = np.argmax(class_score, axis=2) #[1,3,169]
            #最大的分类得分值
            score = np.max(class_score, axis=2)             #[1,3,169]
            #当前格子的置信度
            conf = self.result[:, :, 4, :]                  #[1,3,169]
            #每个格子最后的得分为置信度*分类得分
            prob = score * conf                             #[1,3,169]
            #根据阈值,得到评分
```

```
            score_mask = prob > self.confidence_threshold   #[1,3,169]
            num = np.sum(score_mask.reshape([-1]))
            if num > 0:
                #只有转置成[b * 3* (4+1+num_class) * 169],才能根据 score_mask 进行取
                #值,最后需要得到[b,3,169,7]中的 7
                self.result = np.reshape(self.result, [self.b, 3, self.h * self.w, -1])
                boxes = self.result[score_mask][..., :4]
                                                #根据 score_mask 过滤得到 boxes
                cls_scores = prob[score_mask]  #根据 score_mask 过滤得到 cls_scores
                idx = class_label[score_mask]  #根据 score_mask 过滤得到 labels index
                #将当前 layer 满足 box、cls、idx 进行存储
                all_boxes.append(boxes)
                all_cls_scores.append(cls_scores)
                all_idx.append(idx)

        if len(all_boxes):
            #遍历 3 个 layer 中的预测内容并进行合并
            all_boxes = np.concatenate(all_boxes, axis=0)
            all_cls_scores = np.concatenate(all_cls_scores, axis=0)
            all_idx = np.concatenate(all_idx, axis=0)
            #根据类别进行非极大值抑制
            for label in range(len(self.class_prob)):
                #如果 class_labels_all==label,则取当前 label 中的信息
                mask = all_idx == label
                #如果都不是当前 label,则跳过
                if np.sum(mask) == 0: continue
                #由 cx,cy,w,h 转换成 xmin,ymin,xmax,ymax
                xyxy_box = u.cxcy2xyxy(all_boxes[mask][..., :4])
                cat_boxes = np.concatenate([xyxy_box, all_cls_scores[mask].reshape
([-1, 1])], axis=-1)
                #NMS
                index = u.nms(cat_boxes, self.nms_threshold)
                #绘框
                self.old_img = u.draw_box(self.old_img, cat_boxes[index])
            #最后结果
            u.show(self.old_img)


if __name__ == "__main__":
    d = Detected(r'../weights', input_size=[416, 416])
    d.readImg('../../val_data/pexels-photo-5211438.jpeg')
    d.forward()
    d.classification_filtering()
```

主要变化在__init__()中 self. anchors 变为 9 个锚框，每个预测层分配 3 个锚框。在_generate_anchor(self)中根据 self. currentLayer 的结果取当前预测层的结果，如图 5-64 所示。

在解码_decode(self)方法中，根据 self. output[self. currentLayer]获取当前层的输出，并且解码公式仍然与 YOLOv2 的解码公式相同。

在 classification_filtering(self)方法中，for i in range(3)循环 3 个预测层，然后将 i 值赋给 self. currentLayer 属性并传递给 self. _generate_anchor()方法生成锚框、self. _decode()

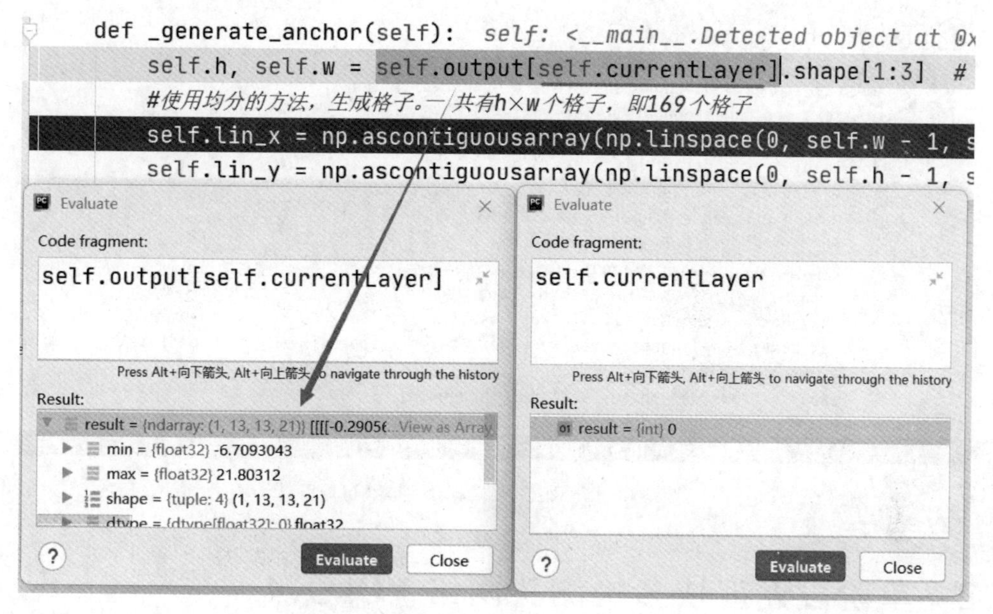

图 5-64 YOLOv3 根据预测层生成锚框

方法进行当前预测 Layer 的解码，将满足 score_mask＝prob＞self. confidence_threshold 阈值的结果存储到 all_boxes、all_cls_scores、all_idx 变量中，如图 5-65 所示。

```
prob = score * conf  # [1,3,169]  prob:
#根据阈值，得到评分
score_mask = prob > self.confidence_threshold  # [1,3,169]  score_mask:
num = np.sum(score_mask.reshape([-1]))  num: 1
if num > 0:
    #只有转置成 [b*3*(4+1+num_class)*169]，才能根据score_mask进行取值，最后需要得到[b,3,169,7]中的7
    self.result = np.reshape(self.result, [self.b, 3, self.h * self.w, -1])
    boxes = self.result[score_mask][..., :4]  # 根据score_mask过滤得到boxes  boxes: [[0.10625909
    cls_scores = prob[score_mask]  # 根据score_mask过滤得到cls_scores  cls_scores: [0.26640323]
    idx = class_label[score_mask]  # 根据score_mask过滤得到labels index  idx: [0]
    #将当前layer满足box、cls、idx进行存储
    all_boxes.append(boxes)
    {list: 1} [array([[0.10625909, 0.09792083, 0.11657397, 0.1282883 ]])]  (cls_scores)
```

图 5-65 YOLOv3 阈值过滤

然后根据 all_boxes 的结果，根据 mask＝all_idx＝＝label 进行 mask 的过滤，最后通过 NMS 算法就可以得到预测结果，如图 5-66 所示。

总结

YOLOv3 有 3 个检测并且每个检测头分配 3 个 Anchor，分别检测大、中、小物体。同时引入 FPN 以加强几何信息与语义信息的融合，这样更有利于检测小目标。

练习

运行并调试本节代码，理解算法的设计与代码的结合，重点梳理本算法的实现方法。

```
#遍历3个layer中的预测内容并进行合并
all_boxes = np.concatenate(all_boxes, axis=0)
all_cls_scores = np.concatenate(all_cls_scores, axis=0)
all_idx = np.concatenate(all_idx, axis=0)
#根据类别进行非极大值抑制
for label in range(len(self.class_prob)):  label: 0
    # 如果class_labels_all==label，则取当前label中的信息
    mask = all_idx == label  mask: [ True  True  True]
    # ...l，则跳过
      + {ndarray: (3,)} [0 0 0]
    if np.sum(mask) == 0: continue
    # 由cx,cy,w,h转换成xmin,ymin,xmax,ymax
    xyxy_box = u.cxcy2xyxy(all_boxes[mask][..., :4])  xyxy_box: [[ 0.0479721   0.03377668
    cat_boxes = np.concatenate([xyxy_box, all_cls_scores[mask].reshape([-1, 1])], axis=-1)
    # nms
    index = u.nms(cat_boxes, self.nms_threshold)  index: [2, 1, 0]
    # 绘框
    self.old_img = u.draw_box(self.old_img, cat_boxes[index])
```

图 5-66 YOLOv3 根据类别索引过滤

5.8 单阶段速度快多检测头网络 YOLOv4

5.8.1 模型介绍

YOLOv4 由 Bochkovskiy 等在 2020 年论文 *YOLOv4：Optimal Speed and Accuracy of Object Detection* 中提出，其主要特点在 YOLOv3 的基础上将主干特征提取网络换成 CSPDarkNet-53，增加了 SPP 和 PANet。引入数据增强 Mosic 和 DropBlock 抑制过拟合。在损失函数方面将位置 BOX 损失更改为 CIOU 损失，其结构如图 5-67 所示。

首先看 CBM 结构，引入了 Mish 激活函数，其公式如下：

$$f(x) \equiv x \times \tanh(\ln(1 + e^x)) \tag{5-13}$$

Mish 函数的优点是无上界、有下界。无上界是任何激活函数都需要的特征，因为它避免了导致训练速度急剧下降的梯度饱和。Mish 函数具有非单调性，这种性质有助于保持小的负值，从而稳定网络梯度流。Mish 函数是光滑函数，具有较好的泛化能力和结果，可以提高训练的准确率，但是 Mish 函数的计算量相比 ReLU 要大，需要的资源更多，其值域如图 5-68 所示。

主干提取网络由 CSP 构成，原作者认为在推理过程中计算量过高的问题是由于网络优化中的梯度信息重复导致的，CSPNet 通过将梯度的变化从头到尾地集成在特征图中，在减少计算量的同时可以保证准确率。进入 CSP 之后，一条分支进行残差并重复指定次数，另外一条分支经过卷积后与其他分支 concat，使用残差可以使网络更稀疏的同时保留更好的特征信息，另一个分支可以保留不同尺度的特征，使 concat 之后能够增强主干特征的提取能力。ShuffLeNet 的思想，分支结构只有两个时也可降低内存消耗。图 5-67 中 CSP 模块残差处重复的次数分别是 1、2、8、8、4 次。

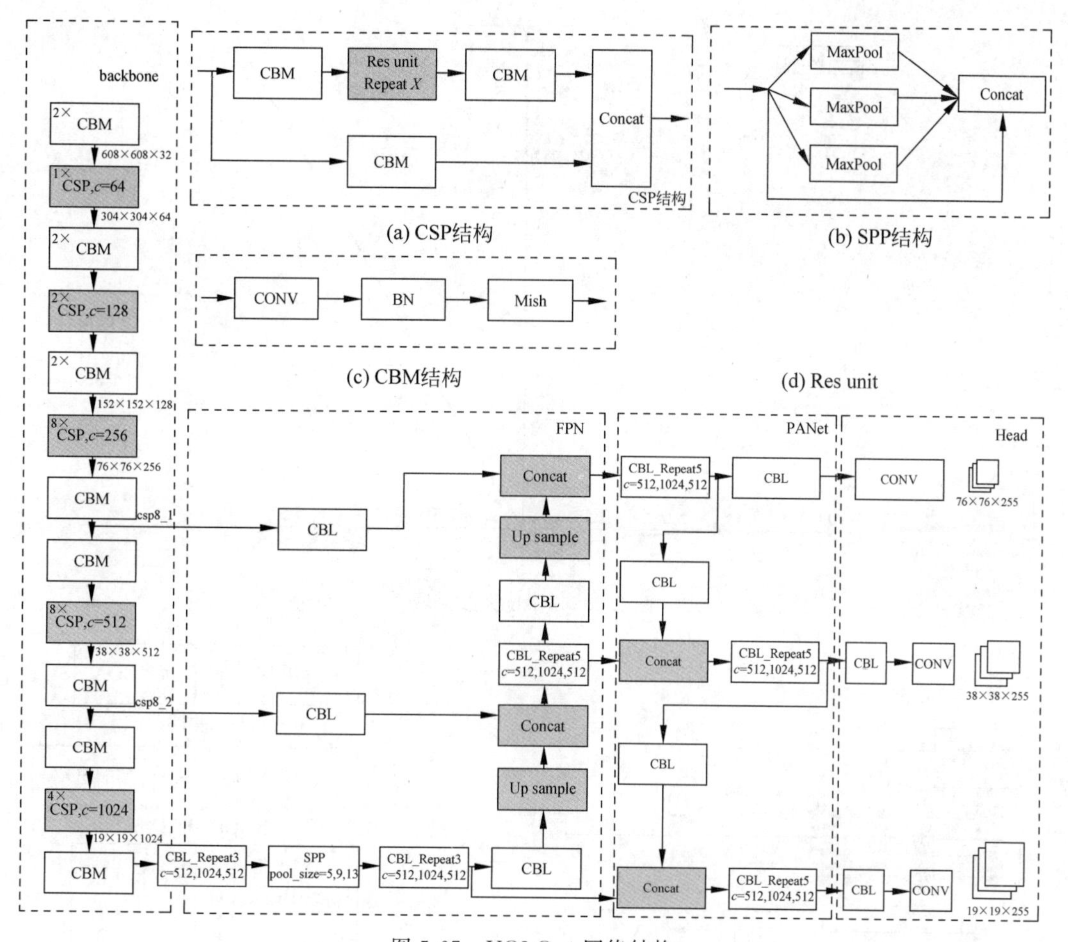

图 5-67 YOLOv4 网络结构

SPP 模块将输入直接作为输出，这为第 1 个分支，第 2 个分支为 5×5 的最大池化，第 3 个分支为 9×9 的最大池化，第 4 个分支为 13×13 的最大池化，其步距都为 1，进行 padding 填充，然后将 4 个分支的特征信息 concat，实现多尺度特征信息的融合，更有助于提升检测的性能。

经过 SPP 模块之后，进行 2 倍上采样为 38×38 并与 csp8_2 的输入进行 concat，再进行 1 次上采样与 csp8_1 的输入进行 concat 得到 76×76 的特征信息。沿 76×76 的特征进行 2 次下采样得到 19×19 的特征，从而用来预测较大目标。FPN 是自底向上的，将深层的语义特征传递过来，对整个特征金字塔进行了增强，不过更多地是增强了语义信息，对定位更有益的几何信息没有传递，PAN 针对这一点，在 FPN 的后面进行了补充，将浅层的定位特征传递到深层，更有助于语义和几何特征信息的融合，如图 5-69 所示。

在计算位置损失时 YOLOv1～3 都使用了均方差损失，这是因为 IOU(GT BOX 与预测 BOX)损失当 IOU=0 时并不能反映两个框的距离，此时损失函数不可导，无法优化两个框不相交的情况；另一种情况虽然 IOU 的值相同，但是从 IOU 的角度也无法区分两者相交情况的不同，如图 5-70 所示。

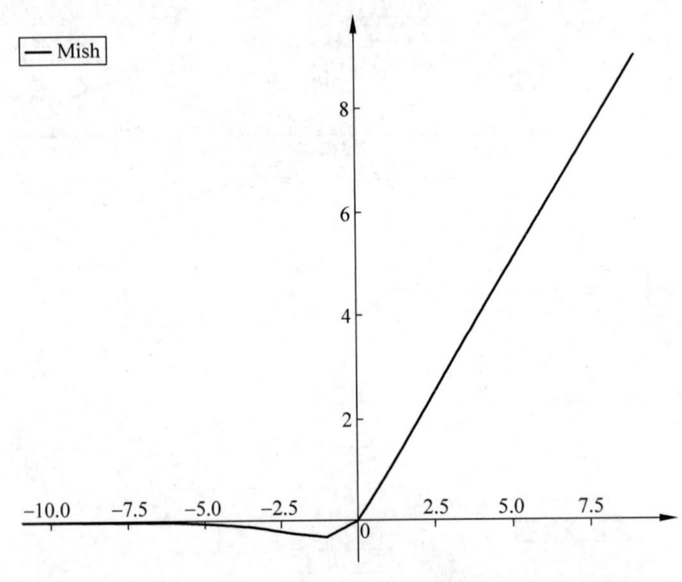

图 5-68　Mish 函数值域

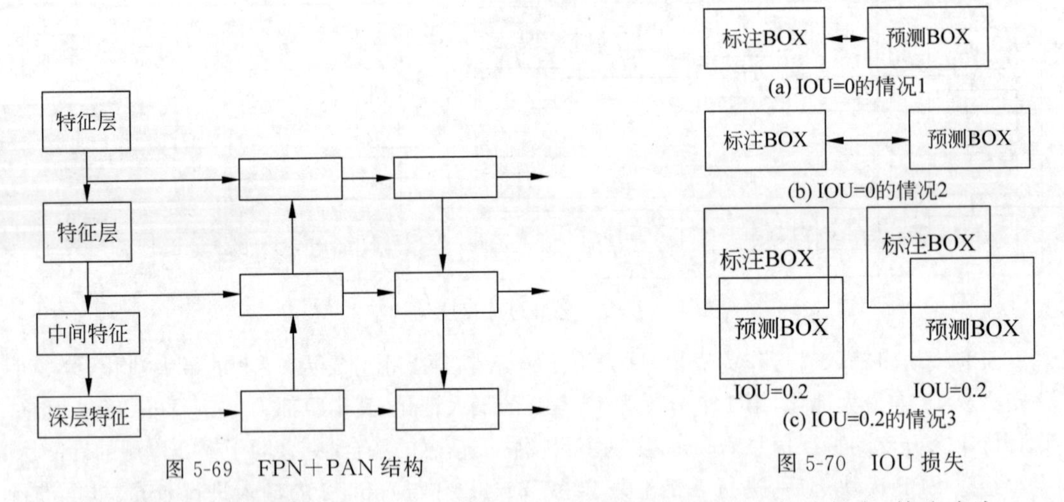

图 5-69　FPN＋PAN 结构

图 5-70　IOU 损失

为了缓解 IOU 损失的不足，在 GIOU 损失中增加相交尺度的衡量方式，其公式为

$$\text{GIOULoss} = 1 - \text{GIOU} = 1 - \left(\text{IOU} - \frac{|D|}{|C|}\right) \tag{5-14}$$

其中，C 代表全集，D 代表差集。当差集 $D=0$ 时，此时为 IOU 损失。当 IOU$=0$ 时，$1-\frac{|D|}{|C|}$仍然能够进行求导计算损失。不过，如果某个 BOX 在另外一个 BOX 的中间时，则差集为 0，此时也就退化成 IOU 损失，仍然无法区分相对位置关系，如图 5-71 所示。

DIOU 损失考虑了重叠面积和中心点距离，其公式：

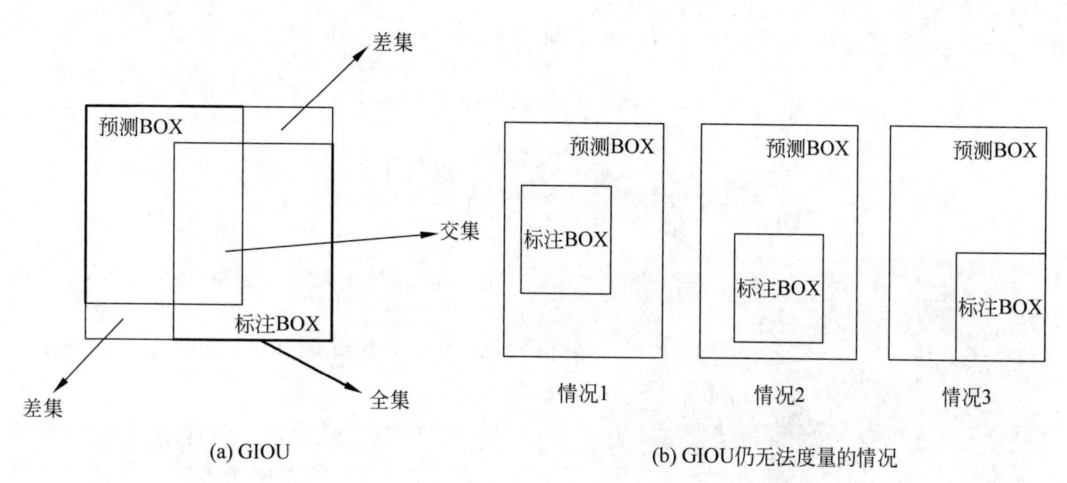

图 5-71 GIOU 损失

$$\text{DIOULoss}=1-\text{DIOU}=1-\left(\text{IOU}-\frac{\text{Distance}_{2^2}}{\text{Distance}_{C^2}}\right) \tag{5-15}$$

其中，Distance_{2^2} 代表两个框中心点的距离，Distance_{C^2} 代表最小外接矩形(对角线)的距离。当两个框重叠时，$\text{Distance}_{2^2}=0$，IOU$=1$，损失为 0；当离得很远时，IOU$=0$，损失为 $1-\left(0-\frac{\text{Distance}_{2^2}}{\text{Distance}_{C^2}}\right)$ 仍然能够进行求导计算损失，如图 5-72 所示。

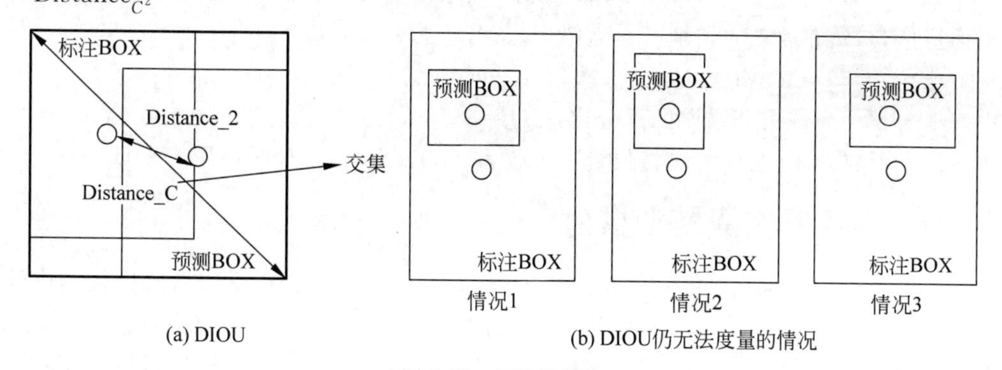

图 5-72 DIOU 损失

DIOU 损失：当目标框包住预测 BOX 时，改为直接度量两个框的距离，但是如图 5-72 中的情况，当预测框的中心点都一样时，没有考虑到宽高比的情况，仍然不能更好地做出选择。

CIOU 损失在 DIOU 损失的基础上考虑了宽高比，其公式如下：

$$\text{CIOULoss}=1-\text{CIOU}=1-\left(\text{IOU}-\frac{\text{Distance}_{2^2}}{\text{Distance}_{C^2}}-\frac{v^2}{(1-\text{IOU})+v}\right) \tag{5-16}$$

$$v=\frac{4}{\pi^2}\left(\arctan\frac{w^{gt}}{h^{gt}}-\arctan\frac{w^p}{h^p}\right)^2$$

其中，v 为衡量宽高比一致性的参数。对 v 进行求导可得

$$\frac{\partial v}{\partial w}=\frac{8}{\pi^2}\left(\arctan\frac{w^{gt}}{h^{gt}}-\arctan\frac{w^p}{h^p}\right)\frac{h}{w^{p^2}+h^{p^2}}$$

$$\frac{\partial v}{\partial h}=\frac{8}{\pi^2}\left(\arctan\frac{w^{gt}}{h^{gt}}-\arctan\frac{w^p}{h^p}\right)\frac{w}{w^{p^2}+h^{p^2}}$$

当预测到 $w^{p^2}+h^{p^2}$ 很小时，$\frac{1}{w^{p^2}+h^{p^2}}$ 的值会很大，为了避免梯度爆炸，在反向传播时此项值设置为 1。

YOLOv4 在主干特征提取时使用了 DropBlock 按空间位置进行丢弃，强迫网络去学习特征图的其他地方的特征。DropBlock 作用在卷积层，而 DropOut 则作用在全连接层，在 RGB 通道进行 DropBlock 的效果如图 5-73 所示。

图 5-73　DropBlock 可视化效果

关于正负样本匹配，YOLOv4 仍然有 3 个检测头，每个检测头有 3 个 Anchor，当 Anchor 与标注 GT BOX 之间的 IOU 大于阈值时都设定为正样本，其他样本为负样本。在设置置信度的值时使用了标签平滑，允许标签的类别有一点错误，其公式为

$$\text{smooth}_{\text{labels}}=y_{\text{true}}\times(1.0-\text{error}_{\text{rate}})+0.5\times\text{error}_{\text{rate}} \tag{5-17}$$

其中，$\text{error}_{\text{rate}}$ 为允许出现错误的百分比，$y_{\text{true}}=1$。

在损失函数方面，在训练时将位置损失改为 CIOU_loss，推理时使用 DIOU_loss。

5.8.2　代码实战模型搭建

从结构图 5-67 中可知，在每次进行 CSP 结果之前有两个 CBM 模块，所以代码中使用 self.split_conv0、self.split_conv1 存储这两个结构。在每次进入 CSP 结构时会经过卷积核为 3×3 且步长 $s=2$ 的下采样。在 1×CSP，$c=64$ 时，输入 channel 与输出 channel 相等，当为其他 CSP 结构时，输出 channel 是输入 channel 的 1 半，关键代码如下：

```
#第 5 章/ObjectDetection/TensorFlow_Yolo_V4_Detected/utils/conv_utils.py

class CSPResBlock(layers.Layer):
    def __init__(self, in_channel, out_channel, num_block, first=False, **kwargs):
        super(CSPResBlock, self).__init__(**kwargs)
        #CSP 中第 1 个 layer1 是用来做下采样的，s=2
        #ConvBNMish 是 Conv->BN->Mish 函数的封装
        self.downSample = ConvBNMish(in_channel, kernel_size=3, strides=2) #32

        if first:
```

```
            #在 CSPRes1 中输入的是 64,残差之后输出的仍然是 64
            self.split_conv0 = ConvBNMish(out_channel, kernel_size=1, strides=1)
            self.split_conv1 = ConvBNMish(out_channel, kernel_size=1, strides=1)
            #CSP 结构中的残差
                self.bocks_conv = Sequential([
                ResBlockMish(in_channel, out_channel),  #残差结构
                ConvBNMish(out_channel, kernel_size=1, strides=1)
            ])
            self.cat_conv = ConvBNMish(out_channel, kernel_size=1, strides=1)
        else:
            #假设:
            #CSPRes2 128->64
            #CSPRes3 256->128
            #CSPRes4 512->256
            #CSPRes5 1024->512
            #从第 2 个 CSP 开始,输出 channel 是输入 channel 的 1 半
            self.split_conv0 = ConvBNMish(out_channel //2, kernel_size=1, strides=1)
                                                            #128 //2
            self.split_conv1 = ConvBNMish(out_channel //2, kernel_size=1, strides=1)
            #*[ResBlockMish for ...] *的作用是将列表反射成参数。num_block 重复的次数
               self.bocks_conv = Sequential([
                *[ResBlockMish(out_channel //2, out_channel //2) for _ in range
(num_block)],
               ConvBNMish(out_channel //2, kernel_size=1, strides=1)
            ])
                #合并
            self.cat_conv = ConvBNMish(out_channel, kernel_size=1, strides=1)

    def call(self, inputs, *args, **kwargs):
        #layer_1 进行下采样
        x = self.downSample(inputs)
        #两个分支
        x0 = self.split_conv0(x)
        x1 = self.split_conv1(x)
        #残差
        x0 = self.bocks_conv(x0)
        out = layers.Concatenate()([x1, x0])
        #再进行一个 1x1
        out = self.cat_conv(out)
        return out
```

然后根据配置文件中的参数进行设置,构建成 cspdarknet 模块,代码如下:

```
#第 5 章/ObjectDetection/TensorFlow_Yolo_V4_Detected/backbone/YOLOv4.py

def csp_darknet_body(input_x):
    """"""
    x = ConvBNMish(32, kernel_size=3, strides=1)(input_x)
    #CPS 模块的输入 channel 和输出 channel,重复次数
    items = [
        [64, 64, 1],
        [128, 128, 2],
```

```
        [256, 256, 8],                    #76 × 76 × 255
        [512, 512, 8],                    #38 × 38 × 255
        [1024, 1024, 4]                   #19 × 19 × 255
    ]
    result = []
    for i, item in enumerate(items):
            #i = 0 first=True
        first = True if i == 0 else False
            #按结构图传递参数
        x = CSPResBlock(item[0], item[1], item[2], first=first)(x)
    return x
```

根据 items 中描述的输入 channel 和输出 channel 循环调用 CSPResBlock(item[0], item[1], item[2], first=first)函数,实现 cspdarknet 特征提取网络的构建。

然后根据结构图构建 CBL_Repeat3 模块,代码如下:

```
#第 5 章/ObjectDetection/TensorFlow_Yolo_V4_Detected/utils/conv_utils.py
class CBL_Repeat3(layers.Layer):
    def __init__(self, list_channels: list):
        super(CBL_Repeat3, self).__init__()
        self.list_channels = list_channels
        self.cbl1 = ConvBnLeakRelu(self.list_channels[0], 1, 1)
        self.cbl2 = ConvBnLeakRelu(self.list_channels[1], 3, 1)
        self.cbl3 = ConvBnLeakRelu(self.list_channels[2], 1, 1)

    def call(self, inputs, *args, **kwargs):
        inputs = self.cbl1(inputs)
        inputs = self.cbl2(inputs)
        inputs = self.cbl3(inputs)
        return inputs
```

根据传入的 3 个 channel 数进行卷积、BN、LeakRelu 的计算。与 CBL_Repeat5 的计算类似,只是增加了 self. cbl4、self. cbl5。

对检测头进行封装,输入[channel,3 * (4+1+class_num)]数进行输出,不同检测头的通道 channel 略有不同,代码如下:

```
#第 5 章/ObjectDetection/TensorFlow_Yolo_V4_Detected/utils/conv_utils.py
class Yolc_V4_Head(layers.Layer):

    def __init__(self, list_channels: list):
        super(Yolo_V4_Head, self).__init__()
        #list_channels=[1024, len(anchors_mask[0]) *(4 + 1 + class_num)]
        self.list_channels = list_channels
        #3×3 卷积
        self.cbl1 = ConvBnLeakRelu(self.list_channels[0], 3, 1)
        #1×1 卷积,输出
        self.conv = layers.Conv2D(self.list_channels[1], 1, 1)
    def call(self, inputs, *args, **kwargs):
        inputs = self.cbl1(inputs)
        inputs = self.conv(inputs)
        return inputs
```

对 CspDarknet、CBL_Repeat3、CBL_Repeat5、Yolo_V4_Head 等结构进行封装，构建完成 YOLOv4 的主体结构，代码如下：

```
#第 5 章/ObjectDetection/TensorFlow_Yolo_V4_Detected/backbone/YOLOv4.py

def YOLOv4_body(input_shape, anchors=None, class_num=80, is_class=False):
    """构建 YOLO 的网络模型"""
    if anchors is None:
        anchors_mask = np.array([
            [[10, 13], [16, 30], [33, 23]],                           #小目标
            [[30, 61], [62, 45], [59, 119]],                          #中目标
            [[116, 90], [156, 198], [373, 326]]                       #大目标
        ])
    else:
        anchors_mask = anchors
    input_x = layers.Input(shape=input_shape, dtype='float32')
    #backbone
    #76 × 76, 38 × 38, 19 × 19
    result = csp_darknet_body(input_x)
    csp8_1, csp8_2, csp4 = result[0]
    #19 × 19 的分支处理
    csp4_19 = CBL_Repeat3(list_channels=[512, 1024, 512])(csp4) #19 *19 *512
    csp4_19_spp = SPP()(csp4_19)                                      #19 × 19 × 2048
    #供 19 × 19 输出头 concatenate
    csp4_19_spp_cbl3 = CBL_Repeat3(list_channels=[512, 1024, 512])(csp4_19_spp)
                                                                      #19 *19 *512
    #上采样,38 × 38 × 256
    csp4_19_spp_cbl3_for_middle_upsample = ConvUpsample(256)(csp4_19_spp_cbl3)
    #csp8_2 分支的处理 38 × 38 × 256
    csp8_2_cbl = ConvBnLeakRelu(filters_channels=256, kernel_size=1, strides=1)
(csp8_2)
    middle_concat1 = layers.Concatenate()([csp4_19_spp_cbl3_for_middle_
upsample, csp8_2_cbl])                                                #38 × 38 × 512
    #38 × 38 × 256
    middle_concat_cbl5 = CBL_Repeat5(list_channels=[256, 512, 256, 512, 256])
(middle_concat1)                                                      #等待第 2 个 concat
    #上采样 76 × 76 × 128
    middle_concat_cbl5_for_76_upsample = ConvUpsample(128)(middle_concat_cbl5)
    #csp8_1 分支的处理 76 × 76 × 128
    csp8_1_cbl = ConvBnLeakRelu(filters_channels=128, kernel_size=1, strides=1)
(csp8_1)
    #76 × 76 × 256
    last_concat = layers.Concatenate()([csp8_1_cbl, middle_concat_cbl5_for_76_
upsample])
    #供中间层进行 middle_concat2，并且也是 76 × 76 的检测头的来源
    #最后一层的输出 76 × 76
    #76 × 76 × 128
    last_concat_cbl5 = CBL_Repeat5(list_channels=[128, 256, 128, 256, 128])(last_
concat)                                                               #p5
    #cbl5 后面还有一个下采样
    #38 × 38 × 256
    last_concat_cbl5_down_sample = ConvBnLeakRelu(filters_channels=256, kernel_
size=3, strides=2)(last_concat_cbl5)
```

```
    #中间层的输出 38 × 38 × 512
    middle_concat2 = layers.Concatenate()([last_concat_cbl5_down_sample, middle_concat_cbl5])
    #38 × 38 的检测头来源,并且也给 19 × 19 进行 concatenate 38 × 38 × 256
    middle_head_cbl5 = CBL_Repeat5(list_channels=[256, 512, 256, 512, 256])(middle_concat2)                    #p4
    #下采样 19 × 19 × 512
    middle_head_cbl5_down_sample = ConvBnLeakRelu(filters_channels=512, kernel_size=3, strides=2)(middle_head_cbl5)
    #第 1 个层的输出 19 × 19 × 1024
    first_concat = layers.Concatenate()([csp4_19_spp_cbl3, middle_head_cbl5_down_sample])
    #19 × 19 × 512
    first_cbl5 = CBL_Repeat5(list_channels=[512, 1024, 512, 1024, 512])(first_concat)                    #p3
    #构建检测头
    #19 × 19 × 255 用于检测大图
    out1 = Yolo_V4_Head([1024, len(anchors_mask[0]) * (4 + 1 + class_num)])(first_cbl5)
    #38 × 38 × 255 用于检测中图
    out2 = Yolo_V4_Head([512, len(anchors_mask[1]) * (4 + 1 + class_num)])(middle_head_cbl5)
    #76 × 76 × 255 用于检测小图
    out3 = Yolo_V4_Head([256, len(anchors_mask[2]) * (4 + 1 + class_num)])(last_concat_cbl5)
    return Model(inputs=input_x, outputs=[out1, out2, out3])
```

csp8_1、csp8_2 和 csp4 为 cspdarknet 中获取的特征输出,经过 SPP、FPN、PAN 之后得到 19×19×255、38×38×255、76×76×255 的 3 个检测头,然后通过 outputs=[out1, out2,out3]进行输出。在 FPN、PAN、Head 部分使用激活函数 LeakRule 是由于 Mish 函数的运算量过大,能够加快模型的收敛。

5.8.3 代码实战建议框的生成

在 YOLOv4 中如果真实 GT BOX 与 Anchor 的 IOU>0.5,则设为正样本,这样有可能会导致某个 GT BOX 均能在 3 个检测头中找到正样本,其实现代码如下:

```
#第 5 章/ObjectDetection/TensorFlow_Yolo_V4_Detected/utils/tools.py
def yolo_v4_true_encode_box(
        true_boxes,
        anchors=None,
        class_num=80,
        input_size=(416, 416),
        ratio=[32, 16, 8],
        iou_threshold=0.213
):
    """当 YOLOv4 使用真值编码时,如果 IOU>0.26,则为正样本。也就说一个 GT 可能会有多个正样本,而 YOLOv3 是一个 GT 只有一个正样本"""
    if anchors is None:
        anchors = np.array([
```

```
        [[10, 13], [16, 30], [33, 23]],                          #小目标
        [[30, 61], [62, 45], [59, 119]],                         #中目标
        [[116, 90], [156, 198], [373, 326]]                      #大目标
    ])
#Anchor 的数量
detected_num = anchors.shape[0]
#用来存储结果
grid_shape = []
true_boxes2 = []
#用来存储结果,对输入的 boxes 进行了升维
true_boxes = np.expand_dims(true_boxes, axis=0)
#因为每个 img 都要转换一下,所以应该是 1。
batch_size = true_boxes.shape[0]
#跟 YOLOv3 一样,初始 3 个检测头的矩阵
for i in range(detected_num):
        #13 × 13
    sw, sh = input_size[0] //ratio[i], input_size[1] //ratio[i]
    grid = np.zeros(
        [batch_size, sw, sh, len(anchors[i]), 4 + 1 + class_num],
        dtype=np.float32)
    grid_shape.append(grid)
    true_boxes2.append(
        np.zeros(
            [batch_size, sw, sh, len(anchors[0]), 4 + 1 + class_num],
            dtype=np.float32))
grid = grid_shape
#遍历每个 BOX 进行 IOU 的选择
for box_index in range(true_boxes.shape[1]):
    #BOX 信息
    box = true_boxes[:, box_index, :]
    box = np.squeeze(box, axis=0)                                #降维
    #类别
    box_class = box[4].astype('int32')
    best_choice = {}
    for index in range(0, 3, 1):
        #anchors = anchors *grid[index].shape[-3]
        #box = (13 × x,13 × y, 13 × w, 13 × h) 换算成相对 grid cell 的值
        sa = grid[index].shape[-3]
        #box[0:4]是归一化后的结果 × 13,将 BOX 信息映射到某个特征层
        box2 = box[0:4] *np.array([sa, sa, sa, sa])
        box_true = box2.copy()
        box2 = box2.copy()
        #在 i,j 个格子
        i = np.floor(box2[1]).astype('int')
        j = np.floor(box2[0]).astype('int')
        if i > grid[index].shape[-3] or j > grid[index].shape[-3]:
            continue
        #遍历每个 Anchor
        for k, anchor in enumerate(anchors[2 - index]):
            #wh
            box_maxes = box2[2:4] *0.5
            box_mines = -box_maxes
            #anchor wh
```

```
            anchor_maxes = anchor * 0.5
            anchor_mines = -anchor_maxes
            #将真实 wh 与 Anchor 之间进行 IOU 的计算,并获取最佳 IOU 是哪个 Anchor
            intersect_mines = np.maximum(box_mines, anchor_mines)
            intersect_maxes = np.minimum(box_maxes, anchor_maxes)
            intersect_wh = np.maximum(intersect_maxes - intersect_mines, 0.)
            intersect_area = intersect_wh[0] * intersect_wh[1]
            #计算 IOU
            box_area = box2[2] * box2[3]
            anchor_area = anchor[0] * anchor[1]
            iou_score = intersect_area / (box_area + anchor_area - intersect_area)
            #只要 IOU 大于 0.26 都认为是正样本
            if iou_score >= iou_threshold:
                #计算偏移值
                adjusted_box = np.array([
                    box2[0] - j, #x 和 y 都是相对于 gird cell 的位置,左上角为[0,0],右
                                 #下角为[1,1]
                    box2[1] - i,
                    np.log(box2[2] / anchors[2 - index][k][0]),
                    np.log(box2[3] / anchors[2 - index][k][1]),
                ], dtype=np.float32)
                #存储偏移值
                grid[index][..., j, i, k, 0:4] = adjusted_box
                #置信度
                grid[index][..., j, i, k, 4] = 1
                #标签平滑处理
                grid[index][..., j, i, k, 5 + box_class] = smooth_labels(1, 0.1)
                #获取真值,方便后面计算损失时做 IOU
                true_boxes2[index][..., j, i, k, 0:4] = box_true
    return true_boxes2, grid
```

与 YOLOv3 的代码类似,for box_index in range(true_boxes.shape[1])循环每个 GT BOX,for index in range(0, 3, 1)循环每个检测头,for k, anchor in enumerate(anchors[2-index])循环每个 Anchor,当 iou_score＞＝iou_threshold(大于设定的阈值)时,grid[index][...,j,i,k,0:4]＝adjusted_box 设置为 j,i 个格子中的第 k 个 Anchor 的偏移值为 adjusted_box。与 YOLOv3 不同之处在于可能有多个正样本,如图 5-74 所示。通常来说提高正样本的数量对于模型的训练有益,正样本越多越容易学到更多的特征。

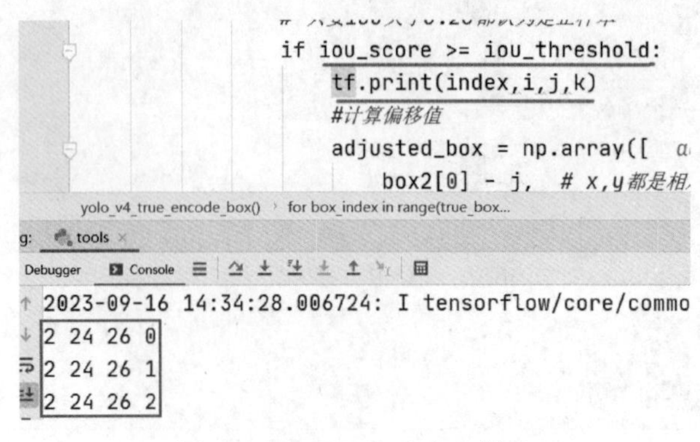

图 5-74　YOLOv4 多个正样本

5.8.4 代码实现损失函数的构建及训练

YOLOv4 损失函数的代码实现与 YOLOv3 损失函数的代码实现基本相同，不同之处仅在于计算 BOX 损失时使用 CIOU 损失，CIOU 的计算代码如下：

```
#第 5 章/ObjectDetection/TensorFlow_Yolo_V4_Detected/utils/iou_help.py
def c_iou(priors_box, true_box):
    """
    不仅考虑到重叠面积和中心点距离，还考虑到纵横比。c_iou 一般用于训练
    :param priors_box: cx, cy, w,h
    :param true_box: cx, cy, w,h
    :return:
    """
    #box1 的信息
    b1_xy = priors_box[..., :2]
    b1_wh = priors_box[..., 2:4]
    b1_wh_half = b1_wh / 2.                    #中心点
    b1_mins = b1_xy - b1_wh_half               #xmin, ymin
    b1_maxes = b1_xy + b1_wh_half              #xmax, ymax
    #box2 的信息
    b2_xy = true_box[..., :2]
    b2_wh = true_box[..., 2:4]
    b2_wh_half = b2_wh / 2.
    b2_mins = b2_xy - b2_wh_half               #xmin, ymin
    b2_maxes = b2_xy + b2_wh_half              #xmax, ymax
    #计算 IOU
    intersect_mins = tf.maximum(b1_mins, b2_mins)
    intersect_maxes = tf.minimum(b1_maxes, b2_maxes)
    intersect_wh = tf.maximum(intersect_maxes - intersect_mins, 0.)
    intersect_area = intersect_wh[..., 0] *intersect_wh[..., 1]
    b1_area = b1_wh[..., 0] *b1_wh[..., 1]
    b2_area = b2_wh[..., 0] *b2_wh[..., 1]
    union_area = b1_area + b2_area - intersect_area
    #calculate IoU, add epsilon in denominator to avoid dividing by 0
    iou = intersect_area / (union_area + tf.keras.backend.epsilon())   #iou

    #计算两个 BOX 的中心点距离
    center_distance = tf.keras.backend.sum(tf.square(b1_xy - b2_xy), axis=-1)
    #获得全集
    enclose_mins = tf.minimum(b1_mins, b2_mins)
    enclose_maxes = tf.maximum(b1_maxes, b2_maxes)
    enclose_wh = tf.maximum(enclose_maxes - enclose_mins, 0.0)
    #计算全集中心点的距离
    enclose_diagonal = tf.keras.backend.sum(tf.square(enclose_wh), axis=-1)

    #       d**2
    #求 Diou = iou - ---------
    #       c**2
    diou = iou - 1.0 *(center_distance)/(enclose_diagonal + tf.keras.backend.epsilon())

    #计算 w、h 宽高比
    v = 4 *tf.keras.backend.square(
       tf.math.atan2(b1_wh[..., 0], b1_wh[..., 1]) - tf.math.atan2(b2_wh[..., 0],
b2_wh[..., 1])
```

```
    ) / (tf.cast(np.pi **2, dtype=tf.float32))
    #公式中的 alpha 为了防止梯度爆炸,alpha 不进行梯度计算
    alpha = v / (1.0 - iou + v)
    K.stop_gradient(alpha) #此句不进行梯度求解
    #得到 CIOU
    ciou = diou - alpha *v
    ciou = tf.expand_dims(ciou, -1)
    return ciou
```

在计算 CIOU 时先计算 iou＝intersect_area/(union_area＋tf.keras.backend.epsilon())，再计算 diou＝iou－1.0 * (center_distance)/(enclose_diagonal＋tf.keras.backend.epsilon())，然后计算 v，最后得 ciou＝diou－alpha * v 的损失结果。计算损失的核心代码如下：

```
#第 5 章/ObjectDetection/TensorFlow_Yolo_V4_Detected/utils/loss.py

class MultiAngleLoss(object):
    def yolo_v4_loss(self, y_true, y_pred, true_box2):
        """整体过程跟 YOLOv3 区别不大。主要是在计算 BOX 时,使用的是 CIOU 进行计算"""
        #网络层数
        num_layers = self.num_layers
        loss = 0
        for index in range(num_layers):
                #每层的预测值
            y_pre1 = y_pred[index]
            y_pre = tf.reshape(y_pre1,
                               [-1, y_pre1.shape[1], y_pre1.shape[2], len(self.
anchor[0]), 4 + 1 + self.num_class])

            #y_pre = tf.nn.sigmoid(y_pre) #把预测出来的值限定在 0~1
            #所有的同层 true_box concat 在一起
            true_boxes = tf.concat([t[index] for t in true_box2], axis=0)

            true_grid = tf.concat([t[index] for t in y_true], axis=0)
                #获取置信度信息,在真值中只有一个检测头是用来预测的
            object_mask = true_grid[..., 4:5]
            num_pos = tf.maximum(K.sum(K.cast(object_mask, tf.float32)), 1)
            #预测返回的是位置信息、置信度信息和分类信息,与 YOLOv3 相同
            pre_boxes, pre_box_confidence, pre_box_class_props = yolo_v4_head_
decode(y_pre, 2 - index,

num_class=self.num_class)
            #平衡系数
            #2 - (w × h) 如果 wh 较小,则惩罚学习小框。也可人为设定
            box_loss_scale = 2 - true_boxes[..., 2] *true_boxes[..., 3]

            #使用 CIOU 直接求 BOX 的 loss
            ciou_score = c_iou(pre_boxes, true_boxes[0:4])
            ciou_loss = object_mask *tf.expand_dims(box_loss_scale, axis=-1) *
(1 - ciou_score)

            #预测值与真值之间的 IOU,如果大于指定阈值,但是又不是最大 IOU 的内容,则会被
            #忽略处理
            iou_score = tf.reduce_max(ciou_score, axis=-1)
            iou_score = tf.expand_dims(iou_score, axis=-1)
```

```
            object_detections = iou_score > self.overlap_threshold
            #正样本的 mask
            object_detections = tf.cast(object_detections, dtype=iou_score.dtype)
            #负样本置信度损失
            no_object_loss = (1 - object_detections) *(
                         1 - object_mask) *tf.expand_dims(tf.keras.losses.binary_
crossentropy(
            object_mask, pre_box_confidence, from_logits=True), axis=-1)
            #正样本损失
                object_loss = object_mask *tf.expand_dims(tf.keras.losses.binary_
crossentropy(
            object_mask, pre_box_confidence, from_logits=True), axis=-1)
            #负样本+正样本的损失
            confidence_loss = object_loss + no_object_loss
            #分类损失
            class_loss = object_mask *tf.expand_dims(tf.keras.losses.binary_
crossentropy(
                     true_grid[..., 5:], y_pre[..., 5:], from_logits=True), axis=-1)
            #位置平均损失
            location_loss = tf.abs(tf.reduce_sum(ciou_loss)) / num_pos
            confidence_loss = tf.reduce_mean(confidence_loss)
            class_loss = tf.reduce_sum(class_loss) / num_pos
                #求合
            loss += location_loss + confidence_loss + class_loss
        return loss
```

跟 YOLOv3 相比区别仅在 ciou_score=c_iou(pre_boxes,true_boxes[0:4])的计算上,即位置损失使用 ciou_score,同时根据 object_detections=iou_score>self.overlap_threshold 的得分计算正样本的 mask,然后由 confidence_loss=object_loss+no_object_loss 计算正样本、负样本的置信度损失,最后再将 location_loss+confidence_loss+class_loss 相加得到总损失。

5.8.5 代码实战预测推理

YOLOv4 的预测推理流程与 YOLOv3 的预测推理流程一致,即需要先遍历 3 个检测头,接着对每个检测头的输出结果进行置信度、分类概率的过滤,然后对 3 个检测头中满足条件的结果进行合并,再通过 NMS 去除重复的框,详细代码可参考 YOLOv3 中的内容。

总结

YOLOv4 在主干特征提取方面使用了 CSP 结构,引用 SPP、FPN 和 PAN 进行更细粒度的特征融合,仍然由 3 个检测头构成多尺度的预测。在损失函数方面,引入 CIOU 来计算位置损失。

练习

运行并调试本节代码,理解算法的设计与代码的结合,重点梳理本算法的实现方法。

5.9 单阶段速度快多检测头网络 YOLOv5

5.9.1 模型介绍

YOLOv5 由 ultralytics 公司开源在 GitHub 上面,到本书成稿时没有发表论文。此模型得以流行，主要在于 GitHub 上代码工程化的易用性，并且精度和速度较佳，其主要结构如图 5-75 所示。

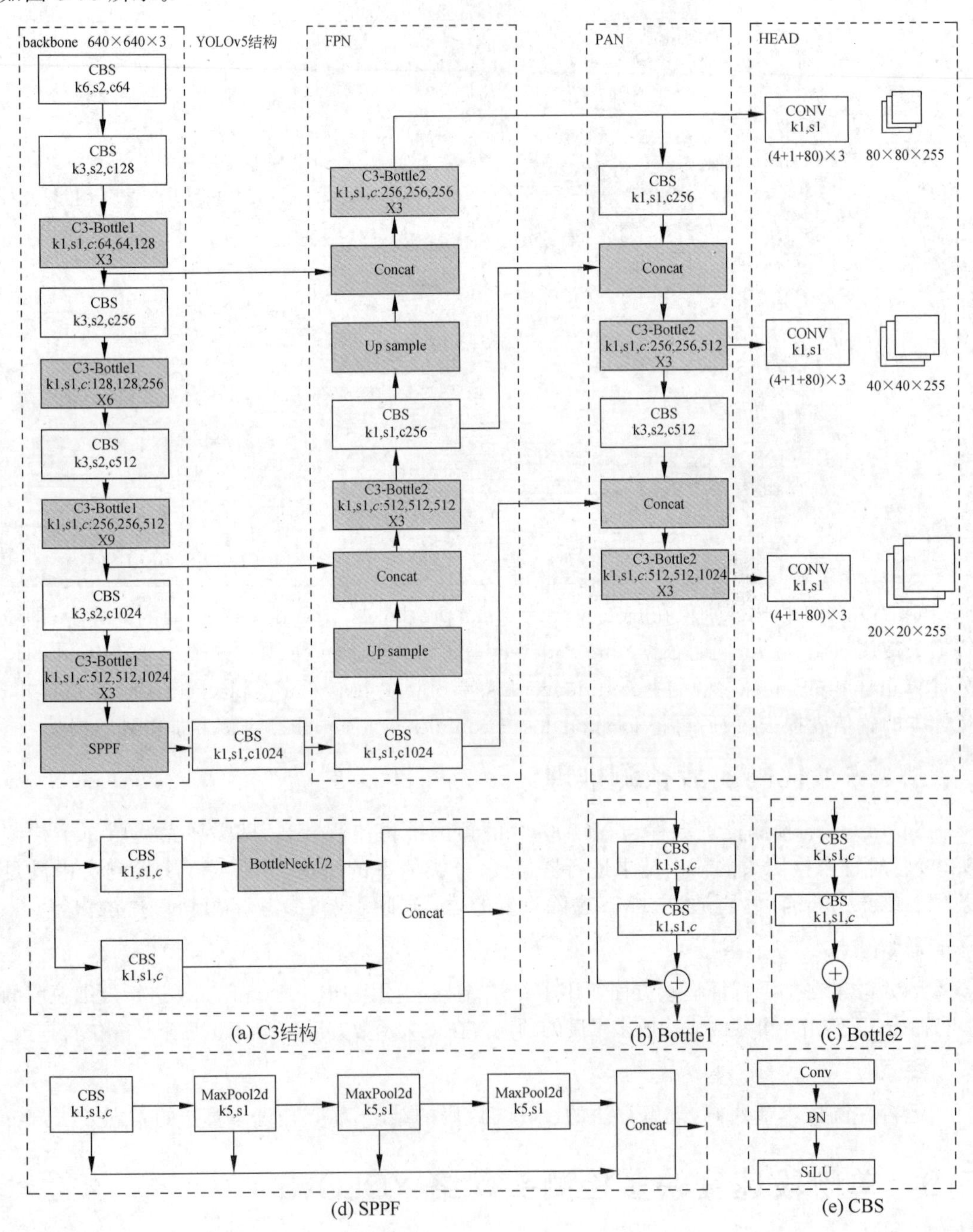

图 5-75 YOLOv5 的结构

图像的输入尺寸设定为640×640,CBS为卷积、池化、SiLU的激活函数,其公式为

$$\mathrm{SiLU}_{(x)}=x*\mathrm{Sigmoid}(x) \tag{5-18}$$

SiLU函数在接近0时具有更平滑的曲线,并且由于Sigmoid(x)函数,可以使网络输出在一个较小值的范围,其函数的值域如图5-76所示。

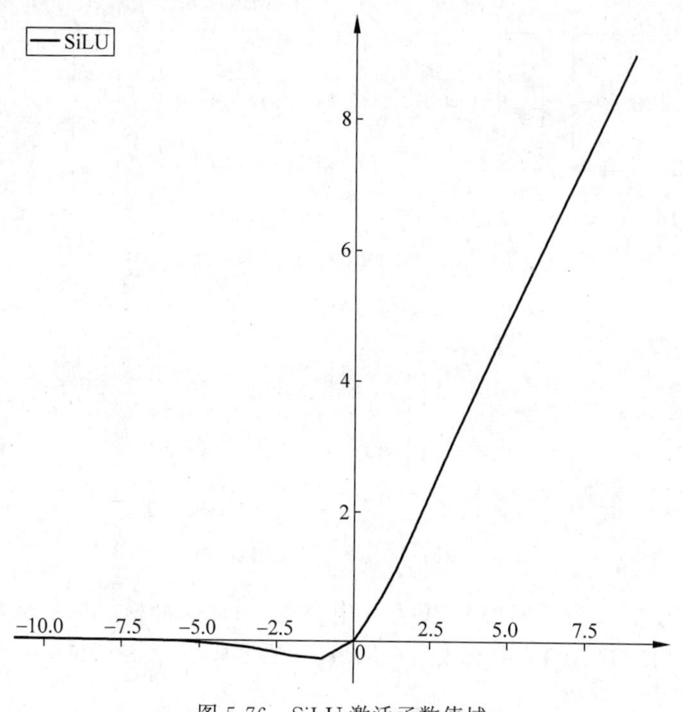

图5-76 SiLU激活函数值域

C3模块分两种情况,在主干特征提取部分为C3-Bottle1,其结构与YOLOv4中的CSP结构相同,并且在YOLOv5中将C3-Bottle2应用于FPN、PAN结构之中(YOLOv4没有)。C3-Bottle2为正常卷积结构,使用C3为一个普通残差结构。将YOLOv4中的SPP结构更换为SPPF,由原来的并联更改为了3个5×5卷积的串联,据作者说能够提高运行速度。

因为输入的分辨率增大了,所以网络的3个检测头的输出也有变化,并且不同尺寸的输出分别用于检测小、中、大目标。输出尺寸为80×80×(4box+1conf+num_class)×3Anchor、40×40×255、20×20×255。

在正负样本匹配方面,YOLOv5改为将真实GT BOX宽和高与Anchor BOX宽和高的比例值小于4设为正样本,如图5-77所示,只要在这个范围内的Anchor均被选取,如果超过这个范围,则不选取。

在计算偏移值时,围绕GT BOX的中心点cx和cy所在的格子选择了附近的3个格子,并且都计算偏移值,并设定增加新的ij作为有目标的格子,由原来1个格子,变为3个格子,结合宽高比小于4的Anchor,极大地增加了正样本的数量,更有利于网络的学习,如图5-78所示。

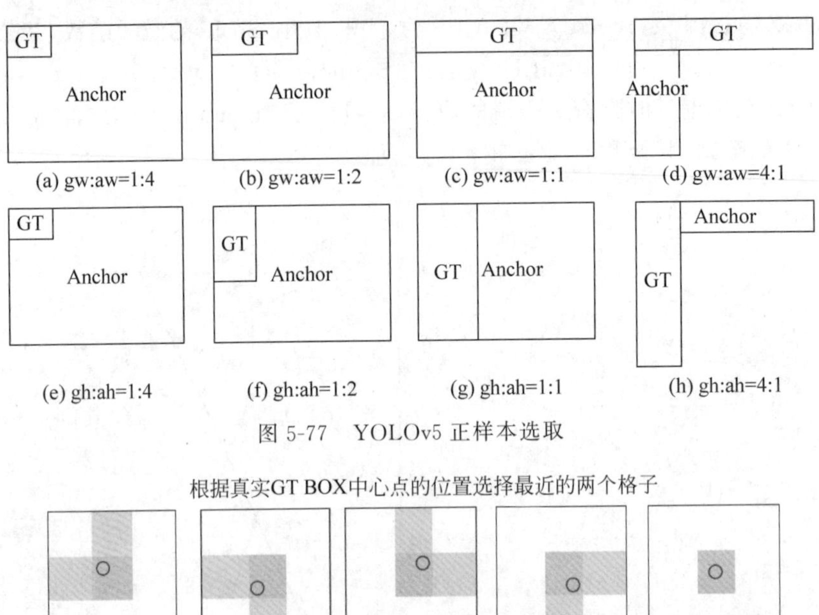

图 5-77 YOLOv5 正样本选取

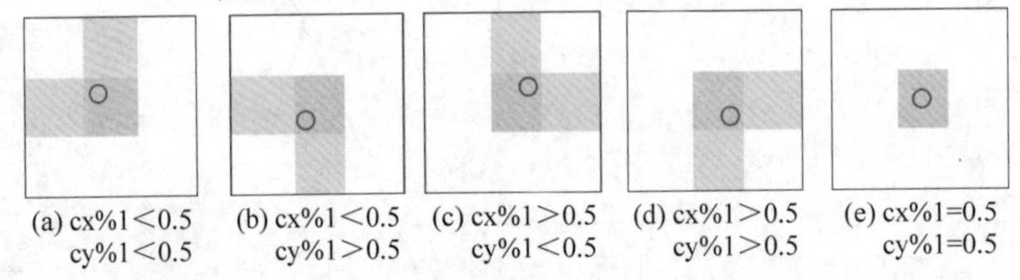

图 5-78 单 GT 多锚点

图 5-78 中 cx 和 cy 为真实 GT BOX 的中心点,因为每个格子的值域为[0,1],所以由 cx%1 和 cy%1 可得 GT BOX 中心点所在区域,然后根据这个区域选择附近的格子来计算偏移值。

根据预测偏移值和 Anchor,计算预测 P_x、P_y、P_w、P_h 时解码式(5-11),YOLOv4 的作者指出当真实目标中心点非常靠近网格的左上角时,Sigmoid(t_x)、Sigmoid(t_y)会趋近于 0,如果在右下角,则会趋近于 1,网络的预测值需要在负无穷或者正无穷时才能取得,而这种极端值网络一般无法达到,GT BOX 在黄色、红色处时 X 值需要很大才能接近,如图 5-79 所示。

为了解决这个问题,YOLOv4 的作者将偏移值从原来的[0,1]缩放到[-0.5,1.5],所以其公式变为

$$
\begin{aligned}
P_x &= (2 * \mathrm{Sigmoid}(t_x) - 0.5) + c_x \\
P_y &= (2 * \mathrm{Sigmoid}(t_y) - 0.5) + c_y \\
P_w &= A_w (2 * \mathrm{Sigmoid}(t_w))^2 \\
P_h &= A_h (2 * \mathrm{Sigmoid}(t_h))^2
\end{aligned} \tag{5-19}
$$

虽然 YOLOv4 提出了这个改进策略,但是代码中没有实现,而这一点在 YOLOv5 中被应用。在 YOLOv5 中除了调整 P_x、P_y 外,还对 P_w、P_h 进行了限制,这样可以避免出现梯度爆炸、训练不稳定的情况。

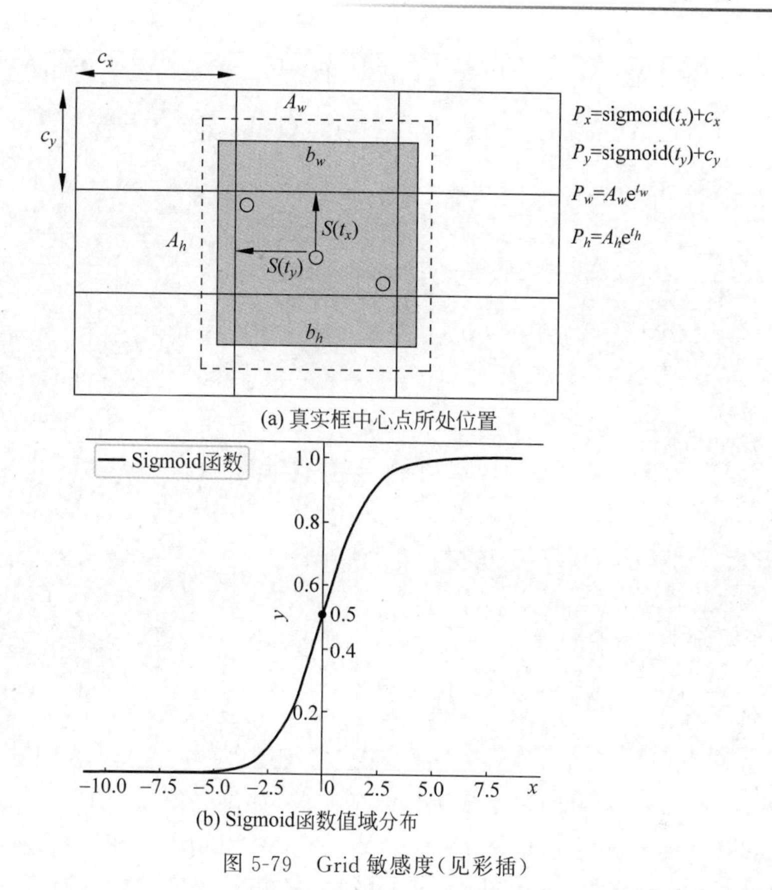

图 5-79 Grid 敏感度(见彩插)

在损失函数方面没有大的变化,位置损失使用 CIOU,而置信度和分类损失则使用交叉熵,不同的检测头有不同的权重比例。

5.9.2 代码实战模型搭建

首先根据如图 5-75 所示的结构图实现 CSP_Bottle1,即 CSP 结构,代码如下:

```
#第 5 章/ObjectDetection/TensorFlow_Yolo_V5_Detected/utils/conv_utils.py

class Yolo5_BottleNeck1(layers.Layer):
    #残差结构,输入 channel,输出 channel
    def __init__(self, input_channels1, input_channels2):
       super(Yolo5_BottleNeck1, self).__init__()
       self.silu1 = ConvBNSiLU(input_channels1, 1, 1)#conv->bn->silu
       self.silu2 = ConvBNSiLU(input_channels2, 3, 1)

    def call(self, inputs, *args, **kwargs):
       return layers.add([inputs, self.silu2(self.silu1(inputs))])

#CSP bottle1 的结构
class Yolo5_CSP_Bottleneck1(layers.Layer):
```

```
    #CSP bottle1 的结构
    def __init__(self,
                 left_channel, right_channel1,
                 bottleneck_channel: list,
                 bottlenect_repeat: int, out_channel):
        super(Yolo5_CSP_Bottleneck1, self).__init__()
        #左侧 channel
        self.left = ConvBNSiLU(left_channel, 1, 1)
        #右侧 channel
        self.right1 = ConvBNSiLU(right_channel1, 1, 1)
        #残差结构
        self.bottle = Yolo5_BottleNeck1(bottleneck_channel[0], bottleneck_
channel[1])
        #输出结构
        self.out = ConvBNSiLU(out_channel, 1, 1)
        self.repeat = bottlenect_repeat

    def call(self, inputs, *args, **kwargs):
        left = self.left(inputs)
        right = self.right1(inputs)
        #重复次数
        for _ in range(self.repeat):
            right = self.bottle(right)
        out = layers.concatenate([left, right], axis=-1)
        return self.out(out)
```

代码中 Yolo5_BottleNeck1()实现了残差的封装，如果是第 1 个 C3-Bottle1，则通道值 input_channels1＝64，input_channels2＝64。Yolo5_CSP_Bottleneck1()实现了 C3-Bottle1 整个的封装，left_channel 和 right_channel1 分别为左侧输入、右侧输入通道数，bottleneck_channel 为残差通道数，bottlenect_repeat 为残差重复的次数，out_channel 为输出的通道数，第 1 个 C3-Bottle1 重复的次数为 3，输入通道为 64，输出为 128，则第 1 个 C3-Bottle1 应传入的参数值为 Yolo5_CSP_Bottleneck1(64,64,[64,64],3,128)。

然后根据结构图实现 C3-Bottle2 的结构，C3-Bottle1 主要用在 FPN、PAN 中，代码如下：

```
#第 5 章/ObjectDetection/TensorFlow_Yolo_V5_Detected/utils/conv_utils.py

class Yolo5_BottleNeck2(layers.Layer):
    #bottle2，没有残差
    def __init__(self, input_channels1, input_channels2):
        super(Yolo5_BottleNeck2, self).__init__()
        self.silu1 = ConvBNSiLU(input_channels1, 1, 1)
        self.silu2 = ConvBNSiLU(input_channels2, 3, 1)

    def call(self, inputs, *args, **kwargs):
        return self.silu2(self.silu1(inputs))

class Yolo5_CSP_Bottleneck2(layers.Layer):
    #neck 部分使用 csp_bottle2 结构
    def __init__(self, left_channel, right_channel1,
```

```
                bottleneck_channel: list,
                bottlenect_repeat: int, out_channel):
        super(Yolo5_CSP_Bottleneck2, self).__init__()
        self.left = ConvBNSiLU(left_channel, 1, 1)
        self.right1 = ConvBNSiLU(right_channel1, 1, 1)
              #此时就是一个串行的卷积
        self.bottle = Yolo5_BottleNeck2(bottleneck_channel[0], bottleneck_
channel[1])
        self.out = ConvBNSiLU(out_channel, 1, 1)
        self.repeat = bottlenect_repeat

    def call(self, inputs, *args, **kwargs):
        left = self.left(inputs)
        right = self.right1(inputs)
            #重复多次
        for _ in range(self.repeat):
            right = self.bottle(right)
            #concat
        out = layers.concatenate([left, right], axis=-1)
        return self.out(out)
```

Yolo5_BottleNeck2()是没有残差的，只是一个串行的卷积结构。按指定的重复次数bottlenect_repeat进行解析。Yolo5_CSP_Bottleneck2()的形参的含义相同，根据传入的参数如Yolo5_CSP_Bottleneck2(512,512,[512,512],3,512)进行解析。

完成了相关模块的封装后，对主干特征提取网络进行实现，代码如下：

```
#第5章/ObjectDetection/TensorFlow_Yolo_V5_Detected/backbone/yolo_v5.py

def yolo5_csp_darknet(input_x):
    """DarkNet部分,也是backbone的部分"""
    x = ConvBNSiLU(64, 6, 2)(input_x)
    x = ConvBNSiLU(128, 3, 2)(x)
    x = Yolo5_CSP_Bottleneck1(64, 64, [64, 64], 3, 128)(x)
    p3 = ConvBNSiLU(256, 3, 2)(x)                          #P3

    neck1 = Yolo5_CSP_Bottleneck1(128, 128, [128, 128], 6, 256)(p3)
    p4 = ConvBNSiLU(512, 3, 2)(neck1)                      #P4

    neck2 = Yolo5_CSP_Bottleneck1(256, 256, [256, 256], 9, 512)(p4)
    p5 = ConvBNSiLU(1024, 3, 2)(neck2)                     #P5
    x = Yolo5_CSP_Bottleneck1(512, 512, [512, 512], 3, 1024)(p5)
    return x, neck2, neck1
```

函数yolo5_csp_darknet()实现了backbone的封装，每个C3-Bottle1结构之后经过ConvBNSiLU()得到结构图中的P3、P4、P5，而neck1、neck2、x会与FPN模块进行结合。对x进行SPPF，以便进行多尺度池化提取，代码如下：

```
#第5章/ObjectDetection/TensorFlow_Yolo_V5_Detected/utils/conv_utils.py

class Yolo5_SPPF(layers.Layer):
```

```
    #SPPF 模块
    def __init__(self, input_channel=512, out_channel=1024):
        super(Yolo5_SPPF, self).__init__()
        self.con_si = ConvBNSiLU(input_channel, 1, 1)
        #都是 5×5 的池化
        self.maxpool = layers.MaxPool2D(5, 1, padding='same')
        self.con_out = ConvBNSiLU(out_channel, 1, 1)

    def call(self, inputs, *args, **kwargs):
        inputs = self.con_si(inputs)  #输入
        pool1 = self.maxpool(inputs)
        pool2 = self.maxpool(pool1)
        pool3 = self.maxpool(pool2)
        #输入与 3 个 5×5 串联的池化 concat
        out1 = layers.concatenate([pool3, pool2, pool1, inputs], axis=-1)
        return self.con_out(out1)
```

Yolo5_SPPF()中对 inputs 进行 pool1，然后对 pool1 最大池化得到 pool2，对 poo2 最大池化得到 pool3，然后将 inputs 与 pool1、pool2、pool3 进行 concat 后由 ConvBNSiLU()输出，从而得到多尺度的特征池化特征信息。

将上面的模块按 YOLOv5 结构图组合得到 YOLOv5 的前向传播，代码如下：

```
#第 5 章/ObjectDetection/TensorFlow_Yolo_V5_Detected/backbone/yolo_v5.py

def yolo_v5(input_shape, anchors=None, class_num=80):
    """YOLOv5 的前向传播"""
    if anchors is None:
        anchors_mask = np.array([
            [[10, 13], [16, 30], [33, 23]],                 #小目标
            [[30, 61], [62, 45], [59, 119]],                #中目标
            [[116, 90], [156, 198], [373, 326]]             #大目标
        ])
    else:
        anchors_mask = anchors
    input_x = layers.Input(shape=input_shape, dtype='float32')
    #neck1 80 × 80 × 256
    #neck2 40 × 40 × 512
    #neck3 20 × 20 × 1024
    x, neck2, neck1 = yolo5_csp_darknet(input_x)
    #SPPF 实现多尺度池化
    neck3 = Yolo5_SPPF(512, 1024)(x)
    #输入 head
    head3 = ConvBNSiLU(512, 1, 1)(neck3)
    #2 倍上采样，这里用转置卷积代替 upsample
    up3 = layers.Conv2DTranspose(512, 2, 2, padding='same')(head3)
    #将 up3 与 neck2 进行 concat
    up3_neck2 = layers.concatenate([neck2, up3], axis=-1)
    #经过 csp2
    up3_neck2_bot1 = Yolo5_CSP_Bottleneck2(512, 512, [512, 512], 3, 512)(up3_neck2)
    #输入 head2
    head2 = ConvBNSiLU(256, 1, 1)(up3_neck2_bot1)
```

```
    up2_head2 = layers.Conv2DTranspose(256, 2, 2, padding='same')(head2)
    #将 up2 与 neck1 进行 concat
    up2_neck1 = layers.concatenate([neck1, up2_head2], axis=-1)

    #输入 head1
    pred1 = Yolo5_CSP_Bottleneck2(256, 256, [256, 256], 3, 256)(up2_neck1)

    #进行 PAN 下采样
    pred1_down1 = ConvBNSiLU(256, 3, 2)(pred1)
    pred1_down1_cat = layers.concatenate([pred1_down1, head2], axis=-1)
    pred2 = Yolo5_CSP_Bottleneck2(256, 256, [256, 256], 3, 512)(pred1_down1_cat)

    pred2_down2 = ConvBNSiLU(512, 3, 2)(pred2)
    pred2_down2_cat = layers.concatenate([pred2_down2, head3], axis=-1)
    pred3 = Yolo5_CSP_Bottleneck2(512, 512, [512, 512], 3, 1024)(pred2_down2_cat)

    #构建检测头
    out1 = layers.Conv2D((4 + 1 + class_num) * len(anchors_mask[0]), 1, 1)(pred1) #小
    out2 = layers.Conv2D((4 + 1 + class_num) * len(anchors_mask[1]), 1, 1)(pred2) #中
    out3 = layers.Conv2D((4 + 1 + class_num) * len(anchors_mask[2]), 1, 1)(pred3) #大

    return Model(inputs=input_x, outputs=[out3, out2, out1]) #大,中,小
```

代码中 layers. Conv2DTranspose()替换了 layers. UpSample(2),输出内容分别为out1、out2、out3。

5.9.3 代码实战建议框的生成

YOLOv5 中正样本选择分为两步,即先对真实 GT BOX 与 Anchor 进行宽高比的计算,当宽高比小于4时选择此 Anchor 作为正样本,同时还实现了单个真实 GT BOX 挑选多个锚点的操作,代码如下:

```
#第 5 章/ObjectDetection/TensorFlow_Yolo_V5_Detected/util/tools.py

def yolo_v5_true_encode_box(
        true_boxes,                     #GT BOX 的值
        anchors=None,                   #传入的 Anchor
        class_num=80,                   #num class
        input_size=(640, 640),          #输入尺度
        ratio=[32, 16, 8],              #特征图的尺寸比
        max_wh_threshold=4.0            #阈值
):
    """
    YOLOv5 不再采用 IOU 匹配,YOLOv5 采用基于宽高比例的匹配策略,GT 的宽和高与 Anchors
的宽和高对应相除得到 ratio1;
    Anchors 的宽和高与 GT 的宽和高对应相除得到 ratio2;
    取 ratio1 和 ratio2 的最大值作为最后的宽高比,该宽高比和设定的阈值(默认为 4)比较,
小于设定阈值的 Anchor 则为匹配到的 Anchor
    """
    if anchors is None:
```

```
    anchors = np.array([
        [[10, 13], [16, 30], [33, 23]],                         #小目标
        [[30, 61], [62, 45], [59, 119]],                        #中目标
        [[116, 90], [156, 198], [373, 326]]                     #大目标
    ]) / 255.
  #Anchor 的数量
  detected_num = anchors.shape[0]
  #存储结果
  grid_shape = []
  true_boxes2 = []
  true_boxes = np.expand_dims(true_boxes, axis=0)
  batch_size = true_boxes.shape[0]
  #初始 3 个检测头的矩阵
  for i in range(detected_num):
          sw,sh = input_size[0] //ratio[i],input_size[1] //ratio[i]
     grid = np.zeros(
        [batch_size, sw, sh, len(anchors[i]), 4 + 1 + class_num],
        dtype=np.float32)
     grid_shape.append(grid)
     true_boxes2.append(
        np.zeros(
           [batch_size, sw, sh, , len(anchors[0]), 4 + 1 + class_num],
           dtype=np.float32))
  grid = grid_shape
  #遍历每个 GT BOX
  for box_index in range(true_boxes.shape[1]):
     #box 信息
     box = true_boxes[:, box_index, :]
     box = np.squeeze(box, axis=0)                                  #降维
         #类别
     box_class = box[4].astype('int32')
     best_choice = {}
     for index in range(0, 3, 1):

        #box = (13 × x,13 × y, 13 × w, 13 × h) 换算成特征图上的值
        box2 = box[0:4] *np.array([
           grid[index].shape[-3], grid[index].shape[-3], grid[index].shape
[-3], grid[index].shape[-3]
        ])
        box_true = box2.copy()
        box2 = box2.copy()
        #GT BOX 本来所处的格子
        i = np.floor(box2[1]).astype('int')                        #y 轴
        j = np.floor(box2[0]).astype('int')                        #x 轴
        if i > grid[index].shape[-3] or j > grid[index].shape[-3]:
           continue
        #循环每个 Anchor
        for k, anchor in enumerate(anchors[2 - index]):
           box_maxes = box2[2:4]
           anchor_maxes = anchor

           #(1)计算每个 GT BOX 与对应的 Anchor Template 模板的高宽比例
```

```
            #(2)统计这些比例和它们倒数之间的最大值
            #这里可以理解成计算 GT BOX 和 Anchor Template 分别在宽度及高度方向的
            #最大差异(当相等时比例为 1,差异最小)
            #假设 r_wh_ratio = 1,则 r_wh_max 就是 max(1, 1),即为 1
            #假设 r_wh_ratio = 0.5, 则 r_wh_max 就是 max(0.5, 1/0.5)--> max(0.5,2)-->2
            #假设 r_wh_ratio = 0.25,则 r_wh_max 就是 max(0.25, 1/0.25)-->
            #max(0.5,4)-->4
            #假设 r_wh_ratio = 0.1, 则 r_wh_max 就是 max(0.1, 1/0.1)-->
            #max(0.1,10) --> 10
            #离得越近,比值越接近 1;离得越远,比值越大(正无穷)

            #每层的 Anchor 相比较
            r_wh_ratio = box_maxes / anchor_maxes #r_w, r_h = w_gt, h_gt / w_ach, h_ach
            r_wh_max = np.maximum(r_wh_ratio, 1 / r_wh_ratio)
            #(3)获取宽度方向与高度方向的最大差异之间的最大值
            r_max = np.max(r_wh_max)
            #如果最大的比例小于 4,则将 GT BOX 分配给该 Anchor Template 模板
            if r_max >= 4:
                continue
            ##################################
            #下一步需要计算 i, j, 2-index(第几组 Anchor),k(第几组的第几个 Anchor)
            #按规则获取偏移范围
            offset = get_near_points(box2[0], box2[1])
            #计算偏移位置,假设 offset 有 3 个偏移位置,则都认为是正样本
            for off_xy in offset:
                #原来的位置 + 偏移后的位置
                off_xy_value = np.array([j, i]) + np.array(off_xy)
                          #新的第 j,i 个格子
                new_j = int(off_xy_value[0])
                new_i = int(off_xy_value[1])

                if (new_j >= grid[index].shape[-3] or new_j <= 0) or (new_i >=
grid[index].shape[-3] or new_i <= 0):
                    continue
                #计算新的偏移
                adjusted_box = np.array([
                    box2[0] - new_j, #x 和 y 都是相对于 grid cell 的位置,左上角为
                                     #[0,0],右下角为[1,1]
                    box2[1] - new_i,
                    np.log(box2[2] / anchor[0]),
                     np.log(box2[3] / anchor[1])
                 ], dtype=np.float32)
                 #正样本赋值
                 grid[index][..., new_j, new_i, k, 0:4] = adjusted_box
                 grid[index][..., new_j, new_i, k, 4] = 1
                 #标签平滑处理
                 grid[index][..., new_j, new_i, k, 5 + box_class] = smooth_labels
(1, 0.1)
                 true_boxes2[index][..., j, i, k, 0:4] = box_true
    return true_boxes2, grid
```

```python
def get_near_points(x, y):
    """根据 cx 和 cy 计算新的锚点"""
    gx = x % 1
    gy = y % 1
    g = 0.5
    #先水平,再垂直
    if gx < g and gy < g:
        return [[0, 0], [-1, 0], [0, -1]]
    elif gx > g and gy > g:
        return [[0, 0], [1, 0], [0, 1]]
    elif gx > g and gy < g:
        return [[0, 0], [1, 0], [0, -1]]
    elif gx < g and gy > g:
        return [[0, 0], [-1, 0], [0, 1]]
    else:
        return [[0, 0]]
```

代码较长但是与 YOLOv3、YOLOv4 类似,重点看不同之处是将 GT BOX 与 Anchor 做宽高比,如图 5-80 所示。

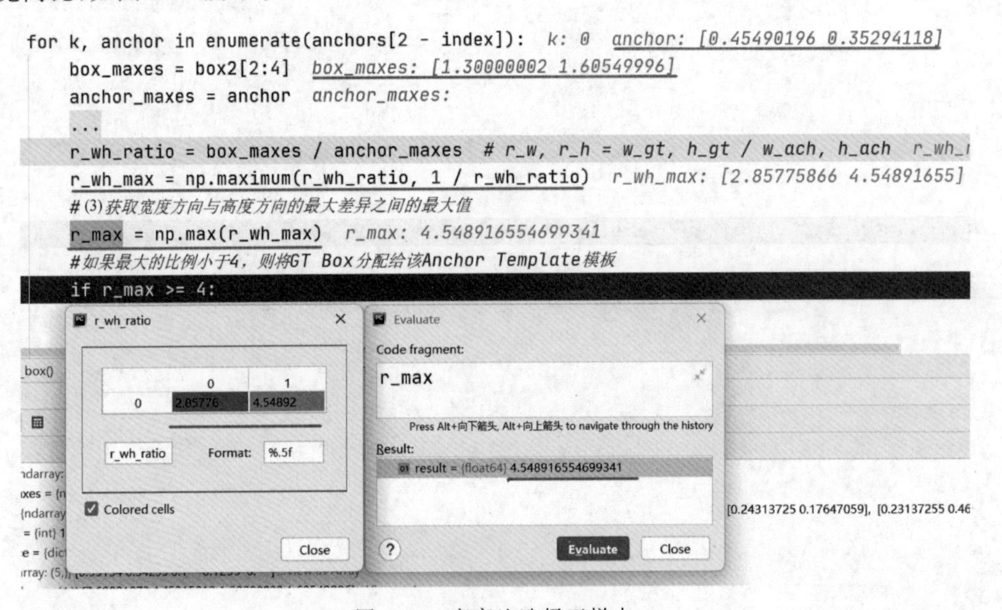

图 5-80 宽高比选择正样本

代码中 box_maxes 的真实 BOX 的宽和高为[1.3,1.6],此时选择的 Anchor 的宽和高为[0.45,0.35],box_maxes/anchor_maxes=[2.85,4.54],然后如果 np.max(r_wh_max)=4.54,宽高比>4 则不会选取,循环下 1 个 Anchor 的值。

多次循环后,当前 GT BOX 所在的格子为 i=4,j=7 且第 3 个检测头中的第 2 个 Anchor 此时 r_max<4,然后根据 cx=7.69 和 cy=4.45 在 get_near_points(box2[0],box2[1])中计算最近的 3 个锚点,如图 5-81 所示。

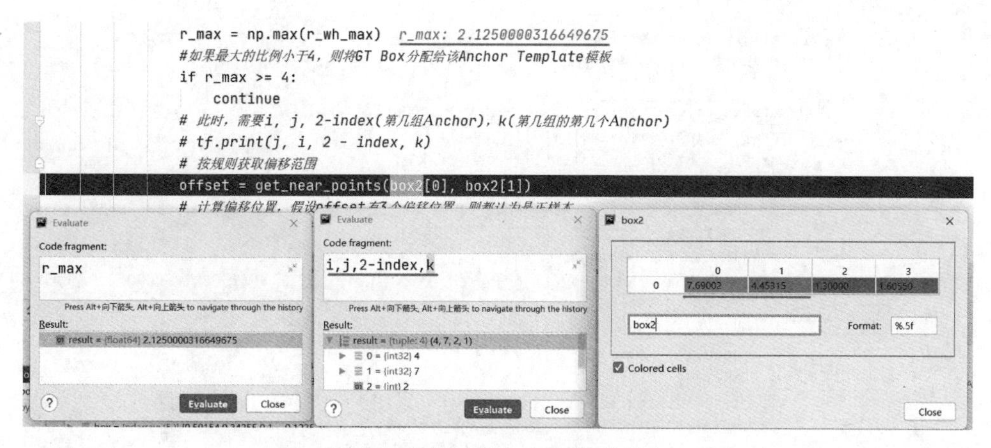

图 5-81　根据 GT BOX 选择锚点

因为 cx=7.69 和 cy=4.45，所以 gx=0.69 和 gy=0.45，其 GT BOX 的锚点在格子的右上方，所以新锚点格子需要[7,4]+[[0,0],[1,0],[0,−1]]，其锚点由原来的 1 个变成了 3 个，如图 5-82 所示。

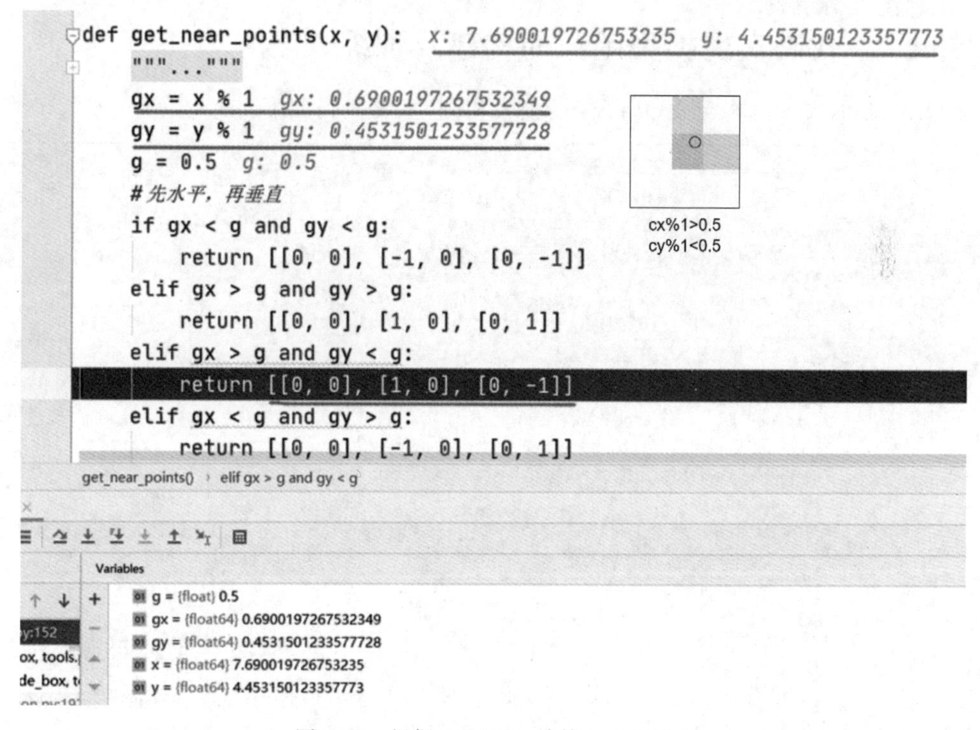

图 5-82　根据 GT BOX 计算 3 个新锚点

然后得到的新的格子为[[7,4],[8,4],[7,3]]，根据新的锚点计算偏移值，box2[0]−new_j、box2[1]−new_i，如图 5-83 所示。

```
for off_xy in offset:  off_xy: [0, -1]
    #原来的位置 + 偏移后的位置
    # 7, 4 | 8, 4 | 7, 3
    off_xy_value = np.array([j, i]) + np.array(off_xy)  off_xy_value: [7 3]
    new_j = int(off_xy_value[0])  new_j: 7
    new_i = int(off_xy_value[1])  new_i: 3
    # tf.print(new_i, new_j)
    if (new_j >= grid[index].shape[-3] or new_j <= 0) or (new_i >= grid[index].shape|
        continue
    # YOLOv4代码好像是用BOX来进行偏移的
    adjusted_box = np.array([  adjusted_box: [0.6900197 1.4531502 1.3000001 1.6055
        box2[0] - new_j,  # x,y都是相对于grid cell的位置，左上角为[0,0]，右下角为[1,1]
        box2[1] - new_i,
        box2[2],  #w和h取的是true值中的内容
        box2[3]
    ], dtype=np.float32)
    #tf.print(f'正样本:index={index},new_j={new_j},new_i={new_i},k={k}')
    grid[index][..., new_j, new_i, k, 0:4] = adjusted_box
    grid[index][..., new_j, new_i, k, 4] = 1
    #标签平滑处理
    grid[index][..., new_j, new_i, k, 5 + box_class] = smooth_labels(1, 0.1)
    true_boxes2[index][..., j, i, k, 0:4] = box_true
```

{ndarray: (4,)} [0.6900197 1.4531502 1.3000001 1.6055]

图 5-83 新锚点的正样本赋值

5.9.4 代码实现损失函数的构建及训练

YOLOv5 的损失基本与 YOLOv4 的损失相同，主要代码如下：

```
#第 5 章/ObjectDetection/TensorFlow_Yolo_V5_Detected/util/tools.py
class MultiAngleLoss(object):
    def yolo_v5_loss(self, y_true, y_pred, true_box2):
        """YOLOv5 的损失的整体计算过程跟 YOLOv4 区别不大"""
        #3 个检测头
        num_layers = self.num_layers
        loss = 0
        for index in range(num_layers):
            y_pre1 = y_pred[index]
            y_pre = tf.reshape(y_pre1, [
                -1, y_pre1.shape[1], y_pre1.shape[2],
                len(self.anchor[0]), 4 + 1 + self.num_class
            ])
            #真实值合并
            true_grid = tf.concat([t[index] for t in y_true], axis=0)
            #获取置信度信息
            object_mask = true_grid[..., 4:5]
            #对每个检测头进行解码,预测返回的是 xmin,ymin,xmax,ymax 位置信息、置信度
            #信息和分类信息
            pre_boxes, pre_box_confidence, pre_box_class_props = yolo_v5_head_decode(
                y_pre, 2 - index, anchor=self.anchor,
                num_class=self.num_class
            )
            #使用 CIOU 直接求 BOX 的 loss
            ciou_score = c_iou(pre_boxes, true_grid)
```

```
            ciou_loss = object_mask *tf.abs((1 - ciou_score))

            #确定正样本
            tobj = tf.where(
                tf.equal(object_mask, 1),
                tf.maximum(ciou_score, tf.zeros_like(ciou_score)),
                tf.zeros_like(ciou_score)
            )
            #置信度损失
            confidence_loss = K.binary_crossentropy(tobj, pre_box_confidence,
from_logits=True)
            #分类损失
            class_loss = object_mask *K.binary_crossentropy(
                true_grid[..., 5:], y_pre[..., 5:], from_logits=True
            )
            #正样本数量
            num_pos = tf.maximum(K.sum(K.cast(object_mask, tf.float32)), 1)
            #位置损失
            location_loss = tf.abs(K.sum(ciou_loss)) *self.box_ratio / num_pos
            #置信度损失
            confidence_loss = K.mean(confidence_loss) *self.balance[index] *
self.obj_ratio
            #分类损失
            class_loss = K.sum(class_loss) *self.cls_ratio / num_pos / self.num_class
            loss += location_loss + confidence_loss + class_loss
        return loss
```

代码中没有求负样本的置信度损失，而是使用 K. binary_crossentropy(tobj,pre_box_confidence)求正样本的置信度损失，位置损失使用 ciou_loss 进行求解。总损失仍然是位置损失、置信度损失、分类损失之和 location_loss＋confidence_loss＋class_loss。

与 YOLOv4 实现不同的是在解码函数 yolo_v5_head_decode()中对 Grid 敏感度进行了限制，代码如下：

```
#第 5 章/ObjectDetection/TensorFlow_Yolo_V5_Detected/util/tools.py
def yolo_v5_head_decode(features, index, anchor=None, num_class=80, scale_x_y=2):
    """对预测出来的值进行解码,得到 xmin,ymin,w,h"""
    #features 1 × 13 × 13 × 3 × 85
    if anchor is None:
        anchor = np.array([
            [[10, 13], [16, 30], [33, 23]],                    #小目标
            [[30, 61], [62, 45], [59, 119]],                   #中目标
            [[116, 90], [156, 198], [373, 326]]                #大目标
        ])/255.
    anchor_size = len(anchor)
    features = tf.reshape(features, [-1, features.shape[1], features.shape[2],
len(anchor[0]), 4 + 1 + num_class])
    conv_height, conv_width = features.shape[-4], features.shape[-3]
    #2 ×sigmoid(xy) - 0.5,值域就变成了-0.5~1.5
    xy_offset = scale_x_y *tf.nn.sigmoid(features[..., 0:2]) - 0.5 *(scale_x_y - 1)
    #2 × pow(sigmoid(wh),2),限定值域
```

```
    wh_offset = tf.square(scale_x_y *tf.sigmoid(features[..., 2:4]))
    #置信度
    box_confidence = tf.sigmoid(features[..., 4:5])
    #
    box_class_props = tf.nn.sigmoid(features[..., 5:])
    #在 feature 上面生成 anchors
    height_index = tf.range(conv_height, dtype=tf.float32)
    width_index = tf.range(conv_width, dtype=tf.float32)
    #得到网格
    x_cell, y_cell = tf.meshgrid(height_index, width_index)
    x_cell = tf.reshape(x_cell, [x_cell.shape[0], x_cell.shape[1], 1])
    y_cell = tf.reshape(y_cell, [x_cell.shape[0], x_cell.shape[1], 1])
    #根据网格求坐标位置,套公式
    bbox_x = (x_cell + xy_offset[..., 0]) / conv_height
    bbox_y = (y_cell + xy_offset[..., 1]) / conv_width
    bbox_w = (anchor[index][:, 0] *wh_offset[..., 0]) / conv_height
    bbox_h = (anchor[index][:, 1] *wh_offset[..., 1]) / conv_width
    boxes = tf.stack(
        [
            bbox_x,
            bbox_y,
            bbox_w,
            bbox_h
        ],
        axis=3
    )
    boxes = tf.reshape(boxes, [boxes.shape[0], boxes.shape[1], boxes.shape[2],
anchor_size, 4])
    return boxes, box_confidence, box_class_props
```

在代码 xy_offset 和 wh_offset 中根据公式使值域为[−0.5,1.5],通过 bbox_x、bbox_y、bbox_w 和 bbox_h 解码得到预测值。

5.9.5 代码实战预测推理

YOLOv5 的预测推理流程仍然与 YOLOv3 一致,不同之处是在将预测值解码时的公式变为公式 5-18,所以在 YOLOv3 的基础上应对解码函数_decode(self)进行修改,修改后的代码如下:

```
def _decode(self):
        #根据当前预测 layer 对预测出来的结果解码
        #YOLOv3 的输出(4+1+num_class) ×3 个 Anchor
        self.b = self.output[self.currentLayer].shape[0]
        #变成[b,3,(4+1+2),169)
        logits = np.reshape(self.output[self.currentLayer], [self.b, 3, -1,
self.h *self.w])
        self.result = np.zeros(logits.shape)
        #套用公式进行解码,使用 Sigmoid()函数对值域进行限制
        self.result[:, :, 0, :] = ((2 *tf.nn.sigmoid(logits[:, :, 0, :]) - 0.5 *
(2 - 1)).NumPy() + self.lin_x) / self.w
```

```
        self.result[:, :, 1, :] = ((2 *tf.nn.sigmoid(logits[:, :, 1, :]) - 0.5 *
(2 - 1)).NumPy() + self.lin_y) / self.h
        self.result[:, :, 2, :] = (tf.square(2 *tf.sigmoid(logits[:, :, 2, :]))).
NumPy() *self.anchor_w) / self.w
        self.result[:, :, 3, :] = (tf.square(2 *tf.sigmoid(logits[:, :, 3, :]))).
NumPy() *self.anchor_h) / self.h
        self.result[:, :, 4, :] = tf.sigmoid(logits[:, :, 4, :]).NumPy()
        self.result[:, :, 5:, :] = tf.nn.softmax(logits[:, :, 5:, :]).NumPy()
```

总结

YOLOv5 没有发表论文(截至本节撰写时),其主要结构与 YOLOv4 类似。在代码中主要使用了 GT 和 Anchor 宽高比小于 4,并且由某个 GT 中心点附近的 3 个格子进行预测,极大地提高了正样本的数量。

练习

运行并调试本节代码,理解算法的设计与代码的结合,重点梳理本算法的实现方法。

5.10 单阶段速度快多检测头网络 YOLOv7

5.10.1 模型介绍

YOLOv7 于 2022 年由 Chien-Yao Wang 等在发表的论文 *YOLOv7: Trainable bag-of-freebies sets new state-of-the-art for real-time object detectors* 中提出。相对于 YOLOv5,主要引入了 ELAN、MP、SPPCSPC、REPConv 等模块,正负样本匹配时使用了 Better simOTA 的计算方法,并在训练时增加了 Aux Head 检测,其网络结构如图 5-84 所示。

图像的输入尺寸设定为 640×640,进行两次 CBS,其中 CBS 分别为卷积、BN、SiLU 激活函数。第 1 个 CBS 的步长为 1,改变通道数,第 2 个 CBS 的步长为 2,进行下采样,重复后得到 160×160×128 的特征图。

ELAN 模块是一个高效的网络结构,它通过控制最短和最长的梯度路径,使网络能够学习到更多的特征,并且具有更强的稳健性。ELAN 有两个分支,第 1 个分支经过 1×1 卷积做通道数的变化;第 2 个分支首先经过 1×1 卷积模块做通道数的变化,接着经过 4 个 3×3 的卷积做特征提取,然后将 4 个特征叠加在一起再经过 CBS 模块输出特征。ELAN-W 模块与 ELAN 模块类似,只是融合的特征层多了两个分支。

MP 模块进一步做下采样,由两个分支构成。第 1 个分支先经过最大池化后接 1×1 卷积以改变通道数。第 2 个分支经过 1×1 卷积做通道数的变化后再经过步长为 2 的 3×3 卷积也进行下采样,然后将两个分支进行 Concat 得到不同尺度,以及不同特征值的下采样特征图。MP1 在 backbone 中输入通道数减半,而 head 时 MP2 的通道数不变。

在 YOLOv4 中 SPP 的作用是能够增大感受野,使算法适应不同分辨率的图像,获取不同的感受野。在 SPPCSPC 中首先将特征分为两部分,第 1 个分支经过 5、9、13 的最大池

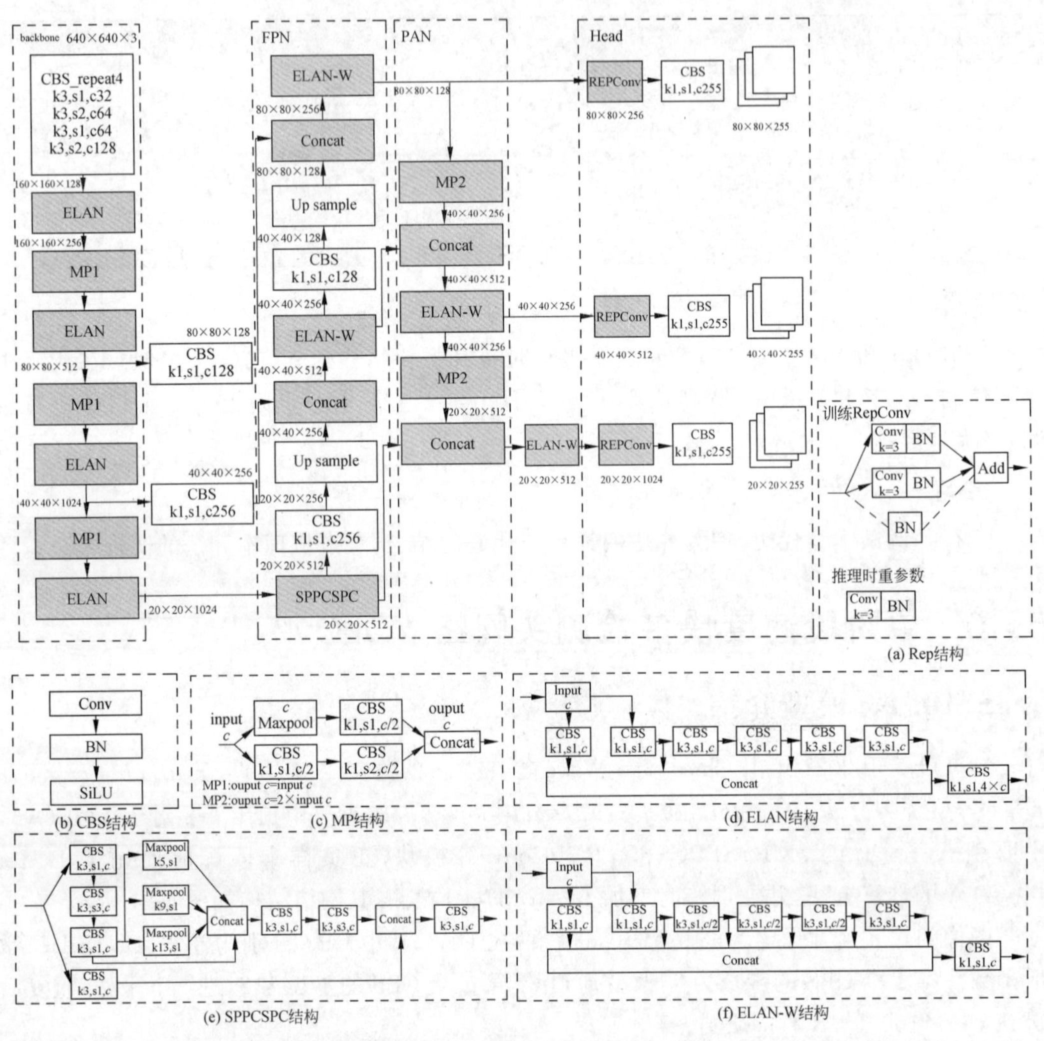

图 5-84 YOLOv7 的结构

化,并与 s=1 的 CBS 模块进行 Concat,使该结构能够处理不同尺度的感受野,利于区分大目标和小目标。另 1 个分支进行常规处理,并与经过最大池化的分支合并,使该结构能够减少计算量,从而使精度得到提升。

模块重参数 RepConv 在训时将一个整体模块分割成多个不同的模块分支,而在推理过程将多个分支模块集成到一个完全等价的模块中,在保证精度的条件下使推理效率更高。

正负样本匹配继承了 YOLOv5 中的宽高比的匹配方法,并根据标注框 GT BOX 的中心位置获取临近的两个网格并作为正样本,即 1 个 GT BOX 由 3 个网格来预测。默认 Anchor 有 9 个,1 个 GT BOX 如果与 Anchor 的宽高比都小于 4,则一共有 3×9=27 个正样本,并且此时正样本将分布在 3 个检测头中。有可能存在一个 GT BOX 对应多个正样本的情况，而且一个正样本有可能对应多个 GT BOX。这种多 Anchor、多 Grid 网格、多检测头 Head 匹配的方法是 YOLOv7 中的初选，如图 5-85 所示。

GT
Anchor
GT BOX
中心点

图 5-85　YOLOv7 正样本选择

然后进入复选 OTA 算法，动态地确定每个 GT BOX 真正需要的正样本数量。首先计算 GT BOX 与预测框的 IOU，然后根据 IOU 的得分从大到小进行排序，取 Top 10 个 IOU 的得分进行求和，求和所得值设为当前动态正样本数 K（K 最小取 1）。这个 K 值就是 GT BOX 所要选取的正样本的数量。

然后计算每个 GT BOX 中的置信度与预测结果中的置信度×分类的交叉熵损失 pair_wise_cls_loss，再将初选框正样本的回归 Reg IOU Loss 加上 pair_wise_cls_loss，并为每个 GT BOX 取 loss 最小的前 K 个样本作为正样本。

因为一个 GT BOX 可以有多个正样本，一个正样本应该只能对应一个 GT BOX，所以如果一个正样本对应多个 GT BOX，就能找到它跟多个 GT BOX 的损失值，用最小的那个损失所在的正样本进行预测。这个过程可以称为精选，更详细的代码实现过程可参考后文。

YOLOv7 论文中使用一辅助头 Aux Head 检测，需要注意的是正样本在粗选时，辅助头中每个网络与 GT BOX 匹配后选择附近的 4 个网格，而 Lead Head 是两个。在精选时 Aux Head 选择 Top 20 个进行 GT BOX 与初选正样本的求和，而 Lead Head 是 10 个，然后在平衡 Aux Head loss 和 Lead Head loss 时，需要按照 0.25：1 的比例进行，否则会导致 Lead Head 精度变低，如图 5-86 所示。

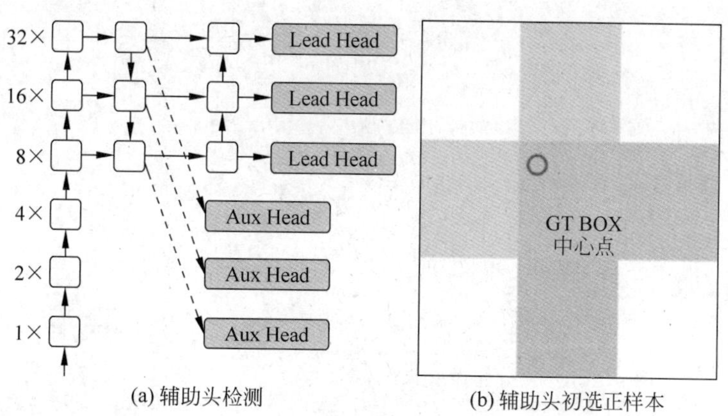

(a) 辅助头检测　(b) 辅助头初选正样本

图 5-86　辅助头检测

在损失函数方面没有大的变化，位置损失使用 CIOU，置信度和分类损失使用交叉熵，不同的检测头有不同的权重比例。

5.10.2　代码实战模型搭建

根据模型结构图 5-84 使用 PyTorch 生成结构 ELAN，代码如下：

```
#第 5 章/ObjectDetection/Pytorch_Yolo_V7_Detected/conv_utils.py
def autopad(k, p=None):
    #根据步长自动补 0
    #Pad to 'same'
    if p is None:
        p = k //2 if isinstance(k, int) else [x //2 for x in k] #auto-pad
    return p
class CBS(nn.Module):
    #标准卷积模块
    #输入 cannel,输出 channel,核长,步长,padding 数量,groups 分组卷积数量
    def __init__(self, c1, c2, k=1, s=1, p=None, g=1, act=True):
        super(CBS, self).__init__()
        #autopad(k,p)根据卷积核自动补 0
        self.conv = nn.Conv2d(c1, c2, k, s, autopad(k, p), groups=g, bias=False)
        self.bn = nn.BatchNorm2d(c2)
        self.act = nn.SiLU() if act is True else (act if isinstance(act, nn.Module)
else nn.Identity())

    def forward(self, x):
        return self.act(self.bn(self.conv(x)))

    def fuseforward(self, x):
        return self.act(self.conv(x))

class ELAN(nn.Module):
    #ELAN 模块
    def __init__(self, c1, c2):
        #C1 为输入通道,C2 为输出通道
        super(ELAN, self).__init__()
        #CBS 为 Conv->BN->SiLU 的封装
        #c1、c2、1、1 分别为输入、输出、kernelSize、strides
        self.cbs1 = CBS(c1, c2, 1, 1)
        self.cbs2 = CBS(c1, c2, 1, 1)
        #cbs3->cbs6 输入与输出通道相同
        self.cbs3 = CBS(c2, c2, 3, 1)
        self.cbs4 = CBS(c2, c2, 3, 1)
        self.cbs5 = CBS(c2, c2, 3, 1)
        self.cbs6 = CBS(c2, c2, 3, 1)
        #因为会将 out1,out2,out3,out4 合并,所以输入/输出通道是 4*c2
        self.cbs7 = CBS(4 *c2, c2 *4, 1, 1)

    def forward(self, x):
        out1 = self.cbs1(x) #假设 x 为 128×160×160,则经 cbs1 后输出为 64×160×160
        out2 = self.cbs2(x) #cbs2 为另一个分支,输出为 64×160×160
        out3 = self.cbs4(self.cbs3(out2)) #out2 进行 cbs3 和 cbs4
        out4 = self.cbs6(self.cbs5(out3)) #out3 进行 cbs5 和 cbs6
        #合并后使用 cbs7
        return self.cbs7(torch.cat([out1, out2, out3, out4], dim=1))
```

PyTorch 在创建子模块时继承了 nn.Module,然后在__init__()中描述算子的属性,在 forward()中实现前向传播。ELAN 模块在 cbs3、cbs4、cbs5、cbs6 时,输入与输出的通道相同。在 forward()中将 out1、out2、out3、out4 进行 cat 操作,通道数变为 4 倍,所以 cbs7 的

输出为 4＊c2。另外，PyTorch 的输入 shape 为[batch_size，channel，height，width]，所以在 cat 操作时 dim＝1（在 1 轴），而 TensorFlow 的输入 shape 为[batch_size，height，width，channel]，所以在 cat 操作时 axis＝－1。

ELAN-W 模块与 ELAN 模块类似，将 out1、out2、out3、out4、out5、out6 的输出特征进行了融合，cbs3、cbs4、cbs5、cbs6 是输入通道数的一半，详细的代码如下：

```
#第 5 章/ObjectDetection/Pytorch_Yolo_V7_Detected/conv_utils.py
class ELAN_W(nn.Module):
    #ELAN_W 模块
    def __init__(self, c1, c2):
        super(ELAN_W, self).__init__()
        self.cbs1 = CBS(c1, c2, 1, 1)
        self.cbs2 = CBS(c1, c2, 1, 1)
        #输出是 c2//2 的一半
        self.cbs3 = CBS(c2, c2 //2, 3, 1)
        self.cbs4 = CBS(c2 //2, c2 //2, 3, 1)
        self.cbs5 = CBS(c2 //2, c2 //2, 3, 1)
        self.cbs6 = CBS(c2 //2, c2 //2, 3, 1)
        self.cbs7 = CBS(4 *c2, c2, 1, 1)

    def forward(self, x):
        out1 = self.cbs1(x) #假设 x 为 512 × 40 × 40,则经 cbs1 后输出为 512 × 40 × 40
        out2 = self.cbs2(x)
        out3 = self.cbs3(out2) #256 × 40 × 40
        out4 = self.cbs4(out3)
        out5 = self.cbs5(out4)
        out6 = self.cbs6(out5)
        #将每个 cbs 的输出都进行了合并
        return self.cbs7(torch.cat([out1, out2, out3, out4, out5, out6], dim=1))
```

接下来对 MP 模块进行实现，代码如下：

```
#第 5 章/ObjectDetection/Pytorch_Yolo_V7_Detected/conv_utils.py
class MP(nn.Module):
    #MP 模块
    def __init__(self, c1, c2, k=2, p=None):
        #C1 为输入通道,C2 为输出通道
        super(MP, self).__init__()
        self.k = k
        self.pool1_1 = nn.MaxPool2d(self.k, self.k)
        #输入与输出通道相同
        self.cbs1_2 = CBS(c1, c2, 1, 1)
        self.cbs2_1 = CBS(c1, c2, 1, 1)
        self.cbs2_2 = CBS(c2, c2, 3, 2)

    def forward(self, x):
        #先池化后 cbs
        x1 = self.cbs1_2(self.pool1_1(x)) #128 × 80 × 80
        #另一个分支进行两次 cbs
        x2 = self.cbs2_2(self.cbs2_1(x)) #128 × 80 × 80
        #池化和 cbs 进行融合
        return torch.cat([x1, x2], dim=1)
```

当采用 MP(2 * c1,c1)时,即 c1 是 c2 的两倍时为 MP1 结构；当采用 MP(c1,c1)时,即 c1==c2 时为 MP2 结构。

对 SPPCSPC 模块进行封装,代码如下：

```
#第 5 章/ObjectDetection/Pytorch_Yolo_V7_Detected/conv_utils.py
class SPPCSPC(nn.Module):
    #SPPCSPC 模块
    def __init__(self, c1, c2, e=0.5, k=(5, 9, 13)):
        #c1 为输入,c2 为输出,e 为通道扩展倍数
        #k=(5, 9, 13)表明进行 SPP 时尺寸的变化
        super(SPPCSPC, self).__init__()
        c_ = int(2 *c2 *e) #输出 channel
        self.cv1 = CBS(c1, c_, 1, 1)
        self.cv2 = CBS(c1, c_, 1, 1)
        self.cv3 = CBS(c_, c_, 3, 1)
        self.cv4 = CBS(c_, c_, 1, 1)
        #对 k=(5, 9, 13)进行 3 个尺度的池化
        self.m = nn.ModuleList([nn.MaxPool2d(kernel_size=x, stride=1, padding=
x //2) for x in k])
        #因为 cv5 是对 cv4 和 m 的 cat,所以输入通道扩展了 4 倍
        self.cv5 = CBS(4 *c_, c_, 1, 1)
        #对 cv5 进行卷积,所以输入、输出又降为 c
        self.cv6 = CBS(c_, c_, 3, 1)
        #对 cv2 和 cv6 进行 cat,所以输入变为 2c
        self.cv7 = CBS(2 *c_, c2, 1, 1)

    def forward(self, x):
        #先进行串列卷积
        x1 = self.cv4(self.cv3(self.cv1(x)))
        #不同尺度池化,进行 cv5、cv6 的卷积
        y1 = self.cv6(self.cv5(torch.cat([x1] + [m(x1) for m in self.m], 1)))
        #对原输入 x 进行卷积
        y2 = self.cv2(x)
        #合并后卷积
        return self.cv7(torch.cat((y1, y2), dim=1))
```

SPPCSPC 模块对于输入 x 进行 3 次串列 cv1、cv3、cv4 的输出,从而得到 x1,然后对 x1 进行 5、9、13 的不同尺度池化 cat,从而得到 x2,将 x1 和 x2 合并进行两次串联 cv5、cv6,从而得到 y1。另 1 个分支对于 x 进行 1 次 cv2,从而得到 y2,将 y1 和 y2 进行 cat 后 cv7 得到最终的输出。该结构能够处理不同尺度的感受野,利于区分大目标和小目标。RepConv 模块可参考 4.12 节重参数网络 RepVGGNet 的实现。

将 CBS、ELAN、MP、ELAN-W、SPPCSPC 模块按模型图进行组装就能实现 YOLOv7 的前向传播,代码如下：

```
#第 5 章/ObjectDetection/Pytorch_Yolo_V7_Detected/yolo7.py
class Yolo7(nn.Module):
    #YOLOv7 模型的实现
    def __init__(self, num_class=80, layer_num_anchors=3):
        #num_class:预测类别数
```

```
        #layer_num_anchors:锚框数量
        super(Yolo7, self).__init__()
        self.num_class = num_class
        ####bakcbone
        #重复 4 次 cbs,即结构图中的 cbs_repeat4
        self.cbs1 = CBS(3, 32, 3, 1)
        self.cbs2 = CBS(32, 64, 3, 2)
        self.cbs3 = CBS(64, 64, 3, 1)
        self.cbs4 = CBS(64, 128, 3, 2)
        #第 1 个 ELAN->MP,输出是 b×(64×4) ×160×160
        #因为 ELAN 会对 out1、out2、out3 和 out4 进行合并,所以输出是 64×4
        self.elan_b1 = ELAN(128, 64)
        #mp1,输出是 b×(128×2) ×80×80,因为 MP 会对 x1 和 x2 合并,所以输出是 128×2
        self.mp1_b1 = MP(64 *4, 128)
        #第 2 个 ELAN->MP
        self.elan_b2 = ELAN(128 *2, 128)
        self.mp1_b2 = MP(128 *4, 256)                    #out b×(256×2) ×40 × 40
        #第 3 个 ELAN->MP
        self.elan_b3 = ELAN(256 *2, 256)
        self.mp1_b3 = MP(256 *4, 512)                    #out b×(512×2) ×20 × 20
        #ELAN->SPPCSP
        self.elan_b4 = ELAN(512 *2, 256)                 #out b×(256×4) ×20 × 20
        self.sppcsp = SPPCSPC(512 *2, 512)               #out b×512×20 × 20

        ####FPN 结构,cbs->upsample->cat->ELAN_W...
        self.cbs_f1 = CBS(512, 256, 1, 1)
        self.up_f1 = nn.Upsample(scale_factor=2)
        self.cbs_route2 = CBS(1024, 256, 1, 1)
        self.cbs_route1 = CBS(512, 128, 1, 1)
        self.elan_w_f1 = ELAN_W(512, 256)
        self.cbs_f2 = CBS(256, 128, 1, 1)
        self.up_f2 = nn.Upsample(scale_factor=2)
        self.elan_w_f2 = ELAN_W(256, 128)
        ####PAN 结构下采样
        self.mp2_p1 = MP(128, 128)
        self.elan_w_p1 = ELAN_W(512, 256)
        self.mp2_p2 = MP(256, 256)
        self.elan_w_p2 = ELAN_W(1024, 512)
        ####head 输出头 4 代表位置,1 为置信度,num_class 为分类的概率,每个检测头有 3 个锚框
        out_c = (4 + 1 + num_class) *layer_num_anchors
        #第 1 个检测头,进行重参数化 255 × 80 ×80
        self.repcov1 = RepConv(128, 256)
        #预测输出
        self.cbs_out1 = CBS(256, out_c, act=False)
        #第 2 个检测头,进行重参数化,预测输出 255 × 40 ×40
        self.repcov2 = RepConv(256, 512)
        self.cbs_out2 = CBS(512, out_c)
        #第 3 个检测头,进行重参数化,预测输出 255 × 20 ×20
        self.repcov3 = RepConv(512, 1024)
        self.cbs_out3 = CBS(1024, out_c)

    def forward(self, x):
```

```
        #cbs repeat4->elan->mp1
        x = self.cbs4(self.cbs3(self.cbs2(self.cbs1(x)))) #1×128×160×160
        x = self.mp1_b1(self.elan_b1(x))                  #256 × 80 × 80

        #route1 会经过 cbs 进入 FPN 结构进行 cat
        route1 = self.elan_b2(x)                          #512 × 80 × 80
        x = self.mp1_b2(route1)                           #512 × 40 × 40

        #route2 会经过 cbs 进入 FPN 结构进行 cat
        route2 = self.elan_b3(x)                          #1024 × 40 × 40
        x = self.mp1_b3(route2)                           #1024 × 20 × 20

        #route3 会经过 cbs 进入 FPN 结构进行 cat
        x = self.elan_b4(x)                               #1024 × 20 × 20
        route3 = self.sppcsp(x)                           #512 × 20 × 20

        #FPN 上采样的构建 512 × 40 × 40
        #上采样,并实现浅深层特征合并,然后 ELAN-W
        x = torch.concat([self.up_f1(self.cbs_f1(route3)), self.cbs_route2
(route2)], dim=1)
        #这一层特征会进入 PAN 进行特征合并
        fpn_route1 = self.elan_w_f1(x)                    #256 × 40 × 40

        #继续上采样,并实现浅深层特征合并,然后 ELAN-W
        x = torch.concat([self.up_f2(self.cbs_f2(fpn_route1)), self.cbs_route1
(route1)], dim=1)                                         #256 × 80 × 80
        #上采样,并实现浅深层特征合并,然后 ELAN-W,并作为第 1 个检测层来检测小目标
        out1 = self.elan_w_f2(x) #128 × 80 × 80

        #PAN 下采样的构建 256 × 40 × 40
        #MP->CAT-ELAN-W 作为第 2 个检测头,检测中目标
        out2 = self.elan_w_p1(torch.concat([self.mp2_p1(out1), fpn_route1], dim=1))
        #继承 MP->CAT-ELAN-W 作为第 3 个检测头,检测大目标
        out3 = self.elan_w_p2(torch.concat([self.mp2_p2(out2), route3], dim=1))
                                                          #512 × 20 × 20

        #根据 out1、out2 和 out3 输出进行重参数化和检测
        p1 = self.cbs_out1(self.repcov1(out1))            #255 × 80 ×80
        p1_shape = p1.shape
        #维度(batch,3,height,width, (4 + 1 + self.num_class))
        #表示每个特征图有 3 个锚框,每个锚框有 4 个位置、1 个置信度和 num_class 个类别
        p1 = p1.view(p1_shape[0], 3, p1_shape[2], p1_shape[3], (4 + 1 + self.num_class))

        p2 = self.cbs_out2(self.repcov2(out2))            #255 × 40 ×40
        p2_shape = p2.shape
        p2 = p2.view(p2_shape[0], 3, p2_shape[2], p2_shape[3], (4 + 1 + self.num_class))

        p3 = self.cbs_out3(self.repcov3(out3))            #255 *20 *20
        p3_shape = p3.shape
        p3 = p3.view(p3_shape[0], 3, p3_shape[2], p3_shape[3], (4 + 1 + self.num_class))
        #返回 3 个输出
        return p1, p2, p3
```

在 YOLOv7 类中对模型进行了实现，__init__()为每个结构类初始化，在 forward()中进行了实现。提取 Backbone 特征，输出 route1、route2、route3，然后在 FPN 中对 route3 进行上采样，并与 self.cbs_route2(route2)进行特征合并，通过 self.elan_w_f1 输出 fpn_route1，并对 fpn_route1 继续上采样，然后与 self.cbs_route1(route1)进行特征合并，经过 self.elan_w_f2()后得到 out1。

在 out1 之后就是 PAN 结构，通过 self.mp2_p1(out1)进行下采样，并跟 fpn_route1 进行特征合并，然后经过 self.elan_w_p1 作为 out2；通过 self.mp2_p2(out2)进行下采样，并与 route3 进行特征合并，通过 self.elan_w_pw()输出 out3；在 out1、out2、out3 之后经过 self.repcov()重参数和 self.cbs_out 卷积，输出 p1、p2、p3 并通过 view(p1_shape[0],3,p1_shape[2],p1_shape[3],(4+1+self.num_class))方法输出(batch,3,height,width,(4+1+self.num_class))，表示每个特征图有 3 个锚框，每个锚框有 4 个位置、1 个置信度和 num_class 个类别。

5.10.3　代码实战建议框的生成

YOLOv7 的正样本提取参考了 YOLOv5 中的方法进行初选，然后由 OTA 算法进行精选，其代码实现过程较复杂，在此参考 YOLOv7 官方代码进行实现。

训练标签数据的文件格式的内容如下：

```
文件路径 xmin,ymin,xmax,ymax,label xmin,ymin,xmax,ymax,label
./face_train/19_Couple_Couple_19_461.jpg 441,432,525,528,0 592,435,676,546,0
```

PyTorch 读取自定义标签文件需要继承 Dataset 类，并在__init__()中进行属性的初始化，在__len__()中返回标签的数量，在__getitem__()中实现按 batch size 生成指定的内容，详细的代码如下：

```python
#第 5 章/ObjectDetection/Pytorch_Yolo_V7_Detected/dataloader.py
class FaceDataSet(Dataset):
    def __init__(self, data_root, transform=None, size=(640, 640)):
        """
        自定义读取数据格式初始化
        :param data_dir: 路径
        :param transform: 数据预处理
        """
        self.transform = transform
        self.data_root = data_root
        self.img_size = size
        #获取训练图片和 label 信息
        self.data_info = self.get_data()

    def __len__(self):
        #类返回数据容量
        return len(self.data_info)

    def __getitem__(self, item):
```

```
        #从所有数据中获取指定长度的内容，只能返回维度相等的内容
        #所以需要指定 collate_fn 进行合并拼接
        datas = self.data_info[item]
        img, true_box, path = self.get_data_rows(datas)
        if self.transform is not None:
            #数据格式转换等
            img = self.transform(img)
        return img, true_box, path

    @staticmethod
    def collate_fn(batch):
        #根据指定 batch 返回 img 和 boxes 的信息
        img, true_box, path = zip(*batch)
        #将 true_box [n,6]中的第 1 位标明是哪一张图片
        for i, l in enumerate(true_box):
            l[..., 0] = i
        img = torch.stack(img, 0)
        true_box = torch.cat(true_box, 0)
        return img, true_box, path

    def get_data(self):
        with open(self.data_root, encoding='utf-8') as f:
            rows = f.readlines()
            return rows

    def get_data_rows(self, row):
        #存储 image path 和 boxes
        #按格式进行解析
        #./face_train/19_Couple_Couple_19_461.jpg 441,432,525,528,0 592,435,
        #676,546,0
        #解析后为['./face_train/19_Couple_Couple_19_461.jpg', '441,432,525,528,0',
        #'592,435,676,546,0']
        data = row.strip().split(' ')
        path = data[0]
        #读图片和 boxes
        img = cv2.imread(path)
        true_box = np.array([np.array(list(map(int, box1.split(',')))) for box1
in data[1:]])
        #转换为指定大小的图片和 boxes 信息
        img, true_box = u.letterbox_image(
            Image.fromarray(np.uint8(img.copy())),
            self.img_size, true_box
        )
        #print(true_box)
        #将 true_box 由 xmin,ymin,xmax,ymax 转换为 cx,cy,w,h
        true_box[..., :4] = u.xyxy2cxcywh(true_box)
        #[len,6]是为了在 collate_fn 中的第 1 个位置标明是第几张图片
        true_box_out = torch.zeros([len(true_box), 6])
        true_box = torch.from_numpy(true_box)
        if true_box[0] != 0:
            true_box_out[..., 1] = true_box[..., -1]
            true_box_out[..., 2:6] = true_box[..., 0:4] / self.img_size[0]
```

```
        return torch.tensor(img / 255., dtype=torch.float32).transpose(0, -1),
true_box_out, path
if __name__ == "__main__":
    data_root = r"../2021_train_yolo.txt"
    t = FaceDataSet(data_root)
    train_loader = DataLoader(
        dataset=t,
        batch_size=4,
        shuffle=True,
        collate_fn=FaceDataSet.collate_fn
    )
    #每次得到的是 batch size 的矩阵。在 data 中应保留 inputs 和 labels
    for i, (img, truebox, path) in enumerate(train_loader):
        #truebox 返回的是[4,6],第 1 位表明是第几张图片
        inputs, labels = img, truebox
```

在自定义 FaceDataSet 类的__init__()中 self.data_info 属性用于返回所有训练标签文件,如图 5-87 所示。

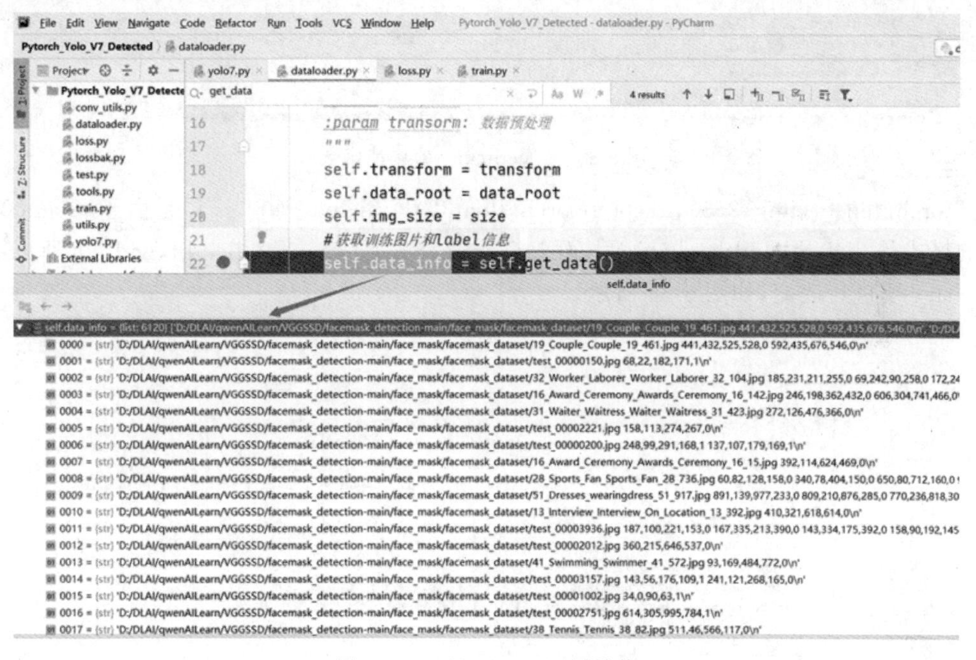

图 5-87 self.data_info 属性值

在__getitem__()中根据随机的 item 从 self.get_data_row 中获取数据,datas 为标签文件中的第 3902 行数据,如图 5-88 所示。

在 get_data_rows(row)中根据传入的标签内容,先对标签的数据进行清洗,然后经过 letterbox_image()对图片和 GT BOX 进行等比例缩放,并将输入的标签格式转换成 cx,cy,w,h,然后使用 true_box_out 矩阵的第 1 位标明当前图片是 batch size 中的第几张图片,使输出格式的矩阵变为[batch index,label,cx,cy,w,h]。

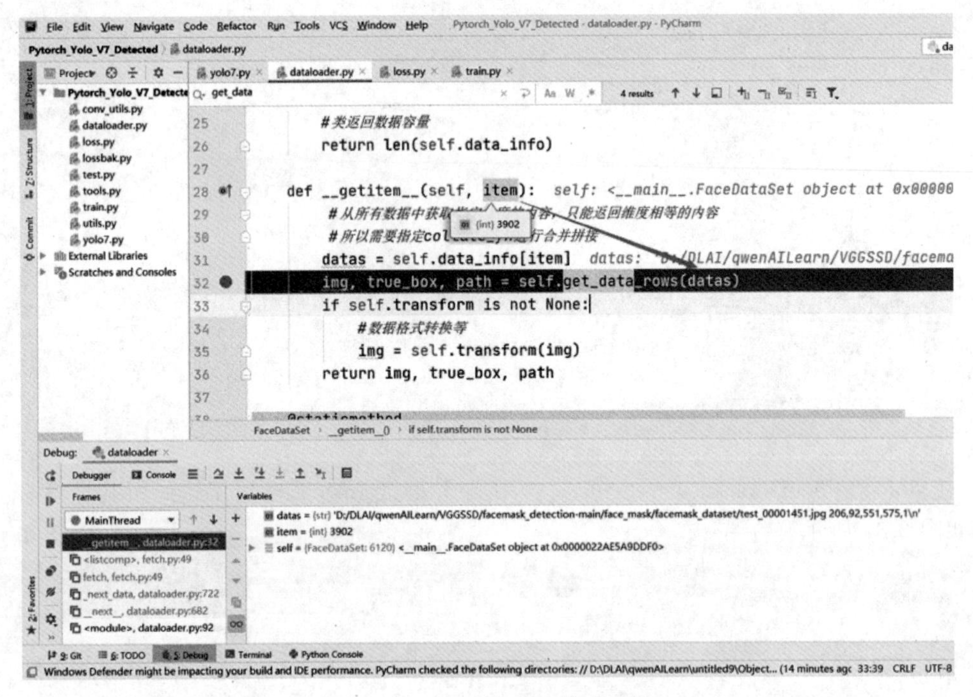

图 5-88 __getitem__()构造函数

torch.tensor(img/255.,dtype=torch.float32).transpose(0, -1)是由于 OpenCV 读取的格式是[height,width,channel],而 PyTorch 格式是[channel,height,width],所以图片的 shape 要进行改变,如图 5-89 所示。

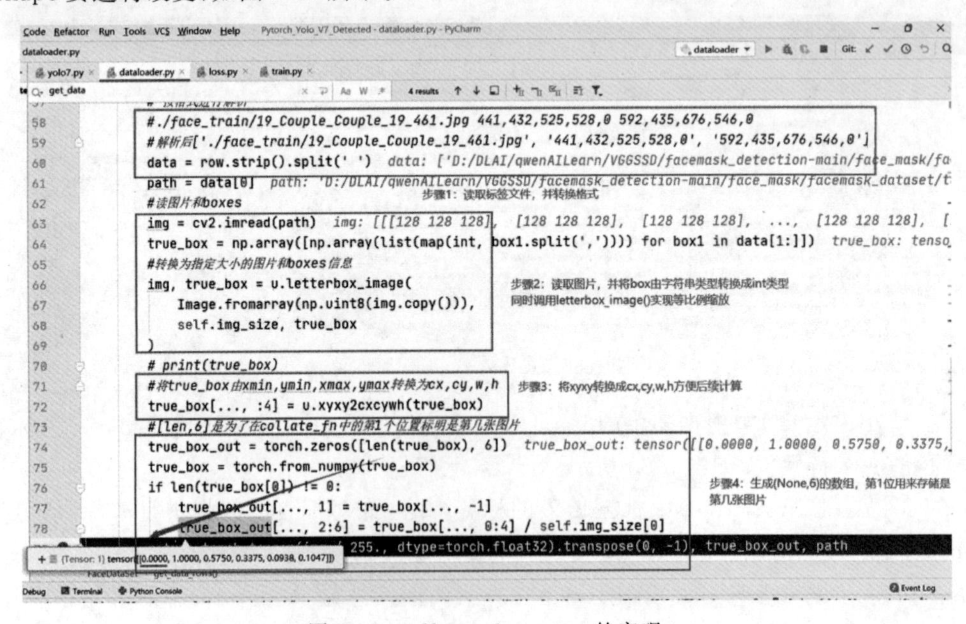

图 5-89 self.get_data_rows 的实现

在 collate_fn(batch)中根据 batch size 返回数据。batch 变量用于存储 img、true_box、path 的信息,如图 5-90 所示。

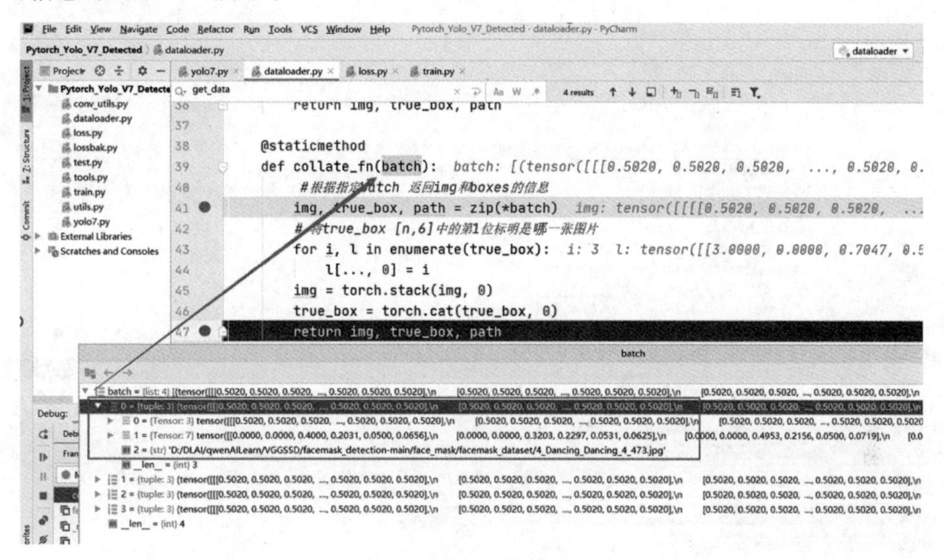

图 5-90 batch 参数的值

然后根据 enumerate(true_box)中的内容,通过 l[...,0]=i 修改第 1 位的值,以此来表明是 batch size 中的第几张图片。最后通过 img=torch.stack(img,0)对图片数据进行合并,true_box=torch.cat(true_box,0)实现标签数据的合并,如图 5-91 所示。

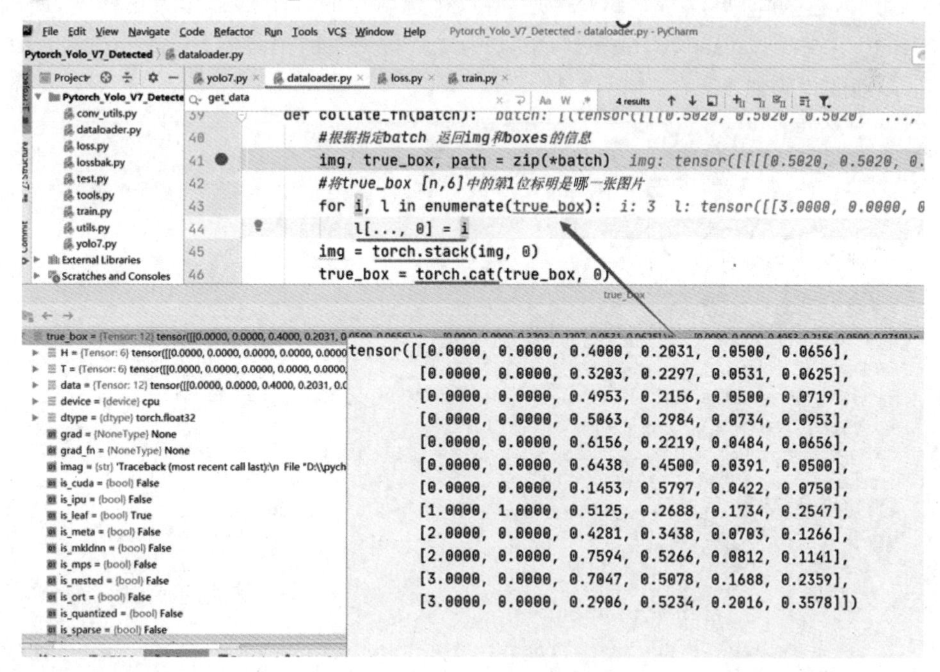

图 5-91 collate_fn 的实现

然后使用 DataLoader 构建生成器 train_loader，经 enumerate(train_loader)之后便可获得指定 batch size 的图片和 label 信息，如图 5-92 所示。

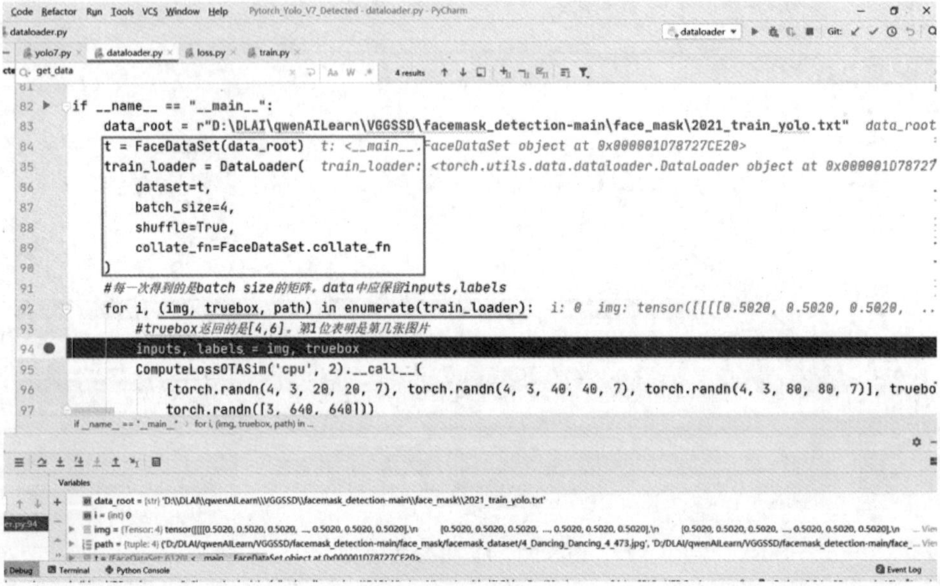

图 5-92 DataLoader 的实现

读取完数据后根据 YOLOv5 中的实现，先将 GT BOX 与 Anchor BOX 进行高宽比，如果小于 4，则入选，同时再以 GT BOX 为中心点挑选附近的 3 个格子作为正样本，此函数被封装在 find_3_positive(self,p,targets)中，p 为 3 个检测头的预测值，targets 为 GT BOX 的标签值，详细的代码如下：

```
#第 5 章/ObjectDetection/Pytorch_Yolo_V7_Detected/loss.py
def find_3_positive(self, p, targets):
    na = len(self.anchor)                          #Anchor 的数量
    nt = targets.shape[0]                          #GT BOX 的数量，4 × 6
    #存放 indices 和 anchors
    indices, anch = [], []
    #序号 2:6 为特征层的高宽
    gain = torch.ones(7, device=targets.device).long()
    #GT BOX 的下标
    ai = torch.arange(na, device=targets.device).float().view(na, 1).repeat(1, nt)
    #实现每个 Anchor 都复制一份 GT 的信息
    targets = torch.cat((targets.repeat(na, 1, 1), ai[:, :, None]), 2)
    #targets [0.0000, 1.0000, 0.4344, 0.2141, 0.7750, 0.6594, 2.0000]
    #第几张图, 类别, gt_cx, gt_cy, gt_h, gt_w, 第几个 Anchor
    g = 0.5 #偏置
    off = torch.tensor([
        [0, 0],
        [1, 0], [0, 1], [-1, 0], [0, -1],          #j,k,l,m
    ], device=targets.device).float() *g           #offsets
    #遍历每个检测头
```

```
    for i in range(len(p)):
        #每个检测头对应的 Anchor
        anchors = self.anchor[2 - i] *p[i].shape[2]
        #4×3×20×20×7-->20×20×20×20
        #gain 本来是[1,1,1,1,1,1,1],此时变为[1,1,20,20,20,20,1]
        gain[2:6] = torch.tensor(p[i].shape)[[3, 2, 3, 2]]   #xyxy gain
        #将 targets 乘以 gain,将真实框映射到特征层上。每个格子都存有 targets 值
        t = targets *gain
        #如果存在 GT
        if nt:
            #4:6 为 GT BOX 的高宽,在 YOLOv5 中根据高宽比来确定正样本
            r = t[:, :, 4:6] / anchors[:, None]
            #高宽比小于 4.0 的 mask
            j = torch.max(r, 1. / r).max(2)[0] < 4.0
            t = t[j]                                     #通过 j 的布尔值过滤出 t 的正样本

            #gxy 用于获得 t 对应的正样本的 x 坐标和 y 坐标
            gxy = t[:, 2:4]
            #gxi 用于获取每个格子的 xy 坐标 20*20
            gxi = gain[[2, 3]] - gxy                     #gxy 离每个格子左上角点的距离
            #根据 gxi 的值,计算附近的格子的 mask
            j, k = ((gxy % 1. < g) & (gxy > 1.)).T
            l, m = ((gxi % 1. < g) & (gxi > 1.)).T
            j = torch.stack((torch.ones_like(j), j, k, l, m))
            #t 重复 5 次,使用满足条件的 j 进行框的提取
            #假设 t 本来有 17 个,则先扩充 5 倍,变成[5,17,4+1+num_class]
            #j 代表当前特征点在[0, 0],[1, 0],[0, 1],[-1, 0],[0, -1]方向是否存在
            t = t.repeat((5, 1, 1))[j] #在[5,17,4+1+num_class]个样本中根据 mask j 提
                                       #取正样本,就变成了[50,7]
            offsets = (torch.zeros_like(gxy)[None] + off[:, None])[j] #附近新正样本
                                                                      #的偏移值
        else:
            #没有目标当负样本
            t = targets[0]
            offsets = 0
        #b 为第几张图片,c 为类别
        b, c = t[:, :2].long().T
        gxy = t[:, 2:4]                                  #正样本 xy
        #gwh = t[:, 4:6]                                 #正样本 wh
        gij = (gxy - offsets).long()                     #偏移值
        gi, gj = gij.T                                   #得到在 gi,gj 个格子中存在目标

        #a 为第几个 Anchor
        a = t[:, 6].long()
        #返回 indices [第几张图片,第几个 anchor,第 j 个列,第 i 行]
        indices.append(
            (b, a, gj.clamp_(0, gain[3] - 1),            #gj.clamp_(0, gain[3] - 1) 限定格
                                                         #子的位置在 0~19
             gi.clamp_(0, gain[2] - 1))
        )                                                #image, anchor, grid indices
        anch.append(anchors[a])                          #此时正样本的 anchors 值是多少
    return indices, anch
```

代码较难理解，先看第1部分，targets 为输入 GT BOX 的信息，因为 batch size=4，所以这里的 shape=4*6。ai 这个变量根据设置的 anchor=3 数量，生成 Anchor 的下标索引值，因为 Anchor 也有4个位置，所以 ai 的值为 tensor([[0.,0.,0.,0.],[1.,1.,1.,1.],[2.,2.,2.,2.]])；targets.repeat(na,1,1)将每个 Anchor 都分配 targets 的内容，所以 shape=3*4*6，然后与 ai[…,None]升维合并后变成 shape=3*4*7，如内容[0.0000,1.0000,0.4344,0.2141,0.7750,0.6594,2.0000]的含义为[第几张图，类别，gt_cx，gt_cy，gt_h，gt_w，第几个 anchor]，如图 5-93 所示。

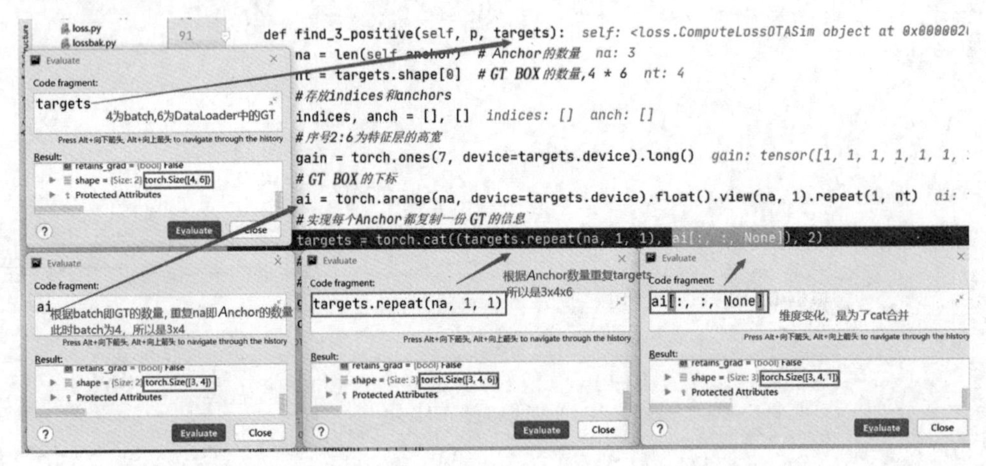

图 5-93 find_3_positive 中的 targets

第2部分，offset 描述了每个当前格子，如果假设为[0,0]，则附近的格式的位置为[1,0]，[0,1]，[-1,0]，[0,-1]。anchors=self.anchor[2-i]*p[i].shape[2]根据当前预测 p 的 shape 计算 Anchor 在当前特征图上的大小。t=targets*gain 为计算 GT BOX 在当前特征图上的大小；gain 变量用来存储计算后的值。r=t[:,:,4:6] / anchors[:,None]用于得到当前特征图上 GT BOX 与 anchors 的高宽比，并通过 j=torch.max(r,1./r).max(2)[0]<4.0 来判断宽高比或者高宽比是否小于4，如果小于4，则 j 为 True，否则为 False。t[j]用于取出满足小于4.0的 GT BOX 在此特征图上的值，关键代码如图 5-94 所示。

第3部分，t[:,2:4]为 targets 的 x 和 y 坐标，假设特征图为 20*20，则 gain[[2,3]]=20*20，则 gxi=gain[[2,3]]-gxy 用于获取 targets 离每个格子左上角所在的格子数。(gxy%1.<g)&(gxy>1.)用于计算当前格子偏离每个格子中心点的位置，得到 j、i 个格子的布尔值，然后将 t=t.repeat((5,1,1))[j]复制5份，通过 j 的布尔值得到离 GT 中心点最近的3个框，假设原有11个正样本 BOX，那么此时就扩展为33个正样本，如图 5-95 所示。

第4部分，根据 offsets 值 gij=(gxy-offsets).long()计算所在格子的位置，根据 t[:,6].long()计算 Anchor 的位置，indices 返回[第几张图片，第几个 anchor，第 j 个列，第 i 行]，anch.append(anchors[a])返回此时正样本的 anchors 值，如图 5-96 所示。

根据上面初选的正样本，进入复选流程，复选的代码在 build_targets()中，详细的代码如下：

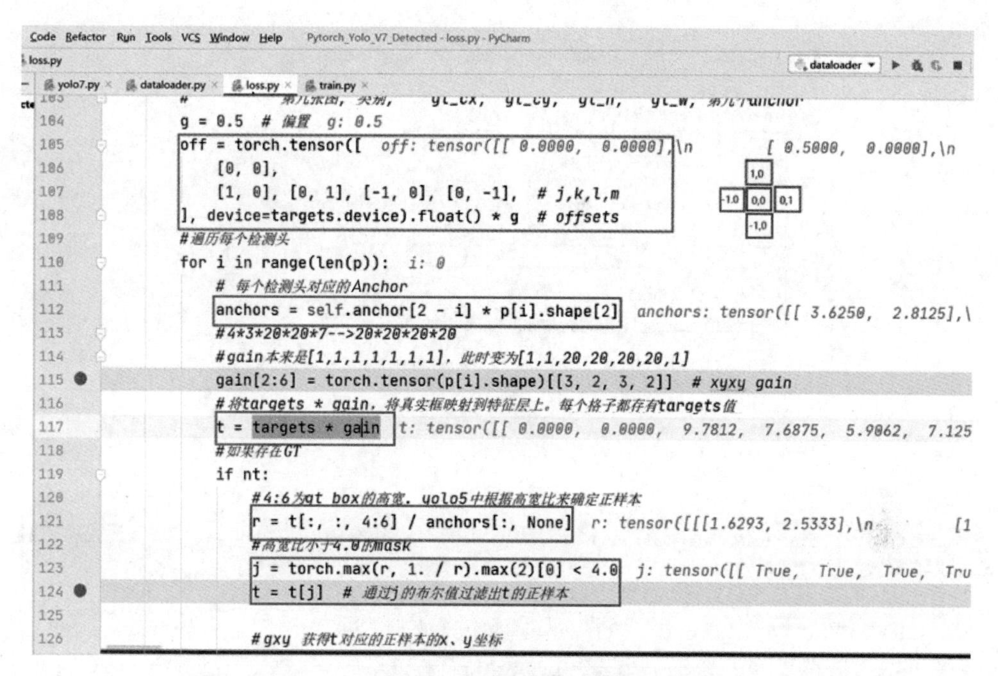

图 5-94 find_3_positive 中的 GT BOX 与 Anchors 的高宽比

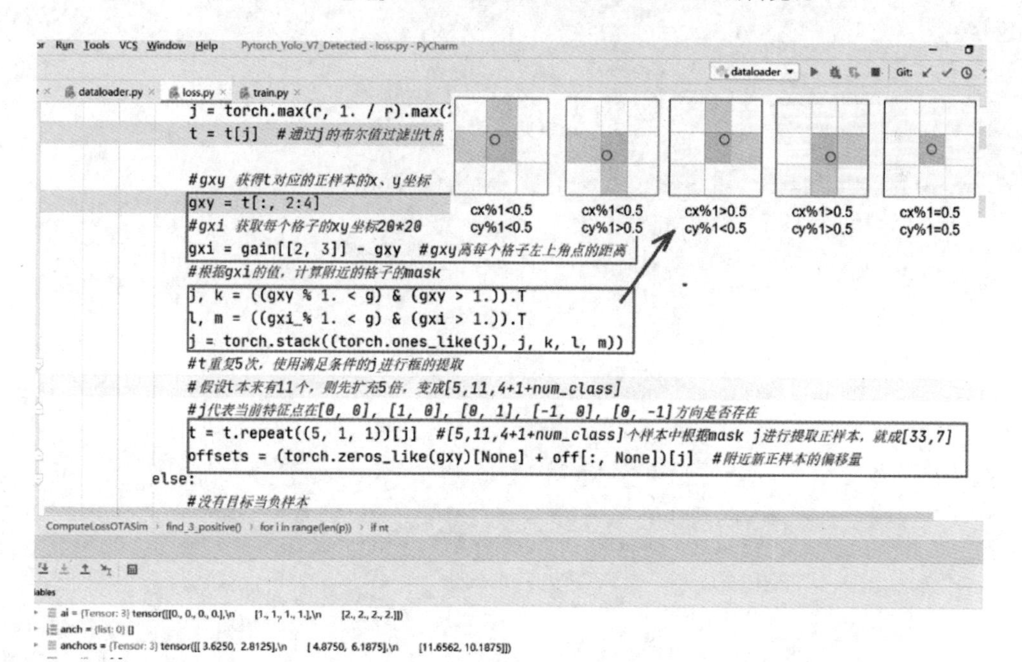

图 5-95 find_3_positive 中正样本的扩展

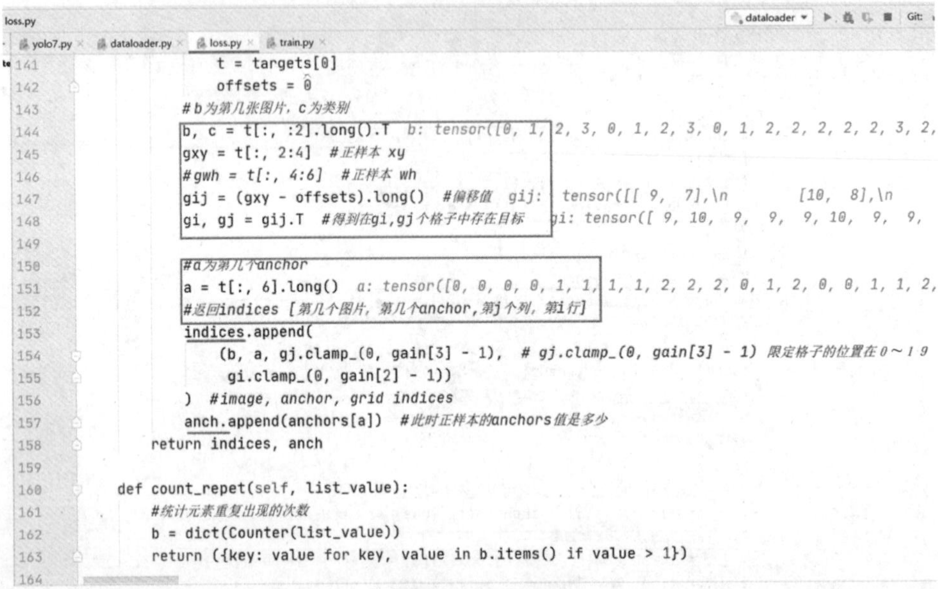

```
141                t = targets[0]
142                offsets = 0
143            #b为第几张图片，c为类别
144            b, c = t[:, :2].long().T  b: tensor([0, 1, 2, 3, 0, 1, 2, 3, 0, 1, 2, 2, 2, 2, 2, 3, 2,
145            gxy = t[:, 2:4]  #正样本 xy
146            #gwh = t[:, 4:6]  #正样本 wh
147            gij = (gxy - offsets).long()  #偏移值  gij:  tensor([[ 9,  7],\n        [10,  8],\n
148            gi, gj = gij.T  #得到在gi,gj个格子中存在目标   gi: tensor([ 9, 10,  9,  9,  9, 10,  9,  9,
149
150            #a为第几个anchor
151            a = t[:, 6].long()  a: tensor([0, 0, 0, 0, 1, 1, 1, 1, 2, 2, 2, 0, 1, 2, 0, 0, 1, 1, 2,
152            #返回indices [第几个图片，第几个anchor,第j个列，第i行]
153            indices.append(
154                (b, a, gj.clamp_(0, gain[3] - 1),  # gj.clamp_(0, gain[3] - 1) 限定格子的位置在0～19
155                 gi.clamp_(0, gain[2] - 1))
156            )  #image, anchor, grid indices
157            anch.append(anchors[a])  #此时正样本的anchors值是多少
158        return indices, anch
159
160    def count_repet(self, list_value):
161        #统计元素重复出现的次数
162        b = dict(Counter(list_value))
163        return ({key: value for key, value in b.items() if value > 1})
164
```

图 5-96　find_3_positive 返回 indices

```
#第 5 章/ObjectDetection/Pytorch_Yolo_V7_Detected/loss.py

def build_targets(self, p, targets, imgs):
    device = torch.device(targets.device)
    #初选,寻找目标 GT 中心点位置附近的 3 个格子作为正样本
    indices, anch = self.find_3_positive(p, targets)

    #匹配,[batch img,anchor,j,i,gt box, anchors 此时的比例]
    matching_bs = [[] for pp in p]
    matching_as = [[] for pp in p]
    matching_gjs = [[] for pp in p]
    matching_gis = [[] for pp in p]
    matching_targets = [[] for pp in p]
    matching_anchs = [[] for pp in p]
    #检测头
    nl = len(p)
    #p[0].shape[0]即 batch size,遍历每幅图来计算
    for batch_idx in range(p[0].shape[0]):
        #targets [0.0000, 1.0000, 0.4344, 0.2141, 0.7750, 0.6594, 2.0000]
        #第几张图, 类别, gt_cx, gt_cy, gt_h, gt_w, 第几个 Anchor
        #只取当前图的 targets 信息
        b_idx = targets[:, 0] == batch_idx
        this_target = targets[b_idx]
        #如果没有 GT BOX,则不处理
        if this_target.shape[0] == 0:
            continue
        #因为 target 是归一化后的值,所以如果乘原图的尺寸,则返回真实的 txywh
        #这样算的原因是后面要计算 IOU
        txywh = this_target[:, 2:6] * imgs[batch_idx].shape[1]
```

```
    #将 xywh 转换成 xmin,ymin,xmax,ymax,也是为了计算 IOU
    txyxy = tools.xywh2xyxy(txywh)
    #预测 xyxy,预测分类,预测置信度
    pxyxys = []
    p_cls = []
    p_obj = []
    #在哪个预测层
    from_which_layer = []
    #存储所有的[batch img,anchor,j,i,gt box, anchors 此时的比例]
    all_b = []
    all_a = []
    all_gj = []
    all_gi = []
    all_anch = []
    #遍历每个预测层
    for i, pi in enumerate(p):
        b, a, gj, gi = indices[i] #根据检测头,从初选中获取
        #获取当前图片
        idx = (b == batch_idx)
        #进一步获取当前图片
        b, a, gj, gi, anchi = b[idx], a[idx], gj[idx], gi[idx], anch[i][idx]
        all_b.append(b)
        all_a.append(a)
        all_gj.append(gj)
        all_gi.append(gi)
        all_anch.append(anchi)
        from_which_layer.append((torch.ones(size=(len(b),)) *i).to(device))

        #在预测结果中根据真实 b,a,gj,gi 进行筛选
        fg_pred = pi[b, a, gj, gi]
        #存储预测结果置信度和分类
        p_obj.append(fg_pred[:, 4:5])
        p_cls.append(fg_pred[:, 5:])
        #所在格子合并
        grid = torch.stack([gi, gj], dim=1)
        #因为预测出来的是偏移值,所以需要解码成 xyxy
        #即式(2×sigmoid(txy)-0.5 + grid) × 32
        pxy = (fg_pred[:, :2].sigmoid() *2. - 0.5 + grid) *self.stride[i]   #32
        pwh = (fg_pred[:, 2:4].sigmoid() *2) **2 *anchi *self.stride[i]   #32
        pxywh = torch.cat([pxy, pwh], dim=-1)
        pxyxy = tools.xywh2xyxy(pxywh)
        pxyxys.append(pxyxy)
    #合并 3 个检测头的预测值
    pxyxys = torch.cat(pxyxys, dim=0)
    if pxyxys.shape[0] == 0:
        continue
    #对每个层预测的相关信息进行合并
    p_obj = torch.cat(p_obj, dim=0)
    p_cls = torch.cat(p_cls, dim=0)
    from_which_layer = torch.cat(from_which_layer, dim=0)
    all_b = torch.cat(all_b, dim=0)
    all_a = torch.cat(all_a, dim=0)
```

```
all_gj = torch.cat(all_gj, dim=0)
all_gi = torch.cat(all_gi, dim=0)
all_anch = torch.cat(all_anch, dim=0)

#每张图片和预测值进行 IOU
pair_wise_iou = tools.box_iou(txyxy, pxyxys) #GT BOX 和预测值进行 IOU
#-log(pair_wise_iou),pair_wise_iou 越大,-log(y)就越小。反之离得越远,重合度越低
pair_wise_iou_loss = -torch.log(pair_wise_iou + 1e-8)
#假设 pair_wise_iou.shape[1]=66,则取 10 个。如果 pair_wise_iou.shape[1]=1,则取 1
top_k, _ = torch.topk(pair_wise_iou, min(10, pair_wise_iou.shape[1]), dim=1)
#得到动态 k,假设为[2,2,3,3],代表 4 个 GT BOX 中的每个与 pre box 的前 10 个 IOU
#得分之和的值
dynamic_ks = torch.clamp(top_k.sum(1).int(), min=1)
#根据标签类别进行 one_hot 编码并升维与 pxyxys 一致[预测 box 的个数与正样本个数相同]
gt_cls_per_image = (
   F.one_hot(this_target[:, 1].to(torch.int64), self.num_class)
      .float()
      .unsqueeze(1)
      .repeat(1, pxyxys.shape[0], 1)
)
#GT 的数量
num_gt = this_target.shape[0]
#预测 置信度 *分类
cls_preds_ = (
      p_cls.float().unsqueeze(0).repeat(num_gt, 1, 1).sigmoid_()
      *p_obj.unsqueeze(0).repeat(num_gt, 1, 1).sigmoid_()
)
#开平方根
y = cls_preds_.sqrt_()
#将 GT BOX 的置信度与预测 log(y/1-y)进行求交叉熵损失,从而得到 pair_wise_cls_
#loss 损失
pair_wise_cls_loss = F.binary_cross_entropy_with_logits(
   torch.log(y / (1 - y)), gt_cls_per_image, reduction="none"
).sum(-1)
del cls_preds_
#精选总损失
cost = (
   #置信度的损失
      pair_wise_cls_loss
      #GT BOX 与 Pre BOX 重合度的损失。3.0 表明 cost 更多地学习 BOX 之间的重叠
      + 3.0 *pair_wise_iou_loss
)
#根据 cost 初始矩阵,假设 pxyxys 是 12 个,则 cost 也是 12 个
#pxyxys 的个数与正样本的 i,j,k 有关。根据正样本的 i,j,k 取对应的预测框
matching_matrix = torch.zeros_like(cost, device=device)
new_pos_idx = []
for gt_idx in range(num_gt): #循环每个 GT BOX,gt_idx 为其下标
   _, pos_idx = torch.topk(
      cost[gt_idx], k=dynamic_ks[gt_idx].item(), largest=False
   ) #dynamic_ks 按 batch size 生成 tensor([2, 2, 3, 3, 3, 3], dtype=torch.int32)
   #如果 gt_idx=0,则 k = 2。cost[0]即损失 12 个预测框中的第 1 组。topk 则取第 1 组
   #的两个损失,并且 largest 是从小到大
```

```
        matching_matrix[gt_idx][pos_idx] = 1.0  #matching_matrix将对应位置设置为1
        #如果 match[0][[17,17]] = 1.0,则表明此时有一个 Anchor 被分配到多个 GT
        new_pos_idx += list(pos_idx.NumPy())

    del top_k, dynamic_ks
    #假设 matching_matrix 4×21,将变为 21
    anchor_matching_gt = matching_matrix.sum(0)
    if (anchor_matching_gt > 1).sum() > 0:     #Anchor 匹配 GT的个数,如果大于 1,
                                                #则需要去重
        print(self.count_repet(new_pos_idx)) #pos_idx 如果有重复项,就会去重
        #anchor_matching_gt > 1,只留 1 个 Anchor 匹配 1 个 GT 的内容。
        #anchor_matching_gt 如果有 2,则会舍弃
        #cost[:, anchor_matching_gt > 1]选出不重复 Anchor 匹配 GT 的 cost 损失
        #torch.min 表明取 cost 里面的最小值。返回最小值_和其索引 cost_argmin_,
        #cost_argmin = torch.min(cost[:, anchor_matching_gt > 1], dim=0)
        #将 anchor_matching_gt>1 在 matching_matrix 中的值设置为 0; 如果原来是 2,
        #则现在变成 0
        matching_matrix[:, anchor_matching_gt > 1] *= 0.0
        #将 anchor_matching_gt>1 且可以令 cost 最小的位置重新设置为 1,表明只取
        #1 个 anchor
        matching_matrix[cost_argmin, anchor_matching_gt > 1] = 1.0
    #matching_matrix 已去掉重复的 Anchor 分配到某个 GT 上。再次求 sum(0)>0.0 是否
    #存在 Anchor
    fg_mask_inboxes = (matching_matrix.sum(0) > 0.0).to(device)
    #最后确定有多少个正样本,并确定其编号
    matched_gt_inds = matching_matrix[:, fg_mask_inboxes].argmax(0)
    #根据最后的正样本的 index 去筛选相关信息
    from_which_layer, all_b, \
    all_a, all_gj, all_gi, all_anch = [
        x[fg_mask_inboxes] for x in
        [from_which_layer, all_b, all_a, all_gj, all_gi, all_anch]
    ]
    #根据正样本对 GT BOX 的对应的位置进行赋值。使两者维度保持一致
    this_target = this_target[matched_gt_inds]
    #遍历 3 个检测头。根据检测头的编号重新整理 all_b 的值。使 b,a,j,i 等值分配到
    #正确的 layer 头上
    #matching_bs,最后正样本分配在 3 个头的存储
    for i in range(nl):
        layer_idx = from_which_layer == i
        matching_bs[i].append(all_b[layer_idx])
        matching_as[i].append(all_a[layer_idx])
        matching_gjs[i].append(all_gj[layer_idx])
        matching_gis[i].append(all_gi[layer_idx])
        matching_targets[i].append(this_target[layer_idx])
        matching_anchs[i].append(all_anch[layer_idx])
    #按 batch size 重新整合
    for i in range(nl):
        if matching_targets[i] != []:
            matching_bs[i] = torch.cat(matching_bs[i], dim=0)
            matching_as[i] = torch.cat(matching_as[i], dim=0)
            matching_gjs[i] = torch.cat(matching_gjs[i], dim=0)
            matching_gis[i] = torch.cat(matching_gis[i], dim=0)
```

```
            matching_targets[i] = torch.cat(matching_targets[i], dim=0)
            matching_anchs[i] = torch.cat(matching_anchs[i], dim=0)
        else:
            matching_bs[i] = torch.tensor([], device=device, dtype=torch.int64)
            matching_as[i] = torch.tensor([], device=device, dtype=torch.int64)
            matching_gjs[i] = torch.tensor([], device=device, dtype=torch.int64)
            matching_gis[i] = torch.tensor([], device=device, dtype=torch.int64)
            matching_targets[i] = torch.tensor([], device=device, dtype=torch.int64)
            matching_anchs[i] = torch.tensor([], device=device, dtype=torch.int64)
    #最后返回 3 个检测头中的相关信息
    return matching_bs, matching_as, matching_gjs, matching_gis, matching_
targets, matching_anchs
```

代码较长且实现较复杂，先看第 1 部分代码中 matching_的相关变量，用来存储经过复选的正样本内容，包括每个 batch 中的图片，anchor 值、第 ji 个格子，GT BOX 的值，以及此时 anchor 值，[[]for pp in p]是由于有 3 个检测头，所以需要循环接收。在循环每个 batch 的图片中，通过 b_idx＝targets[:,0]＝＝batch_idx 获取当前图片的 b_idx，按每张图片的 b_idx 去获取 this_target[:,2:6]在当前图片中的 txywh 值并转换成 xmin，ymin，xmax，ymax 值，以此来计算 GT BOX 与 Pre BOX 预测值的 IOU，如图 5-97 所示。

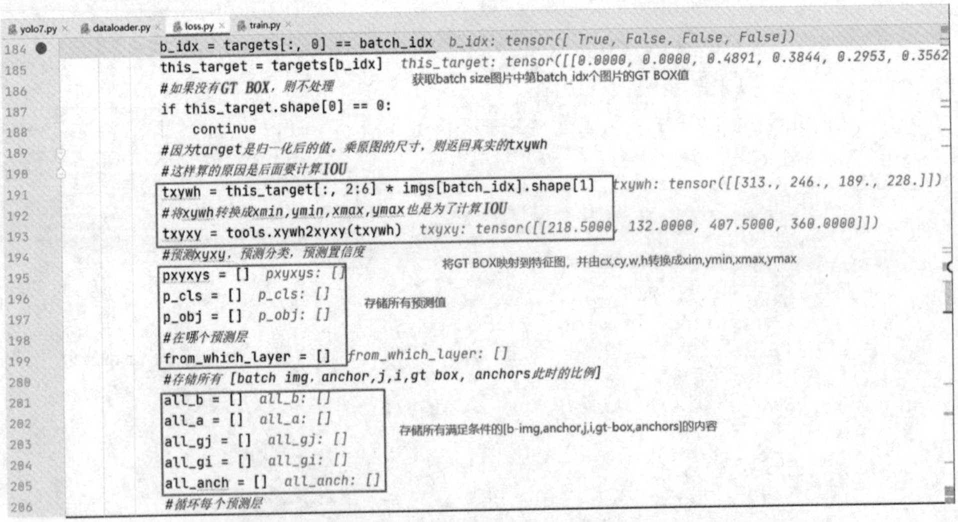

```
b_idx = targets[:, 0] == batch_idx
this_target = targets[b_idx]
#如果没有GT BOX，则不处理
if this_target.shape[0] == 0:
    continue
#因为target是归一化后的值。乘原图的尺寸，则返回真实的txywh
#这样算的原因是后面要计算IOU
txywh = this_target[:, 2:6] * imgs[batch_idx].shape[1]
#将xywh转换成xmin,ymin,xmax,ymax也是为了计算IOU
txyxy = tools.xywh2xyxy(txywh)
#预测xyxy，预测分类，预测置信度
pxyxys = []
p_cls = []
p_obj = []
#在哪个预测层
from_which_layer = []
#存储所有 [batch img, anchor,j,i,gt box, anchors此时的比例]
all_b = []
all_a = []
all_gj = []
all_gi = []
all_anch = []
#循环每个预测层
```

获取batch size图片中第batch_idx个图片的GT BOX值
将GT BOX映射到特征图，并由cx,cy,w,h转换成xim,ymin,xmax,ymax
存储所有预测值
存储所有满足条件的[b-img,anchor,j,i,gt-box,anchors]的内容

图 5-97 build_targets 获取当前 batch idx 的真实值

第 2 部分，根据预测值当前所在 batch 及 find_3_positive()返回的[batch img，anchor，j，i，gt box，anchors 值]去预测值 fg_pred 中获取预测的置信度和分类信息的概率，然后使用 pxy＝(fg_pred[:,:2]. sigmoid()＊2.－0.5＋grid)＊self. stride[i]对于预测的偏移值进行解码操作，转换成 cx，cy，w，h。tools. xywh2xyxy(pxywh)将预测值又转换成 xmin，ymin，xmax，ymax 的内容，循环 enumerate(p)，从而得到 3 个检测头的预测值，如图 5-98 所示。

第 3 部分，计算正样本 txyxy 与预测 pxyxys 的 IOU，存储在 pair_wise_iou 中，然后由 torch. topk(pair_wise_iou，min(10，pair_wise_iou. shape[1])，dim＝1)计算 pair_wise_iou

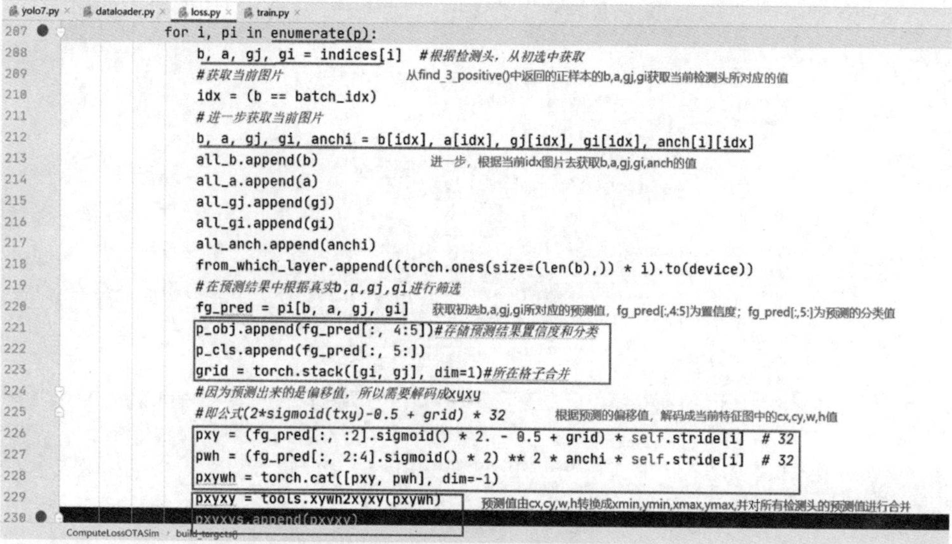

```
287     for i, pi in enumerate(p):
288         b, a, gj, gi = indices[i]  #根据检测头，从初选中获取
289         #获取当前图片              从find_3_positive()中返回的正样本的b,a,gj,gi获取当前检测头所对应的值
210         idx = (b == batch_idx)
211         #进一步获取当前图片
212         b, a, gj, gi, anchi = b[idx], a[idx], gj[idx], gi[idx], anch[i][idx]
213         all_b.append(b)          进一步，根据当前idx图片去获取b,a,gj,gi,anch的值
214         all_a.append(a)
215         all_gj.append(gj)
216         all_gi.append(gi)
217         all_anch.append(anchi)
218         from_which_layer.append((torch.ones(size=(len(b),)) * i).to(device))
219         #在预测结果中根据真实b,a,gj,gi进行筛选
220         fg_pred = pi[b, a, gj, gi]   获取初选b,a,gj,gi所对应的预测值，fg_pred[:,4:5]为置信度；fg_pred[:,5:]为预测的分类值
221         p_obj.append(fg_pred[:, 4:5])#存储预测结果置信度和分类
222         p_cls.append(fg_pred[:, 5:])
223         grid = torch.stack([gi, gj], dim=1)#所在格子合并
224         #因为预测出来的是偏移值，所以需要解码成xyxy
225         #即公式(2*sigmoid(txy)-0.5 + grid) * 32      根据预测的偏移值，解码成当前特征图中的cx,cy,w,h值
226         pxy = (fg_pred[:, :2].sigmoid() * 2. - 0.5 + grid) * self.stride[i]  # 32
227         pwh = (fg_pred[:, 2:4].sigmoid() * 2) ** 2 * anchi * self.stride[i]  # 32
228         pxywh = torch.cat([pxy, pwh], dim=-1)
229         pxyxy = tools.xywh2xyxy(pxywh)    预测值由cx,cy,w,h转换成xmin,ymin,xmax,ymax,并对所有检测头的预测值进行合并
230         pxyxys.append(pxyxy)
```

图 5-98 build_targets 中 pxyxys 的获取

中最大 IOU 值前 10 个 top_k. sum(1). int()之和并作为正样本的 dynamic_ks 个数。gt_cls_per_image 根据标签类别进行 one_hot 编码，升维后跟 pxyxys 一致。cls_preds_是预测置信度 p_obj * 分类 p_cls 的概率。－torch. log(pair_wise_iou＋1e-8)表明 pair_wise_iou 值越大，－torch. log(pair_wise_iou)就越小，反之离得越远，重合度越低，如图 5-99 所示。

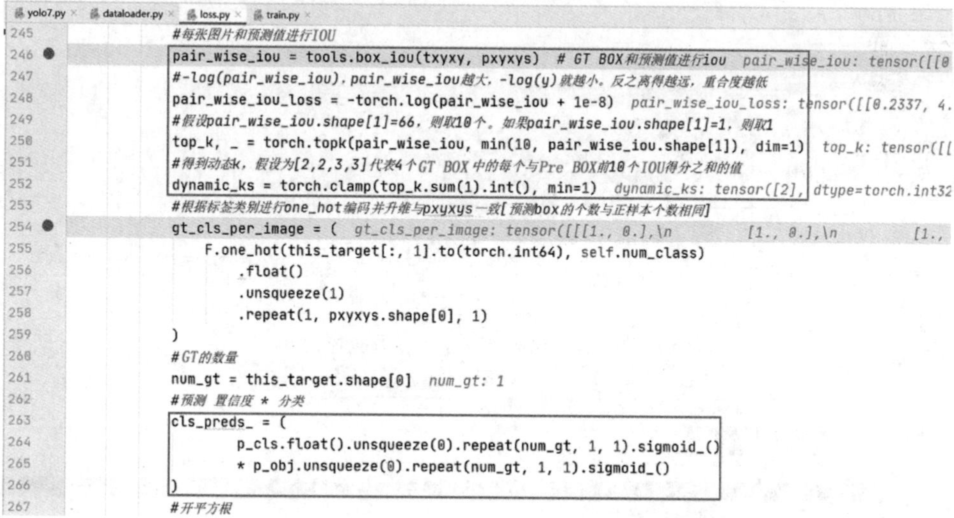

```
245     #每张图片和预测值进行IOU
246     pair_wise_iou = tools.box_iou(txyxy, pxyxys)  # GT BOX和预测值进行iou  pair_wise_iou: tensor([[0
247     #-log(pair_wise_iou)，pair_wise_iou越大，-log(y)就越小。反之离得越远，重合度越低
248     pair_wise_iou_loss = -torch.log(pair_wise_iou + 1e-8)  pair_wise_iou_loss: tensor([[0.2337, 4.
249     #假设pair_wise_iou.shape[1]=66，则取10个。如果pair_wise_iou.shape[1]=1，则取1
250     top_k, _ = torch.topk(pair_wise_iou, min(10, pair_wise_iou.shape[1]), dim=1)  top_k: tensor([[
251     #得到动态k，假设为[2,2,3,3]代表4个GT BOX中的每个与Pre BOX前10个IOU得分之和的值
252     dynamic_ks = torch.clamp(top_k.sum(1).int(), min=1)  dynamic_ks: tensor([2], dtype=torch.int32
253     #根据标签类别进行one_hot编码并升维与pxyxys一致[预测box的个数与正样本个数相同]
254     gt_cls_per_image = (  gt_cls_per_image: tensor([[[1., 0.],\n        [1., 0.],\n        [1.,
255         F.one_hot(this_target[:, 1].to(torch.int64), self.num_class)
256             .float()
257             .unsqueeze(1)
258             .repeat(1, pxyxys.shape[0], 1)
259     )
260     #GT的数量
261     num_gt = this_target.shape[0]  num_gt: 1
262     #预测 置信度 * 分类
263     cls_preds_ = (
264             p_cls.float().unsqueeze(0).repeat(num_gt, 1, 1).sigmoid_()
265             * p_obj.unsqueeze(0).repeat(num_gt, 1, 1).sigmoid_()
266     )
267     #开平方根
```

图 5-99 从 build_targets 中获取动态 k 的值决定正样本

第 4 部分，将预测置信度 * 分类概率 cls_preds_与每张图片中分类的概率 gt_cls_per_image 做 binary_cross_entropy_with_logits()交叉熵损失，并对 pair_wise_cls_loss＋3.0 * pair_wise_iou_loss 进行求和，得到可以令置信度损失和回归损失最小的 cost，也就是

dynamic_ks 个能够令选出的正样本的置信度损失与回归损失最小,如图 5-100 所示。

```
yolo7.py × dataloader.py × loss.py × train.py ×
260                 # GT的数量
261                 num_gt = this_target.shape[0]  num_gt: 1
262                 #预测 置信度 * 分类
263                 cls_preds_ = (
264                         p_cls.float().unsqueeze(0).repeat(num_gt, 1, 1).sigmoid_()
265                         * p_obj.unsqueeze(0).repeat(num_gt, 1, 1).sigmoid_()
266                 )
267                 #开平方根
268                 y = cls_preds_.sqrt_()  y: tensor([[[0.6106, 0.6010],\n          [0.4963, 0.3841],\n          [0
269                 #将GT BOX的置信度与预测log(y/1-y)进行求交叉熵损失,从而得到pair_wise_cls_loss损失
270                 pair_wise_cls_loss = F.binary_cross_entropy_with_logits(  pair_wise_cls_loss: tensor([[1.4122,
271                     torch.log(y / (1 - y)), gt_cls_per_image, reduction="none"
272                 ).sum(-1)                        每个图片中分类的概率
273                 del cls_preds_
274                 #精选总损失
275                 cost = (  cost: tensor([[ 5.7162, 11.4856,  5.8173, 56.5331,  5.6705,  6.0423, 11.7819,  6.7300
276                     #置信度的损失
277                         pair_wise_cls_loss
278                         #GT BOX与Pre BOX重合度的损失。3.0表明cost更多地学习 box之间的重叠
279                         + 3.0 * pair_wise_iou_loss
280                 )
281                 #根据cost初始矩阵, 假设pxyxys是12个, 则cost也是12个
282                 #pxyxys的个数与正样本的i,j,k有关. 根据正样本的i,j,k取对应的预测框
283                 matching_matrix = torch.zeros_like(cost, device=device)
```

图 5-100 build_targets 中的 dynamic_ks 决定 cost 最小

第 5 部分,根据 cost 值,遍历每个 GT BOX,取出 dynamic_ks 动态 K 能够令 cost 最小的正样本索引 pos_idx,并且如果此时存在,则将 matching_matrix[gt_idx][pos_idx]=1.0,表明当前 gt_idx 和 pos_idx 都进行选择。new_pos_idx 用来统计每个 pos_idx 的索引号,如图 5-101 所示。

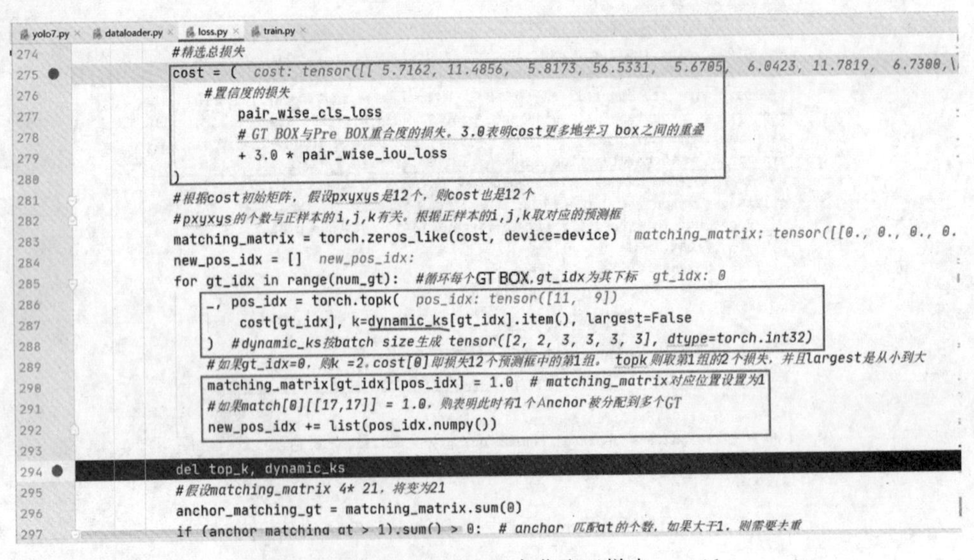

```
yolo7.py × dataloader.py × loss.py × train.py ×
274                 #精选总损失
275                 cost = (  cost: tensor([[ 5.7162, 11.4856,  5.8173, 56.5331,  5.6705,  6.0423, 11.7819,  6.7300,\
276                     #置信度的损失
277                         pair_wise_cls_loss
278                         # GT BOX与Pre BOX重合度的损失。3.0表明cost更多地学习 box之间的重叠
279                         + 3.0 * pair_wise_iou_loss
280                 )
281                 #根据cost初始矩阵, 假设pxyxys是12个, 则cost也是12个
282                 #pxyxys的个数与正样本的i,j,k有关. 根据正样本的i,j,k取对应的预测框
283                 matching_matrix = torch.zeros_like(cost, device=device)  matching_matrix: tensor([[0., 0., 0., 0.
284                 new_pos_idx = []  new_pos_idx:
285                 for gt_idx in range(num_gt):  #循环每个GT BOX,gt_idx为其下标  gt_idx: 0
286                     _, pos_idx = torch.topk(  pos_idx: tensor([11,  9])
287                         cost[gt_idx], k=dynamic_ks[gt_idx].item(), largest=False
288                     )  #dynamic_ks按batch size生成 tensor([2, 2, 3, 3, 3, 3], dtype=torch.int32)
289                     #如果gt_idx=0, 则k =2.cost[0]即损失12个预测框中的第1组, topk则取第1组的2个损失, 并且largest是从小到大
290                     matching_matrix[gt_idx][pos_idx] = 1.0  # matching_matrix对应位置设置为1
291                     #如果match[0][[17,17]] = 1.0, 则表明此时有1个Anchor被分配到多个GT
292                     new_pos_idx += list(pos_idx.numpy())
293
294                 del top_k, dynamic_ks
295                 #假设matching_matrix 4* 21, 将变为21
296                 anchor_matching_gt = matching_matrix.sum(0)
297                 if (anchor_matching_gt > 1).sum() > 0:  # anchor 匹配gt的个数, 如果大于1, 则需要去重
```

图 5-101 从 build_targets 中获取正样本 pos_idx

第 6 部分,如果 pos_idx 存在重复,则表明某个 Anchor 被分配给多个 GT BOX。在理想情况下,一个 GT BOX 可以有多个正样本,但是一个 Anchor 应该只分配 1 个 GT BOX。

在 matching_matrix 中存储 gt_idx 和 pos_idx 的值，假设 matching_matrix[0][[17,17]]=1.0，则表明 pos_idx 被分配给多个 GT BOX，所以如果 matching_matrix.sum(0)>1，则表明有重复 Anchor 被分配给 GT BOX，所以 matching_matrix[:,anchor_matching_gt>1] *= 0.0 可以令 anchor_matching_gt>1 的位置在 matching_matrix 中将值设置为 0，从而剔除重复项，并得到最后的正样本 fg_mask_inboxes，如图 5-102 所示。

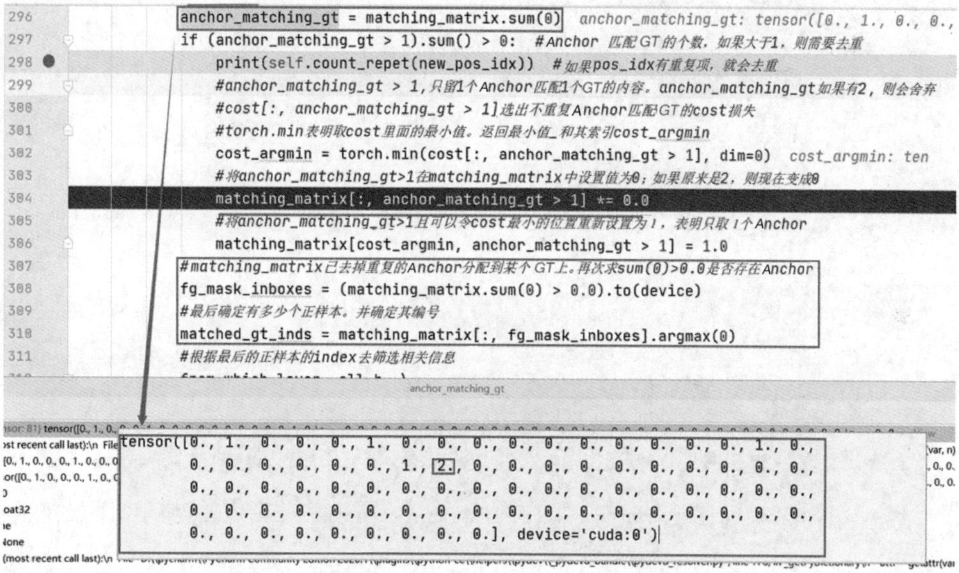

图 5-102 从 build_targets 中去除重复 pos_idx

最后遍历 3 个检测头，根据检测头的编号重新整理 all_b 的值，使 b,a,j,i 值分配到正确的 layer 头上，并使 matching_bs 分配正确，至此 build_targets()完成正样本的提取。

整个正样本的代码比较复杂，建议下载源码后根据注释和解释进行调试。

5.10.4 代码实现损失函数的构建及训练

损失函数的构建根据 self.build_targets(p,targets,imgs)返回的 bs、as_、gjs、gis、targets 和 anchors 值遍历 3 个检测头进行位置损失计算，分类和置信度损失进行计算，并对 3 个检测头的损失按指定的比例进行相加，详细的代码如下：

```
#第 5 章/ObjectDetection/Pytorch_Yolo_V7_Detected/loss.py
class ComputeLossOTASim(object):
    def __call__(self, p, targets, imgs):
        device = targets.device
        #分类、位置和置信度的初始值
        lcls, lbox, lobj = [torch.zeros(1, device=device) for _ in range(3)]
        #在 build_targets 中进行正样本的匹配，分为精选和复选
        bs, as_, gjs, gis, targets, anchors = self.build_targets(p, targets, imgs)
        #因为 p 是 3 个检测头的输出，而 pp 则是 1*3anchor*80*80*num_class 等
```

```
        #所以[3, 2, 3, 2]得到的 Tensor 为 80*80*80*80
        pre_gen_gains = [torch.tensor(pp.shape, device=device)[[3, 2, 3, 2]] for
pp in p]
        #i 表明是第几个检测头。pi 为取的特征
        for i, pi in enumerate(p):
            #b 为第几张图,a 表示第几个 Anchor,gj 表示第 j 个格子,gi 表示第 i 格子
            b, a, gj, gi = bs[i], as_[i], gjs[i], gis[i]
            #初始 GT 的矩阵跟预测出来的结果保持一致
            tobj = torch.zeros_like(pi[..., 0], device=device)
            n = b.shape[0]
            if n:
                ps = pi[b, a, gj, gi]        #根据 build_targets 返回的 index 在预测值
                                             #中取对应位置的值
                #在哪个格子
                grid = torch.stack([gi, gj], dim=1)
                #对预测值限制值域
                pxy = ps[:, :2].sigmoid() *2. - 0.5
                pwh = (ps[:, 2:4].sigmoid() *2) **2 *anchors[i]
                pbox = torch.cat((pxy, pwh), 1) #合并
                #将真实框映射到特性图上
                selected_tbox = targets[i][:, 2:6] *pre_gen_gains[i]
                #计算真实框的偏移值
                selected_tbox[:, :2] -= grid
                #求预测 BOX 与真实 BOX 之间的 loss,这里使用的是 CIOU 损失

                iou = tools.bbox_iou(pbox.T, selected_tbox, x1y1x2y2=False, CIoU=True)
                lbox += (1.0 - iou).mean() #
                #tobj 置信度损失,self.gr obj loss 的权重。原作者默认写为 1
                #tobj[b, a, gj, gi],即对应 build_targets 的位置的置信度值
                tobj[b, a, gj, gi] = (1.0 - self.gr) + self.gr *iou.detach().clamp(0).
type(tobj.dtype)
                #GT 的分类值
                selected_tcls = targets[i][:, 1].long()
                if self.num_class > 1:
                    t = torch.full_like(ps[:, 5:], self.cn, device=device)
                    t[range(n), selected_tcls] = self.cp
                    #分类的损失
                    lcls += self.BCEcls(ps[:, 5:], t)
                #置信度损失
                obji = self.BCEobj(pi[..., 4], tobj)
                lobj += obji *self.balance[i]     #不同检测头的权重不同
        lbox *= self.hyp['box']
        lobj *= self.hyp['obj']
        lcls *= self.hyp['cls']
        bs = tobj.shape[0] #batch size
        #总损失
        loss = lbox + lobj + lcls
        return loss *bs, torch.cat((lbox, lobj, lcls, loss)).detach()
```

第 1 部分代码根据 self.build_targets(p,targets,imgs)获得 bs、as_、gjs、gis、targets 和 anchors,然后遍历 3 个检测头,得到当前检测头中的 grid=torch.stack([gi,gj],dim=1)所在的格子,限制预测值的值域后计算真实框的偏移值 selected_tbox[:, :2]-=grid,然后将

预测框 pbox 与真实框 selected_tbox 做 CIOU 损失，如图 5-103 所示。

```
yolo7.py  dataloader.py  loss.py  train.py
48          for i, pi in enumerate(p):   i: 2  pi: tensor([[[[[-8.4929e-02, -9.0205e-02, -9.9048e-02,  ..., -1.03
49              #b为第几张图, a表示第几个Anchor, gj表示第j个格子, gi表示第i格子
50              b, a, gj, gi = bs[i], as_[i], gjs[i], gis[i]   b: tensor([], device='cuda:0', dtype=torch.int64)
51              #初始GT的矩阵跟预测出来的结果保持一致
52              tobj = torch.zeros_like(pi[..., 0], device=device)  tobj: tensor([[[[0., 0., 0.,  ..., 0., 0., 0
53              n = b.shape[0]  n: 0
54              if n:
55                  ps = pi[b, a, gj, gi]  #根据build_targets返回的index在预测值中取对应位置的值  ps: tensor([[ 0.1844,
56                  #在哪个格子
57                  grid = torch.stack([gi, gj], dim=1)  grid: tensor([[20, 16]], device='cuda:0')
58                  #对预测值限制值域
59                  pxy = ps[:, :2].sigmoid() * 2. - 0.5  pxy: tensor([[0.5920, 0.5434]], device='cuda:0', grad_
60                  pwh = (ps[:, 2:4].sigmoid() * 2) ** 2 * anchors[i]  pwh: tensor([[3.6290, 8.3894]], device='
61                  pbox = torch.cat((pxy, pwh), 1)  # 合并  pbox: tensor([[0.5920, 0.5434, 3.6290, 8.3894]], dev
62                  #将真实框映射到特性图上
63                  selected_tbox = targets[i][:, 2:6] * pre_gen_gains[i]  selected_tbox: tensor([[-0.1875,  0.3
64                  #计算真实框的偏移值
65                  selected_tbox[:, :2] -= grid
66                  #求预测BOX与真实BOX之间的loss,这里使用的是CIOU 损失
67
68                  iou = tools.bbox_iou(pbox.T, selected_tbox, x1y1x2y2=False, CIoU=True)  iou: tensor([0.3291]
69                  lbox += (1.0 - iou).mean()  #
70                  # tobj置信度损失,self.gr obj loss的权重。原作者默认写死为1
ComputeLossOTASim > __call__()
```

图 5-103　CIOU 位置损失

然后对分类损失、置信度损失、回归损失进行求和，从而得到总损失，如图 5-104 所示。

```
yolo7.py  dataloader.py  loss.py  train.py
69                  lbox += (1.0 - iou).mean()  #
70                  #tobj置信度损失,self.gr obj loss的权重。原作者默认写死为1
71                  #tobj[b, a, gj, gi],即对应build_targets的位置的置信度值
72                  tobj[b, a, gj, gi] = (1.0 - self.gr) + self.gr * iou.detach().clamp(0).type(tobj.dtype)
73                  #GT的分类值
74                  selected_tcls = targets[i][:, 1].long()  selected_tcls: tensor([1], device='cuda:0')
75                  if self.num_class > 1:
76                      t = torch.full_like(ps[:, 5:], self.cn, device=device)  t: tensor([[0., 1.]], device='cuda
77                      t[range(n), selected_tcls] = self.cp
78                      #分类损失
79                      lcls += self.BCEcls(ps[:, 5:], t)
80                  #置信度损失
81                  obji = self.BCEobj(pi[..., 4], tobj)  obji: tensor(0.7873, device='cuda:0',\n        grad_fn=<B
82                  lobj += obji * self.balance[i]  #不同检测头的权重不同
83          lbox *= self.hyp['box']
84          lobj *= self.hyp['obj']
85          lcls *= self.hyp['cls']
86          bs = tobj.shape[0]  # batch size
87          #总损失
88          loss = lbox + lobj + lcls  loss: tensor([0.5895], device='cuda:0', grad_fn=<AddBackward0>)
89          return loss * bs, torch.cat((lbox, lobj, lcls, loss)).detach()
90
91      def find_3_positive(self, p, targets):
ComputeLossOTASim > __call__()
```

图 5-104　总损失（求和）

5.10.5　代码实战预测推理

YOLOv7 的预测推理流程与 YOLOv5 一致，即在 YOLOv3 的基础上只变更了解码公式，其流程仍然是先遍历 3 个检测头，接着对每个检测头的输出结果进行置信度、分类概率

的过滤，然后对 3 个检测头中满足条件的结果进行合并，再通过 NMS 去除重复的框，详细代码可参考 YOLOv5 中的解码代码和 YOLOv3 的推理代码。

注意：YOLOv7 使用 PyTorch 框架进行了模型的搭建、训练。

总结

YOLOv7 在 YOLOv4 的基础上引入了 MP、ELAN、SPPCSPC 模块，保护 FPN、PAN、3 个检测头的预测，同时在损失函数方面使用了 OTASim 进行更精细化的正样本提取。

练习

运行并调试本节代码，理解算法的设计与代码的结合，重点梳理本算法的实现方法。

5.11 数据增强

5.11.1 数据增强的作用

数据增强是一种基于原有数据，通过一些技术手段生成新数据的方法，通过数据增强可以提高模型的泛化能力，缓解过拟合，提高模型的稳健性。

目标检测中常见的数据增强有旋转、翻转、HSV 等，而在 YOLOv4、YOLOv5 中使用 Mosic 数据增强提高了模型的精确率。

本节重点介绍 CutOut、MixUp、Mosic、随机复制 label 等增强手段，其他数据增强如图 5-105 所示。

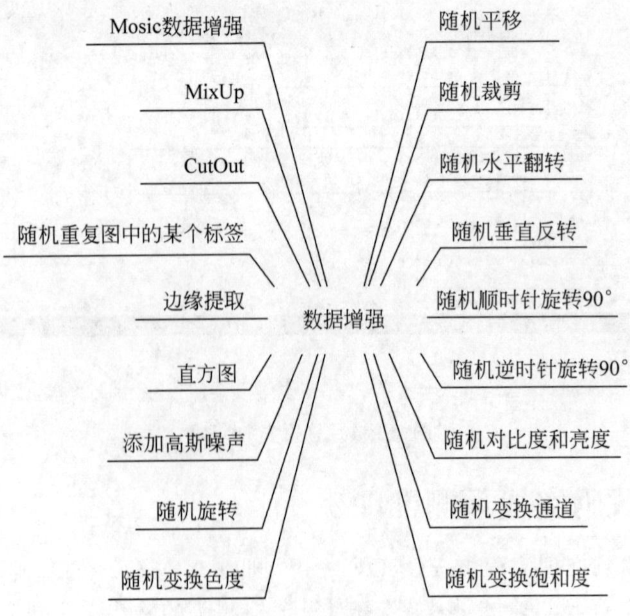

图 5-105 数据增强相关方法

5.11.2 代码实现 CutOut 数据增强

CutOut 是在 2017 年提出的一种数据增强方法，即在训练时随机裁剪掉图像的一部分，起到类似 DropOut 正则化的效果，在论文 *Improved Regularization of Convolutional Neural Networks with CutOut* 中表明在原有数据的基础上精度均有提高，如图 5-106 所示。

Model	STL10	STL10+
WideResNet	23.48±0.68	14.21±0.29
WideResNet+CutOut	**20.77±0.38**	**12.74±0.23**

图 5-106 CutOut 论文效果

原论文表明，增加 CutOut 数据增强在 STL10 上面精度提高了 0.38。CutOut 的代码如下：

```
#第 5 章/ObjectDetection/TensorFlow_Yolo_V5_Detected/data/enhancement.py
def CutOut(img, gt_boxes, amount=0):
    '''CutOut 数据增强
    img: image
    gt_boxes: 格式[[x1 y1 x2 y2,obj],...]
    amount: 概率
    '''
    out = img.copy()
   #随机选择 CutOut 区域
    ran_select = [random.randint(0, int(len(gt_boxes) - 1 * amount)) for i in
range(int(len(gt_boxes) - 1 * amount))]
    #根据区域进行操作
    for i in ran_select:
    #选择哪个 GT BOX 进行 CutOut
        box = gt_boxes[i]
        x1 = int(box[0])
        y1 = int(box[1])
        x2 = int(box[2])
        y2 = int(box[3])
    #在原有 GT BOX 的基础上裁一定的 BOX
        mask_w = int((x2 - x1) * 0.5)
        mask_h = int((y2 - y1) * 0.5)
        mask_x1 = random.randint(x1, x2 - mask_w)
        mask_y1 = random.randint(y1, y2 - mask_h)
        mask_x2 = mask_x1 + mask_w
        mask_y2 = mask_y1 + mask_h
        #绘框
        cv2.rectangle(out, (mask_x1, mask_y1), (mask_x2, mask_y2), (0, 0, 0),
thickness=-1)
        #位置 CutOut
        gt_boxes[i][0:4] = [mask_x1, mask_y1, mask_x2, mask_y2]
    return out, gt_boxes
```

传入图片和 BOX 信息调用后的效果如图 5-107 所示。

图 5-107　CutOut 示意效果图

5.11.3　代码实现 MixUp 数据增强

MixUp 是在论文 *MixUp*：*BEYOND EMPIRICAL RISK MINIMIZATION* 中提出的，其实际上是将两张图片按一定透明度进行叠加的操作。在论文中在 ERM 数据集中使用 VGG-11，使用 MixUp 数据增强对于分类网络错误率约降低了 0.1，如图 5-108 所示。

Model	Method	Validation set	Test set
LeNet	ERM	**9.8**	**10.3**
	MixUP(α=0.1)	10.1	10.8
	MixUP(α=0.2)	10.2	11.3
VGG-11	ERM	5.0	4.6
	MixUP(α=0.1)	4.0	3.8
	MixUP(α=0.2)	**3.9**	**3.4**

图 5-108　MixUp 论文效果

MixUp 的实现代码如下：

```
#第 5 章/ObjectDetection/TensorFlow_Yolo_V5_Detected/data/enhancement.py

def mixup(im, labels, im2, labels2):
    #传入两张图的 image 和 label 信息
    r = np.random.beta(32.0, 32.0) #mixup ratio, alpha 和 beta=32.0
    #resize 到相同的大小
    if im.shape[0] > im2.shape[0]:
        im2 = cv2.resize(im2, (im.shape[1], im.shape[0]))
    else:
        im = cv2.resize(im, (im2.shape[1], im2.shape[0]))
    #两张图按一定比例进行融合
    im = (im *r + im2 *(1 - r)).astype(np.uint8)
```

```
    #对两个 BOX 信息进行融合
    if len(labels) != 0 and len(labels2) != 0:
        labels = np.concatenate((labels, labels2), 0)
    elif len(labels) == 0:
        labels = labels2
    elif len(labels2) == 0:
        labels = labels
    return im, labels
```

调用运行代码后其效果如图 5-109 所示。

图 5-109 MixUp 示意效果图

5.11.4 代码实现随机复制 Label 数据增强

在完成某些任务(例如缺陷检测)时,由于某些类别的缺陷数量较少,所以可以通过复制标注 BOX 实现特定目标分类的重采样,从而提高网络的稳健性,代码如下:

```
#第 5 章/ObjectDetection/TensorFlow_Yolo_V5_Detected/data/enhancement.py
def replicate(im, labels, label_index=0, repetitions=1):
    """标签复制"""
    #得到当前图像的宽和高
    h, w = im.shape[:2]
    #得到 BOX 的位置
    boxes = labels[:, 0:4].astype("int")
    x1, y1, x2, y2 = boxes.T
    #计算所处坐标的位置
    s = ((x2 - x1) + (y2 - y1)) / 2
    for i in s.argsort()[:round(s.size * 0.5)]:
        #如果指定的类别不为空
        if labels[i][-1] != None:
            #判断当前的标签是否与指定的类别一致
            if labels[i][-1] == label_index:
                #重复标签的次数
```

```
                for x in range(repetitions):
                    #得到 4 个坐标点
                    x1b, y1b, x2b, y2b = boxes[i]
                    #得到 BOX 的高、宽
                    bh, bw = y2b - y1b, x2b - x1b
                    #随机原位置偏移位置
                    yc, xc = int(random.uniform(0, h - bh)), int(random.uniform(0, w - bw))
                    x1a, y1a, x2a, y2a = [xc, yc, xc + bw, yc + bh]
                    #从当前图像中切图并改变透明度,赋给新指定的区域
                    im[y1a:y2a, x1a:x2a] = im[y1b:y2b, x1b:x2b]*rand()
                    #保存 BOX 信息
                    labels = np.append(labels, [[x1a, y1a, x2a, y2a, labels[i, -1]]],
axis=0)
    return im, labels
```

调用运行后红色为原目标、绿色为重复目标,如图 5-110 所示,实现目标框增加两次。

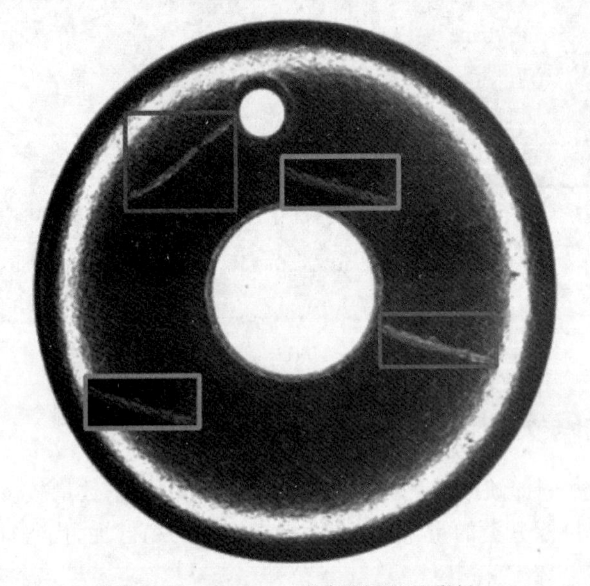

图 5-110　Replicate 数据增强(见彩插)

5.11.5　代码实现 Mosic 数据增强

Mosic 数据增强在 YOLOv4、YOLOv5、YOLOv7 中均有使用,使用 Mosic 数据增强可以涨点,其实现思路是将经过翻转、随机裁剪、翻转、色域变换等其他数据增强的 4 张图合并成 1 张新的图,代码如下:

```
#第 5 章/ObjectDetection/TensorFlow_Yolo_V5_Detected/data/enhancement.py
def getList_img_box(rnd_all_lines: list, num: int = 4):
    """
    获得随机的图片和地址
    :param rnd_all_lines: 所有训练的 lines 中的地址,包括图片和位置
    :param num:
```

```
    :return: 返回对应 lines 的下标
    """
    idx = random.sample(range(len(rnd_all_lines)), num)
    return idx


def rand():
    np.random.seed(1000)
    return np.random.uniform()

#合并 4 张图片
def mosic_join_img(rnd_all_lines: list, output_size=None, scale_range=None):
    """Mosic 合并 4 张图片"""
    if output_size is None:
        output_size = [1024, 1024] #设定图像尺寸
    if scale_range is None:
        scale_range = [0.5, 0.5]
    #新建 1 个为 0 的图像
    output_img = np.zeros([output_size[0], output_size[1], 3], dtype=np.uint8)
    #图像缩小的比例
    scale_x = scale_range[0] + random.random() * (scale_range[1] - scale_range[0])
    scale_y = scale_range[0] + random.random() * (scale_range[1] - scale_range[0])
    #贴的图像大小
    point_x = int(scale_x *output_size[1])
    point_y = int(scale_y *output_size[0])
    new_bbox = []
    #从所有的 lines 中获取随机的 4 张图片进行 Mosic
    idx = getList_img_box(rnd_all_lines, 4)
    for i, ix in enumerate(idx):
        #对选择的 4 张照片进行处理,得到 boxes 信息
        line = rnd_all_lines[ix].split()
        img_path = line[0]
        img_boxes = np.array([np.array(list(map(int, box1.split(',')))) for box1
in line[1:]])

        #读图片
        img = cv2.imread(img_path)
        #对图像进行加工操作,调用已有函数
        #翻转
        if rand() < 0.5:
            img, img_boxes = random_horizontal_flip(img, img_boxes)
        #色域变换
        if rand() < 0.3:
            img, img_boxes = random_hue(img, img_boxes)
        #CutOut
        if rand() < 0.1:
            img, img_boxes = CutOut(img, img_boxes)
        #随机裁剪
        if rand() < 0.2:
            img, img_boxes = random_crop(img, img_boxes)

        #左上角图片的处理
```

```
        if i == 0:
            #用 letter_box 进行替换,得到指定大小的图片和 boxes
            img2, img_boxes = letterbox_image(Image.fromarray(np.uint8(img.
copy())), (point_x, point_y),
                                              np.array(img_boxes.copy()))
            #更新到要保存的图像中
            output_img[:point_x, :point_y, :] = img2
            #处理 bbox
            for bbox in img_boxes:
                xmin = bbox[0]                       #第 1 张图的位置不变
                ymin = bbox[1]
                xmax = bbox[2]
                ymax = bbox[3]
                new_bbox.append([xmin, ymin, xmax, ymax, bbox[-1]])
        elif i == 1:
            #第 2 张图的 x 轴的位置发生了变化
            img2, img_boxes = letterbox_image(
                Image.fromarray(np.uint8(img.copy())), (output_size[1] - point_x,
point_y),
                np.array(img_boxes.copy())
            )
            #更新到要保存的图像中
            output_img[:point_y, point_x:output_size[1], :] = img2
            for bbox in img_boxes:
                xmin = point_x + bbox[0]             #第 2 张图的 x 发生了变化
                ymin = bbox[1]
                xmax = point_x + bbox[2]
                ymax = bbox[3]
                new_bbox.append([xmin, ymin, xmax, ymax, bbox[-1]])
        elif i == 2:
            #第 3 张图的 x 轴的位置发生了变化
            img2, img_boxes = letterbox_image(
                Image.fromarray(np.uint8(img.copy())), (point_x, output_size[0] -
point_y),
                np.array(img_boxes.copy())
            )

            output_img[point_y:output_size[0], :point_x, :] = img2
            #x 不变,y 轴变了
            for bbox in img_boxes:
                xmin = bbox[0]                       #第 3 张图的 y 发生了变化
                ymin = point_y + bbox[1]
                xmax = bbox[2]
                ymax = point_y + bbox[3]
                new_bbox.append([xmin, ymin, xmax, ymax, bbox[-1]])
        elif i == 3:
            img2, img_boxes = letterbox_image(
                Image.fromarray(np.uint8(img.copy())), (output_size[1] - point_x,
output_size[0] - point_y),
                np.array(img_boxes.copy())
            )
            output_img[point_y:output_size[0], point_x:output_size[1], :] = img2
```

```
            for bbox in img_boxes:
                xmin = point_x + bbox[0]          #第 4 张图的 x、y 方向同时变化
                ymin = point_y + bbox[1]
                xmax = point_x + bbox[2]
                ymax = point_y + bbox[3]
                new_bbox.append([xmin, ymin, xmax, ymax, bbox[-1]])
    return output_img, np.array(new_bbox, dtype=np.int)
```

代码实现思路是先通过 np.zeros()得到一张 1024×1024 全黑的图片，然后根据传入的 lines 信息随机获得第 idx 张图片，然后解析出 img_boxes。根据 rand()随机值调用翻转、色域变换、CutOut、随机裁剪的数据增强函数，以 1024×1024 的中心点为坐标，在第 1 个位置将图像更新上去，即通过代码 output_img[:point_x,:point_y,:]=img2 实现，同时计算 bbox 的位置变化，调用该代码实现的效果如图 5-111 所示。

图 5-111　Mosic 数据增强

更多数据增强的代码，可参考随书代码。

总结

数据增强不仅能提高算法的稳健性，同时也能有效地缓解过拟合，是模型训练调优的重要组成方法。

练习

动手实现 Mosic 和标签随机复制的数据增强代码。

第6章 项 目 实 战

本章主要结合项目应用场景和研发流程展示算法项目工作流程、数据分析、模型训练和优化技巧，同时实现预测代码的部署，通过阅读本章读者将会对算法在工业界中的应用有所了解。

6.1 计算机视觉项目的工作流程

算法类工作满足软件开发的基本流程，即立项、需求分析、概要设计、编码、测试、发布的流程，如图6-1所示。

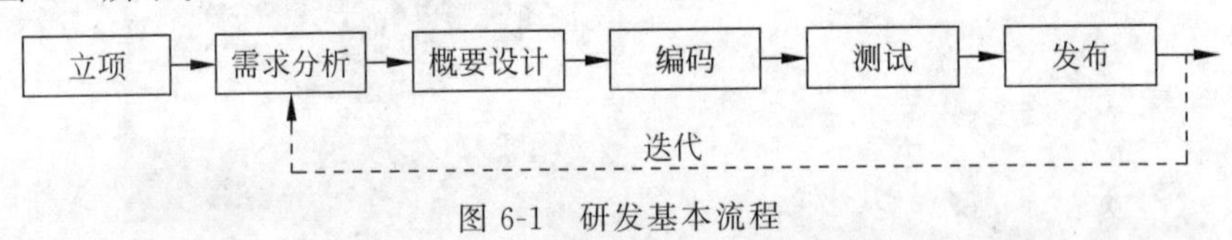

图6-1 研发基本流程

(1) 立项：公司高层或市场相关人员从客户获取初步合作意向，或者根据市场情况进行调研，确定某软件开发的需求，相关人员(如产品经理、技术负责人等)从业务、产品规划、成本、技术等角度分析项目的可行性，然后开展立项决策以确定是否立项。

(2) 需求分析：根据立项确定的客户需求，进一步确定产品的功能性需求及非功能性需求，这是整个软件在开发过程中各个团队的实现标准，在文档中会详细描述某个功能的实现流程、约束条件、期望结果、异常处理、性能指标等内容。

(3) 概要设计：设计阶段主要包括软件的架构设计、数据设计、系统概要设计、详细设计、测试设计、UI设计等内容，设计阶段主要描述如何实现需求中的功能性和非功能性需求。编码过程需要按照与设计相关的内容开展工作。

(4) 编码：根据概要设计、详细设计、UI、需求内容开展编码相关工作。

(5) 测试：对开发出来的产品按照测试设计对软件开展测试工作，发现存在的问题，通过修复问题提高软件质量。

(6) 发布：测试通过后，软件产品可正式交付用户使用。

发布之后通常会有一些缺陷需要修复或者用户会有新的需求，将再次进入需求分析阶段进行多轮迭代。

算法类项目通常来说需要先行，其流程归属于研发主体流程，如图 6-2 所示。

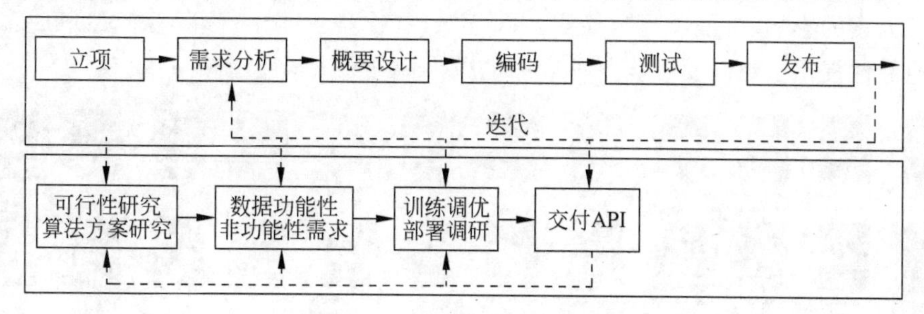

图 6-2 算法研发基本流程

在立项阶段算法工作需要研究其可行性，并对算法实现方案进行初步研究，供立项决策作为参考。在需求分析阶段，需要分析算法能够提供的功能及根据整体项目的非功能性需求，确定算法的非功能性需求（精度、实时性、算力等），并且还需要确定相关数据采集、制作、数据质量等内容。在软件的设计阶段此时就需要模型的训练、调优，快速制作样例，并根据样例进行部署平台的研究和实施，尽早将算法影响项目交付等不利因素排除。当整个软件进入编码阶段时，算法功能相关接口此时应初步具备，以便整个项目进行联调和实施。

算法有一定的不确定性，图 6-2 中只是理想状态，实际情况可能会延后，也有可能存在多次推翻算法实现方案的情况，需要根据项目情况具体问题具体分析。

总结

算法工作是整个研发工作的重要组成部分，算法工作需要与其他研发工作进行紧密的配合。

练习

理解各个研发工作的主要工作内容，并梳理其工作流程和工作内容。

6.2 条形码项目实战

6.2.1 项目背景分析

随着电子商务的快速发展，条形码在零售和物流行业中被广泛应用。条形码是一种以图像形式表示商品信息的编码标识，它能够帮助快速准确地识别商品。目前，随着条形码的广泛使用，条形码目标检测技术变得越来越重要。条形码目标检测技术可以自动识别图像中的条形码，并提取其中的信息，实现商品的追踪、库存管理、商品验证等功能。

在传统的条形码目标检测方法中，通常使用基于特征工程的方法来提取条形码的特征，并结合传统机器学习的算法进行分类。这种方法需要手动设计特征，并且对于不同类型的条形码需要调整特征提取和分类器参数，导致工作量大且算法的稳健性不佳。

因此，鉴于深度学习在目标检测领域取得的显著突破，使用 CNN 通过训练数据集，自动学习条形码的特征和模式，并实现准确的条形码检测和提取是一项重要工作。

在实际应用中，条形码目标检测也面临一些挑战，例如条形码图像常常受到光照、噪声和变形等因素的影响，可能会导致检测精度下降，其次条形码出现在多种尺度和姿态下，对算法的稳健性提出了更高要求。各种商品的条形码如图 6-3 所示。

图 6-3　各种商品的条形码

为了解决这些问题，本项目旨在基于深度学习技术，通过构建合适的数据集、选择合适的网络结构和优化算法参数，提高条形码目标检测的稳健性。

6.2.2　整体技术方案

在数据采集上一方面使用网络爬虫技术获取一部分数据，另一方面从商超实拍数据，考虑光线、角度、褶皱、模糊度方面的影响，共采集了 6075 张图片。在数据标注方面使用 labelimg 进行 Voc 格式的标注，并且同时标注条形码区域和下面的数字，如图 6-4 所示。

图 6-4　标注方法

在对数据进行分析后，根据数据的特点选择模型进行训练和调优，然后根据模型的推理结果对图片进行推理。推理时对条形码和文字区域进行推理。此时只能获得目标的区域，接下来还需要更进一步地调用条形码识别程序，以及 OCR 识别程序对条形码做进一步的提取。如果两者返回其一，则选择有内容的那个，如果两者都返回，则判断是否一致，如果不一致，则优先返回条形码识别内容。整个算法的实现流程如图 6-5 所示。

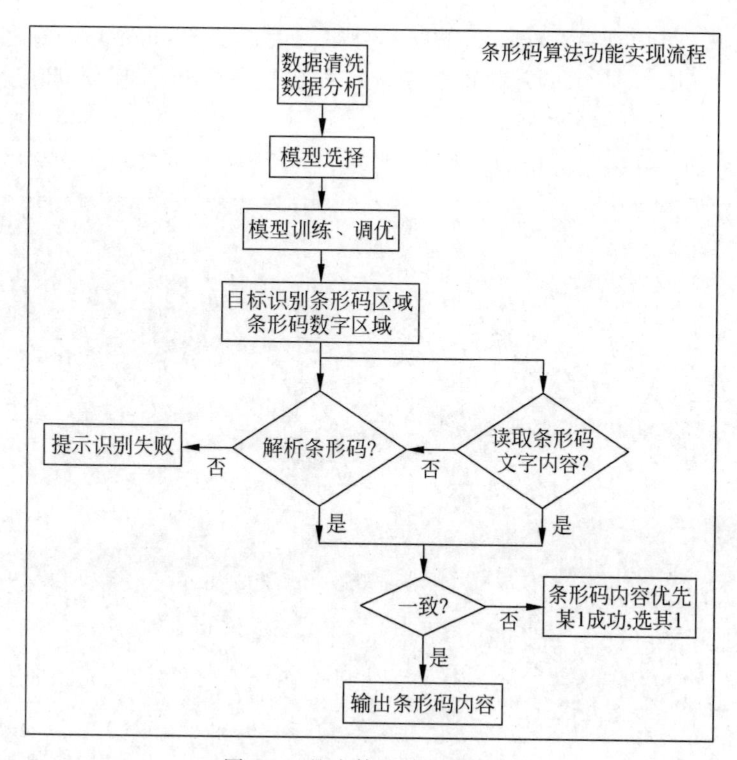

图 6-5　整个算法的实现流程

6.2.3　数据分布分析

在深度学习项目中对数据足够了解是成功的前提条件，在获取数据源后的第 1 步不是进行训练，而是对数据进行分析，其中主要分析以下内容：

(1) 原有图像尺寸的分布。由于采集数据的来源多样，可能某张图片过小或者过大，此时通过图像缩放后会失真，有时会失去某些重要的特征信息，从而影响检测结果。

(2) 标注类别数量的分布。当有多个类别分布时，某些类别的标注 BOX 可能过大，也可能过小，有时会出现数据长尾分布等不平衡问题。当数据不平衡时会使神经网络更倾向于学习数据多的类别，而数据少的类别的精度会相应地降低。

(3) 标注类别 BOX 的面积分布。在不同数据集下标注 BOX 的面积分布不一样，例如当标注 BOX 小于 16×16 的面积较多时，小目标识别将成为一个难点。

(4) 标注类别 BOX 的宽和高分布。目标检测生成的区域选择框是按一定预设比例设置的，当检测目标 BOX 的宽高比极其不规则时会影响目标检测的结果。

(5) 在指定尺寸下标注 BOX 聚类，从而得到 Anchor。论文中给出的 Anchor 通常是基于 COCO 数据集的，与自有数据集分布不一致，所以需要手动聚类。同时由于聚类是针对所有数据集进行的，所以在某些特殊情况可以针对子数据集(如小于 16×16 的 BOX 图片)进行单独聚类，效果往往较佳。

(6) Anchor与标注BOX的最大IOU的分布。设置了Anchor后与标注BOX进行最大IOU的分布统计(YOLOv5为宽和高分布),有利于训练和推理的阈值设定,同时对于Anchor的设置也有更进一步的认识。

对原图尺寸的分布采集分析的代码如下:

```
#第5章/数据分析/分析原图像尺寸分布.py
def get_all_files(path):
    #获取某个目录下的所有文件
    return os.listdir(path)

def get_image_wh(jpg_name):
    #获取图片的宽和高
    img = Image.open(jpg_name)
    return img.size

if __name__ == "__main__":
    path = "../barcodeDataset/images"
    #各种标准图像分辨率的分布,并用来进行计数,初始值为0
    size = {k: [0, []] for k in (
        '<10w-390×260', '>=10w-390×260-640×480',
        '>=30w-640×480-800×600', '>=50w-800×600-1024×768',
        '>=80w-1024×768-1280×960', '>=130w-1280×960-1600×1200',
        '>=310w-1600×1200-2048×1536', '>=400w-2048×1536-2400×1800',
        '>=500w-2560×1920-2560×1920', '>=600w-2560×1920-3000×2000',
        '>=600w-3000×2000', 'other',
    )}
    #对某张图片进行不同尺寸的统计
    for file in get_all_files(path):
        file_name = f"{path}/{file}"
        w, h = get_image_wh(file_name)
        img_area = w * h
        if img_area < 390 * 260:
            size['<10w-390×260'][0] += 1
            size['<10w-390×260'][1].append(file_name)
        elif img_area > 390 * 260 and img_area < 640 * 480:
            size['>=10w-390×260-640×480'][0] += 1
            size['>=10w-390×260-640×480'][1].append(file_name)
        elif img_area > 640 * 480 and img_area < 800 * 600:
            size['>=30w-640×480-800×600'][0] += 1
            size['>=10w-390×260-640×480'][1].append(file_name)
        elif img_area > 800 * 600 and img_area < 1024 * 768:
            size['>=50w-800×600-1024×768'][0] += 1
            size['>=10w-390×260-640×480'][1].append(file_name)
        elif img_area > 1024 * 768 and img_area < 1280 * 960:
            size['>=80w-1024×768-1280×960'][0] += 1
            size['>=10w-390×260-640×480'][1].append(file_name)
        elif img_area > 1280 * 90 and img_area < 1600 * 1200:
            size['>=130w-1280×960-1600×1200'][0] += 1
```

```
        size['>=130w-1280×960-1600×1200'][1].append(file_name)
    elif img_area > 1600 * 1200 and img_area < 2048 * 1536:
        size['>=310w-1600×1200-2048×1536'][0] += 1
        size['>=310w-1600×1200-2048×1536'][1].append(file_name)
    elif img_area > 2048 * 1536 and img_area < 2400 * 1800:
        size['>=400w-2048×1536-2400×1800'][0] += 1
        size['>=400w-2048×1536-2400×1800'][1].append(file_name)
    elif img_area > 2400 * 1800 and img_area < 2560 * 1920:
        size['>=500w-2560×1920-2560×1920'][0] += 1
        size['>=500w-2560×1920-2560×1920'][1].append(file_name)
    elif img_area > 2560 * 1920 and img_area < 3000 * 2000:
        size['>=600w-2560×1920-3000×2000'][0] += 1
        size['>=600w-2560×1920-3000×2000'][1].append(file_name)
    elif img_area > 3000 * 2000:
        size['>=600w-3000×2000'][0] += 1
        size['>=600w-3000×2000'][1].append(file_name)
    else:
        size['other'][0] += 1
        size['other'][1].append(file_name)
#展示数据
show_hist(size)
```

标准图像按像素进行划分，当通过代码计算像素值时按相应像素值进行比较划分，同时也对与 size['other'][1].append(file_name)相关的文件名称进行保存，以便下一步处理。从运行结果图 6-6 中可以发现 130 万像素的图片有 1057 张，大于 600 万像素的图片有 2255 张，图像尺寸较大，如果将输入目标检测网络缩放至 640×640，则标注 BOX 会变小，从而导致特征丢失，影响检测精度。

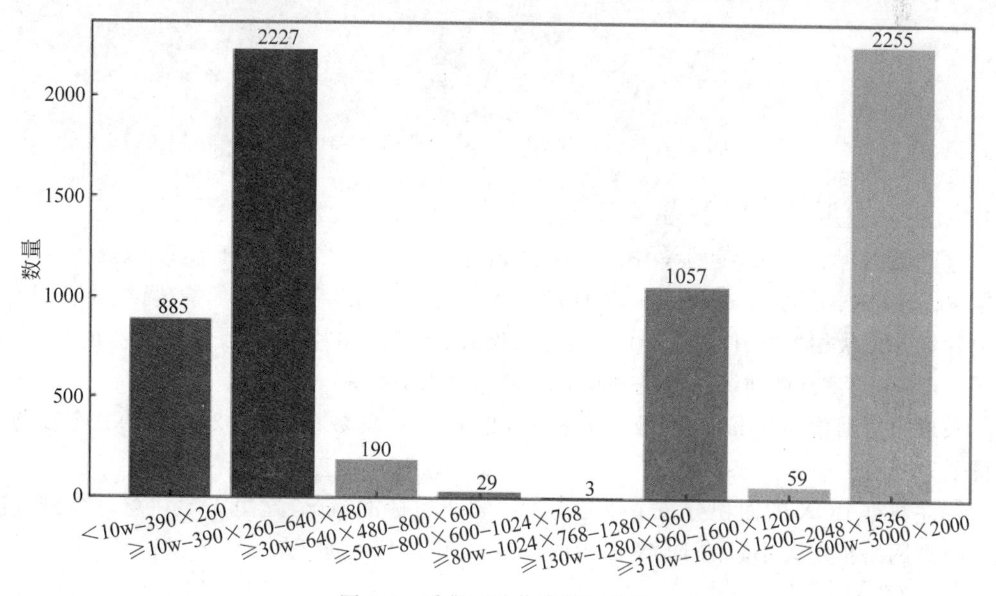

图 6-6 采集图片像素数据分布

根据 Voc 数据集中的分布，统计每个标签的数据分类情况，代码如下：

```
#第 5 章/数据分析/分析标注 BOX 中的 Label 分类.py
#voc2yolo 该库为 "5.1 标签处理及代码"中的代码，主要包括读 Voc 文件，格式转换等程序
import voc2yolo as v

def get_all_files(path):
    #获取某个目录下的所有文件
    return os.listdir(path)


if __name__ == "__main__":
    path = "../barcodeDataset/labels"
    #类别
    class_name = ['barcode', 'number', 'noLabel']
    #构建计数字典
    total = {i: [0, []] for i, k in enumerate(class_name)}
    #对某张图片进行不同尺寸的统计
    for file in get_all_files(path):
        file_name = f"{path}/{file}"
        #调用读取 Voc 文件的函数
        boxes = v.read_annotations(file_name, class_name)
        #如果没有 BOX，则在原有分类的基础上+1，否则根据类别号+1
        if len(boxes):
            boxes_index = boxes[..., 0]
            for index in boxes_index:
                total[index][0] += 1
                #将不同 label 的文件存储起来
                total[index][1].append(file_name)
        else:
            total[len(total) - 1][0] += 1
            total[len(total) - 1][1].append(file_name)
    #重组类别名
    total = {name: v for (k, v), name in zip(total.items(), class_name)}
    #展示图表
    show_hist(total)
```

这里调用了已经写好的 v.read_annotations()函数，主要用于读取 Voc 标签中的[[index,xmin,ymin,xmax,ymax]]的信息，然后根据 total 字典格式的构建，统计不同 label 中的数量并记录相关文件。因为可能会存在没有 label 的图片，所以在原有分类的基础上增加了 noLabel 类别，在训练时该类别将会作为背景来训练。

条形码数据集条码 label 的数量为 8151，数字 label 的数量为 6802，数据的分布较均衡，如图 6-7 所示。

标注数据 BOX 的面积分布主要是为了了解小目标的情况，这里需要注意，此时应统计输入模型网络指定大小时的 BOX 面积的分布，详细的代码如下：

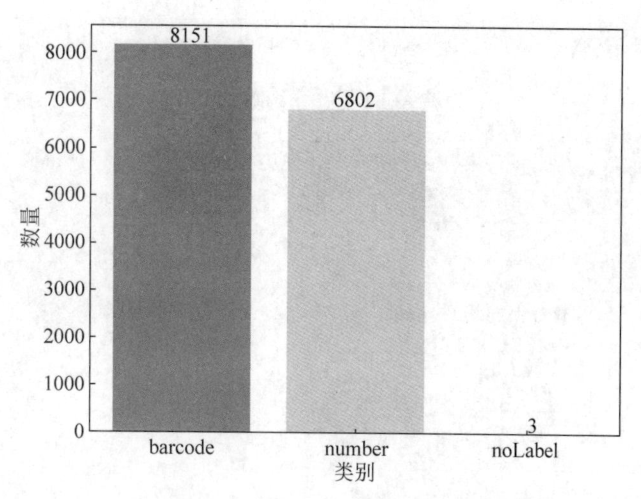

图 6-7　不同 label 的数据分布

```
#第 5 章/数据分析/分析模型网络大小 BOX 的面积.py
import voc2yolo as v
import enhancement as e

if __name__ == "__main__":
    path = "../barcodeDataset/labels"
    jpgDir = "../barcodeDataset/images"
    #类别
    class_name = ['barcode', 'number', 'noLabel']
    is_resize = True
    #构建计数字典 k 为类别;[0, [], []]分别用于存放计数、图片地址、wh 比值
    total = {}
    for i, n in enumerate(class_name):
        size = {k: [0, [], []] for k in (
            '<8×8', '>=8×8-16×16',
            '>=16×16-32×32', '>=32×32-64×64',
            '>=64×64-128×128', '>128×128',
            'other', 'ratio',
        )}
        total[i] = size

    for file in os.listdir(path):
        name = file.split('.')[0]        #获取文件名称
        file_name = f"{path}/{name}.xml"
        #调用读取 Voc 文件的函数,得到 xmin,ymin,xmax,ymax
        old_boxes = v.read_annotations(file_name, class_name)
        if len(old_boxes):
            jpg_name = f"{jpgDir}/{name}.jpg"
            img = Image.open(jpg_name)
            mode_size = (640, 640) if is_resize else img.size
            #将图片和 BOX 缩放到指定的尺寸,调用已有函数
```

```
        new_image, model_boxes = e.letterbox_image(img.copy(), mode_size, old_
boxes)
        #组成 width,height,width,height 的数组,方便 xyxy2cxcywh 归一化计算
        image_shape = np.array(mode_size * 2)
        model_boxes = v.xyxy2cxcywh(model_boxes.copy(), image_shape)
        #统计每个 label 的 area
        for bbox in model_boxes:
            index = bbox[0]
            wh = bbox[-2:] * image_shape[0:2]
            #label 面积的统计
            area = wh[0] * wh[1]
            if area < 8 * 8:
                total[index]["<8×8"][0] += 1
                total[index]["<8×8"][1].append(jpg_name)
            elif area >= 8 * 8 and area < 16 * 16:
                total[index][">=8×8-16×16"][0] += 1
                total[index][">=8×8-16×16"][1].append(jpg_name)
            elif area >= 16 * 16 and area < 32 * 32:
                total[index][">=16×16-32×32"][0] += 1
                total[index][">=16×16-32×32"][1].append(jpg_name)
            elif area >= 32 * 32 and area < 64 * 64:
                total[index][">=32×32-64×64"][0] += 1
                total[index][">=32×32-64×64"][1].append(jpg_name)
            elif area >= 64 * 64 and area < 128 * 128:
                total[index][">=64×64-128×128"][0] += 1
                total[index][">=64×64-128×128"][1].append(jpg_name)
            elif area >= 128 * 128:
                total[index][">128×128"][0] += 1
                total[index][">128×128"][1].append(jpg_name)
            else:
                total[index]["other"][0] += 1
                total[index]["other"][1].append(jpg_name)
            #将 w/h 记录下来
            total[index]['ratio'][2].append(round(wh[0] / wh[1], 1))

    #图像展示
    for i in total.keys():
        if i != len(total) - 1:
            word = "640×640 时" if is_resize else "原图大小时"
            #展示图像
            show_hist(total[i], title=f"{word}{class_name[i]}类标注 BOX 面积大小分布")
            show_ratio(total[i], title=f"{word}{class_name[i]}类标注 BOX 宽高比分布")
```

构建 total 字典，k 为类别，[0,[],[]]分别用于存放计数、图片地址、wh 比值信息，根据 is_resize 参数来设定是否将图片 resize 到指定大小，然后按 BOX 预测面积的当前类别进行计数，最后调用 show_hist()、show_ratio()绘图方法，将图表展示出来。

当将图像重置为 640×640 大小时，其分布情况如图 6-8 和图 6-9 所示。

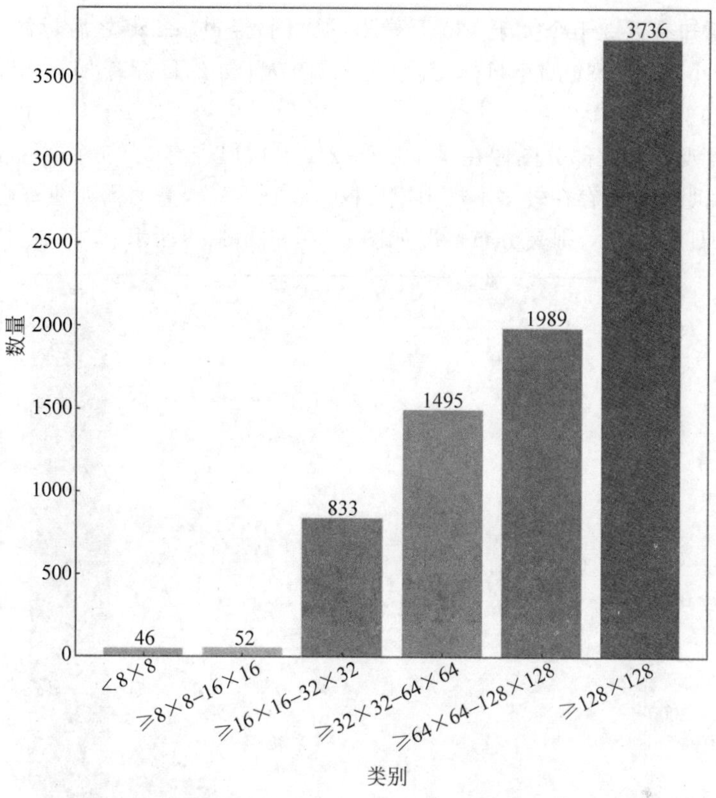

图 6-8　640×640 时 barcode 类标注 BOX 面积的分布

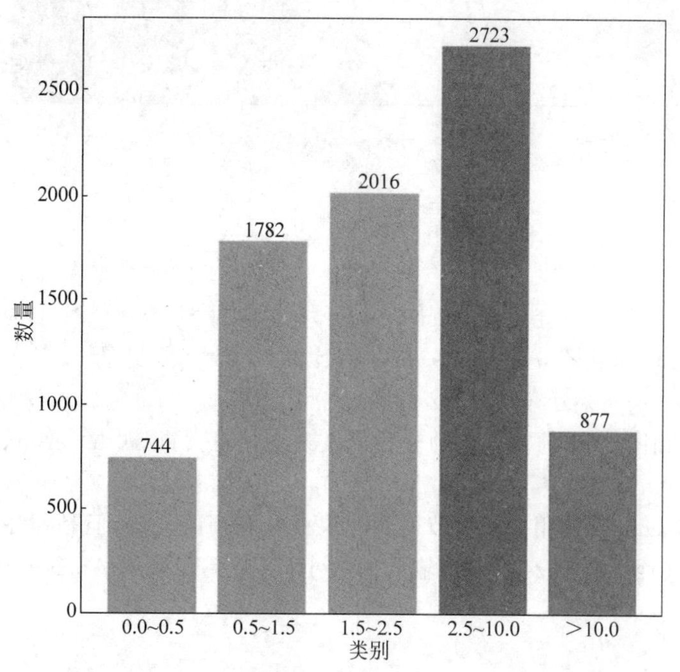

图 6-9　640×640 时 barcode 类标注 BOX 宽和高的分布

从图 6-6 可知有近 3 千个大于 130 万像素的大图，经过 resize 之后标注 BOX，如图 6-8 所示，有 931 个小于 32×32 的小目标，其中大于 128×128 的目标有 3736 个，小目标占比为 931/8151≈11.4%。

从图 6-9 可观察到大部分图像在 1∶2.5～2.5∶1，但也有 3600 个标注 BOX 的宽高比大于 2.5∶1，说明本数据存在较多不规则矩形框，在 Anchor 设置时需要注意这种情况。

如果保留原图像大小，则其分布情况如图 6-10 和图 6-11 所示。

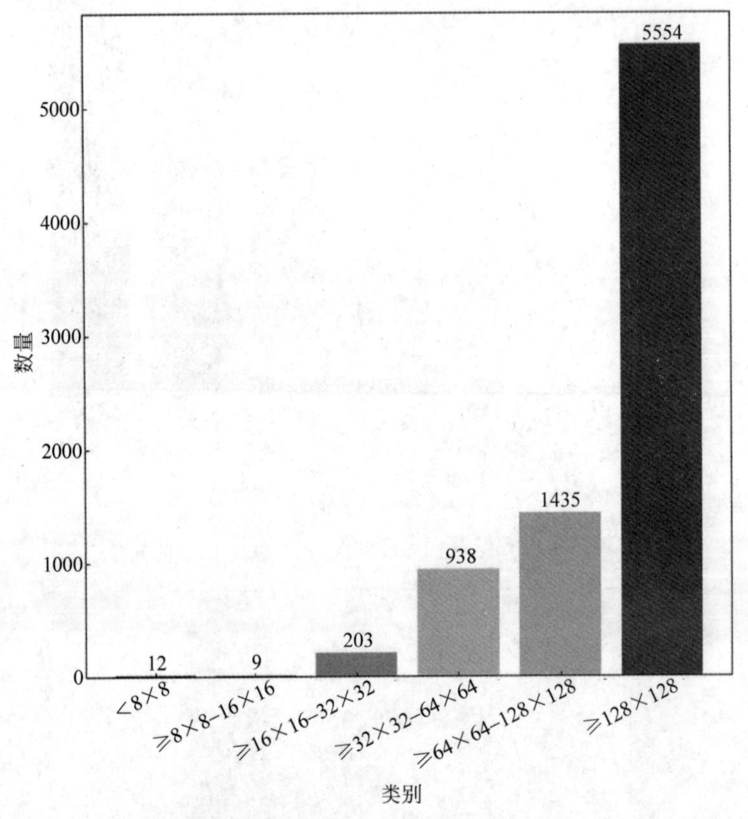

图 6-10　原图大小时 barcode 类标注 BOX 面积的分布

对比图 6-10 和图 6-8 会发现，resize 到 640×640 时小于 32×32 目标 BOX 的数量从图 6-10 中的 224 增加到图 6-8 中的 931，从占比约 2.7%增长到约 11.4%，如果精度不佳，则可能需要处理 resize 后小于 32×32 的图像。

对比图 6-9 和图 6-11 的宽高比的分布可发现变化不大，说明 letterbox_image()函数功能实现正确。

聚类得到 Anchor 可使用 5.6.3 节得到建议框宽和高的代码，但是要注意此时应先使用 letterbox_image()函数 resize 到模型输入的大小，以避免获取的 Anchor 不准确，如图 6-12 所示。

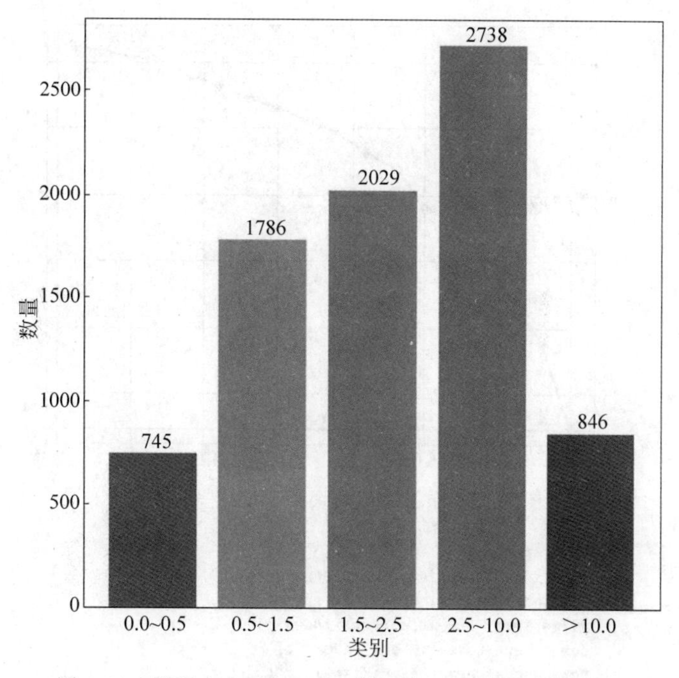

图 6-11 原图大小时 barcode 类标注 BOX 宽和高的分布

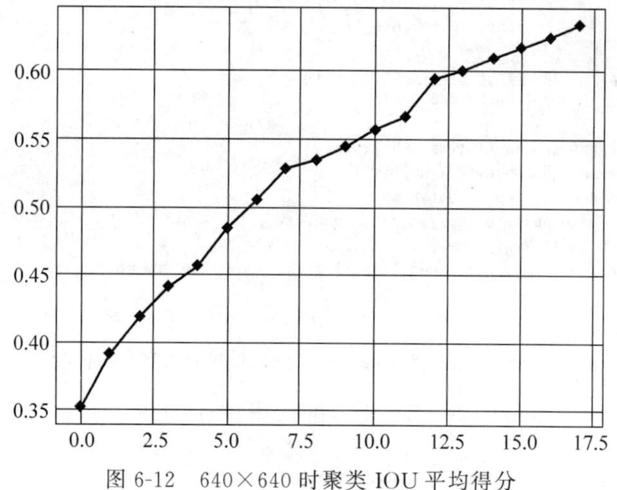

图 6-12 640×640 时聚类 IOU 平均得分

从图 6-12 中可知当且仅当 Anchor 数量超过 6 个时聚类得到的 Anchor 与标注 BOX 的 IOU 得分大于 0.5，当设置为 12 个时平均 IOU 得分超过 0.6，而当为原图大小时约 4 个 Anchor 时 IOU 得分大于 0.5，大于 0.6 只要 6 个，如图 6-13 所示。从两张图可知图像缩放对于 Anchor 的数量设置有影响。

最后在模型训练编码函数（如 yolo_v5_true_encode_box()函数处）统计 Anchor 与标注 BOX 的最大 IOU 值，以便进一步了解 Anchor 设置的 IOU 分布，增加如图 6-14 所示的代码。

如图 6-14 所示，增加 get_max_wh_list()方法，将每个 GT BOX 与 Anchor 宽高比值记

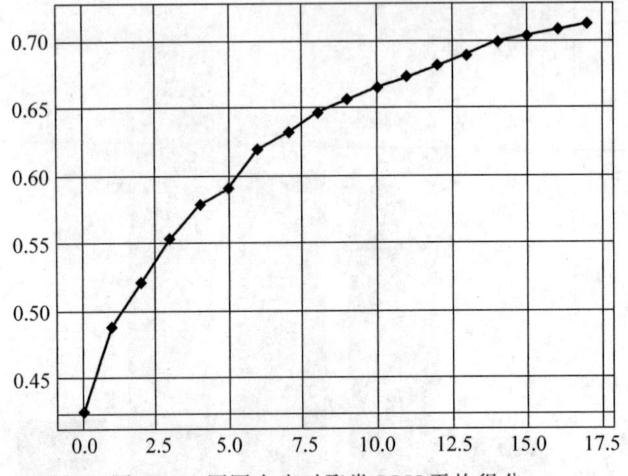

图 6-13 原图大小时聚类 IOU 平均得分

```
    max_wh_list = []
    #for box_index, box in enumerate(true_boxes):
    for box_index in range(true_boxes.shape[1]):
        #tf.print(f'box_index={box_index}')
        box = true_boxes[:, box_index, :]
        box = np.squeeze(box, axis=0)  # 降维
        box_class = box[4].astype('int32')
        best_choice = {}
        for index in range(0, 3, 1):...
        #1个GT至少应该有1个小于4
        get_max_wh_list(max_wh_list)
    return true_boxes2, grid

def get_max_wh_list(max_wh_list):
    wh = np.array(max_wh_list)
    new_wh = np.round(wh[wh <= 4], 2)
    with open('gt_box_wh_ratio_bak.txt', 'a') as f:
        if len(new_wh):
            f.writelines([str(i) + "\n" for i in new_wh])
        else:
            f.write('0\n')  #0表示某个GT没有找到任何1个宽高比小于4的Anchor
```

图 6-14 获取每个 GT 与 Anchor 的宽高比

录到文件中,然后对文件进行如图 6-15 所示的代码统计。

```
with open('gt_box_wh_ratio_bak.txt') as f:
    data = [float(i.strip()) for i in f.readlines()]
    r = {'GT_Anchor_WH>4': 0, 'GT_Anchor_WH<=4': 0}
    for d in data:
        if d == 0:
            r['GT_Anchor_WH>4'] += 1
        else:
            r['GT_Anchor_WH<=4'] += 1
    show_hist(r, "每个GT找到对应Anchor的数量")
```

图 6-15 统计每个 GT 与 Anchor 的宽高比代码

运行程序后得到图 6-16,由此图可知,在 YOLOv5 中现有 Anchor 设置绝大多数满足 GT 与 Anchor 宽高比小于 4 的设置,也就保证了 Anchor 设置在正样本的正确性。

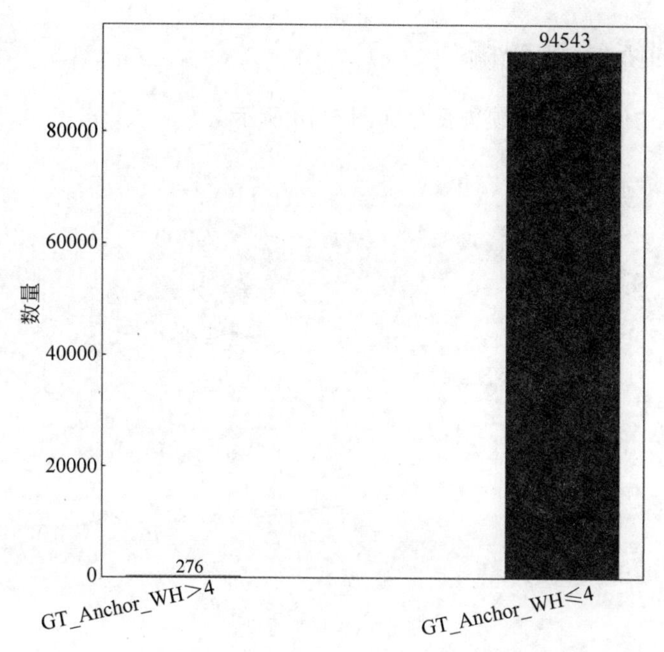

图 6-16　统计每个 GT 与 Anchor 的宽高比小于 4 的数量

6.2.4　参数设置

对于 Anchor 的设置采用默认的 Anchor 值，即[[10,13],[16,30],[33,23]],[[30,61],[62,45],[59,119]],[[116,90],[156,198],[373,326]]；初始学习率 init_lr=1e-5，并采用余弦退化的方法进行学习率的控制；学习输入图像尺寸为 640×640；优化器为 Adam；EPOCH=50 轮；Batch_Size=12，数据增强参数如图 6-17 所示。

```
is_enhancement = True  #是否进行数据增强
#0.3 指的是一个随机小数，当小于0.3时，执行相关的数据增强
enhancement_ratio = {
    'random_translate': 0.3,  #随机平移
    'random_crop': 0.2,  #随机裁剪
    'random_horizontal_flip': 0.5,  #随机水平翻转
    'random_vertical_flip': 0.1,  #随机垂直反转
    'random_rot90_1': 0.3,  #随机顺时针旋转90°
    'random_rot90_2': 0.0,  #随机逆时针旋转
    'random_bright': 0.1,  #随机对比度和亮度
    'random_swap': 0.2,  #随机变换通道
    'random_saturation': 0.1,  #随机变换饱和度
    'random_hue': 0.2,  #随机变换色度(HSV空间下(-180, 180))
    'random_rot': 0.0,  #随机旋转，默认为45°
    'cutout': 0.3,  # cutout
    'add_blur_filters': 0.2,  #添加高斯噪声
    'add_noise_noisy': 0.2,
    'replicate': 0.4,  # 随机重复图中的某个标签，如果数据样本不足，则可以指定某个样本进行多次复制
    'mixup': 0.4,  #mixup
    'mosic_v4': 0.5,  #mosic数据增强
    'cartoonise': 0.3,  #提取物体的主要几位信息
    'equalizeHist': 0.2,  #直方图
}
```

图 6-17　数据增强参数

6.2.5 训练结果分析

经过 50 轮的迭代学习，其损失变化如图 6-18 所示。

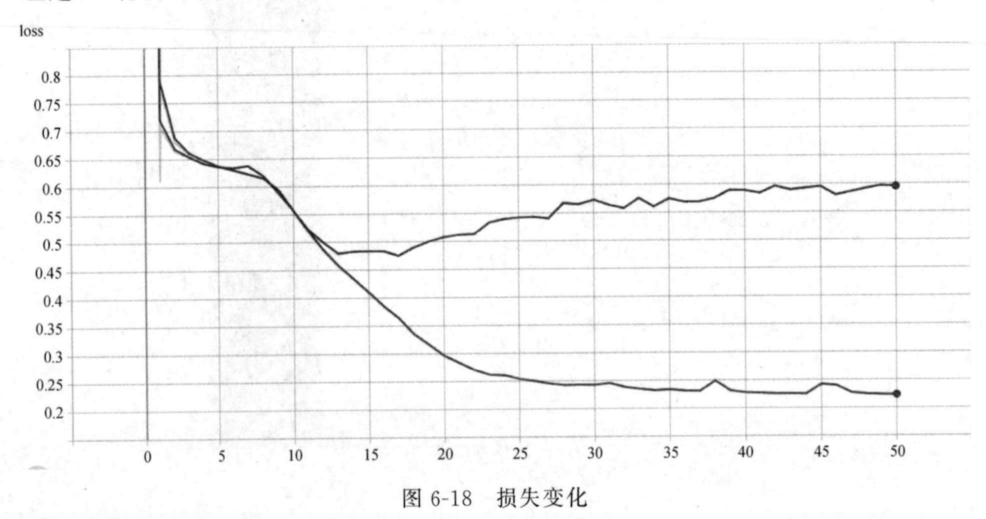

图 6-18 损失变化

在大约第 17 轮训练时，训练损失在下降，而验证损失几乎没有变化，甚至出现略有回升的趋势。同时查看准确率指标，其结果如图 6-19 所示。

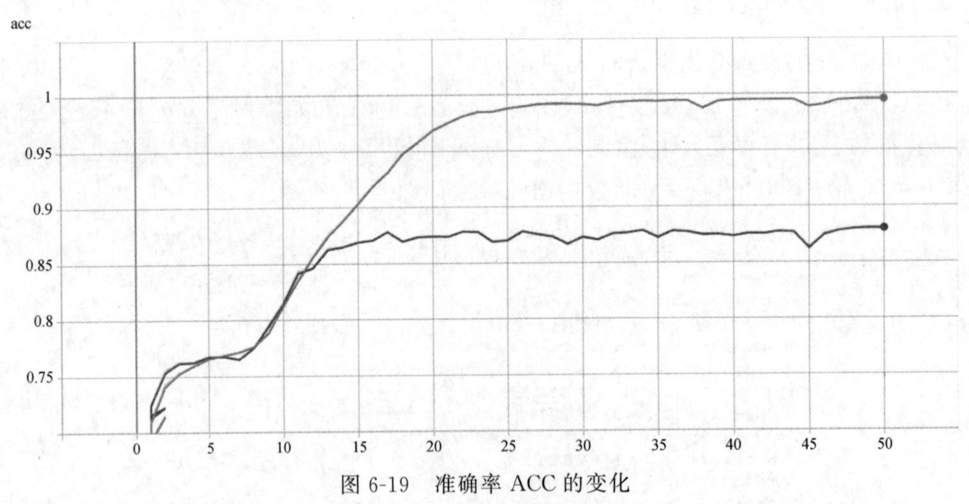

图 6-19 准确率 ACC 的变化

从 ACC 准确率来看，大约在第 17 轮验证准确率在 0.87 左右之后几乎没有增长，而训练准确率在第 22 轮达到 0.98。

解析日志文件将每个 GT BOX 的位置损失、置信度损失、分类损失可视化，如图 6-20～图 6-22 所示。

从图 6-22 可观察，分类损失在训练的中后期呈波动状态，从而导致模型中后期的准确率没有提高并导致模型出现一定的过拟合现象。

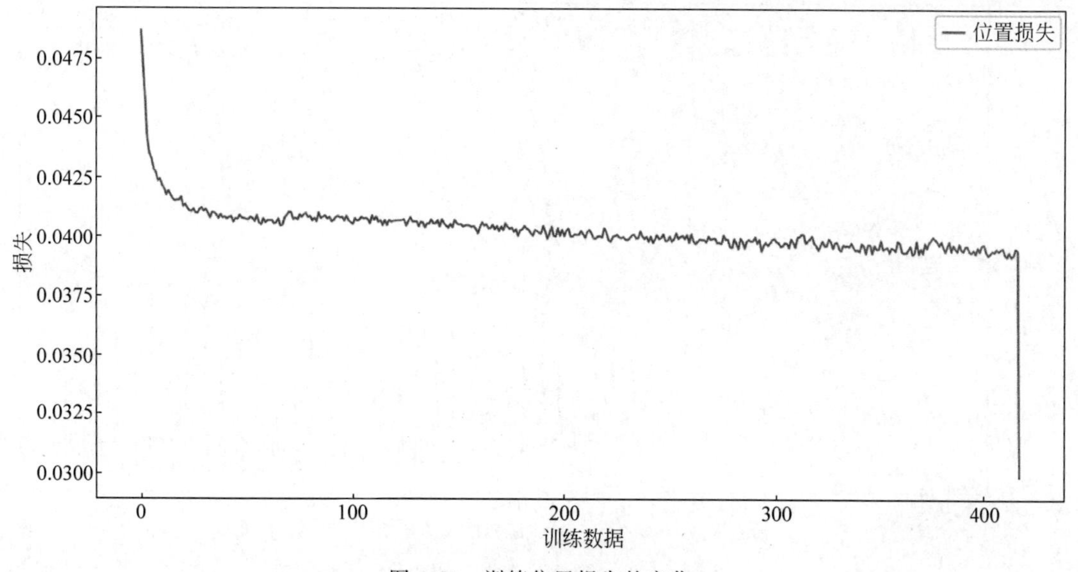

图 6-20　训练位置损失的变化

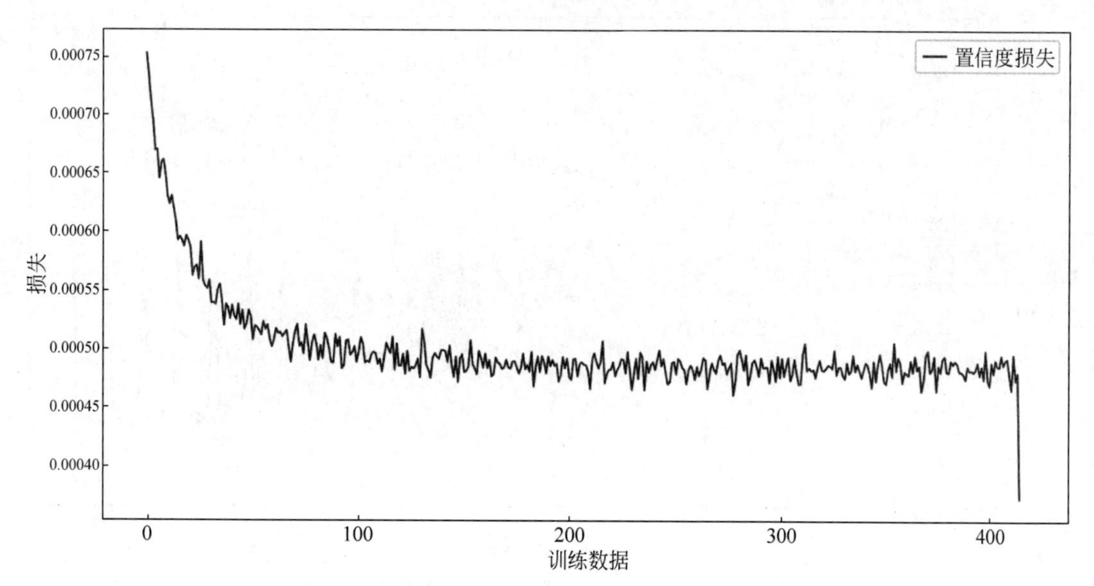

图 6-21　训练置信度损失的变化

添加日志信息，分别观察条码分类、数字分类、背景分类的损失变化，如图 6-23 所示。

从图 6-23 中观察可知，在训练前期模型分类损失下降较快，但到中后期条码类别和数字类别呈波动状态，数字类别的波动更剧烈。观察训练图片可知由于两个类别有遮挡情况，并且存在部分条形码及其数字存在模糊的情况，可考虑调节数据增强的比例继续训练，观察其收敛的变化。

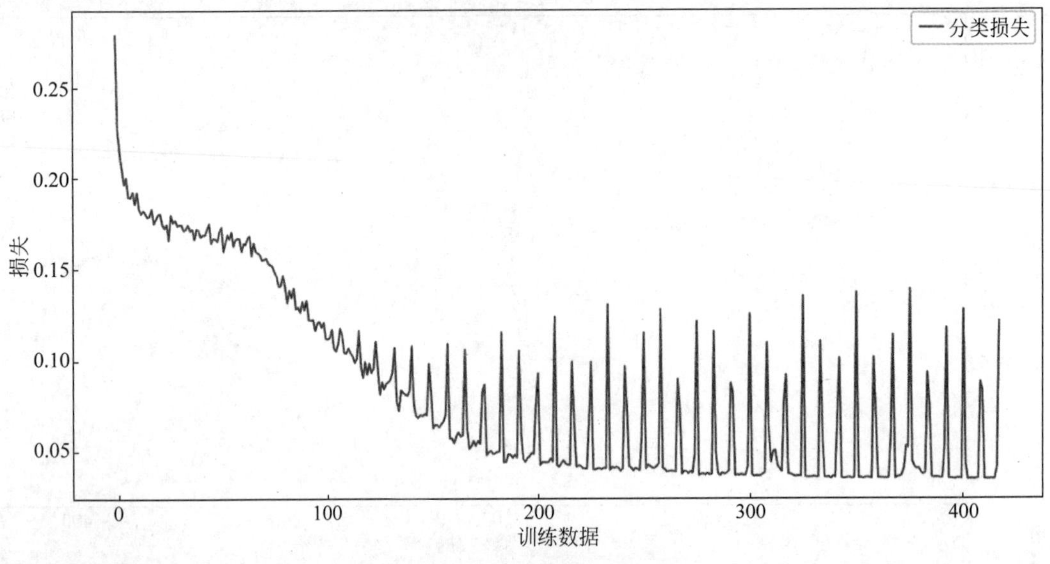

图 6-22 训练分类损失的变化

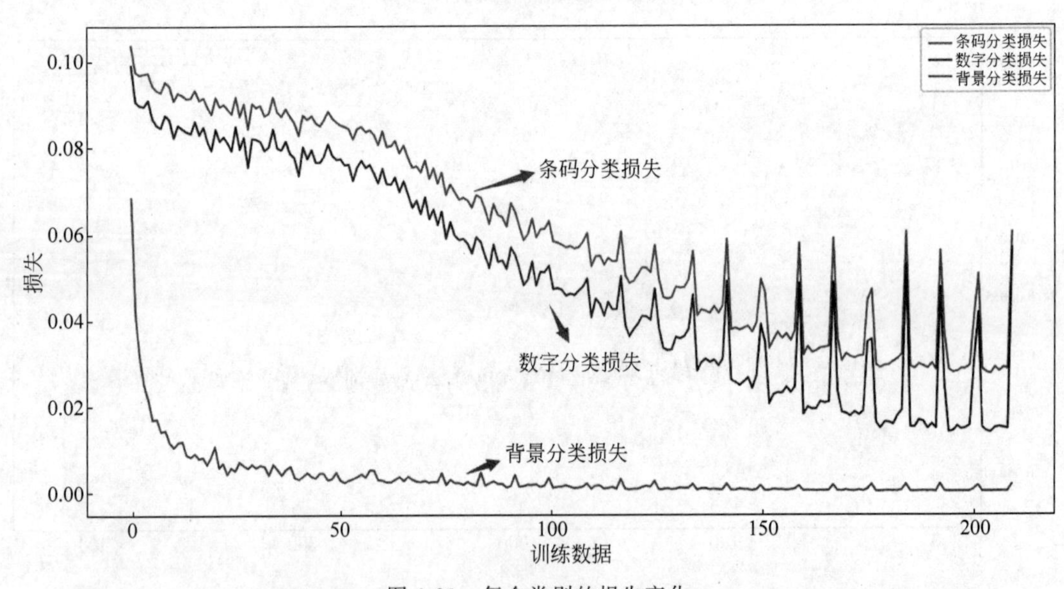

图 6-23 每个类别的损失变化

6.2.6 OpenCV DNN 实现推理

OpenCV 提供了深度学习推理框架接口 DNN，它支持 TensorFlow、Caffe、Keras、PyTorch 等框架训练模型的推理工作，同时支持 CPU 和 GPU 运行模型。

首先将训练保存的 saved_model.pb 文件转换成通用 ONNX 格式。ONNX（Open Neural Network Exchange）开放神经网络交换，它定义了一组和环境、平台无关的标准格式，用来增强各种 AI 模型的可交互性。

转换命令如下：

```
#pip install tf2onnx
#pip install onnx==1.14.1
python -m tf2onnx.convert --saved-model ./last --output ./yolo5.onnx

--saved-model：保存 save_model.pb 文件的路径
--output：生成格式的路径
```

命令内容如图 6-24 所示。

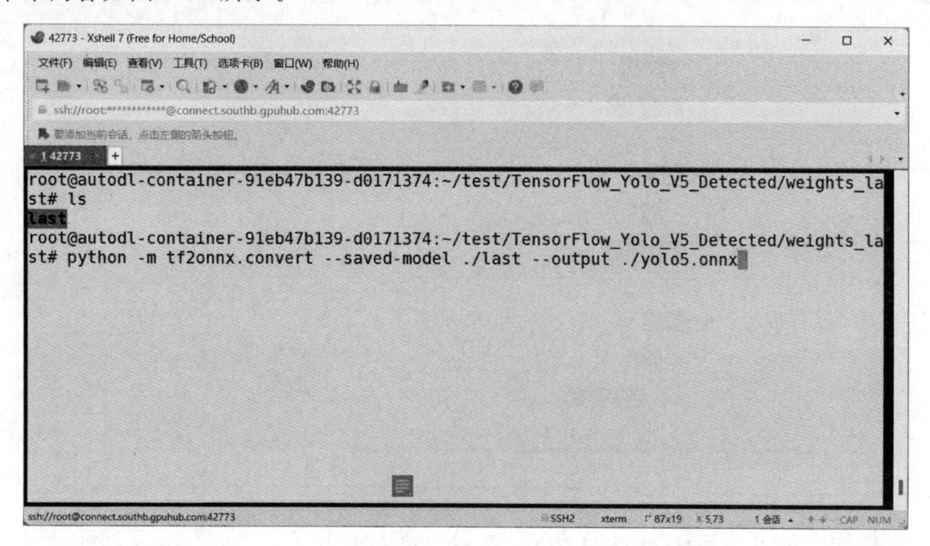

图 6-24　转 ONNX 命令

转换成功的日志如图 6-25 所示。

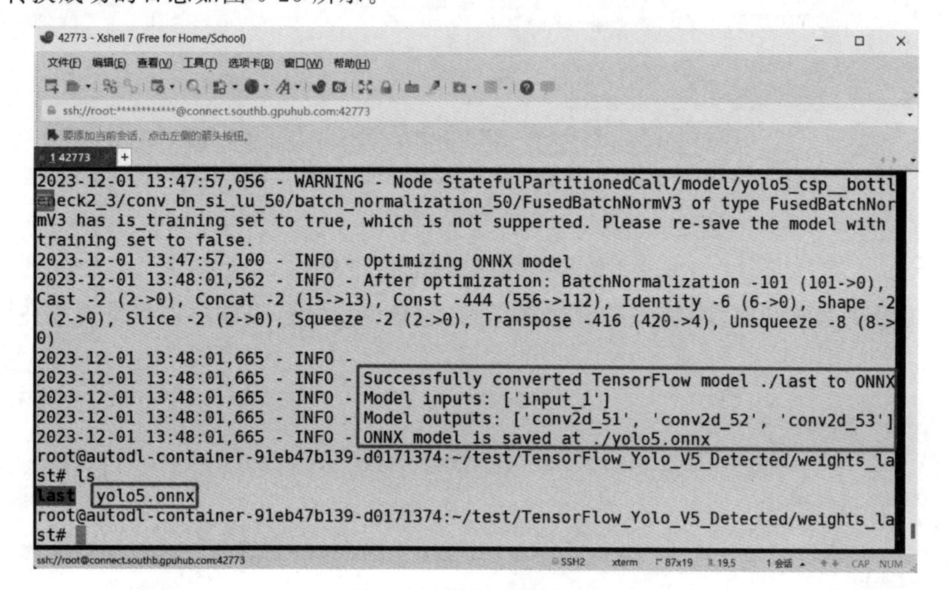

图 6-25　转换成功的日志

注意：如果训练是在 GPU 环境下进行的，则转换成 ONNX 时也需要在 GPU 环境下进行，转换版本与训练 TensorFlow 版本相同，同时需要使用 onnx==1.14.1 版本。

转换成功后可在网站 https://netron.app/中打开，观察转换后的网络结构图，如图 6-26 所示。

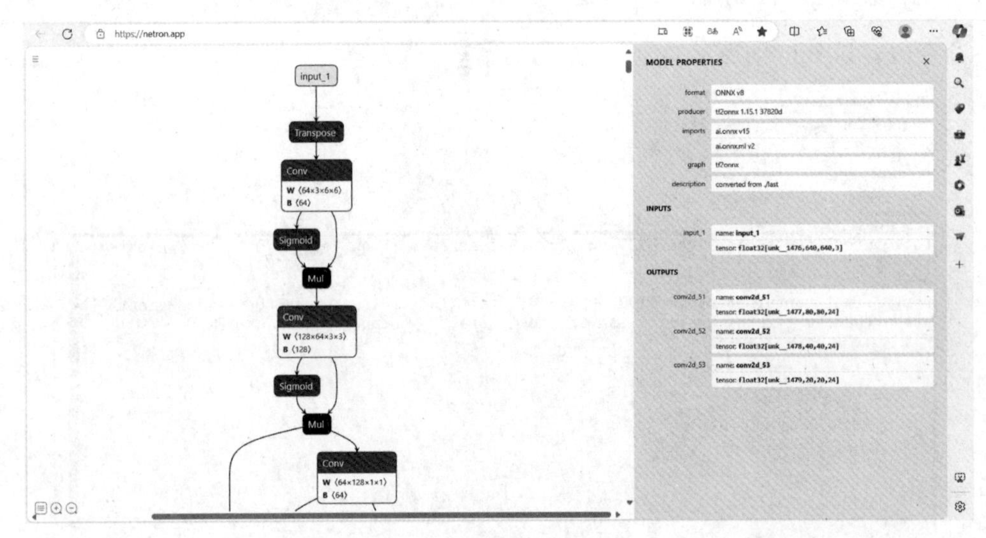

图 6-26　ONNX 可视化

然后使用 OpenCV 的 readNetFromONNX()读取模型文件，如图 6-27 所示。

```
opencvYolo5.py
1  import cv2
2
3  cv2.dnn.readNet()
      readNetFromONNX(onnxFile)                 cv2.dnn
      readNetFromModelOptimizer(xml, …          cv2.dnn
   Press Ctrl+. to choose the selected (or first) suggestion and insert a dot afterwards  Next Tip
```

图 6-27　OpenCV 读取 ONNX 函数

然后编写推理代码，相对于历史推理代码，加载模型并实现前向传播略有变化，代码如下：

```
class Detected:
    def __init__(self, model_path, input_size):
            #加载 ONNX 模型
        self.net = cv2.dnn.readNetFromONNX(model_path)
        self.confidence_threshold = 0.5                #有无目标置信度
```

```
        self.class_prob = [0.5, 0.5]          #两个类别的分类阈值
        self.nms_threshold = 0.5              #NMS 的阈值
        self.input_size = input_size
            #预设 anchors
        self.anchors = np.array([
            [[10, 13], [16, 30], [33, 23]],
            [[30, 61], [62, 45], [59, 119]],
            [[116, 90], [156, 198], [373, 326]]
        ]) / 640
    #NumPy 版本的 Sigmoid
    def sigmoid(self, x):
        return 1.0 / (1 + np.exp(-x))
    #前向传播
    def forward(self):
        #因为 OpenCV 默认为三维,所以需要调用此函数升成 4 维,并归一化
        frame = cv2.dnn.blobFromImage(self.img, 1 / 255.0, [640, 640], swapRB=True)
            #获取输出层的名称
        ln = self.net.getLayerNames()
        #获取输出层的名称
        output_name = [ln[n - 1] for n in self.net.getUnconnectedOutLayers()][::-1]
            #网络的输入要求为 1,640,640,3。转换为 ONNX 格式后默认为 1,3,640,640
        frame = np.reshape(frame, [1, 640, 640, 3])
        #将 frame 对象送入网络
        self.net.setInput(frame)
        #前向传播得到预测值
        self.output = self.net.forward(output_name)
```

代码中 self. net. getLayerNames()用于获取所有 ONNX 中的网络层的名称，output_name 是输出层的名称，ln[n－1]是由于 ONNX 的层数比实际网络层多 1 个，如图 6-28 所示。

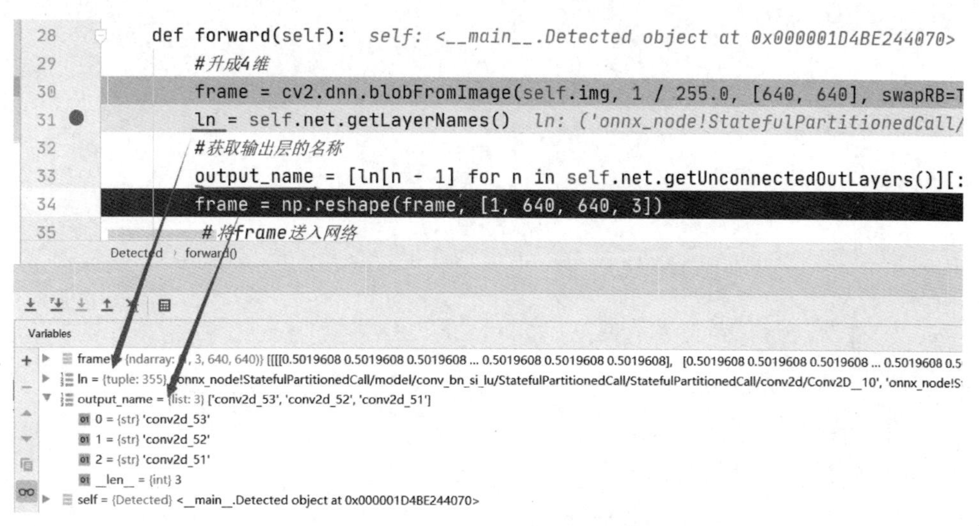

图 6-28　从 ONNX 中获取输出名称

完善代码推理后的结果如图 6-29 所示。

图 6-29　OpenCV ONNX 预测结果

总结

先分析数据的构成，选择合适的模型进行训练调优，并将模型整合到应用程序中才能完成整个算法项目的主要工作。

练习

选择 GitHub YOLOv5 官方代码，选择某个数据集进行模型的训练与调优，可尝试完成 OpenCV DNN 推理代码的实现。

参考文献

[1] 阿斯顿·张,李沐,何孝霆,等. 动手学深度学习[M]. 北京：人民邮电出版社,2021.
[2] 文青山. Python软件测试实战宝典[M]. 北京：机械工业出版社,2022.

图书推荐

书　　名	作　　者
Diffusion AI 绘图模型构造与训练实战	李福林
图像识别——深度学习模型理论与实战	于浩文
HuggingFace 自然语言处理详解——基于 BERT 中文模型的任务实战	李福林
动手学推荐系统——基于 PyTorch 的算法实现(微课视频版)	於方仁
TensorFlow 计算机视觉原理与实战	欧阳鹏程、任浩然
自然语言处理——原理、方法与应用	王志立、雷鹏斌、吴宇凡
人工智能算法——原理、技巧及应用	韩龙、张娜、汝洪芳
跟我一起学机器学习	王成、黄晓辉
深度强化学习理论与实践	龙强、章胜
Java+OpenCV 高效入门	姚利民
Java+OpenCV 案例佳作选	姚利民
计算机视觉——基于 OpenCV 与 TensorFlow 的深度学习方法	余海林、翟中华
深度学习——理论、方法与 PyTorch 实践	翟中华、孟翔宇
Flink 原理深入与编程实战——Scala+Java(微课视频版)	辛立伟
Spark 原理深入与编程实战(微课视频版)	辛立伟、张帆、张会娟
PySpark 原理深入与编程实战(微课视频版)	辛立伟、辛雨桐
Python 预测分析与机器学习	王沁晨
Python 人工智能——原理、实践及应用	杨博雄 等
Python 深度学习	王志立
编程改变生活——用 Python 提升你的能力(基础篇·微课视频版)	邢世通
编程改变生活——用 Python 提升你的能力(进阶篇·微课视频版)	邢世通
编程改变生活——用 PySide6/PyQt6 创建 GUI 程序(基础篇·微课视频版)	邢世通
编程改变生活——用 PySide6/PyQt6 创建 GUI 程序(进阶篇·微课视频版)	邢世通
Python 量化交易实战——使用 vn.py 构建交易系统	欧阳鹏程
Python 从入门到全栈开发	钱超
Python 全栈开发——基础入门	夏正东
Python 全栈开发——高阶编程	夏正东
Python 全栈开发——数据分析	夏正东
Python 编程与科学计算(微课视频版)	李志远、黄化人、姚明菊 等
Python 游戏编程项目开发实战	李志远
Python 数据分析实战——从 Excel 轻松入门 Pandas	曾贤志
Python 概率统计	李爽
Python 数据分析从 0 到 1	邓立文、俞心宇、牛瑶
Python Web 数据分析可视化——基于 Django 框架的开发实战	韩伟、赵盼
Python 玩转数学问题——轻松学习 NumPy、SciPy 和 Matplotlib	张骞
AR Foundation 增强现实开发实战(ARKit 版)	汪祥春
AR Foundation 增强现实开发实战(ARCore 版)	汪祥春
ARKit 原生开发入门精粹——RealityKit+Swift+SwiftUI	汪祥春
HoloLens 2 开发入门精要——基于 Unity 和 MRTK	汪祥春
Octave GUI 开发实战	于红博

续表

书　名	作　者
Octave AR 应用实战	于红博
HarmonyOS 移动应用开发(ArkTS 版)	刘安战、余雨萍、陈争艳 等
openEuler 操作系统管理入门	陈争艳、刘安战、贾玉祥 等
JavaScript 修炼之路	张云鹏、戚爱斌
深度探索 Vue.js——原理剖析与实战应用	张云鹏
前端三剑客——HTML5+CSS3+JavaScript 从入门到实战	贾志杰
剑指大前端全栈工程师	贾志杰、史广、赵东彦
HarmonyOS 应用开发实战(JavaScript 版)	徐礼文
HarmonyOS 原子化服务卡片原理与实战	李洋
鸿蒙操作系统开发入门经典	徐礼文
鸿蒙应用程序开发	董昱
鸿蒙操作系统应用开发实践	陈美汝、郑森文、武延军、吴敬征
HarmonyOS 移动应用开发	刘安战、余雨萍、李勇军 等
HarmonyOS App 开发从 0 到 1	张诏添、李凯杰
从数据科学看懂数字化转型——数据如何改变世界	刘通
JavaScript 基础语法详解	张旭乾
5G 核心网原理与实践	易飞、何宇、刘子琦
恶意代码逆向分析基础详解	刘晓阳
深度探索 Go 语言——对象模型与 runtime 的原理、特性及应用	封幼林
深入理解 Go 语言	刘丹冰
Vue+Spring Boot 前后端分离开发实战	贾志杰
Spring Boot 3.0 开发实战	李西明、陈立为
Flutter 组件精讲与实战	赵龙
Flutter 组件详解与实战	[加]王浩然(Bradley Wang)
Dart 语言实战——基于 Flutter 框架的程序开发(第 2 版)	亢少军
Dart 语言实战——基于 Angular 框架的 Web 开发	刘仕文
IntelliJ IDEA 软件开发与应用	乔国辉
FFmpeg 入门详解——音视频原理及应用	梅会东
FFmpeg 入门详解——SDK 二次开发与直播美颜原理及应用	梅会东
FFmpeg 入门详解——流媒体直播原理及应用	梅会东
FFmpeg 入门详解——命令行与音视频特效原理及应用	梅会东
FFmpeg 入门详解——音视频流媒体播放器原理及应用	梅会东
Power Query M 函数应用技巧与实战	邹慧
Pandas 通关实战	黄福星
深入浅出 Power Query M 语言	黄福星
深入浅出 DAX——Excel Power Pivot 和 Power BI 高效数据分析	黄福星
从 Excel 到 Python 数据分析:Pandas、xlwings、openpyxl、Matplotlib 的交互与应用	黄福星
云原生开发实践	高尚衡
云计算管理配置与实战	杨昌家
虚拟化 KVM 极速入门	陈涛
虚拟化 KVM 进阶实践	陈涛
Octave 程序设计	于红博